Current Research in Astrophysics

Current Research in Astrophysics

Edited by Jun Yu

SYRAWOOD
PUBLISHING HOUSE

New York

Published by Syrawood Publishing House,
750 Third Avenue, 9th Floor,
New York, NY 10017, USA
www.syrawoodpublishinghouse.com

Current Research in Astrophysics
Edited by Jun Yu

International Standard Book Number: 978-1-68286-682-5 (Hardback)

Cataloging-in-Publication Data

Current research in astrophysics / edited by Jun Yu.
    p. cm.
Includes bibliographical references and index.
ISBN 978-1-68286-682-5
1. Astrophysics. 2. Cosmic physics. I. Yu, Jun.
QB461 .C87 2019
523.01--dc23

TABLE OF CONTENTS

PREFACE

In my initial years as a student, I used to run to the library at every possible instance to grab a book and learn something new. Books were my primary source of knowledge and I would not have come such a long way without all that I learnt from them. Thus, when I was approached to edit this book; I became understandably nostalgic. It was an absolute honor to be considered worthy of guiding the current generation as well as those to come. I put all my knowledge and hard work into making this book most beneficial for its readers.

The study of astronomical objects through the analysis of characteristics like luminosity, density, chemical composition and temperature, by employing principles of physics and chemistry, falls under the domain of astrophysics. It also aims to understand and explain blackhole dynamics and dark matter phenomenology. Modern astrophysics implements a multidisciplinary approach by incorporating principles of varied areas of physics such as electromagnetism, thermodynamics, nuclear and particle physics. This book elucidates the concepts and innovative models around prospective developments with respect to astrophysics. It strives to provide a fair idea about this discipline and to help develop a better understanding of the latest advances in this field. This book is a complete source of knowledge on the present status of this important discipline.

I wish to thank my publisher for supporting me at every step. I would also like to thank all the authors who have contributed their researches in this book. I hope this book will be a valuable contribution to the progress of the field.

Editor

On the reliability of N-body simulations

Tjarda Boekholt[*] and Simon Portegies Zwart

Abstract

The general consensus in the N-body community is that statistical results of an ensemble of collisional N-body simulations are accurate, even though individual simulations are not. A way to test this hypothesis is to make a direct comparison of an ensemble of solutions obtained by conventional methods with an ensemble of true solutions. In order to make this possible, we wrote an N-body code called `Brutus`, that uses arbitrary-precision arithmetic. In combination with the Bulirsch-Stoer method, `Brutus` is able to obtain converged solutions, which are true up to a specified number of digits.

We perform simulations of democratic 3-body systems, where after a sequence of resonances and ejections, a final configuration is reached consisting of a permanent binary and an escaping star. We do this with conventional double-precision methods, and with `Brutus`; both have the same set of initial conditions and initial realisations. The ensemble of solutions from the conventional simulations is compared directly to that of the converged simulations, both as an ensemble and on an individual basis to determine the distribution of the errors.

We find that on average at least half of the conventional simulations diverge from the converged solution, such that the two solutions are microscopically incomparable. For the solutions which have not diverged significantly, we observe that if the integrator has a bias in energy and angular momentum, this propagates to a bias in the statistical properties of the binaries. In the case when the conventional solution has diverged onto an entirely different trajectory in phase-space, we find that the errors are centred around zero and symmetric; the error due to divergence is unbiased, as long as the time-step parameter, $\eta \leq 2^{-5}$ and when simulations which violate energy conservation by more than 10% are excluded. For resonant 3-body interactions, we conclude that the statistical results of an ensemble of conventional solutions are indeed accurate.

Keywords: methods: numerical; methods: N-body simulations; stars: dynamics; binaries: formation

1 Introduction

Analytical solutions to the N-body problem are known for $N = 2$, which are the familiar conic sections. Also, for several systems possessing symmetries, analytical solutions have been found, for example the equilateral triangle (Lagrange 1772). For a more general initial configuration, solutions have to be obtained by means of numerical integration. Given an initial N-body realisation, one can calculate all mutual forces and subsequently the net acceleration of each particle. Different integration methods exist which take the accelerations, and update the positions and velocities to a time $t + \Delta t$, with Δt the time-step size. This process is repeated until the end time is reached.

Miller (1964) recognised that obtaining the solution to an N-body problem by numerical integration is difficult. This is caused by exponential divergence. Consider a certain N-body problem, i.e. N point-particles, each with a given mass, position and velocity. This system evolves with time in a definite and unique way. If one goes back to the initial state and slightly perturbs only one coordinate of a single particle, the perturbed N-body problem will also have a definite and unique but different solution than the original one. When the two solutions are compared as a function of time, it is observed that differences can grow exponentially (Miller 1964; Dejonghe and Hut 1986; Goodman et al. 1993; Hut and Heggie 2002). If the initial perturbation is due to a numerical error, the calculated solution will also diverge away from the true solution.

Several authors have estimated the time-scale of this divergence (Goodman et al. 1993; Hut and Heggie 2002), and arrived at an e-folding time-scale of the order a dynam-

[*]Correspondence: boekholt@strw.leidenuniv.nl
Leiden Observatory, Leiden University, PO Box 9513, Leiden, 2300 RA, The Netherlands

ical, crossing time. Simulation times of interest are typically much longer than a crossing time and therefore staying close to the true solution is numerically challenging.

If the result of a direct N-body simulation of for example a star cluster, has diverged away from the true solution, the result may well be meaningless (Goodman et al. 1993). The general consensus however, is that statistically the results are representative for the true solution to the N-body problem (Smith 1979; Heggie 1991; Goodman et al. 1993). The underlying idea is that the statistics of an ensemble of N-body simulations are representative for the true statistics, obtained by an ensemble of true solutions, with the same set of initial conditions. We regard this the hypothesis we want to test.

One way to test this hypothesis is to directly compare statistics obtained by conventional methods, with the statistics obtained from an ensemble of true solutions. To obtain true solutions, we wrote an N-body code which can solve the N-body problem to arbitrary precision.

Such a code can be realised if the different sources of error are controlled. The error has contributions from the time discretisation of the integrator and the round-off due to the limited precision of the computer (Zadunaisky 1979). Another possible source of error is in the initial conditions, for example the configuration of the Solar System is only approximately known (Ito and Tanikawa 2002). However, if the initial condition is a random realisation of a distribution function, this is less often a problem. Using the Bulirsch-Stoer method (Bulirsch and Stoer 1964; Gragg 1965), the discretisation error can be controlled to stay within a specified tolerance. Using arbitrary-precision arithmetic instead of conventional double-precision or single-precision, the round-off error can be reduced by increasing the number of digits.

We obtain converged solutions to the N-body problem by decreasing the Bulirsch-Stoer tolerance and increasing the number of digits systematically. We define a converged solution in our experiments as a solution for which the first specified number of decimal places of every phase-space coordinate in our final configuration in the N-body experiment becomes independent of the length of the mantissa and the Bulirsch-Stoer tolerance. We explain the method of convergence in Section 2 and we give examples of the procedure in Section 3.

Using this new, brute force N-body code which we call Brutus, we test the reliability of N-body simulations by a controlled numerical experiment which we describe in Section 4. In this experiment we perform a series of resonant 3-body simulations, where the term resonant implies a phase or multiple phases during the interaction where the stars are more or less equidistant (Hut and Bahcall 1983). These phases are intermingled by ejections, where a binary and single star are clearly separated. We perform the simulations with conventional double-precision, and

with arbitrary-precision to reach the converged solution. In Section 5, the solutions are compared individually to investigate the distribution of the errors. We also compare the global statistical distributions using two-sample Kolmogorov-Smirnov tests (Kolmogorov 1933; Smirnov 1948).

2 Methods

2.1 The benchmark integrator

The gravitational N-body problem aims to solve Newton's equations of motion under gravity for N stars (Newton 1687). A popular integrator to perform this task is the fourth-order Hermite predictor-corrector scheme (Makino and Aarseth 1992), using double-precision arithmetic. The experiments we discuss in Section 4 will use this integrator as a benchmark test. We adopt a shared, adaptive time-stepping scheme with the following criterion:

$$\Delta t = \eta \min \sqrt{\Delta r_{ij}/\Delta a_{ij}}. \tag{1}$$

Here η is the time-step parameter and Δr_{ij} and Δa_{ij} are the relative distance and acceleration for the pair of particles i and j. We implement no further constraints on the time-step size.

To test how inaccurate we are allowed to integrate while still obtaining accurate statistics (Smith 1979; Quinlan and Tremaine 1992) we vary the time-step parameter η, to obtain statistics from conventional simulations with different precision.

2.2 The Brutus N-body code

The results obtained with the benchmark integrator are compared to those obtained with Brutus, which uses an arbitrary-precision library.[a] With this library we can specify the number of bits, L_w, used to store the mantissa, while the exponent has a fixed word-length. The length of the mantissa can be specified and increased, with the aim of controlling the round-off error.

The integration of the equations of motion is realised using the Verlet-leapfrog scheme (Verlet 1967). The time-step is shared among all particles, but varies for every step according to equation (1).

To control the discretisation error, we implemented the Bulirsch-Stoer (BS) method, which uses iterative integration and polynomial extrapolation to infinitesimal time-step size (Bulirsch and Stoer 1964; Gragg 1965). An integration step is accepted, when two subsequent BS iterations have converged to below the BS tolerance level, ϵ.

The time-step parameter η and the BS tolerance ϵ, both influence the performance. If η is too big, convergence may not be achieved for any tolerance. If η is too small, the many integration steps will render the integration too expensive. There is an optimal value for η as a function of ϵ.

We measured this relation empirically, which results in:

$$\log_{10} \eta = A \log_{10} \epsilon + B. \tag{2}$$

For $\epsilon < 10^{-50}$ the powerlaw converges to $A = 0.029$ and $B = 0.45$. Extrapolating this relation to $\epsilon > 10^{-50}$ will cause the time-step size to become larger than the time scale for the closest encounter in the system. Therefore this relation saturates to a flatter powerlaw for $\epsilon > 10^{-50}$ with $A = 0.012$ and $B = -0.40$. Compared to a fixed value for η, this relation speeds up the iterative procedure by about a factor three or more. The code is implemented as a community code in the AMUSE framework (Portegies Zwart et al. 2012) under the name `Brutus`.

2.3 Method of convergence

For every simulation we have to define the BS tolerance parameter ϵ and the word-length L_w. In an iterative procedure we vary both parameters systematically, each time carrying out a simulation until $t = t_{end}$. We subsequently calculate the phase space distance, $\delta_{A,B}^2$, between two solutions A and B:

$$\delta_{A,B}^2 = \frac{1}{6N} \sum_{i=1}^{N} \sum_{j=1}^{6} (q_{A,i,j} - q_{B,i,j})^2. \tag{3}$$

The first summation is over all particles and the second summation is over the six phase-space coordinates denoted by q (Miller 1964). We normalise by $6N$, so that δ represents the average difference per phase-space coordinate between two solutions A and B. In our experiments we adopt Hénon units[b] (Hénon 1971; Heggie and Mathieu 1986), in which the typical values for the distance and velocity are of the same order. We will also use the distance in just position or just velocity space as they might behave differently.

We consider the solutions A and B to be converged when $\delta_{A,B} < 10^{-p}$ at all times during the simulation. Note that converged in this case means convergence of the total solution, contrary to convergence per integration step as in the previous section. This criterion for convergence is roughly equivalent to comparing the first p decimal places of the positions and velocities for all N stars, in two subsequent calculations A, B. In most of our experiments we adopt $p = 3$, i.e. all coordinates have to converge to about three decimal places or more. We perform a subset of simulations with $p = 15$ to investigate the effect of small errors (see Section 5.4.3).

Each simulation starts by specifying the initial positions and velocities of N stars in double-precision (see Section 4). The simulation is carried out with the parameter set (ϵ, L_w). We start each simulation with $\epsilon = 10^{-6}$ and $L_w = 56$ bits. This corresponds to a level of accuracy similar to what we reach with the conventional `Hermite` integrator. After this simulation, we increase L_w, for example to

72 bits (~ 22 decimal places), redo the simulation and calculate δ^2. We repeat this procedure until $\delta < 10^{-p}$. When this is achieved, we have obtained a solution in which the round-off error is below a specified number of decimal places for this particular value of ϵ.

We now reduce the tolerance parameter ϵ, for example by a factor of 100, and repeat the procedure of increasing L_w. This series will again lead to a converged solution, but this time it is obtained using a smaller ϵ, and is likely to be different than the previous converged solution. We continue decreasing the value of ϵ by factors of 100 and repeat the procedure, until two subsequent iterations in ϵ lead to a converged solution with a value of $\delta < 10^{-p}$. By this time we are assured of having a solution to the gravitational N-body problem, that is accurate up to at least p decimal places.

In practice, we speed up the procedure by writing the word-length as a function of BS tolerance. Consider for example a BS tolerance of 10^{-20}. To reach convergence up to this level, we need at least 20 decimal places. Adding an extra buffer of 10 digits gives a total of 30 digits, or equivalently a word-length of about 112 bits. For this example, 112 bits turns out to be a good minimum word-length. For a first estimate of the word-length, we use:

$$L_w = 4|\log_{10} \epsilon| + 32 \text{ bits.} \tag{4}$$

With this relation, we will only have to specify a single parameter ϵ, which directly controls the discretisation error and indirectly controls the round-off error. For most of the systems in our experiment the discretisation error turns out to be the dominant source of error and as a consequence ϵ has to decrease quite drastically. When ϵ decreases, L_w increases, even up to the point that there are many more digits available than really needed to control the round-off error. In the case when the discretisation error dominates, the above defined minimum word-length for a given BS tolerance will result in the converged solution. When the round-off error dominates the word-length should be varied independently.

3 Validation and performance
3.1 The Pythagorean problem

To show that our method works, we adopt the Pythagorean 3-body system (Burrau 1913). Previous numerical studies have shown that this system dissolves into a binary and an escaper (Szebehely and Peters 1967; Aarseth et al. 1994). After many complex, close encounters the dissolution happens at about $t = 60$ time units (Dejonghe and Hut 1986), or about 16 crossing times.

We adopt the initial conditions for the Pythagorean problem and integrate up to $t = 100$. To illustrate how the method works we start with a high tolerance and short word-length ($\epsilon = 10^{-2}$, $L_w = 40$ bits), which is less precise

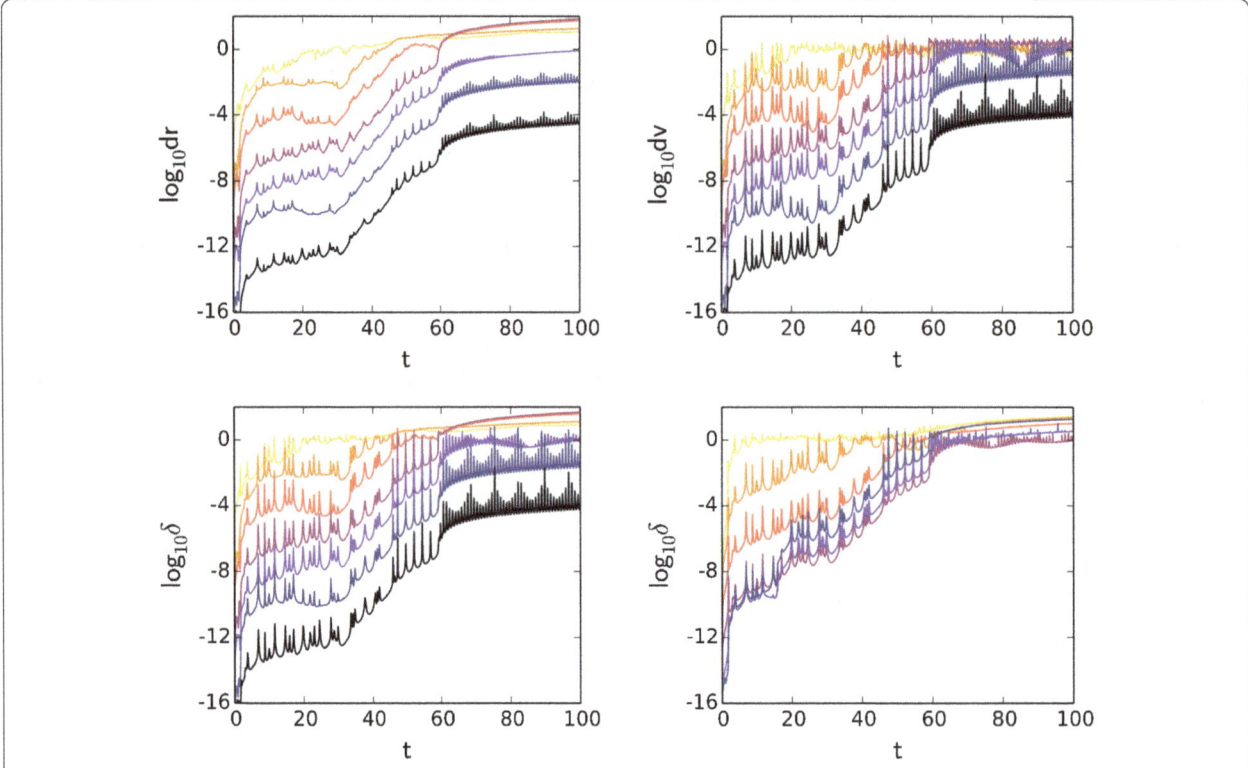

Figure 1 Exponential divergence in the Pythagorean problem. In the top two panels and the lower left panel, Brutus is compared with Brutus with increasing precision. The yellow curves (curves at the top) compare a tolerance of 10^{-2} with 10^{-4}, the orange curves (second curve from the top) compare 10^{-4} with 10^{-6} and so on. The word-length is a function of the tolerance as in equation (4). In the top left panel we show the distance in position-space, in the top right panel in velocity-space and in the bottom left panel in the full phase-space (all normalized by the number of stars and coordinates). The lower right panel compares the converged solution (black and lowest curve in the other plots), with Hermite solutions with time-step parameters $\eta = 2^{-3}, 2^{-5}, 2^{-7}$ up to 2^{-13}, with a color sequence similar as in the other panels.

than double-precision. In Figure 1, this calculation is compared to a simulation with ($\epsilon = 10^{-4}, L_w = 48$ bits), through the yellow (upper) curves in the first three panels. After the first BS integration step, δ obtains a value of the order of the BS tolerance, and continues to increase due to exponential divergence, to eventually exceed $\delta \sim 10^{-1}$, after which the errors become on the order of the typical distance and speed in the system.

In the following step, we repeat the calculation with a precision of ($\epsilon = 10^{-6}, L_w = 56$ bits), and compare the result with the calculation using ($\epsilon = 10^{-4}, L_w = 48$ bits). The comparison is represented by the orange curves (second from above) in Figure 1. The overall behaviour of δ is similar, but the system diverges at a later time due to a higher initial precision.

We continue the iterative procedure until a converged solution has been obtained. In the first three panels of Figure 1, it can be seen that subsequent simulations with higher precision shift the curve to lower values of δ. Superposed on the steady growth of the error are sharp spikes, where the error grows by several orders of magnitude, after which the error restores again (Miller 1964). These spikes

are dominated by errors in the velocity, as can be deduced by comparing the magnitude of the spikes in position and velocity-space. Eccentric binaries which are out of phase when comparing two solutions cause large, periodic errors in the velocity. We finish the procedure when a solution is obtained for which the criterion for convergence is fulfilled, considering the magnitude of the error between the sharp spikes (bottom, black curves).

In the bottom right panel of Figure 1, we compare solutions obtained by the Hermite integrator to the converged solution. The different curves belong to different time-step parameters; $\eta = 2^{-3}, 2^{-5}, 2^{-7}$ up to 2^{-13}. Note that for a time-step parameter $\eta < 2^{-9}$, the curve is not shifted to lower values of δ, but even increases again. At this point round-off error becomes important, making the solution less accurate. The final close encounter in the Pythagorean problem occurs around 60 time units, after which a permanent binary and an escaper are formed. The Hermite integrator is able to accurately reproduce the evolution up to this point, but not subsequently, because δ has increased to values of order unity or higher. This can be explained by a small error in the final close encounter between all three stars, such that the direction of the escaper is slightly different.

To obtain the converged solution up to the first three decimal places, a tolerance of 10^{-14} and a word-length of 88 bits were needed. The simulation was about twice as slow compared to the Hermite simulation with $\eta = 2^{-9}$. The Hermite simulation, however, had a slightly different solution and a final, relative energy conservation of 10^{-8}. Decreasing the value of η will improve the level of energy conservation, but due to round-off error δ will not decrease.

3.2 The equilateral triangle
As a second test case, we adopt the 3-body equilateral triangle as an initial condition (Lagrange 1772). In the exact solution this configuration remains intact, but small perturbations, such as produced by numerical errors, quickly cause the triangle to fall apart. For this problem, we also have a source of error in the initial conditions. Whereas the Pythagorean problem can be set up using integers, the initial condition for the equilateral triangle contains irrational numbers. To control the error in the initial condition, we calculate the initial coordinates with the same word-length as used for the simulation.

In the left panel of Figure 2, a similar diagram is shown as for the Pythagorean problem in the lower left panel of Figure 1. The starting precision is $\epsilon = 10^{-10}$ and the word-length is a function of ϵ as in equation (4). Subsequent simulations are performed with a 10 orders of magnitude higher precision. For a short initial phase of 5 time units, the rate of divergence follows a power law. At later time, the solutions start to diverge exponentially with a characteristic rate independent of the tolerance and word-length. To investigate this transition, we redo the simulations with a large, fixed word-length of 512 bits (green dotted curves). This way, we reduce the amount of round-off error. As a consequence the rate of divergence is first dominated by the accumulation of discretisation errors and this phase lasts for a longer time, until the transition in the behaviour of the divergence, is reached, but now at ~45 time units. The time of the transition depends on word-length. Why the exponential divergence starts once the round-off error has kicked in, is a question that is still under investigation.

The red dashed curves in the same diagram in Figure 2 give the results of the fourth-order Hermite, which are compared with the most precise Brutus simulation (with $\epsilon = 10^{-80}$, $L_w = 352$ bits). The time-step parameter $\eta = 10^{-1}, 10^{-2}, 10^{-3}$ and 10^{-4} for subsequent curves. The Hermite integrations show a similar behaviour as the Brutus results, which could imply that the rate of divergence is a physical property of the configuration, rather than a property of the integrator.

In the right panel of Figure 2 we show the duration for which the triangular configuration remains intact as a function of BS tolerance. For this experiment we halt the simulation when the distance between any two particles has increased or decreased by 10%, after which the triangle falls apart quickly. This diagram also illustrates the linear relation between accuracy and time in this system, which is caused by the constant number of digits being lost during every unit of time. The small scatter is due to the discrete times at which we check the triangular configuration. The solid, blue line is a fit to the data and its slope is $-0.52(3)$, which is equivalent to a loss of 1.9(1) digits per cycle.

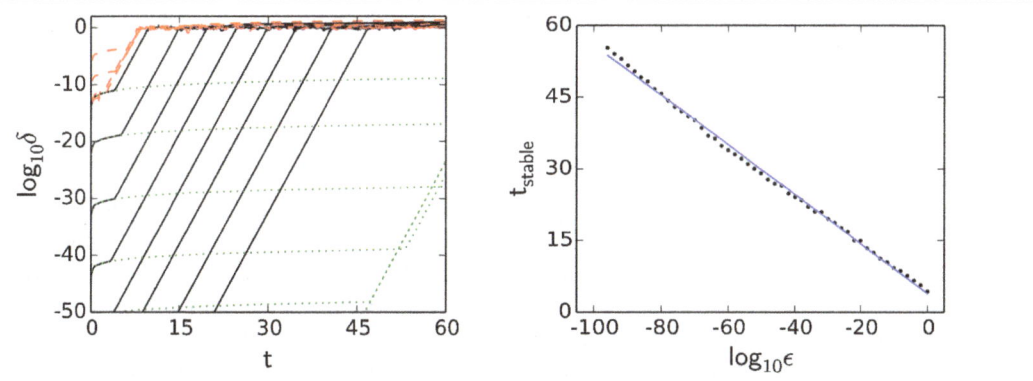

Figure 2 Divergence for the equilateral triangle configuration. In the left panel we show the divergence as a function of time. The solid, black curves compare Brutus solutions with increasing precision, where subsequent precisions are increased by 10 orders of magnitude and where the word-length is a function of tolerance as in equation (4). The dotted, green curves show results for similar simulations, but with a much longer, fixed word-length of 512 bits. The initial power law phase of divergence lasts longer in this case. The exponential divergence becomes dominant when the round-off error has had time to accumulate to become of the order the discretisation error. The dashed, red curves compare the highest precision Brutus solution with Hermite solutions with time-step parameters $10^{-1}, 10^{-2}, 10^{-3}$ and 10^{-4}. In the right panel we show for Brutus, the duration for which the triangular configuration remains intact as a function of Bulirsch-Stoer tolerance ϵ. Note that the time is in units of the period of one complete rotation of the system. The small scatter in the data is due to the discrete times at which we check the triangular configuration.

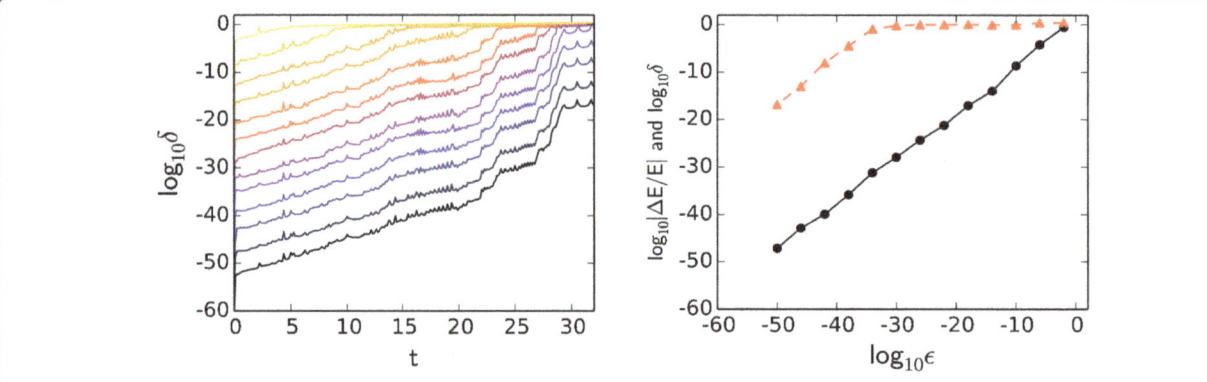

Figure 3 Exponential divergence in a 16-body cluster. In the left panel we illustrate the exponential divergence between Brutus simulations with increasing precision. In the right panel we show the final relative energy conservation (black bullets, solid line) and the final normalized phase space distance between two subsequent simulations (red triangles, dashed line) versus the Bulirsch-Stoer tolerance parameter ϵ. The solution starts to converge at a level of final relative energy conservation of $\sim 10^{-34}$.

3.3 A Plummer distribution with N = 16

As a third test we simulate the dynamical formation of the first hard binary in a small star cluster. We select a moderate number of sixteen equal mass stars and draw them randomly from a Plummer distribution (Plummer 1911). We integrate this system for about ten crossing times and apply the method of convergence. In Figure 3 we present how two solutions with the same initial conditions, but different precisions, diverge as a function of time. The rate of exponential divergence, on average, starts rather constant, with a loss of $\sim 2/3$ digits per time unit. This is equivalent to an e-folding time of $t_e = 0.65$, which is consistent with the results of Goodman et al. (1993) (see their Figure 8). From $t = 20$ onwards, the rate of divergence experiences systematic changes, in particular a steep rise of the error of about 10 orders of magnitude between $t = 26$ and $t = 29$. Such rises are a signature for the presence of a hard binary interacting with surrounding stars.

The right panel in Figure 3 shows the energy conservation (black bullets, solid line) and the normalized phase space distance (red triangles, dashed line) versus ϵ. Energy conservation is proportional to ϵ, but the solutions only start to converge for $\epsilon < 10^{-34}$. More generally, even if conserved quantities like total energy are conserved to machine-precision or better, it is not guaranteed that the solution itself has converged.

The highest precision Brutus simulation in this example ($\epsilon = 10^{-50}$, $L_w = 232$ bits), took about a day of wall-clock time, which is about 7,000 times slower than a simulation with Hermite using $\eta = 2^{-9}$.

3.4 Scaling of the wall-clock time

The use of arbitrary-precision arithmetic dramatically increases the CPU time of N-body simulations. Also the BS method, which performs integration steps iteratively, makes an integration scheme more expensive by at least a

factor two or more. To investigate for example how feasible it would be to run a converged N-body simulation for 10^3 stars through core collapse, we perform a scaling test in which we vary the number of particles and the precision, ϵ and L_w.

We randomly select positions and velocities for N equal mass stars from the virialised Plummer distribution (Plummer 1911), for $N = 2, 4, 8, \ldots$, up to 1,024. The BS tolerance is fixed at a level of 10^{-6} and the word-length at 64 bits. We integrate the systems for one Hénon time unit and measure the wall-clock time. In the top left panel in Figure 4 we show the wall-clock time as a function of N, which fit the relation $t_{CPU} \propto N^{2.6}$.

For $N > 32$, it becomes efficient to parallellise the code. Our version implements i-parallellisation (Portegies Zwart et al. 2008) in the calculation of the accelerations. In the top right panel of Figure 4, we plot the speed-up, S, against the number of cores. For $N = 1,024$, we obtain a speed up of a factor 30 using 64 cores.

In the lower panels of Figure 4 we present the scaling of the wall-clock time with BS tolerance and word-length. To measure the dependence on BS tolerance, we simulated a 16-body cluster for 1 Hénon time unit. We varied the BS tolerance while keeping the word-length fixed at $L_w = 1,024$ bits. The relation obtained converges to $t_{CPU} \propto \epsilon^{-0.032}$. A similar experiment was performed to measure the dependence on word-length. This time we fixed the BS tolerance at $\epsilon = 10^{-10}$ and varied the word-length. For $L_w < 1,024$, the relation can be estimated as $t_{CPU} \propto L_w^{0.33}$, while for $L_w > 1,024$, $t_{CPU} \propto L_w$. This transition depends on the internal workings of the arbitrary-precision library which we will not discuss here.

Using a very long word-length of 4,096 bits, i.e. $\sim 10^3$ digits, results in a slowdown of a factor $f_s \sim 16$ compared to 64 bits. But for some simulations a BS tolerance smaller than 10^{-50} can easily be required to reach convergence, and

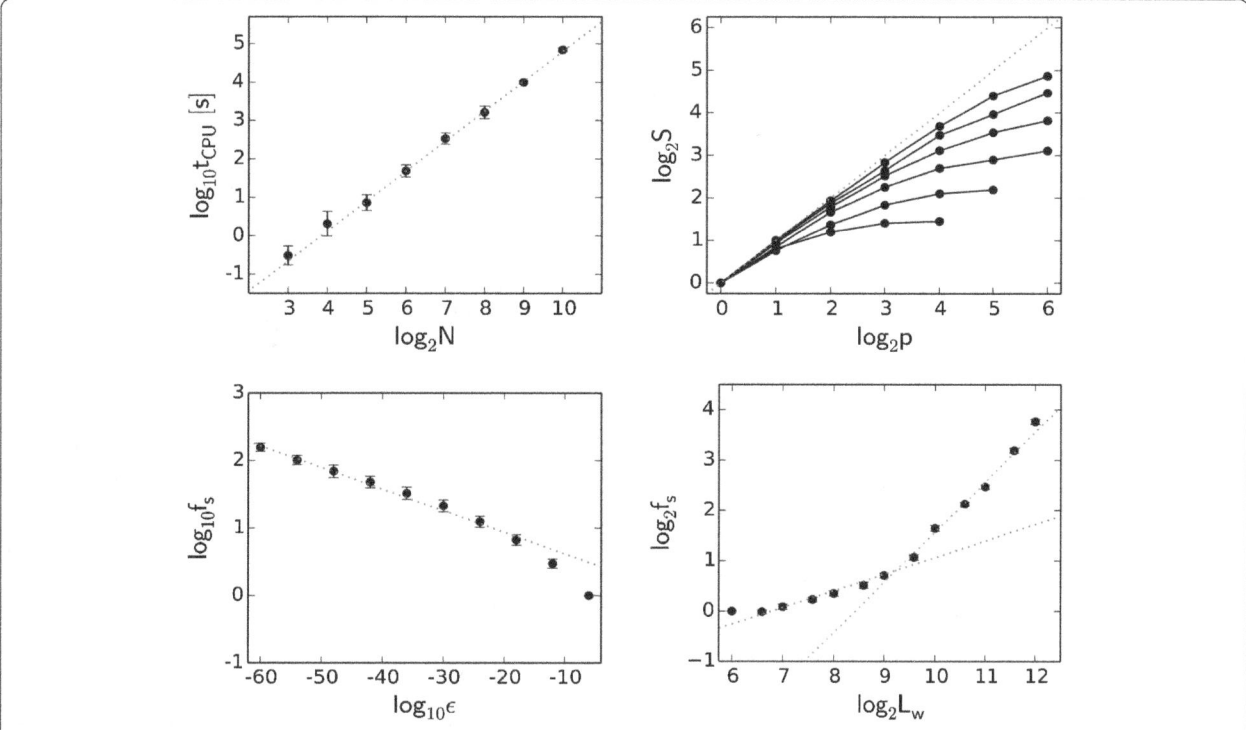

Figure 4 Scaling of Brutus. In the top left panel we show the scaling of the wall-clock time that Brutus needs as a function of number of stars N. The dotted curve is a fit to the data given by $t_{CPU} \propto N^{2.6}$. In the top right panel we show the speed-up when the number of cores, p, is increased. The bottom, solid curve represents $N = 32$ and each curve above has an N a factor two higher than the previous curve. The dotted curve represents ideal scaling. In the bottom left panel we plot the slowdown factor as a function of the Bulirsch-Stoer tolerance ϵ, for a fixed word-length of 1,024 bits. In the bottom right panel we plot the slowdown factor as a function of word-length L_w, for a fixed tolerance of 10^{-10}. The slowdown of the simulations is mainly caused by the very small Bulirsch-Stoer tolerances required.

this will result in a slowdown of a factor $f_s > 100$. The very small BS tolerance is often the main cause for the slowdown of the simulations, instead of the increased word-length.

Using the above results, we can construct the following model to estimate the wall-clock time for integrating 1 Hénon time unit with $L_w < 1,024$ bits:

$$t_{CPU} = \left(\frac{N}{512}\right)^{2.6} \left(\frac{\epsilon}{10^{-6}}\right)^{-0.032} \left(\frac{L_w}{64}\right)^{0.33} 10^4 \ [s]. \quad (5)$$

Integrating $N = 1,024$ with standard precision ($\epsilon = 10^{-6}$, $L_w = 64$ bits), up to core collapse at $\sim$300 time units, and taking into account a speed up of a factor 30 due to parallellisation, we estimate a total wall-clock time of a week. Increasing the precision to ($\epsilon = 10^{-20}$, $L_w = 112$ bits), will take about a month. A precision of ($\epsilon = 10^{-50}$, $L_w = 232$ bits) will take roughly a year. To estimate how much precision is needed, we will assume that the rate of exponential divergence before the formation of the first hard binary is approximately constant. In the left panel of Figure 3, the initial slopes correspond to a loss of $\sim$2/3 digits per time unit. We construct the following approximate model for

the initial BS tolerance needed to end up with a converged solution:

$$\log_{10} \epsilon = \log_{10} \delta_{final} - R_{div} t_{cc}. \quad (6)$$

Here ϵ is the BS tolerance parameter, δ_{final} is the final precision of all the coordinates in the system, R_{div} is the approximately constant rate of divergence, e.g. the number of accurate digits lost per unit of time, and t_{cc} is the core collapse time. We set the final precision to 10^{-6}, i.e. convergence to the first 6 decimal places, and we set the core collapse time to $\sim$300 as before. If we adopt $R_{div} = 2/3$, we estimate that we need an $\epsilon \sim 10^{-206}$. This would take about 10^5 years to finish. It would be more practical to simulate a 256-body cluster. If we set the core collapse to 100 time units we estimate $\epsilon \sim 10^{-73}$, which would take about a month on a cluster of 64 Intel® Xeon® E5530 cores.

For direct N-body codes, the time for integrating up to core collapse usually scales as $\mathcal{O}(N^3)$. Using the analysis above, we estimate that the time for converged core collapse simulations scales approximately exponentially. This is effectively caused by the exponential divergence.

4 Precision of statistical results: experimental setup

In the previous section we demonstrated that it is possible to obtain a converged solution for a particular initial condition. We have also shown that a solution obtained by `Hermite` diverges from the converged solution, even up to the point that the microscopic solution given by `Hermite` is beyond recognition. We now perform a statistical study, to examine the hypothesis that double-precision N-body simulations produce statistically indistinguishable results, from those obtained from an ensemble of converged solutions with the same set of initial conditions. Because it is computationally expensive to reach convergence, we start investigating the hypothesis above by exploring the accuracy of 3-body statistics.

The $N = 3$ experiment is inspired by the Pythagorean problem, where after a complex 3-body interaction, a binary and an escaper are formed. As a variation to this, we define four different sets of initial conditions as follows:

1. Plummer distribution equal mass.
2. Plummer distribution with masses 1:2:4.
3. Plummer distribution equal mass with zero velocities.
4. Plummer distribution with masses 1:2:4 and zero velocities.

The positions and velocities of the three stars are selected randomly from a virialised Plummer distribution (Plummer 1911; Aarseth et al. 1974). For the cold collapse systems, we set the velocities to zero. Then we rescale the positions and velocities to virialise the systems if the initial velocities are non-zero, or we set the total energy equal to $E = -0.25$ if the system starts out cold. We adopt standard Hénon units (Hénon 1971; Heggie and Mathieu 1986) throughout.

In the case of the cold initial conditions, the systems start democratically, i.e. the minimal distance between each pair of particles is greater than N^{-1}. We reject initial conditions in which this criterion is not satisfied. This is to prevent initial realisations where two stars which are very near, fall to each other radially causing very long wall-clock times for the integration. When starting with a democratic configuration, there will also be an initial close triple encounter (Aarseth et al. 1994), which is hard to integrate accurately and is therefore a good test. A total of 10,000 random realisations are generated for each set of initial conditions and can be found in Additional files 1, 2, 3 and 4.

We stop the simulations when the system is dissolved into a permanent binary and an escaper. The criteria used to detect an escaper are the following:

1. escaper has a positive energy, $E > 0$,
2. is a certain distance away from the center of mass, $r > 2r_{virial}$,
3. is moving away from the center of mass, $r \cdot v > 0$.

The energy of the escaper is calculated in the barycentric frame of the three particles and r_{virial} is the virial radius of the system, which is of the order unity in Hénon units.

There may be situations in which a star is ejected without actually escaping from the binary. After a long excursion the star turns around and once again engages the binary in a 3-body resonance (Hut and Bahcall 1983). Because these systems need to be integrated for a longer time, they also require higher precision to reach convergence, which takes a long time to integrate [see also Hut (1993)]. To deal with this issue, we perform the simulations iteratively by increasing the final integration time t_{end}. Starting with $t_{end} = 50$ Hénon time units, we evolve every system and detect those that are dissolved. Then we increase t_{end} to 100, 150, 200, etc., but only for those systems which have not yet dissolved. A complete ensemble of solutions is obtained up to $t_{end} \sim 500$, or equivalently $\sim$180 crossing times where the crossing time has a value of $2\sqrt{2}$ in Hénon units (Hénon 1971; Heggie and Mathieu 1986). Systems which take a longer time to integrate are not taken into account in this research. The fraction of long-lived systems is however a statistic we measure. We gathered the final, converged configurations in Additional files 1, 2, 3 and 4.

Each initial realisation is run with the `Hermite` code, using standard double-precision, and with `Brutus`, using arbitrary-precision until a converged solution is obtained. At the end of each simulation, we investigate the nature of the binary and the escaper. In addition to the BS tolerance, word-length, CPU time and dissolution time, we record the mass, speed and escape direction of the escaping single star, and the semimajor axis, binding energy and eccentricity of the binary. In this way, we obtain statistics for $N = 3$ generated by a conventional N-body solver and by `Brutus`.

5 Results

Before we perform a detailed comparison between results obtained by `Hermite` and `Brutus`, we first compare the `Brutus` results with analytical distributions from the literature in order to relate to previous studies. We compare `Hermite` and `Brutus` on a global level by performing two-sample Kolmogorov-Smirnov tests (Kolmogorov 1933; Smirnov 1948) to see whether global distributions are statistically indistinguishable. We also compare the distribution of lifetimes of triples to see whether precision influences the stability and we measure the typical CPU time and BS tolerance needed to obtain a converged solution. After this, we compare `Hermite` and `Brutus` per individual system, with the aim of investigating the nature of the differences of every individual outcome. Finally, we define categories which classify a conventional simulation as a preservation or exchange, depending on whether the identity of the escaping star is consistent between `Hermite` and `Brutus`.

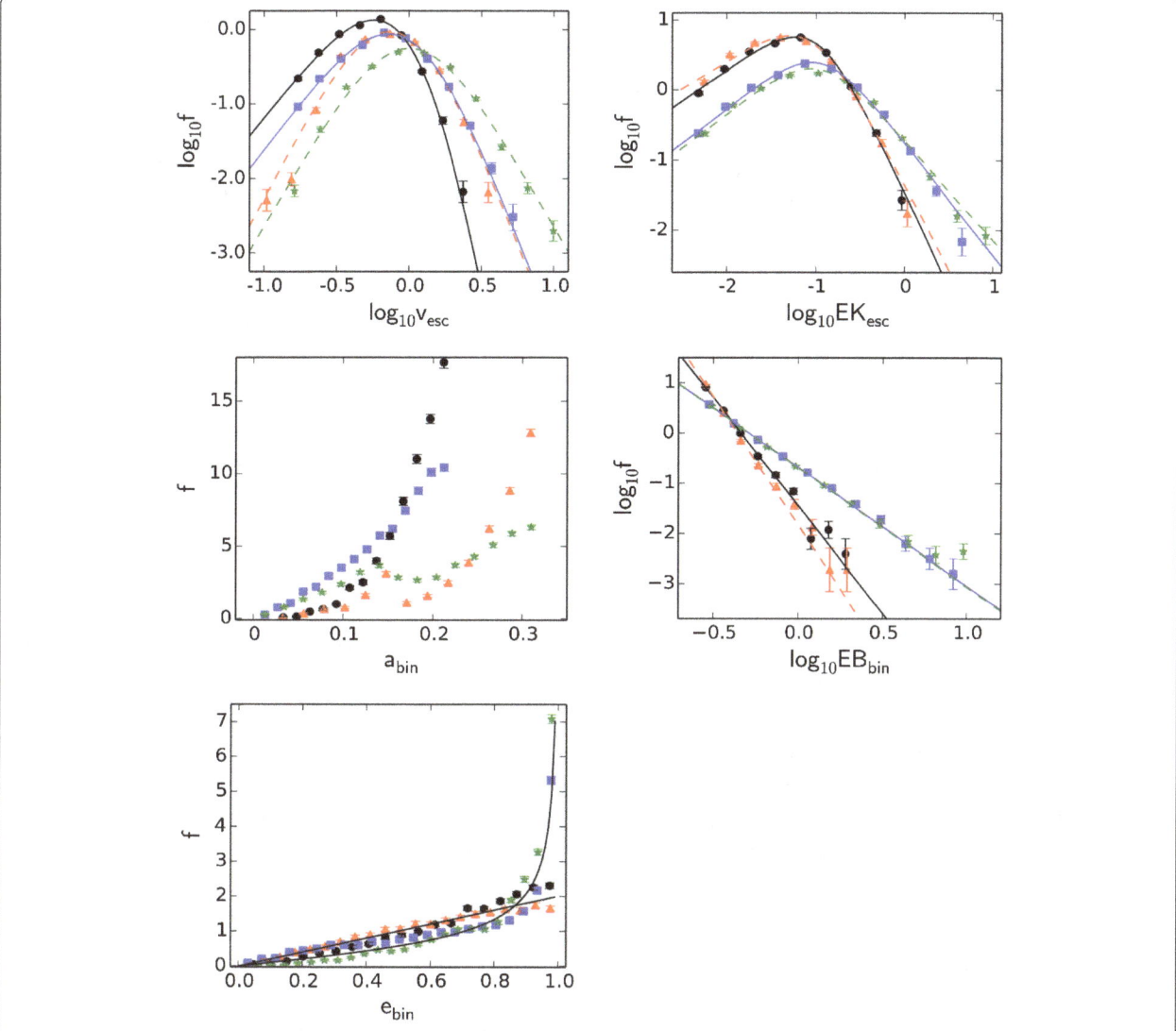

Figure 5 Comparison of Brutus results and analytical distributions. Distributions are given for the escaper speed (top left) and kinetic energy (top right), binary semimajor axis (middle left), binding energy (middle right) and binary eccentricity (bottom). The results from the Brutus simulations are represented by the data points, for each of the four sets of initial conditions: Plummer equal mass (black bullets), Plummer with different masses (red triangles), cold Plummer equal mass (blue squares) and cold Plummer with different masses (green stars). Note that we use standard Hénon units (Hénon 1971; Heggie and Mathieu 1986). Analytical models from the literature are fitted to the empirical distributions represented by the curves. For the eccentricities we plot the thermal distributions.

5.1 Brutus versus analytical distributions

In Figure 5, the distributions obtained by converged solutions are given for the following quantities: velocity and kinetic energy of the escaper in the barycentric reference frame, and semimajor axis, binding energy and eccentricity of the binary. We start by looking at the eccentricity distributions (bottom panel in Figure 5). These distributions can be estimated analytically by assuming that the probability of a certain configuration is proportional to the associated volume in phase space (Monaghan 1976; Valtonen and Karttunen 2006) or by considering an equi-

librium distribution of binary stars in a cluster (Heggie 1975). The resulting thermal distribution in the three-dimensional case is given by

$$f(e) = 2e, \tag{7}$$

and in the two-dimensional case by

$$f(e) = \frac{e}{\sqrt{1 - e^2}}. \tag{8}$$

The 3-body cold collapse problem is essentially a two-dimensional problem. We compare the empirical and theoretical distributions by means of the K-S test (see also next section). It turns out that the distributions in eccentricity are statistically distinguishable. By inspection by eye we observe that in the virialised case, there are slight deviations at high eccentricities. In the case of the equal-mass, cold systems, there are more low eccentricity binaries compared to the theoretical prediction. They coincide at an eccentricity of about 0.7, after which they deviate again. For the cold systems with unequal masses, this behaviour is the other way around. The analytical predictions are able to capture the empirical distributions only in a qualitative manner.

The velocity distribution of the single escaping star can be estimated analytically in a similar way as was done for the eccentricities. The resulting distribution is predicted to be a double powerlaw given by (Monaghan 1976; Valtonen and Karttunen 2006):

$$f(v) \propto \frac{v^{\alpha}}{(1 + \gamma v^2)^{\beta}}. \tag{9}$$

We fit this model to the data (see Figure 5, first panel) and obtain values for α and β which are given in Table 1. The powerlaw indices vary with mass ratio and total angular momentum. To remove the dependence on mass ratio, we plot the kinetic energy of the escaper (see Figure 5, top right panel). Again, we fit a double powerlaw of a similar form as equation (9), and the powerlaw indices are given in Table 1. Both the escaper velocity and kinetic energy are consistent with a double powerlaw distribution.

The binary semimajor axis and binding energy are related quantities. We fit the binding energy distribution (see Figure 5, middle right panel) to a powerlaw (Heggie 1975; Monaghan 1976; Valtonen and Karttunen 2006):

$$f(E_B) \propto E_B^{-\alpha}. \tag{10}$$

The fitted powerlaw indices are given in Table 1. The empirical distributions are consistent with a powerlaw, although somewhat steeper than predicted (Heggie 1975; Monaghan 1976; Valtonen and Karttunen 2006). The slopes do tend to vary somewhat as a function of angular momentum (Monaghan 1976; Valtonen and Karttunen 2006).

The empirical distributions obtained by Brutus are in qualitative agreement with the analytical estimates present in the literature (Heggie 1975; Monaghan 1976; Valtonen and Karttunen 2006). Slight variations are present due to the dependence on total angular momentum, a limited statistical sampling and assumptions made in the derivation of the analytical distributions. Nevertheless, a similar qualitative agreement has been obtained between the analytical distributions discussed above and empirical distributions from an ensemble of conventional numerical solutions, e.g. not converged (Valtonen and Karttunen 2006, Chapters 7-8 and references therein). The question remains to what extend conventional and converged solutions agree quantitatively.

5.2 Brutus versus Hermite: global comparison

A quantitative way to compare global distributions is by performing two-sample Kolmogorov-Smirnov tests (K-S tests) (Kolmogorov 1933; Smirnov 1948). The K-S test gives the likelihood that two samples are drawn from the same distribution, quantified by the value called p. When the p-value is below five percent, the distributions are considered to be significantly different.

In Figure 6 we plot the p-value obtained by comparing the Brutus distribution with the Hermite distribution versus time-step parameter η used for Hermite. In the panel showing the data for the binary semimajor axis, the distributions of the cold systems become significantly different for $\eta > 2^{-6}$. The distributions from the initially virialised systems start to differ for $\eta > 2^{-4}$. The cold systems are harder to model accurately, because of the close encounters that occur shortly after the start. The reason the distributions start to become significantly different at large time-steps is because at these large time-steps most simulations violate energy conservation by $|\Delta E/E| > 0.1$. When this occurs, solutions might reach regions in $6N$-dimensional phase-space, which theoretically are forbidden. The distribution then becomes biased by these outlier solutions.

5.3 Lifetime of triple systems

In Figure 7, we present the fraction of triple systems which are undissolved, i.e. still interacting, as a function of time.

Table 1 Fitted powerlaw indices for the velocity and kinetic energy distributions of the escaping stars and for the binding energy distribution of the binary stars

	α	β
Velocity		
Plummer equal mass	2.5 ± 0.09	6.7 ± 1.02
Plummer mass ratio	3.8 ± 0.16	4.4 ± 0.43
Cold Plummer equal mass	2.6 ± 0.19	3.8 ± 0.28
Cold Plummer mass ratio	3.4 ± 0.45	3.4 ± 0.19
Kinetic energy		
Plummer equal mass	0.9 ± 0.02	1.8 ± 0.04
Plummer mass ratio	0.8 ± 0.02	1.6 ± 0.04
Cold Plummer equal mass	0.99 ± 0.02	1.3 ± 0.03
Cold Plummer mass ratio	0.98 ± 0.03	1.2 ± 0.02
Binding energy		
Plummer equal mass	4.31 ± 0.13	
Plummer mass ratio	5.12 ± 0.32	
Cold Plummer equal mass	2.37 ± 0.11	
Cold Plummer mass ratio	2.38 ± 0.12	

Note that we use equal intervals in logarithmic space.

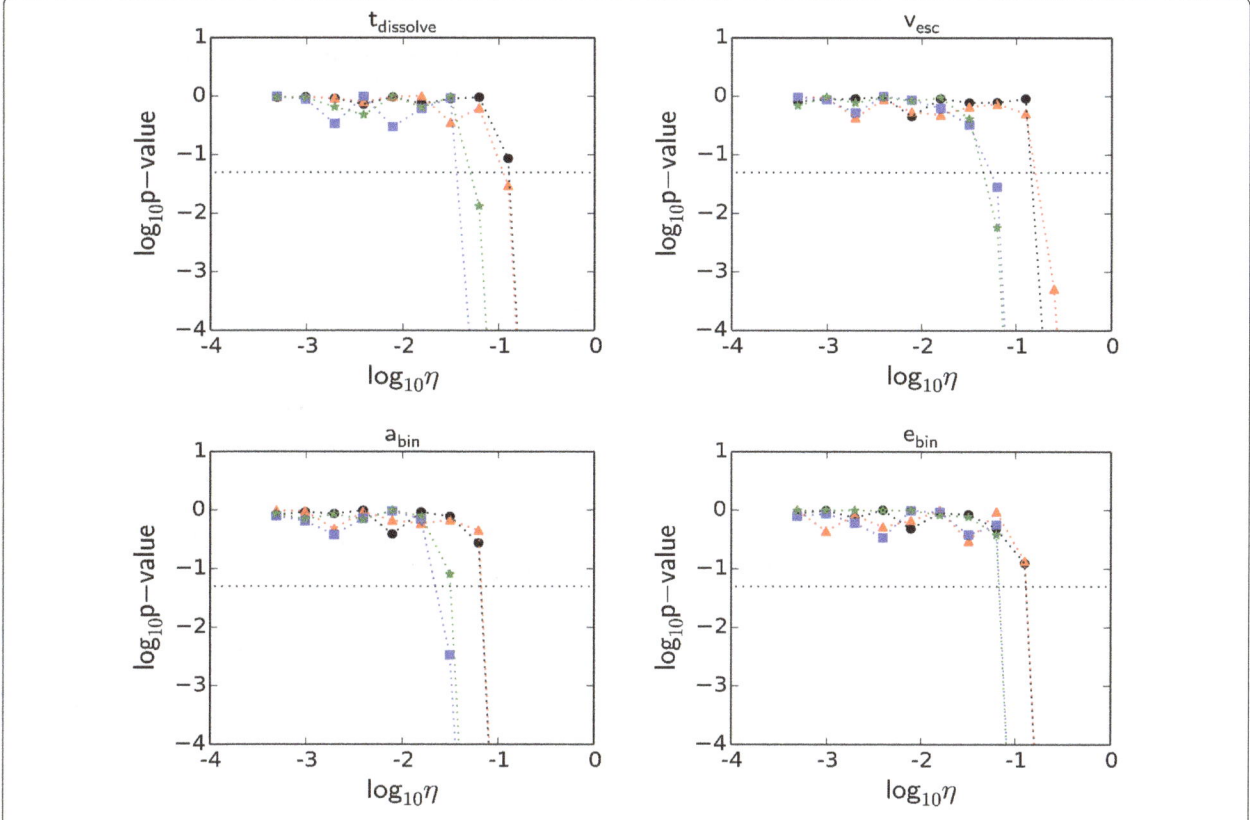

Figure 6 Two-sample K-S tests on distributions obtained by Hermite and Brutus. We compare distributions of dissolution time (top left), escaper speed (top right), binary semimajor axis (bottom left) and binary eccentricity (bottom right)). The color coding is the same as in Figure 5. Two-sample K-S tests are performed and the p-value is plotted versus Hermite time-step parameter η. The dashed line represents the 5% significance level. For $\eta < 2^{-5}$, the distributions are not significantly different.

The results by Brutus are represented by the data points: equal-mass Plummer (black bullets), Plummer with different masses (red triangles), equal-mass cold Plummer (blue squares) and cold Plummer with different masses (green stars). The results by Hermite for a time-step parameter $\eta = 2^{-5}$ are represented by the curves appearing to go through the data points.

The initially cold systems dissolve faster than the initially virialised systems. This is somewhat expected due to the close triple encounter resulting from the initial cold collapse: the rate of energy exchange can be very high for these encounters (Johnstone and Rucinski 1991). After ~180 crossing times, about 40% of the systems which started with an equal-mass Plummer initial configuration, are undissolved, compared to about 10% for the cold Plummer with different masses. Systems which include stars with different masses dissolve faster than their equal mass counterparts. Energy equipartition tends to cause the lightest particle to quickly reach the escape velocity.

In Figure 7, the grey curves through the data points represent the interpolated Hermite results. Even though Hermite and Brutus use different algorithms and preci-

sions to solve the equations of motion, we find that the lifetime of an unstable triple is statistically indistinguishable between converged Brutus and non-converged Hermite solutions (but see also Section 6.3).

In Figure 8, we plot the maximum CPU time and minimum BS tolerance, both as a function of dissolution time. This is shown for the Brutus simulations, for the four different initial conditions. The longer it takes for a system to dissolve, the longer the CPU time and the higher the precision needed to reach a converged solution. To reach ~180 crossing times, there are systems which require a BS tolerance of the order 10^{-100}, with the final converged run taking of the order a few days. The average CPU time as a function of time is about an order of magnitude smaller than the maximum CPU time. The average BS tolerance ranges from ~10^{-20} to 10^{-30}. For systems which dissolve within 100 crossing times, Brutus is on average about a factor 120 slower than Hermite.

We were able to obtain a complete ensemble of systems dissolving within ~180 crossing times. Simulations which take longer than this are not taken into account in this experiment. The fraction of long-lived systems as obtained

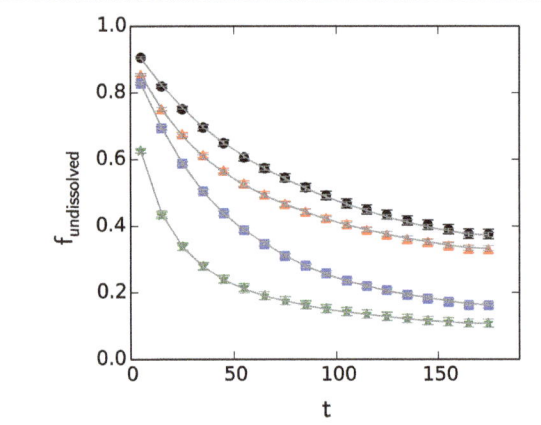

Figure 7 Lifetime of triple systems. We plot the fraction of triple systems that have not dissolved yet into a permanent binary and escaping single star configuration, as a function of simulation time (in units of crossing time). The color coding is the same as in Figure 5. The grey curves through the data points represent the interpolated Hermite results with a time-step parameter $\eta = 2^{-5}$.

by Hermite and Brutus are consistent. For our purpose of comparing results from conventional integrators with the converged solution, integrating up to ~180 crossing times is sufficient, in the sense that there is enough time for conventional solutions to diverge from the true solution (see Section 5.4.1). Including the long-lived triple systems may however influence the statistical distributions and biases on the long term.

5.4 Brutus versus Hermite: individual comparison

For the individual comparison, we take a certain initial realisation and compare the solutions of Hermite and Brutus. In Figure 9 we show scatter plots of the Hermite solution (with time-step parameter $\eta = 2^{-5}$) versus the converged Brutus solution for the equal-mass Plummer data set.

Data points on the diagonal represent accurate solutions, whereas the scatter around it represents inaccurate Hermite solutions. The diagonal is present in each panel and extends throughout the range of possible outcomes. The width of the diagonal is very narrow. When the normalized phase-space distance between the Hermite and Brutus solution $\delta < 10^{-1}$, then the coordinates are accurate enough to produce derived quantities accurate to at least one decimal place and Hermite and Brutus will give similar results. Once $\delta > 10^{-1}$, the solution has diverged to a different trajectory in phase-space leading to a different outcome. This outcome could in principle be any of the possible outcomes as can be derived from the amount of scatter in the Hermite solutions at a fixed Brutus solution.

In the scatter plot of the dissolution time, we observe that for small times ($t < 10$), Hermite and Brutus agree on the solution in the sense that the data points lie on the diagonal. Systems which dissolve after a short time don't have sufficient time to accumulate enough error to diverge to another trajectory in phase-space. Once however this level of divergence is reached, the scatter immediately covers the entire, available outcome space. This randomisation is also observed in the other panels.

5.4.1 The fraction of accurate solutions

In Figure 10 we estimate the fraction of data points on the diagonal as a function of the Hermite time-step parameter, η. We only include the data points for which the normalized phase-space distance $\delta < 10^{-1}$. For the largest time-step parameters used ($\eta > 10^{-1}$) the fraction on the diagonal, or the accurate fraction, varies from zero to about 0.2. By reducing the time-step parameter, the accurate fraction increases until it saturates at about 0.4 to 0.7 depending on the initial conditions. Even though by reducing η, the discretisation error decreases, the number of integration steps increases, which then increases the round-off error. For the data sets with zero angular momentum, the

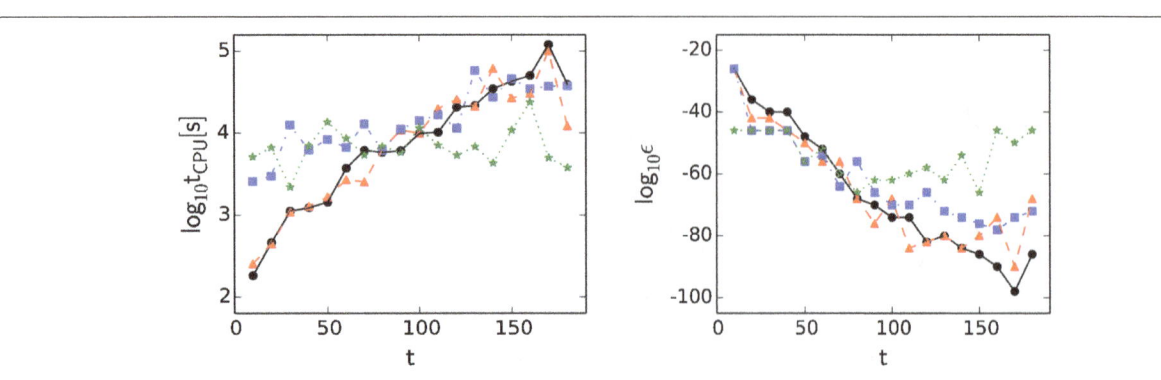

Figure 8 CPU time and precision as a function of time for Brutus. On the left, we plot the CPU time of the simulation which took the longest, as a function of dissolution time. On the right, we plot the Bulirsch-Stoer tolerance of the simulation which needed the highest precision, as a function of dissolution time. The different curves represent the four sets of initial conditions as in the previous plots.

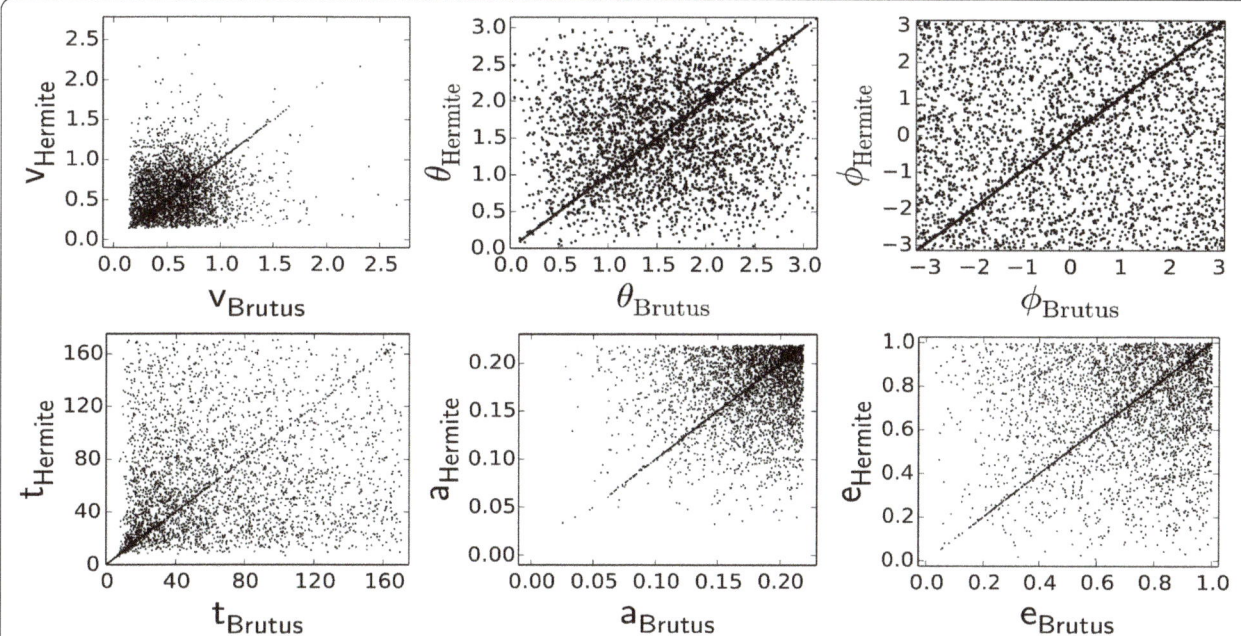

Figure 9 Direct comparison of Brutus and Hermite results per individual simulation. The results are shown only for the $N = 3$ equal mass Plummer data set and for a Hermite time-step parameter $\eta = 2^{-5}$. Each dot in a panel represents a different initial realisation. The value on the ordinate is the value obtained using Hermite and the value on the abscissa the value obtained by Brutus. We compare the escaper velocity (top left), direction of the escaper: polar angle (top middle) and azimuthal angle (top right) (with respect to the plane of the binary and pericentre direction), dissolution time (bottom left), binary semimajor axis (bottom middle) and binary eccentricity (bottom right). The diagonal represents accurate Hermite solutions. The scatter around it represents solutions where Hermite and Brutus have diverged.

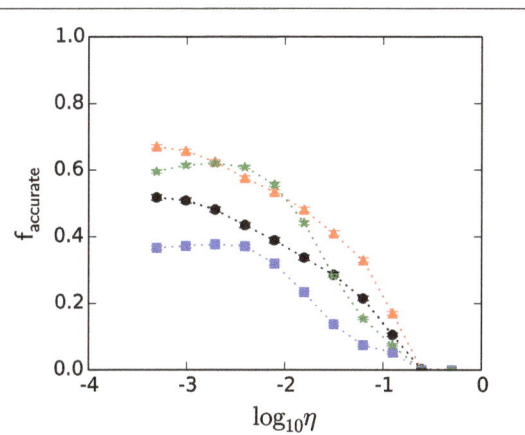

Figure 10 The fraction of accurate Hermite simulations as a function of Hermite time-step parameter η. The different curves represent the different data sets: equal mass Plummer (black bullets), Plummer with different masses (red triangles), equal mass cold Plummer (blue squares) and cold Plummer with different masses (green stars). As η decreases, the accurate fraction increases. However, for $\eta < 2^{-7}$, the fraction starts to saturate, more so for the cold data sets. At this point the effect of round-off error becomes important.

maximum accurate fraction is obtained for $\eta \sim 2^{-9}$. For the initially virialised systems this seems to occur between $\eta \sim 10^{-3}$-10^{-4}, although the actual saturation point is not visible yet. This dependence on angular momentum is due to the initial cold collapse and subsequent close encounters, which increases the round-off error.

5.4.2 The error distribution
In Figure 11 we present statistics on the distribution of the errors, i.e. $S_{\text{Hermite}} - S_{\text{Brutus}}$, with S a statistic. For the dissolution time and the eccentricity, the average error converges to zero for $\eta < 10^{-1}$. For larger time-steps, simulations which grossly violate energy conservation ($|\Delta E / E| > 0.1$) cause biases in the average error. For the binary semimajor axis however, the data representing the cold collapse simulations also seem to be systematically biased for small time-steps, in the sense that Hermite makes fewer tight binaries.

The width of the error distributions converge to a nonzero value. This can be understood because with decreasing time-step, round-off errors will become more important so that the standard deviation of the errors will never reach zero. For the dissolution time, the width of the error distribution for the smallest time-step parameter adopted, varies from 60 to 100 crossing times. For the eccentricities the width is on average ~ 0.2. For the semimajor axis the width approaches ~ 0.05 (in Hénon units). In the case of the semimajor axis, the data representing the cold collapse simulations behave differently, because the width is much

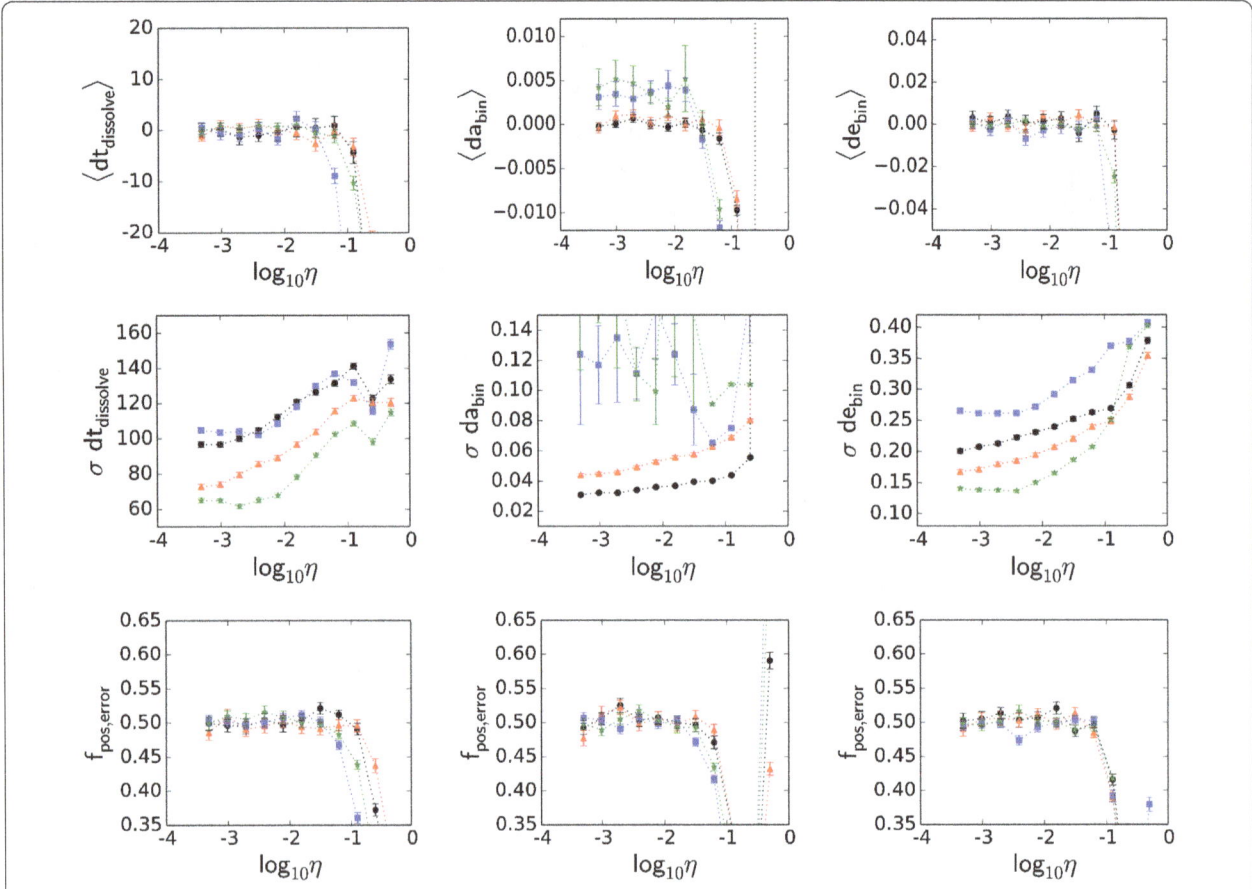

Figure 11 Statistics on the error distribution of Hermite results. We present the average error (top row), the standard deviation of the error distribution (middle row) and the fraction of errors which are positive (bottom row). The errors are given for the dissolution time (left column), binary semimajor axis (middle column) and eccentricity (right column). The different curves represent the different data sets similar as in Figure 10.

larger than the width for the data representing the initially virialised systems.

If we regard the results given by Brutus and Hermite as random variables drawn from the same distribution, then we can write the variance in a certain statistic, in this example the eccentricity, as:

$$\left\langle (e_H - e_B)^2 \right\rangle = \left\langle e_H^2 \right\rangle + \left\langle e_B^2 \right\rangle - 2\left\langle e_H \right\rangle\left\langle e_B \right\rangle. \quad (11)$$

Here e stands for eccentricity and the subscripts for Brutus and Hermite. For a thermal eccentricity distribution (equation (7)), we obtain a standard deviation of 1/3. However, this only applies to inaccurate Hermite results, which had enough time to diverge through outcome space. If we multiply the theoretical standard deviation calculated above by the inaccurate fraction, we obtain a range in the standard deviation from 0.17 to 0.27, as η ranges from the most precise value to $\eta = 10^{-1}$.

5.4.3 Symmetry of the error distribution

To measure the symmetry of the error distribution, we count the fraction of positive errors (Figure 11, bottom panels). Again for an $\eta < 10^{-1}$, this fraction converges to 0.5. A more detailed comparison is given in Figure 12, where we compare distribution functions of positive and negative errors. In Section 2.3, we mentioned that in our experiment we define the Brutus solution to be converged when at least 3 decimal places of every coordinate have converged. To investigate the symmetry up to higher precision, we repeated a subset of 1,000 simulations. We did this only for the initial conditions with equal-mass stars picked randomly from a virialised Plummer distribution and this time we obtain solutions converged up to the first 15 decimal places.

We observe that the majority of errors are larger than $\sim 10^{-3}$ and within the statistical error, the positive and negative errors have a similar distribution. For the smallest errors however, we observe an asymmetry in the sense that there are more negative, small errors. The magnitude of the error where this excess occurs is determined by the

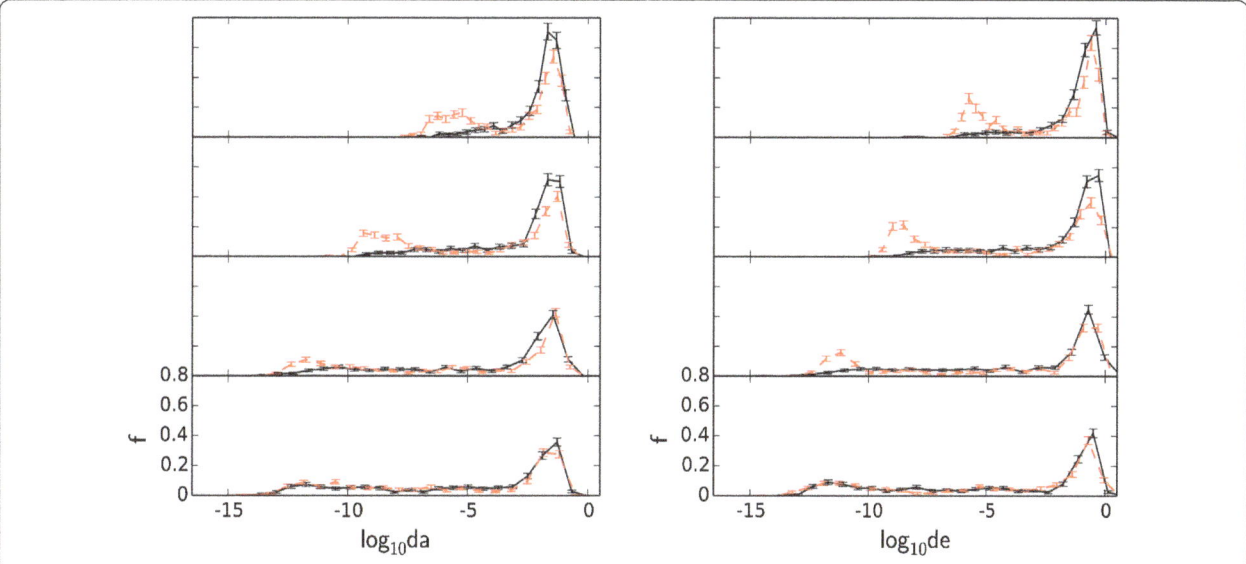

Figure 12 Symmetry of the error distributions. We show distributions of the errors in semimajor axis (left column) and eccentricity (right column) of the binaries formed in the equal-mass Plummer data set. This is shown separately for the positive errors (solid, black) and negative errors (dashed red), to investigate the symmetry of the error distribution. From the panels at the top to the bottom, the time-step parameter for `Hermite` varies as 2^{-5}, 2^{-7}, 2^{-9} and 2^{-11}. An asymmetry can be observed at the smallest errors.

precision of the integration. For the smallest η, the excess is below double-precision and thus not observable anymore (see Section 6.2 for more explanation).

5.5 Escaper identity

In this section we compare the solutions obtained with `Hermite` and `Brutus` individually, by looking at which star eventually becomes the escaper and which form the binary. We define preservation if the `Hermite` and the `Brutus` solution both have the same star as the escaper. We define it as exchange if the escaping star is different. A further distinction can be made in the preservation category, if the `Hermite` simulation is also accurate. We can typify each `Hermite` simulation as follows:

- *Accurate*: The coordinates are accurate, up to at least two digits.
- *Preservation*: The coordinates are inaccurate, but same star escapes.
- *Exchange*: Different star escapes.

In Figure 13 we present the fraction of each category as a function of time. As expected, systems which dissolve quickly, hardly have time to develop errors and are categorized as accurate simulations. In time however, because errors grow exponentially, the solutions become inaccurate. The fractions of preservation and exchange start to grow. For a small time-step parameter ($\eta = 2^{-11}$, top row in Figure 13), this growth starts after ~20 crossing times for the initially virialised systems. For the initially cold systems, the inaccurate fractions already start to grow after a single crossing time.

The cold collapse with equal-mass stars is the hardest problem to integrate as the accurate fraction is of comparable magnitude as the preservation and exchange fractions. The accurate fraction generally remains dominant, with a final fraction varying from about 0.4 for the equal-mass cold Plummer to about 0.7 for the Plummer with different masses. For the lesser precision ($\eta = 2^{-3}$, bottom row in the figure), the accurate fractions decrease to below 0.2.

In the panels in Figure 13, which include the data for the systems with different masses, preservation is more common than exchange. This can be understood, because due to energy equipartition, the lightest particle will be more likely to escape and therefore the identity is more often correct than in the equal mass case. For the equal mass case, the fraction of preservation and exchange is comparable, except in the case of the equal-mass cold Plummer with the low precision ($\eta = 2^{-3}$, the bottom row). If we regard the identity of the escaping star to be completely random once the solution has become inaccurate, we would expect the fraction of exchange to be twice the fraction of preservation. This is roughly what we observe in the equal mass cold collapse case with low precision. Because of the low precision and the initial close encounter, solutions will diverge very quickly. In the panel with the higher precision this trend is not observed because the solutions are less randomised. The preservation category includes solutions which slightly differ from the converged solution only in the escape angle of the escaper. Also the long-lived triples are not taken into account here, which will alter these fractions.

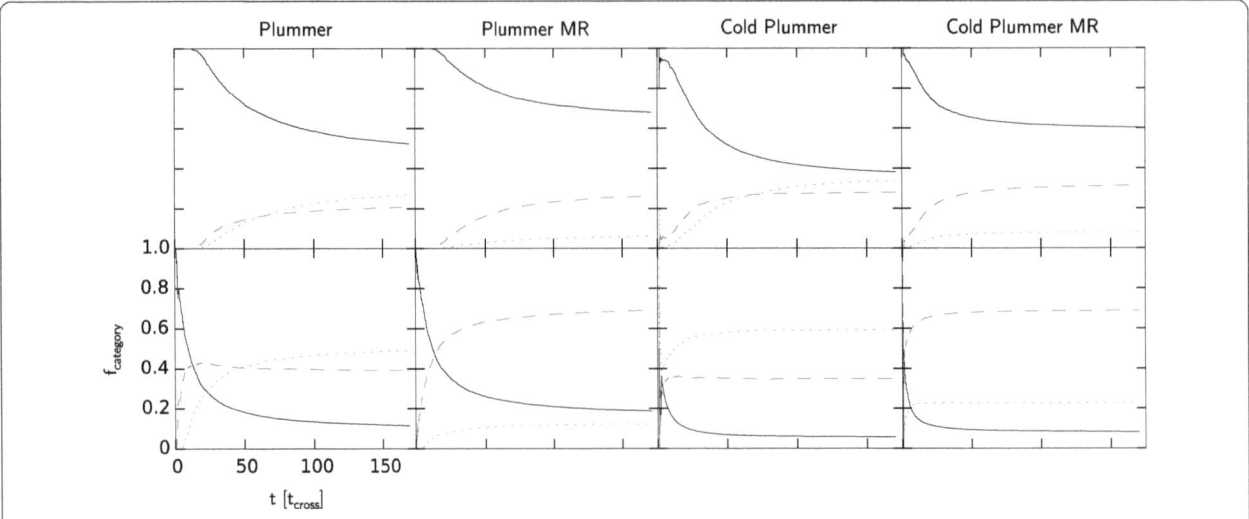

Figure 13 The evolution of the relative fractions of categories. The different curves represent the different categories: accurate (solid, black curves), preservation (dashed, red curves) and exchange (dotted, blue curves). These three categories are defined in the text. From left to right, the data are from the Plummer, Plummer with different masses, cold Plummer and cold Plummer with different masses data sets. In the top panels we show the results for a `Hermite` time-step parameter $\eta = 2^{-11}$ and in the bottom for $\eta = 2^{-3}$.

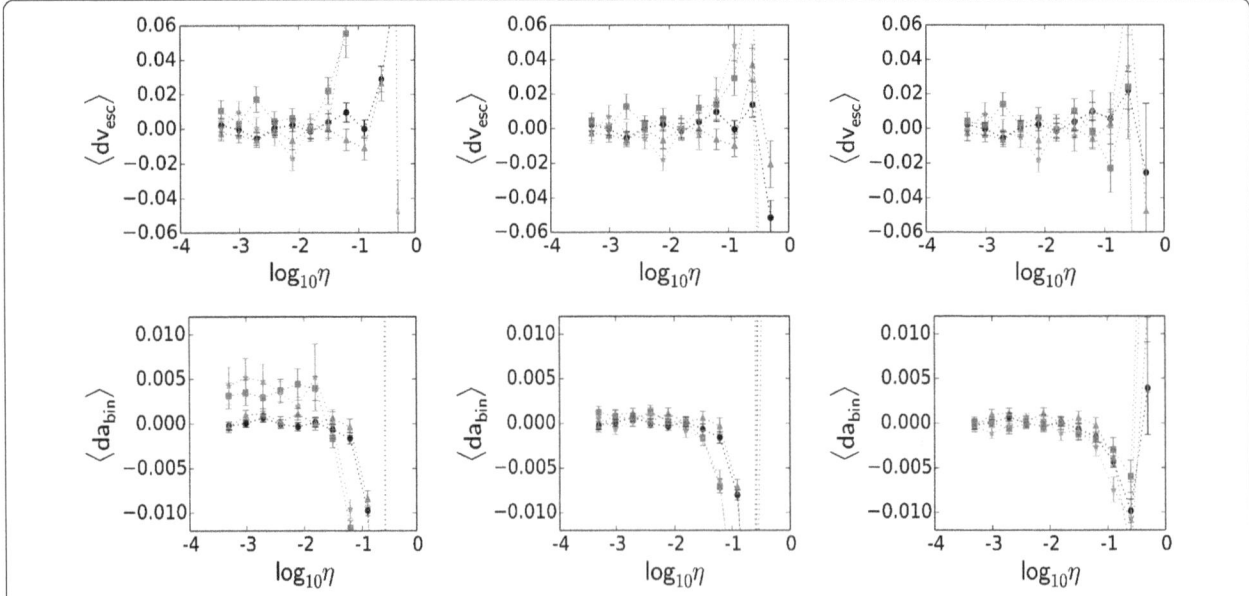

Figure 14 The effect of cuts in final relative energy conservation. We plot the average error in the velocity of the escaping star (top row) and the error in the binary semimajor axis (bottom row) as a function of `Hermite` time-step parameter η (with same color coding as in Figure 10). The three columns differ in the maximum allowed level of relative energy conservation. In the left column we show the results for the total ensemble of solutions, in the middle column for a maximum level of unity and in the right column for 10^{-1}. The bias in the left column for the binary semimajor axis is caused by solutions which grossly violate energy conservation. Note that this only happens for the cold collapse simulations. When these outliers are taken out of the ensemble, the bias vanishes.

6 Discussion

6.1 Energy conservation

In every ensemble of `Hermite` solutions there are some that grossly violate conservation of energy $|\Delta E/E| > 0.1$. This deformation of the energy hyper-surface in phase-space can allow solutions to reach parts of phase-space which are theoretically forbidden. This affects the global statistical distributions. In Figure 14, we replot the average error in the binary semimajor axis as a function of the time-step parameter. We produce similar diagrams as presented in Figure 11, but this time we introduce a maximum allowed error in the energy. If we filter out simulations with a relative energy conservation $|\Delta E/E| > 1$, or $|\Delta E/E| > 0.1$, we observe that the bias in the average error of the semi-major axis of the binaries vanishes. We conclude that this bias is caused by a few simulations which grossly violate energy conservation. A similar bias in the velocity of the escaping star is less pronounced.

Time-reversible, symplectic integrators should in principle conserve energy to a better level than non-symplectic integrators, since there is no drift present in the energy error. Therefore, by using a symplectic integrator, the number of simulations with large energy error could be reduced. Using a Leapfrog integrator with constant time-steps, we tested this assumption and we find that for resonant 3-body interactions, it is challenging to obtain accurate solutions. The main reason is that, contrary to regular systems like, for example, the Solar System, resonant 3-body interactions often include very close encounters, which need a very small time-step size to be resolved accurately. This is especially the case for the initially cold systems. Adopting such a small time-step size for the whole simulation, will increase the wall-clock time to that of Brutus or beyond.

6.2 Asymmetry at small errors

In Section 5.4.3, we discussed an asymmetry at small errors. In Figure 15, we present similar diagrams as in Figure 12 for the positive and negative errors. This time we add the errors in the total energy and angular momentum of the system and the error in the velocity of the escaper.

We also vary the integration method because different methods produce different (biased) error distributions in energy and angular momentum. We use a standard Leapfrog integrator, a standard Hermite integrator and a Hermite integrator which uses the $P(EC)^n$ method (we adopted $n = 3$) (Kokubo et al. 1998). This last method adds an iterative procedure to the algorithm to improve the predictions and corrections, which improves the time-symmetry. For each method we implement a shared, adaptive time-step criterion as in equation (1), with a time-step parameter $\eta = 2^{-7}$. As a consequence they will not be time-symmetric nor symplectic.

We first look at the error distributions in the total energy and angular momentum. We observe that none of them are symmetric, in the sense that the positive and negative errors have identical distributions, except for the angular momentum in the Leapfrog simulations. The Leapfrog solutions tend to gain energy, whereas the standard Hermite loses energy. The Hermite with the $P(EC)^n$ method produces both positive and negative errors in the energy, but not in a symmetric manner.

To investigate whether the bias in energy and angular momentum conservation propagates to a bias in the binary

and escaper properties, we estimate what the errors should be if we regard the error in the energy and angular momentum as a small perturbation to the converged solution. For the error in the velocity of the escaper, using the derivative of the kinetic energy with respect to velocity, we obtain the following expression:

$$\delta v = \frac{1}{mv}\delta E. \tag{12}$$

Here m is the mass of a star, v the velocity as obtained by Brutus, δE the energy error and δv the error in the velocity due to this energy error. For the binary semimajor axis we obtain:

$$\delta a = \frac{2}{m^2}a^2\delta E. \tag{13}$$

Here a is the semimajor axis from the Brutus solution. For the eccentricity we obtain:

$$\delta e = \frac{1}{\sqrt{1 + \frac{2\epsilon l^2}{\mu^2}}}\left(\frac{l^2}{\mu^2}\delta\epsilon + \frac{2\epsilon l}{\mu^2}\delta l\right). \tag{14}$$

Here μ is the total mass of the binary, ϵ and l the specific energy and specific angular momentum of the binary as obtained by Brutus. The error in the eccentricity δe has contributions from errors in the energy $\delta\epsilon$ and angular momentum δl.

If we compare the resulting error distributions to the actual error distributions, we find that the approximated error distribution is positioned at the asymmetry in the empirical error distribution. This is most clearly seen for the semimajor axis and eccentricity (see Figure 15).

The reason why the approximated error distribution overestimates the excess, is because not all errors are solely due to an error in the energy and angular momentum. In time, the numerical solution diverges from the true solution and this error due to divergence will become more dominant. With this in mind, we can approximate the error in a statistic as follows:

$$\delta S = \delta S_{\text{conservation}} + \delta S_{\text{divergence}}. \tag{15}$$

Here S is a statistic that is related to energy and/or angular momentum, $\delta S_{\text{conservation}}$ is the error due to a small perturbation in the energy and/or angular momentum and $\delta S_{\text{divergence}}$ is the error due to divergence of the solution. When the solution has not diverged appreciably yet, the first type of error will dominate and possible biases can be observed. When the second type of error dominates, we observe that the symmetry is restored to within the statistical error.

Upon inspection of the velocity data, we observe no asymmetry in the Hermite results. When we measure

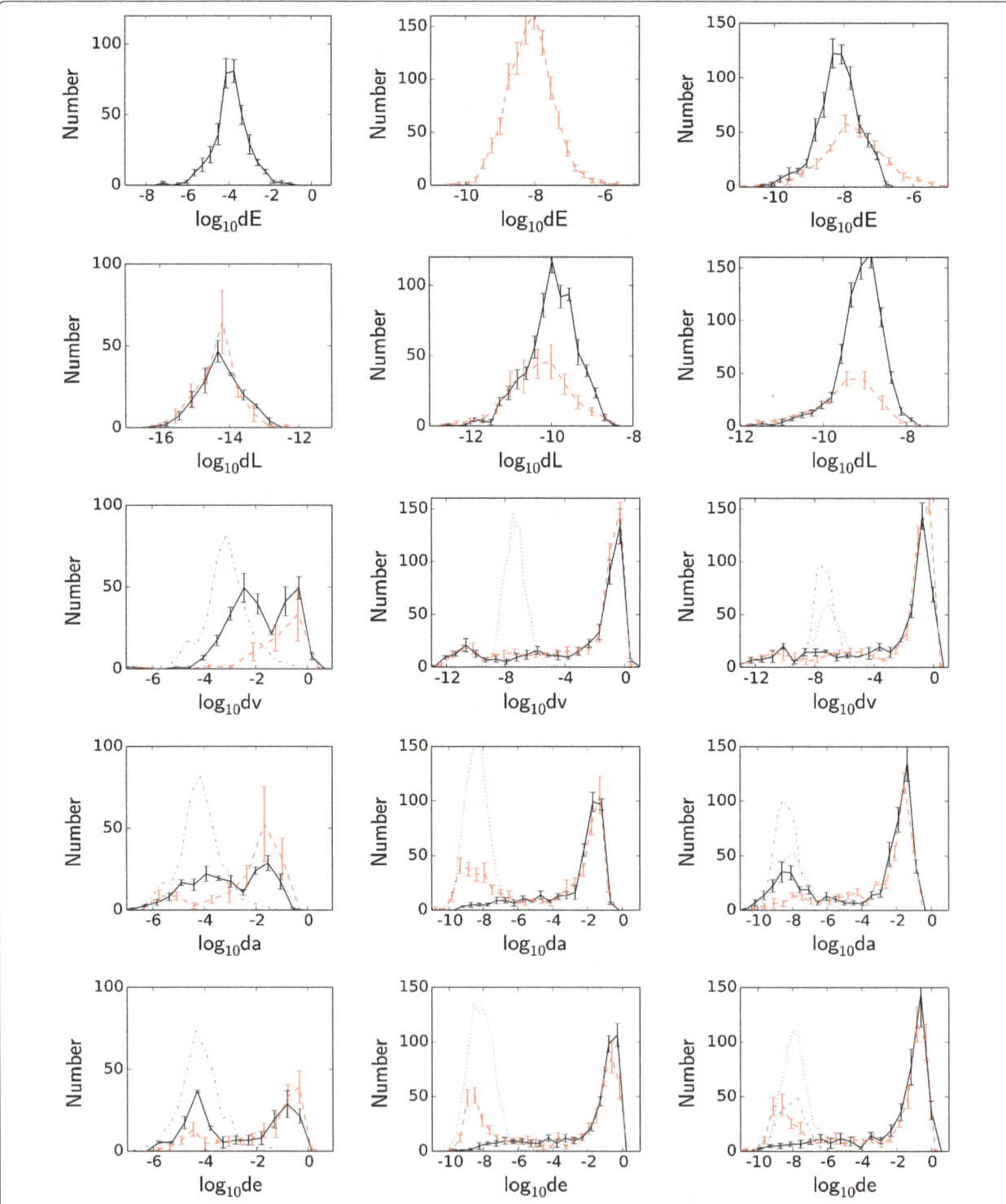

Figure 15 Explanation of the asymmetry at small errors. We show distributions of the positive (solid, black) and negative (dashed, red) errors in the total energy (top row), total angular momentum (second row), escaper velocity (third row), binary semimajor axis fourth row) and eccentricity (bottom row). This is shown for different algorithms: Leapfrog (left column), standard `Hermite` (middle column) and `Hermite` with $P(EC)^n$ method (right column, $n = 3$). Each method implements a shared, adaptive time-step criterion according to equation (1), with a time-step parameter $\eta = 2^{-7}$. Each of these three integrators has a different asymmetry in the conservation of energy and angular momentum. By propagating these asymmetric errors as a small perturbation to the converged solution, we can estimate the resulting asymmetry in the derived quantities. These estimated error distributions are also given separately for the positive (dot-dash, blue) and negative (dotted, green) errors. We observe that the estimated error distributions are located at the asymmetry in the empirical error distributions. The asymmetry at small errors is caused by a bias in the integrator.

which fraction of the energy error is reserved for the binary and which fraction for the escaper, we find that in most cases the error propagates to the binary. For the Leapfrog however, the asymmetry is still present.

6.3 Preservation of the macroscopic properties

Valtonen et al. (2004) state that the final statistical distributions forget the specific initial conditions and only depend on globally conserved quantities. This assumption makes predictions which are verified by our experiment. The results show that for a time-step parameter $\eta < 2^{-5}$, the distributions are statistically indistinguishable, even though at least half of the solutions diverged from the converged solution. If however, energy conservation is grossly violated, biases are introduced in the statistics. In our experiment, a maximum level of relative energy conservation of $|\Delta E/E| = 0.1$ was sufficient to remove the biases. This is a much milder constraint than the $|\Delta E/E| \sim 10^{-6}$ usually adopted in collisional simulations. Whether 0.1 is also sufficient for systems with more stars, should be verified experimentally. Heggie (1991) for example, finds that the energy of escaping stars in higher-N systems, depends sensitively on integration accuracy. The maximum required level of energy conservation should be such that it is below the energy taken away from the cluster by the escaping stars.

The chaoticity of the 3-body problem is illustrated by the scatter diagrams in Figure 9. For a certain value of a statistic obtained by Brutus, any other value in the allowed outcome space is reachable for the Hermite integrator. For example, if the converged solution gives an eccentricity for the binary of 0.6, a diverged solution can produce any eccentricity between 0 and 1. Once the solution has diverged from the true solution, it will start a random walk through or near the allowed phase-space until the 3-body system has dissolved. We observed that this randomisation happens in such a way that the available outcome space is still completely sampled and that it preserves global statistical distributions.

In Section 4, we discussed that the lifetime of an unstable triple does not depend on the integrator used nor on the accuracy of that integrator. This last point should be interpreted in the sense that when more effort is put into performing simulations with higher precision, that this does not change the global statistics, even though individual solutions will change with precision (see for example the Hermite results in Figure 1). If instead we continue to decrease the precision, there will be a point where biases start to appear. Urminsky (Urminsky 2008) analysed the 3-body Sitnikov problem and showed that the precision of the integration influences the average lifetime of triple systems, contrary to our results. The integration times in our experiment however, are much shorter. Obtaining a converged solution for a resonant 3-body system for longer than 200 crossing times, is still computationally challenging. Therefore any statistical difference on the long term will not be visible in our experiment.

7 Conclusion

Brutus is an N-body code that uses the Bulirsch-Stoer method to control discretisation errors, and arbitrary-precision arithmetic to control round-off errors. By using the method of convergence, where we systematically vary the Bulirsch-Stoer tolerance parameter and the word-length, we can obtain a solution for a particular N-body problem, for which the first p digits in the mantissa are independent of the time-step size and word-length. We call this solution converged to p decimal places.

Obtaining the converged solution is computationally expensive, mainly because of the exponential divergence of the solution. In some cases, Bulirsch-Stoer tolerances of 10^{-100} are needed to reach convergence. We estimate that the time for simulating a star cluster up to core collapse, until convergence, scales approximately exponentially with the number of stars. Simulations with 256 stars however, may be performed within a year of computing time.

The motivation to obtain expensive, converged solutions is to test the assumption that the statistics of an ensemble of approximate solutions, are indistinguishable from the statistics of an ensemble of true solutions. To put this assumption to the test, we have investigated the statistics on the breakup of 3-body systems. In our experiment, a bound triple system will eventually dissolve into a binary and an escaping star. Solutions to every initial realisation were obtained using the standard Hermite integrator and using Brutus.

For systems with a long lifetime it is challenging to obtain the converged solution. Due to repeated ejections and resonances, many accurate digits will be lost and so a very small Bulirsch-Stoer tolerance is required. Therefore, we have set an integration limit at $\sim$180 crossing times. For equal-mass, virialised systems, $\sim$40% of the random initial realisations were not dissolved by this time. For the initially cold systems with different masses this was $\sim$10%. Hermite and Brutus are consistent on the average lifetime of an unstable triple system. However, possible differences on the long term are not visible in this experiment.

When we compare the results on an individual basis, we find that on average about half of the Hermite solutions give accurate results, i.e. at most a 1% relative difference compared to Brutus. For the inaccurate results, the error distribution becomes unbiased and symmetric for a time-step parameter $\eta \leq 2^{-5}$ and implementing a maximum level of relative energy conservation of $|\Delta E/E| < 0.1$.

Once the conventional solution has diverged from the converged solution, it will start a random walk through or near the allowed region in phase space. such that any allowed outcome of a statistic is reachable. This randomisation process completely samples the available outcome space of a statistic and it also preserves the global statistical distributions.

Kolmogorov-Smirnov tests were performed to compare the global distributions produced by `Hermite` and `Brutus`. No significant differences were detected when using the criteria mentioned above for the time-step parameter η and relative energy conservation. This research for the 3-body problem supports the assumption that results from conventional N-body simulations are valid in a statistical sense. We observed however that a bias is introduced for the smallest errors, if the algorithm used to solve the equations of motion, is biased in the conservation of energy and angular momentum. In this research however, this bias did not have an appreciable effect. It is important to see whether this remains true for statistics of higher-N systems or systems with a dominant mass. An example of a higher-N system where precision might play a role is a young star cluster (without gas) going through the process of cold collapse (Caputo et al. 2014). At the moment of deepest collapse, a fraction of stars will obtain large accelerations, so that a small error in the acceleration can cause large errors in the position and velocity. The rate of divergence can increase up to about 5 digits per Hénon time unit for 128 particles and it increases with N.

Additional material

Additional file 1: Initial and final configurations for the equal-mass Plummer. This table consists of 10,000 initial configurations for three equal-mass stars drawn randomly from a Plummer distribution, together with the final configurations as obtained by `Brutus`. Additional information is given on the dissolution time, the Bulirsch-Stoer tolerance and word-length. For the configurations which took longer than 500 Hénon time units to dissolve, we give the last configuration of the simulation. For the simulations where the CPU time was very high, we set the final coordinates equal to zero.
Additional file 2: Initial and final configurations for the Plummer with different masses. Similar as the previous additional file, but for the virialised Plummer initial condition with different masses.
Additional file 3: Initial and final configurations for the cold Plummer. Similar as the previous additional file, but for the equal-mass Plummer starting with zero velocities.
Additional file 4: Initial and final configurations for the cold Plummer with different masses. Similar as the previous additional file, but for the Plummer with different masses, starting with zero velocities.

Competing interests
The authors declare that they have no competing interests.

Authors' contributions
TB wrote the `Brutus` N-body code, participated in designing the experiments, performed the N-body simulations, gathered the results from the simulations, interpreted the results and wrote the major part of the manuscript. SPZ thought of the concept of the `Brutus` code, participated in designing the experiments, interpreted the results and helped to draft the manuscript. All authors read and approved the final manuscript.

Acknowledgements
We thank Douglas Heggie, Piet Hut, Michiko Fujii and Guilherme Gonçalves Ferrari for useful discussions and comments on the manuscript. TB would also like to thank Ann Young and Lucie Jíková for carefully reading the manuscript and improving the presentation. The authors also thank the referees for providing useful improvements to our manuscript. This work was supported by the Netherlands Research Council NWO (grants #643.200.503, #639.073.803 and #614.061.608) and by the Netherlands Research School for Astronomy (NOVA). Part of the numerical computations were carried out on the Little Green Machine at Leiden University and on the Lisa cluster at SURFSara in Amsterdam.

Endnotes
[a] We use the open-source library GMP: http://gmplib.org/.
[b] Formerly known as N-body units. Introduced by D. Heggie at MODEST14.

References
Aarseth, SJ, Anosova, JP, Orlov, VV, Szebehely, VG: Global chaoticity in the Pythagorean three-body problem. Celest. Mech. Dyn. Astron. **58**, 1-16 (1994)
Aarseth, SJ, Anosova, JP, Orlov, VV, Szebehely, VG: Close triple approaches and escape in the three-body problem. Celest. Mech. Dyn. Astron. **60**, 131-137 (1994)
Aarseth, SJ, Henon, M, Wielen, R: A comparison of numerical methods for the study of star cluster dynamics. Astron. Astrophys. **37**, 183-187 (1974)
Bulirsch, R, Stoer, J: Fehlerabschätzungen und extrapolation mit rationalen funktionen bei verfahren vom richardson-typus. Numer. Math. **6**, 413-427 (1964)
Burrau, C: Numerische Berechnung eines Spezialfalles des Dreikörperproblems. Astron. Nachr. **195**, 113 (1913)
Caputo, DP, de Vries, N, Portegies Zwart, S: On the effects of subvirial initial conditions and the birth temperature of R136. Mon. Not. R. Astron. Soc. **445**, 674-685 (2014)
Dejonghe, H, Hut, P: Round-off sensitivity in the N-body problem. In: Hut, P, McMillan, SLW (eds.) The Use of Supercomputers in Stellar Dynamics. Lecture Notes in Physics, vol. 267, p. 212. Springer, Berlin (1986)
Goodman, J, Heggie, DC, Hut, P: On the exponential instability of N-body systems. Astrophys. J. **415**, 715 (1993)
Gragg, WB: On extrapolation algorithms for ordinary initial value problems. SIAM J. Numer. Anal. **2**, 384-403 (1965)
Heggie, DC: Binary evolution in stellar dynamics. Mon. Not. R. Astron. Soc. **173**, 729-787 (1975)
Heggie, DC: Chaos in the N-body problem of stellar dynamics. In: Roeser, S, Bastian, U (eds.) Predictability, Stability, and Chaos in N-Body Dynamical Systems, pp. 47-62 (1991)
Heggie, DC, Mathieu, RD: Standardised units and time scales. In: Hut, P, McMillan, SLW (eds.) The Use of Supercomputers in Stellar Dynamics. Lecture Notes in Physics, vol. 267, p. 233. Springer, Berlin (1986)
Hénon, MH: The Monte Carlo method (Papers appear in the Proceedings of IAU Colloquium No. 10 Gravitational N-Body Problem (ed. by Myron Lecar), R. Reidel Publ. Co., Dordrecht-Holland.) Astrophys. Space Sci. **14**, 151-167 (1971)
Hut, P: Binary-single-star scattering. III - Numerical experiments for equal-mass hard binaries. Astrophys. J. **403**, 256-270 (1993)
Hut, P, Bahcall, JN: Binary-single star scattering. I - Numerical experiments for equal masses. Astrophys. J. **268**, 319-341 (1983)
Hut, P, Heggie, DC: Orbital divergence and relaxation in the gravitational N-body problem. J. Stat. Phys. **109**, 1017-1025 (2002)
Ito, T, Tanikawa, K: Long-term integrations and stability of planetary orbits in our Solar system. Mon. Not. R. Astron. Soc. **336**, 483-500 (2002)
Johnstone, D, Rucinski, SM: Statistical properties of planar zero-angular-momentum equal-mass triple systems. Publ. Astron. Soc. Pac. **103**, 359-367 (1991)
Kokubo, E, Yoshinaga, K, Makino, J: On a time-symmetric Hermite integrator for planetary N-body simulation. Mon. Not. R. Astron. Soc. **297**, 1067-1072 (1998)
Kolmogorov, A: Sulla determinazione empirica di una legge di distribuzionc. 1st. Ital. Attuari. G. **4**, 1-11 (1933)

Lagrange, JL: Essai sur le Problème des Trois Corps. Prix de l'Académie Royale des Sciences de Paris **6**, 292 (1772)

Makino, J, Aarseth, SJ: On a Hermite integrator with Ahmad-Cohen scheme for gravitational many-body problems. Publ. Astron. Soc. Jpn. **44**, 141-151 (1992)

Miller, RH: Irreversibility in small stellar dynamical systems. Astrophys. J. **140**, 250 (1964)

Monaghan, JJ: A statistical theory of the disruption of three-body systems. I - Low angular momentum. Mon. Not. R. Astron. Soc. **176**, 63-72 (1976)

Newton, I: Philosophiae Naturalis Principia Mathematica (1687)

Plummer, HC: On the problem of distribution in globular star clusters. Mon. Not. R. Astron. Soc. **71**, 460-470 (1911)

Portegies Zwart, S, McMillan, S, Groen, D, Gualandris, A, Sipior, M, Vermin, W: A parallel gravitational *N*-body kernel. New Astron. **13**, 285-295 (2008)

Portegies Zwart, S, McMillan, S, Pelupessy, I, van Elteren, A: Multi-physics simulations using a hierarchical interchangeable software interface. In: Capuzzo-Dolcetta, R, Limongi, M, Tornambè, A (eds.) Advances in Computational Astrophysics: Methods, Tools, and Outcome. Astronomical Society of the Pacific Conference Series, vol. 453, p. 317 (2012)

Quinlan, GD, Tremaine, S: On the reliability of gravitational *N*-body integrations. Mon. Not. R. Astron. Soc. **259**, 505-518 (1992)

Smirnov, N: Table for estimating the goodness of fit of empirical distributions. Ann. Math. Stat. **19**(2), 279-281 (1948)

Smith, H Jr.: The dependence of statistical results from *N*-body calculations on *N*. Astron. Astrophys. **76**, 192-199 (1979)

Szebehely, V, Peters, CF: Complete solution of a general problem of three bodies. Astron. J. **72**, 876 (1967)

Urminsky, D: On the calculation of average lifetimes for the 3-body problem. In: Vesperini, E, Giersz, M, Sills, A (eds.) IAU Symposium, vol. 246, pp. 235-236 (2008)

Valtonen, M, Karttunen, H: The Three-Body Problem. Cambridge University Press, Cambridge (2006)

Valtonen, M, Mylläri, A, Orlov, V, Rubinov, A: Statistical approach to the three-body problem. In: Byrd, GG, Kholshevnikov, KV, Myllri, AA, Nikiforov, II, Orlov, VV (eds.) Order and Chaos in Stellar and Planetary Systems. Astronomical Society of the Pacific Conference Series, vol. 316, p. 45 (2004)

Verlet, L: Computer 'Experiments' on classical fluids. I. Thermodynamical properties of Lennard-Jones molecules. Phys. Rev. **159**, 98-103 (1967)

Zadunaisky, PE: On the accuracy in the numerical solution of the *N*-body problem. Celest. Mech. **20**, 209-230 (1979)

2

Riemann solvers and Alfven waves in black hole magnetospheres

Brian Punsly[1,2]* (iD), Dinshaw Balsara[3], Jinho Kim[3] and Sudip Garain[3]

Abstract

In the magnetosphere of a rotating black hole, an inner Alfven critical surface (IACS) must be crossed by inflowing plasma. Inside the IACS, Alfven waves are inward directed toward the black hole. The majority of the proper volume of the active region of spacetime (the ergosphere) is inside of the IACS. The charge and the totally transverse momentum flux (the momentum flux transverse to both the wave normal and the unperturbed magnetic field) are both determined exclusively by the Alfven polarization. Thus, it is important for numerical simulations of black hole magnetospheres to minimize the dissipation of Alfven waves. Elements of the dissipated wave emerge in adjacent cells regardless of the IACS, there is no mechanism to prevent Alfvenic information from crossing outward. Thus, numerical dissipation can affect how simulated magnetospheres attain the substantial Goldreich-Julian charge density associated with the rotating magnetic field. In order to help minimize dissipation of Alfven waves in relativistic numerical simulations we have formulated a one-dimensional Riemann solver, called HLLI, which incorporates the Alfven discontinuity and the contact discontinuity. We have also formulated a multidimensional Riemann solver, called MuSIC, that enables low dissipation propagation of Alfven waves in multiple dimensions. The importance of higher order schemes in lowering the numerical dissipation of Alfven waves is also catalogued.

Keywords: black hole physics; magnetohydrodynamics (MHD); galaxies: jets; galaxies: active; accretion; accretion disks

1 Introduction

In this century, there has been great progress in 3-D magnetohydrodynamic (MHD) simulations of black hole magnetospheres (De Villiers et al. 2003; Komissarov 2004; Komissarov 2005; Hawley and Krolik 2006; Fragile et al. 2007; McKinney and Blandford 2009; McKinney et al. 2012). To varying degrees, each of these simulations require knowledge of the 1-D characteristics of the MHD system in order to time evolve the magnetosphere. Specifically, the polarization properties of the waves determine the changes in the fields that can be propagated at the appropriate speed along a particular characteristic direction. In single fluid ideal MHD there are three plasma modes in the system, the fast mode, the Alfven (or intermediate) mode and the slow mode. Black hole magnetospheres

have the property that all plasma must pass progressively through the slow, Alfven and fast critical surfaces before reaching the event horizon (Punsly 2008). As each critical surface is crossed, the unique information associated with each wave mode is unable to be communicated upstream to an outgoing wind or jet. The event horizon wind system has no boundary conditions at its terminus, there are asymptotic infinities both at the event horizon and at large radial coordinate (Punsly 2008). There are only lateral boundary conditions imposed by accreting gas. Thus, the wind system itself and the lateral boundary conditions determine 3-D single fluid perfect MHD wind solutions. Furthermore, due to the paired wind nature of the event horizon wind system (an ingoing accretion inner wind and the outgoing outer wind or jet), plasma is always drained off of the field lines and auxiliary physics (mass floors) must be injected by hand in order to keep numerical simulations from crashing at low density. Mass floors are a source of MHD waves and are generally chosen to enhance dissipation. Consequently, there are many unique

*Correspondence: brian.punsly@cox.net
[1] 1415 Granvia Altamira, Palos Verdes Estates, CA 90274, USA
[2] ICRANet, Piazza della Repubblica 10, Pescara, 65100, Italy
Full list of author information is available at the end of the article

aspects to the application of MHD that can influence the final steady state of the wind system. Describing the evolution of the event horizon magnetosphere with single fluid MHD is wrought with non trivial subtleties.

These subtleties relate to the numerical determination of the field line angular velocity as viewed from asymptotic infinity, Ω_F. This is of primary interest since the Poynting flux of the wind scales as Ω_F^2 (Punsly 2008). First, contrary to previous claims of early simulations, newer simulations indicate that Ω_F can be altered significantly by the auxiliary method of injecting plasma (McKinney et al. 2012; Beskin and Zheltoukhov 2103). We consider the unique role of the oblique Alfven wave in this process. A unique component of the momentum flux is propagated along the Alfven characteristics and this momentum flux is a component of the MHD equations written in conservative form. It is also the only isolated discontinuity that propagates a physical charge. Black hole magnetospheres that support an outgoing relativistic jet, rotate and have a Goldreich-Julian charge density. The Alfven critical surface for the inflow (IACS) is quite far from the event horizon. For rapidly rotating black holes and the most recent Ω_F values from numerical simulations, the proper distance is 2 to 3 black hole radii from the event horizon. For proper evolution of the magnetic field rotation rate and the induced charge density, one must be able to simulate the role of Alfven waves with high fidelity both globally and inside of the IACS. Thusly motivated, we discuss in this article new numerical methods that are designed to accurately characterize the Alfven wave numerically in the rarefied environment of black hole magnetosphere.

An accurate depiction of the time evolution of a black hole magnetosphere and the global considerations germane to the IACS are intimately related to minimizing the numerical dissipation of Alfven waves. The IACS is a one-way surface as far as t he propagation of Alfven waves is concerned. In other words, at the IACS and within it, all Alfven waves should propagate inwards and only inwards. The propagation of waves in a higher order Godunov code is modulated by the Riemann solver. It is, therefore desirable if the Riemann solver can mimic this one-way propagation property for Alfven waves. Alas, whether a Riemann solver does so or not, depends crucially on the design of the Riemann solver. Some Riemann solvers which retain the substructure associated with Alfven waves within the Riemann fan, can indeed represent such a one-way propagation of Alfven waves. Other Riemann solvers, like the HLL Riemann solver, do not retain the substructure associated with the Alfven waves. At extraordinarily high resolutions, any well-designed code will of course minimize this dissipation. However, the present generation of simulations have all been done with low or modest resolutions. Furthermore, they have mostly used the HLL Riemann solver which, we argue, applies

a maximal, and deleterious, dissipation to Alfven waves. To appreciate this point, realize that the HLL Riemann solver is based on a wave model that has just two extremal waves. These two extremal waves determine the ends of the Riemann fan in one-dimension. The speed of these extremal waves is usually set to the extremal signal speeds in the physical problem. For a relativistic MHD (RMHD) simulation of highly magnetized event horizon magnetospheres, these extremal signal speeds are usually set to approximately the speed of light propagating in either direction at a zone interface where the Riemann solver is applied. The HLL Riemann solver does not incorporate any further sub-structure associated with intermediate waves. Consequently, the HLL Riemann solver maximizes the dissipation of Alfven waves even near the IACS. This is the very location where the dissipation of these waves has to be minimized. Introducing the intermediate waves in the Riemann fan reduces the dissipation, but that effect is not incorporated in the HLL Riemann solver.

There is another issue that increases the dissipation of Alfven waves, in more than one dimension. Riemann solvers applied to black hole magnetospheres have been treated as 1-D in each direction. However, a true multidimensional Riemann solver has a strongly interacting region in which the numerical fluxes in orthogonal directions become intertwined (Balsara 2012). Careful treatment of the strongly-interacting region results in far less dissipation (as we show for RMHD in Section 4). The increase in computational complexity associated with a multidimensional Riemann solver is handily offset by larger timesteps and greater code robustness. The accuracy of the numerical depiction of the role of the Alfven waves near a black hole during jet production is facilitated by a true multidimensional scheme that incorporates the strongly interacting region.

Our ultimate goal is to understand the detailed time evolution of black hole magnetospheres. This is subtle because one must try to understand in each time step how the mass floor is affecting the time evolution. Thus, the transient structure is essential to monitor in order to see how transients associated with plasma injection are altering the time evolution of the system. This requires an inherently very stable numerical scheme ('well-balanced', which we describe below). Furthermore, we note that the simulations of Komissarov (2004) and Komissarov (2005) did tend to a steady state and this made it possible to carry out convergence testing for those simulations. However, several simulations seemed to never reach a steady state and, therefore, one cannot carry out convergence studies for them. In particular, an unexpected finding of Krolik et al. (2005) was that the event horizon magnetospheres in numerical simulations are very unsteady and appear to be more like a cauldron of strong MHD waves rather than a force-free structure. 'For example, although the funnel region is magnetically-dominated, it is not in general in a state of force-free equilibrium. Indeed, the very large fluctuations that continually occur in the outflow show that it is never in any state of equilibrium, force-free or otherwise.' This also appears to be the case in the simulations of McKinney et al. (2012) based on the supporting online movies. This suggests that wind formation in event horizon magnetospheres might be subject to large numerically induced transients which would mask the kinds of effects that we are looking for. Alternatively, if these large transients are integral to jet formation it indicates a dynamic in which strong waves from the lateral boundaries created by the accretion flow scatter off the event horizon magnetosphere producing strong gusts of energy in a jetted outflow.

It is important to separate these potential physical effects from numerical effects. However, schemes that are based on higher order Godunov methods, especially those that are based on the HLL Riemann solver, are notorious for not achieving steady state even when the physical problem admits such a steady state! This was first observed when higher order Godunov schemes with Riemann solvers were first applied to metreological simulations (which simulate wind flow in the earth's atmosphere) or to the shallow water equations (which simulate lake and ocean circulation) (Parés 2006). Unless the numerical scheme has a special property called well-balancing, it usually does not find a stationary state even when one exists. Instead, the simulation generates 'numerical storms' - high velocity flows that are entirely a numerical artefact. The issue of well-balancing has recently become more compelling within the context of astrophysical flows with the work of Kappeli and Mishra (2014, 2016). Within the context of Type IIb supernova simulations, it has been found that the proto-neutron stars refuse to reach steady state if the scheme is not well-balanced, i.e., the numerical method has to be specially engineered to that it can find a steady-state proto-neutron star solution if one exists. Kappeli and Mishra have also explored time-dependent simulations that are not in steady state. Their very interesting result is that even for simulations that have no reason to be in steady state, the inclusion of well-balancing provides a substantial improvement in the accuracy of the simulation. This result has bearing on the black hole magnetospheres problem because it suggests that even when the simulations are far from steady state they might indeed be helped by well-balancing.

In order to verify that one has a well-balanced scheme, one has to know what the steady states of the system are and make sure that the simulated system is driven towards that steady state. This does not mean that every system must reach a steady state; it only means that when such a steady state exists, the numerical code is equipped to find

it. For a scheme to be truly well-balanced, it needs to have two attributes. First, the reconstruction procedure has to be modified so that any potential steady state solution is subtracted off from the reconstructed solution. Second, the scheme must use a Riemann solver that can capture intermediate waves - specifically the contact discontinuity in classical Euler flow. We admit, that in certain circumstances it may not be possible to identify the steady state solution that has to be subtracted off. In such situations, the numerical scheme should at least be well-balanced up to second order. In practical terms, this means that a modest resolution simulation will not find the steady state solution to machine accuracy. However, it will nevertheless find the steady state solution with accuracies that are proportional to the size of the mesh squared. Being well-balanced up to second order is a weaker notion of well-balancing compared to being truly well-balanced. In order for a scheme to be well-balanced up to second order, it is imperative that the Riemann solver should at least capture stationary contact discontinuities in a self-gravitating situation involving Euler flow. By extension, any MHD code that is capable of capturing stationary equilibria up to second order should at least be based on a Riemann solver that captures contact discontinuities and Alfven waves. The discussion in this paragraph has made it clear that in order to disentangle a flow that is chaotic because of the physics of the situation from a flow that is chaotic because of simple numerical effects, one must pay attention to the form of the Riemann solver.

In summary, for the purposes of exploring the time evolution of event horizon magnetospheres, we require a well-balanced scheme to second order that will eliminate recurrent large transients. This is facilitated by preserving the Alfven wave as discussed above, but also requires that the inclusion of the contact discontinuity in the Riemann fan. Furthermore, without the contact discontinuity, the huge density gradient between accretion flow and the evacuated funnel of the event horizon magnetosphere means that plasma flows into the funnel by numerical diffusion. This is far from ideal if we want to explore the role of mass injection on the final solutions. To improve this situation, one desires a Riemann solver that at least minimizes the dissipation at contact discontinuities. (For more helpful information on numerical schemes and Riemann solvers of relevance to astrophysics, please visit http://www.nd.edu/~dbalsara/Numerical-PDE-Course.)

Designing low-dissipation Riemann solvers for RMHD is a challenging enterprise. An exact RMHD Riemann solver exists (Giacomazzo and Rezzolla 2006; Giacomazzo and Rezzolla 2007), but it is not practicable for use in numerical codes. HLLC Riemann solvers for RMHD exist (Mignone and Bodo 2006; Honkkila and Janhunen 2007; Kim and Balsara 2014). They enable a stationary contact discontinuity to be captured on a mesh. However, they dissipate stationary Alfven waves just like an HLL Riemann solver. HLLD Riemann solvers for RMHD (Mignone et al. 2009), do exist, which enable stationary contact discontinuities as well as stationary Alfvenic discontinuities to be captured on a computational mesh. Unfortunately, the method is iterative, which makes it computationally very expensive. Furthermore, when an iteration fails to converge, the method becomes brittle. With the emergence of the HLLI Riemann solver (Dumbser and Balsara 2016), it has become possible to capture stationary contact discontinuities as well as stationary Alfvenic discontinuities using a Riemann solver that is non-iterative and computationally inexpensive. The first goal of this paper is to document this capability in the astrophysical literature.

RMHD simulations also have to maintain the divergence-free structure of the magnetic field. This necessitates the use of a Yee-type mesh where the magnetic field components are specified at the faces of the mesh and the electric fields are to be evaluated at the edges of the mesh. It was claimed by Gardiner and Stone (2005), Gardiner and Stone (2008), Beckwith and Stone (2011), White and Stone (2015) that stabilizing the evolution of the magnetic field requires that one should always double the dissipation in the electric field at every timestep. Unfortunately, that approach has been used in the RMHD literature with the result that the already excessive dissipation of the HLL Riemann solver is increased even further in simulations. Such explorations ignore recent advances in multidimensional Riemann solver technology (Balsara 2010; Balsara 2012; Balsara 2014; Balsara 2015; Balsara et al. 2014; Balsara and Dumbser 2015; Balsara et al. 2016a). In a recent paper, Balsara and Kim (2016), showed that an exact analogue of the HLLI Riemann solver in multidimensions can be designed for RMHD. Their work is based on the original paper by Balsara et al. (2016b). Such multidimensional Riemann solvers go under the name of MuSIC Riemann solvers. Here the MuSIC acronym stands for Multidimensional, Self-similar strongly-Interacting Riemann solver that is based on Consistency with the conservation law. By introducing substructure associated with the multidimensional propagation of Alfven waves, the MuSIC Riemann solver reduces the dissipation of Alfven waves that propagate at any angle with respect to the mesh. The second goal of this paper is to catalogue the advantages of the MuSIC Riemann solver in reducing the dissipation involved in the multidimensional propagation of Alfven waves in RMHD.

In this article, we claim that modern 5-wave Riemann solvers can now be implemented that can allow a systematic assessment of the issues related to determination of Ω_F. This would require the formulation and simulation of simple magnetospheres, lateral boundary conditions and plasma injection mechanisms. Proper time evolution of the 3-D magnetosphere requires two main aspects of the solver, low dissipation and well balancing. We have motivated both these issues in the previous paragraphs. We demonstrate that the new Riemann solvers described here are capable of delivering on these goals in the following.

Riemann solvers offer one way of reducing numerical dissipation and Riemann solvers that preserve essential features of the flow are certainly central to many aspects of jet simulation. Recent advances in higher order schemes has made it possible to go beyond the traditional second order Godunov scheme that is commonplace in computational astrophysics. The third goal of this paper is to show that higher order schemes for RMHD do exist (Balsara and Kim 2016; Del Zanna et al. 2007; Zanotti and Dumbser 2016), which make it possible to go beyond second order of accuracy. We show that the combination of higher order schemes and appropriate Riemann solvers can go a long way towards enabling almost dissipation-free propagation of Alfven waves.

The paper is organized as follows. In Section 2, we discuss dissipation inside the IACS in Riemann solvers in general terms. It is shown that the HLLI Riemann solver provides the theoretical minimum dissipation that is consistent with a stable numerical scheme. In Section 3, we demonstrate by explicit examples that the HLLI RMHD Riemann solver preserves the Alfven wave with high accuracy and respects the contact discontinuity. In Section 4, we incorporate the important aspects of multidimensionality with the MuSIC RMHD Riemann solver. This is truly a multi-dimensional scheme and we demonstrate that its ability to resolve the strongly interacting region substantially reduces Alfvenic dissipation compared to higher dimensional schemes that utilize 1-D Riemann solvers in each direction. In our final discussion section, we describe how a numerical scheme that utilizes the MuSIC Riemann solver would be suitable for specialized simulations that would shed light on the causal physics of the time evolution of event horizon magnetospheres and help define the full panoply of physically allowed and disallowed solutions.

2 Dissipation in Riemann solvers

In this section, we illustrate the mathematical implications of the IACS in conservative upwind schemes that utilize Riemann solvers. (Please also note that schemes that do not use Riemann solvers necessarily have to introduce even higher levels of dissipation. This is because they cannot discriminate between wave families in the way that some of the better Riemann solvers can.) For simplicity and without loss of generality, consider a one-dimensional grid. The conservation law that must be solved in each direction and at each time step can be formally written as

$$\frac{\partial U}{\partial x^{0''}} + \frac{\partial F}{\partial x^{1''}} = 0. \tag{1}$$

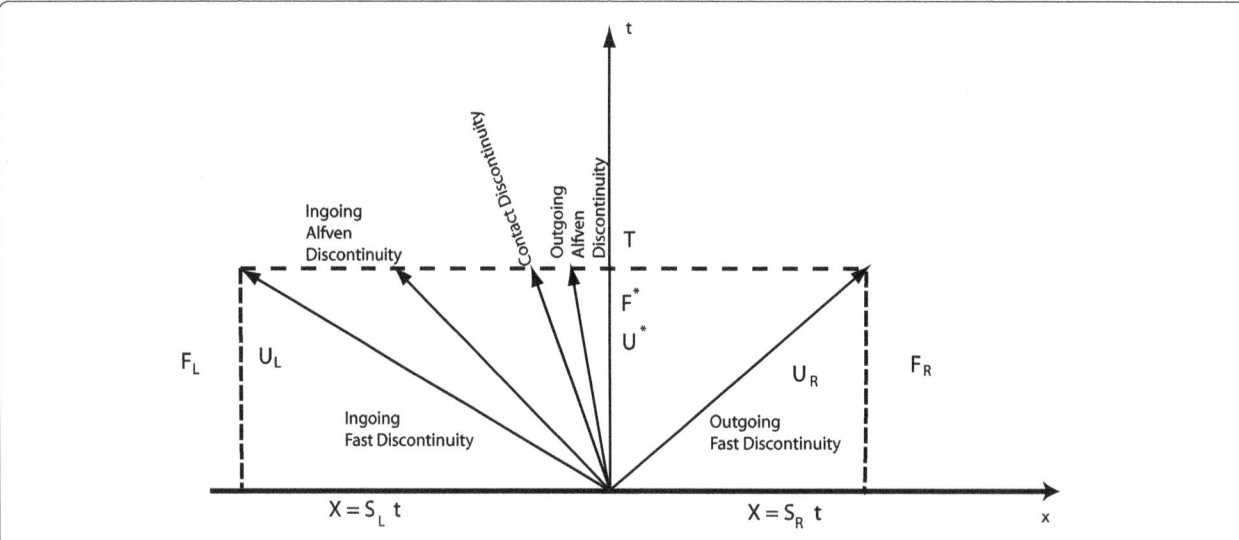

Figure 1 Super Alfvenic Riemann fan. An example of a one dimensional Riemann problem in which the flow is super-Alfvenic inward (to the left). The flow is not super-fast inward. This figure is used to illustrate what happens in a Riemann problem at the interface between two cells in a numerical scheme inside of the IACS. The 5-wave Riemann fan is illustrated (the slow waves are ignored without loss of generality) in order to show the difference of the resolved flux at the interface calculated with a 2-wave HLL Riemann solver and the same calculation performed with a 5-wave HLLI Riemann solver. Outgoing and ingoing are defined in the frame of reference of the plasma.

For a fine enough mesh and a well behaved coordinate system, the covariant derivatives can be replaced with ordinary derivatives in the conservation equation. In fact, RMHD Riemann solvers are used in GRMHD (general relativistic MHD) simulations (Komissarov 2004; Gammie et al. 2003; Etienne et al. 2015). In an integral (weak solution) solution of the Riemann problem, the higher order corrections due to connection coefficients will be small (bilinear) corrections compared to the integral of the derivative terms which are linear in the space-time mesh size. This is the essence of the validity of ignoring the source (connection coefficient) terms in the GRMHD Riemann solvers and is a manifestation of the equivalence principle. We note that in the GRMHD conservation law the connection coefficient terms (source) terms occur. The error induced by these terms can be made arbitrarily small on a fine enough mesh compared to the differential terms. However, in practice the mesh might be coarse enough that the connection terms represent source terms that modify the solution of the conservation law in each time step (Del Zanna et al. 2007). This is not discussed further in this section which is concerned only with the Riemann solvers in GRMHD. The present paper does not focus on a consideration of stiff source terms.

Consider the nature of the flow at the IACS. We are especially interested in the propagation of different RMHD wave families relative to the IACS which, in principle, could be stationary relative to the computational mesh. The space-time diagram is indicated in Figure 1. The flow is super-Alfvenic inward (to the left). The flow is not super-

fast inward. This figure is a spacetime diagram of the MHD characteristics at an interface between cells inside of the IACS. The 5-wave Riemann fan is illustrated (the slow waves are ignored without loss of generality) in order to show the difference in the resolved flux that is produced by an HLL Riemann solver and a 5-wave based Riemann solver. The spacetime in Figure 1 is split into various zones. The world lines of isolated discontinuities that emanate from the zone boundary are shown in Figure 1. The resolved flux is the numerical flux in the zone that straddles the time axis. Inside of the IACS, the resolved flux is the numerical flux in the zone bounded on the left by Alfven wave that is outgoing in the proper frame (but ingoing on the computational mesh) and on the right by the outgoing fast wave. Physically, since the outgoing Alfven wave overlies the time axis in Figure 1, we are interested in capturing the stationary Alfvenic surface with maximum precision and the least possible numerical diffusion.

Consider a one-dimensional Riemann problem with left and right states $\mathbf{U}_L$ and $\mathbf{U}_R$ that are separated by a Riemann fan with extremal speeds that span $[S_L, S_R]$. Let the fluxes that correspond to the left and right states be given by $\mathbf{F}_L$ and $\mathbf{F}_R$. The numerical flux from practically any one-dimensional Riemann solver can be formally written as

$$\mathbf{F}^* = \frac{1}{2}(\mathbf{F}_L + \mathbf{F}_R) - \frac{1}{2}\boldsymbol{\Theta}(\mathbf{U}_R - \mathbf{U}_L). \tag{2}$$

The first term, which is the average of the left and right fluxes in the above expression, simply provides a centered, non-dissipative flux. The second term in the above expression is known as the dissipation term. The matrix, Θ, in the second term is the viscosity matrix, it regulates the dissipation of the Riemann solver. Further details on the ensuing mathematics can be found in Dumbser and Balsara (2016), Appendix B. Here we provide just enough results to show the difference between the dissipation from the HLL Riemann solver and the HLLI Riemann solver.

The viscosity matrix is usually expressed in terms of the right and left eigenvectors of the Roe matrix. Denoting the Roe matrix by $\mathbf{A}(\mathbf{U}_L, \mathbf{U}_R)$, we will refer to its left and right eigenvectors by $\mathbf{L}$ and $\mathbf{R}$. The eigenvalues of the Roe matrix will be denoted by a diagonal matrix, $\mathbf{\Lambda}$. The viscosity matrix for the HLLI Riemann solver can be written as

$$\Theta = \mathbf{R}\mathbf{\Sigma}\mathbf{L} \tag{3}$$

with

$$\mathbf{\Sigma} = \frac{S_R + S_L}{S_R - S_L}\mathbf{\Lambda} - 2\frac{S_R S_L}{S_R - S_L}\mathbf{I} + 2\frac{S_R S_L}{S_R - S_L}\boldsymbol{\delta}. \tag{4}$$

Here $\mathbf{I}$ is the identity matrix and (for our purposes) $\boldsymbol{\delta}$ is a special diagonal matrix that is introduced into the HLLI Riemann solver to judiciously reduce dissipation. Notice, therefore, that $\mathbf{\Sigma}$ is also a diagonal matrix. Please observe from the previous equation that the viscosity matrix introduces dissipation on a wave-by-wave basis, i.e., if the diagonal term corresponding to a particular wave becomes zero, the dissipation that is provided to that wave will also become zero. If we choose the diagonal terms in $\boldsymbol{\delta}$ just right, we can minimize the dissipation and even guarantee that the dissipation of standing Alfven waves is exactly zero. This is exactly what has been done in Dumbser and Balsara (2016). Those authors provide precise expressions for the diagonal matrix, $\boldsymbol{\delta}$, which ensure that stationary waves (whether they are Alfven waves or the entropy wave) have zero dissipation. For the sake of completeness, we catalogue their specification of the ith term of the diagonal matrix, $\boldsymbol{\delta}$, as

$$\delta_i = 1 - \frac{\min(\lambda_i, 0)}{S_L} - \frac{\max(\lambda_i, 0)}{S_R}. \tag{5}$$

Here λ_i is the ith eigenvalue of the Roe matrix corresponding to the wave that we are interested in. We see that the dissipation is finely tuned so that a moving Alfven wave gets just the minimum amount of dissipation that it needs, consistent with numerical stability. For example, a wave that propagates slowly relative to the computational mesh is given smaller dissipation compared to a wave that is propagating at high speed on the mesh. This decision to regulate the dissipation according to the wave speed is also what is demanded by numerical stability.

The viscosity matrix for the HLL Riemann solver is retrieved by setting $\boldsymbol{\delta} = 0$. In that case, a standing Alfven wave has non-zero dissipation which means that the Alfven waves at the IACS surface will dissipate. Consequently, a numerical code that is based on the HLL Riemann solver (especially if it is operated at low to modest resolution) will not treat the IACS as a one-way surface with respect to the propagation of Alfven waves. We feel that this is a very important observation. Furthermore, with $\boldsymbol{\delta} = 0$, it is easy to see that the HLL Riemann solver gives all waves a non-zero dissipation regardless of their wave speed. To see this, let λ_i be a specific eigenvalue. Then the ith term for the diagonal matrix, $\mathbf{\Sigma}$ can be written as

$$\Sigma_i = \frac{S_R(\lambda_i - S_L) - S_L(S_R - \lambda_i)}{S_R - S_L}. \tag{6}$$

For the sub-sonic case shown in Figure 1 we have $S_L < 0 < S_R$. We see that $\Sigma_i > 0$ for all intermediate eigenvalues, λ_i, with $S_L < \lambda_i < S_R$. Consequently all intermediate waves, like Alfven waves or contact discontinuities, will always be dissipated by the HLL Riemann solver.

In this section, we have only given a flavor of the dissipation characteristics of the HLLI Riemann solver and how it offers a substantial improvement over the HLL Riemann solver. The reader who is interested in details should please read Dumbser and Balsara (2016). The eigenvectors for RMHD that were used in this paper can all be obtained from Balsara (2001) or Antón et al. (2010).

3 One dimensional Riemann solvers in RMHD

The Introduction has shown that it is very desirable to have Riemann solvers that can capture stationary, isolated contact discontinuities as well as stationary, isolated Alfven waves. Indeed the first of the one-dimensional Riemann solvers for RMHD by Komissarov (1999) and Balsara (2001) were Roe-type Riemann solvers. Because such Riemann solvers retain the entire set of eigenvectors for the RMHD system, they can indeed capture stationary, isolated contact discontinuities as well as stationary, isolated Alfven waves on a mesh. There has been a recent effort by Antón et al. (2010) to revive the use of Roe-type Riemann solvers in RMHD simulations, but the effort has met with limited success owing to the exorbitant computational cost of such Riemann solvers, especially for RMHD. The Roe-type Riemann solvers also have an inherent deficiency. This has to do with their loss of positivity of density and pressure in certain circumstances (Einfeldt 1988; Einfeldt et al. 1991). It is easy to find mentions in the early literature on RMHD (Komissarov 1999, 2004, 2005), showing that the early RMHD simulation codes struggled with positivity issues and, therefore, reverted to the use of the HLL Riemann solver. This issue is relevant to the extremely low density environment of event horizon magnetospheres.

HLLD Riemann solvers. Mignone et al. (2009), also enable a code to capture stationary, isolated contact discontinuities as well as stationary, isolated Alfven waves. But they have their own set of attendant problems, as discussed in the Introduction. HLLC Riemann solvers (Mignone and Bodo 2006; Honkkila and Janhunen 2007; Kim and Balsara 2014), represent a compromise position where they enable a code to capture stationary, isolated contact discontinuities, but not Alfven waves. The HLLI Riemann solver of Dumbser and Balsara (2016) is built on top of an HLL Riemann solver, so it inherits all the beneficial positivity properties of the HLL Riemann solver. However, it introduces sub-structure in the Riemann fan. Typically, that substructure includes the contribution from eigenvectors of the contact discontinuity and eigenvectors associated with Alfven waves. The eigenvectors for the fast and slow magnetosonic waves are very expensive to evaluate computationally, and their evaluation is avoided in the HLLI Riemann solver. As a result, the HLLI Riemann solver enables a code to capture stationary, isolated contact discontinuities as well as stationary, isolated Alfven waves at a very low computational cost. Unlike HLLC and HLLD, the HLLI Riemann solver does not require an iterative solution, thereby ensuring that it has even lower computational cost. In the next few paragraphs we demonstrate this facet of the HLLI Riemann solver for RMHD.

Figure 2 shows two simulations of an isolated, stationary contact discontinuity as suggested by Honkkila and Janhunen (2007). The density variable is shown. We use the same parameters as the previous authors and we run the simulation to a final time that is ten times larger than the one suggested by Honkkila and Janhunen. The result from the HLLI Riemann solver is shown with filled dots, the result from the HLL Riemann solver is shown with crosses.

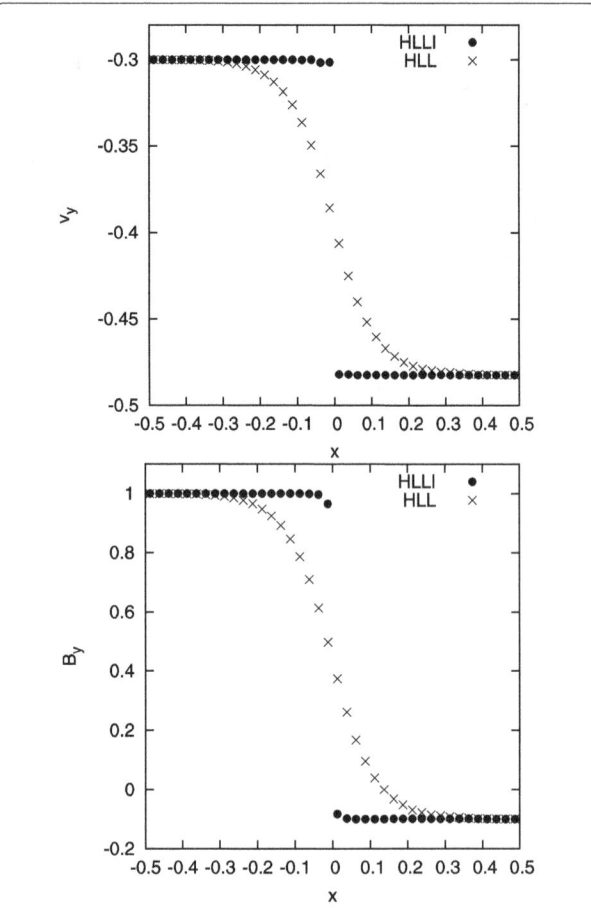

Figure 3 HLL Riemann solver vs. HLLI Riemann solver contact discontinuity. Figure 3 shows two simulations of an isolated, stationary Alfven wave. The transverse velocity and magnetic field are shown. The result from the HLL Riemann solver is shown with crosses while the result from the HLLI Riemann solver is shown with dots.

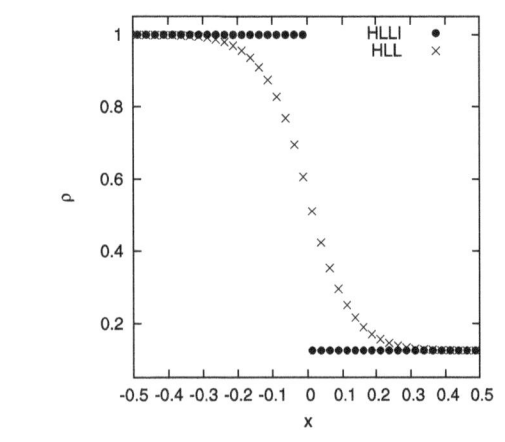

Figure 2 HLL Riemann solver vs. HLLI Riemann solver stationary Alfven discontinuity. Figure 2 shows two simulations of an isolated, stationary contact discontinuity. The density variable is shown. The result from the HLL Riemann solver is shown with crosses while the result from the HLLI Riemann solver is shown with dots.

We see that the HLL Riemann solver has produced significant dissipation of the contact discontinuity, while the HLLI Riemann solver has captured the contact discontinuity exactly.

Figure 3 shows two simulations of an isolated, stationary Alfven wave. The transverse velocity and magnetic field are shown. This problem was suggested by Mignone et al. (2009) and we use the same parameters as the previous authors but we run the simulation to a final time that is four times larger than the final time quoted by Mignone et al. (2009). The result from the HLLI Riemann solver is shown with filled dots, the result from the HLL Riemann solver is shown with crosses. As before, we see that the HLL Riemann solver has produced significant dissipation of the Alfven wave discontinuity, while the HLLI Riemann solver has captured the Alfven wave discontinuity exactly. Figures 2 and 3 both used a standard, second order scheme, the only difference being the use of the HLLI Riemann solver.

4 Multidimensional propagation of Alfven waves

Figures 2 and 3 showed that the one-dimensional HLLI Riemann solver can dramatically reduce the dissipation compared to the HLL Riemann solver. Multidimensional treatment of Alfven waves on a computational mesh requires a multidimensional Riemann solver. In Balsara (2004), we were able to formulate a test problem that measures the ability of a multidimensional MHD code to propagate Alfven waves with the least amount of dissipation. It has been shown that this test problem is very important in benchmarking the dissipation characteristics of multidimensional codes for classical MHD. The analogous problem which benchmarks the low dissipation propagation of Alfven waves in RMHD has been recently presented in Balsara and Kim (2016). It consists of torsional Alfven waves propagating obliquely to the mesh lines of a two-dimensional mesh. The mesh has 120×120 zones. We do not repeat the problem description. Instead, we show the results and intercompare with older methods that involve dissipation doubling from Gardiner and Stone (2005, 2008).

Figure 4 shows the results of the torsional Alfven wave dissipation test from Balsara and Kim (2016). In Figure 4, we use the same second order reconstruction algorithm from the RMHD version of the RIEMANN code. Figure 4a shows the decay in the z-component of the velocity of the Alfven wave as a function of time. Figure 4b shows the same for the z-component of the magnetic field of the Alfven wave. The vertical axis is logarithmically scaled. A faster rate of decline in Figure 4 indicates that the associated numerical scheme has higher dissipation. The curve that is labeled 'MuSIC + 1DHLLI' uses the one-dimensional HLLI Riemann solver at the zone faces and the MuSIC Riemann solver with sub-structure at the zone edges. We see that it displays minimal dissipation. This is because the MuSIC Riemann solver is designed to be the exact, multidimensional analogue of the one-dimensional HLLI Riemann solver. The curve that is labeled 'MuSIC-NoSS + 1DHLLI' uses the same algorithmic combination with one simple exception. The MuSIC Riemann solver is prevented from endowing sub-structure to the strongly-interacting state. We see that when the sub-structure in the MuSIC Riemann solver is artificially removed, the dissipation of Alfven waves increases. This makes the very nice point that all facets of the newly designed MuSIC Riemann solver play a role in reducing dissipation. It is very useful to cross-compare with the dissipation doubling ideas from Gardiner and Stone (2005, 2008). The curve that is labeled '1D HLLC (dissipation doubling)' doubles the dissipation in the HLLC Riemann solver using the ideas from Gardiner and Stone (2005, 2008). Despite the one-dimensional HLLC Riemann solver being an able performer, we see that it dramatically increases the dissipation that is provided to the torsional Alfven waves.

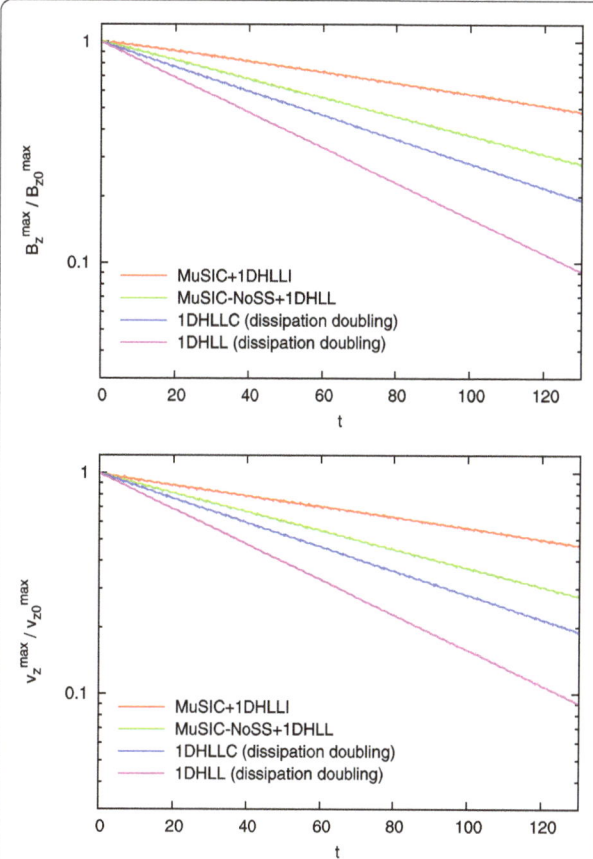

Figure 4 Torsional Alfven wave dissipation test. Figure 4 shows the results of the torsional Alfven wave dissipation test. A second order WENO reconstruction was used in all these tests. Figure 4a shows the decay in the z-component of the velocity as a function of time. Figure 4b shows the same for the z-component of the magnetic field.

Lastly, one is most interested in understanding what happens when the dissipation doubling ideas from Gardiner and Stone (2005, 2008) are applied to the one-dimensional HLL Riemann solver. This is shown in Figure 4 by the curve labeled '1D HLL (dissipation doubling)'. We see that Alfven waves are strongly dissipated.

It should also be emphasized that the reconstruction that was used in Figure 4 is the linear part of the $r = 3$ WENO reconstruction. This is already a very superior reconstruction strategy. It is almost as superior as a true third order reconstruction strategy. It is quite possible that reconstruction is done with a second order TVD limiter, like the MC limiter. In that case, the analogous results are shown in Figure 5. We see considerably increased dissipation in Figure 5 compared to Figure 4.

In Balsara and Kim (2016), ADER-WENO schemes were designed that go all the way up to fourth order of accuracy. It is, therefore, very interesting to ask whether improved accuracy gives us an improved result for the prop-

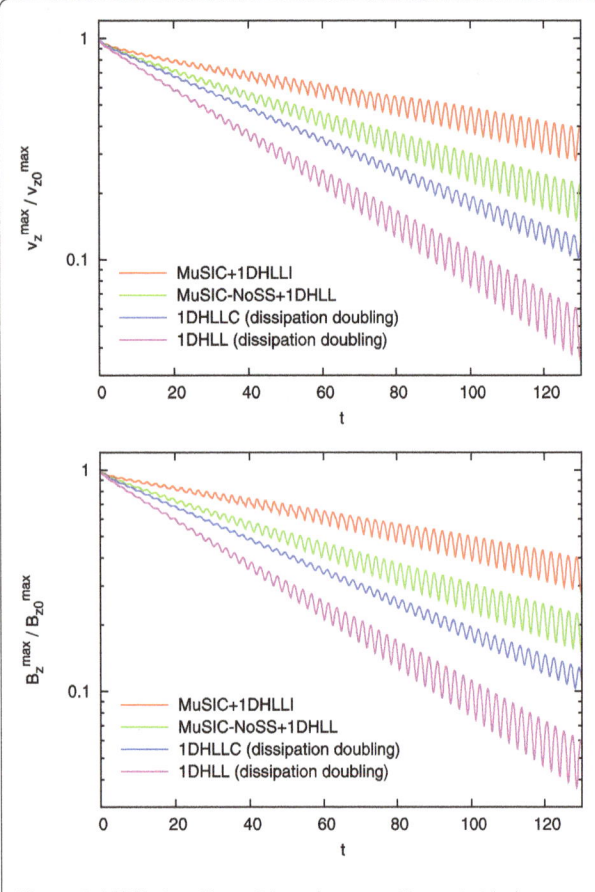

Figure 5 MC limiter. Figure 5 is analogous to Figure 4 with the exception than an MC limiter was used. The MC limiter is considered inferior to a good WENO scheme. Comparing Figures 4 and 5, this observation is apparent in the figures.

agation of torsional Alfven waves. Figure 6 shows the propagation of torsional Alfven waves when second, third and fourth order ADER-WENO schemes are used. All these schemes used the one-dimensional HLLI Riemann solver at the zone faces and the MuSIC Riemann solver with sub-structure at the zone edges. We clearly see that the higher order schemes show vastly reduced dissipation. In Balsara and Kim (2016), we also demonstrate that modern high order schemes perform robustly even in the vicinity of strong shocks. Thus the barrier to their use in astrophysics is dramatically reduced by this work.

Figure 6 clearly shows us that the combination of a high accuracy method and 1D and 2D Riemann solvers that preserve sub-structure produces very low dissipation. This is especially apparent in the fourth order simulation shown in Figure 6. Out of curiosity, we can always ask what fraction of the improvement in Alfven wave propagation stems from the use of a higher order scheme and what fraction of the improvement in Alfven wave propagation stems from the use of Riemann solvers that retain substructure? For that reason, the same simulations from Figure 6 were run

again with 1D and 2D Riemann solvers that do not preserve sub-structure. The results are shown in Figure 7. In other words, for Figure 7, the 1D Riemann solver was an HLL Riemann solver and the 2D Riemann solver was a 2D analogue of an HLL Riemann solver. Consequently, Figure 7 shows the result of Alfven wave propagation when lower quality Riemann solvers are used. We see that the second order result in Figure 7 is substantially more dissipative than the second order result from Figure 6. We also see that the second order result from Figure 6 has a dissipation that is comparable to the third order result from Figure 7. In other words, using a Riemann solver with sub-structure produces a very palpable improvement in second and third order schemes. When we compare the fourth order results from Figures 6 and 7, we see that they are indeed quite comparable. In other words, we suggest that at fourth and higher orders of accuracy the value of a Riemann solver that preserves sub-structure is diminished because the fourth order reconstruction itself is so very accurate! Note though that a third order scheme will typically be two to three times more expensive compared to a second order scheme. Similarly, a fourth order accurate scheme will be about three times more expensive compared to a third order scheme. For that reason, it is very profitable to try and extract as much performance and quality from a lower order scheme, especially if one does not have access to a higher order scheme.

Figures 6 and 7 show that there is always a small wiggle in the maximum amplitude of the Alfven waves. This wiggle stems from the fact that the Alfven waves in our test problem have large amplitude and are, therefore, prone to small initialization errors or errors in start-up transients that never go away. The non-relativistic analogue of this test problem also shows the same issue. The wiggles make the test problem used in Figure 6 an inadequate test problem for demonstrating order of accuracy. That is especially true at fourth order where the wiggles have an amplitude that is comparable to the decay of the Alfven wave. (In Balsara and Kim (2016) we do present better test problems for demonstrating the accuracy of an RMHD scheme.) However, please observe from Figure 6 that at second and third orders of accuracy, the rate of decay in the Alfven wave amplitude is much larger than the wiggles, so that it can be used to document the second and third orders of accuracy of the schemes used here. Table 1 shows the L_1 error in the z-momentum density and the z-component of the magnetic field as a function of mesh resolution at the second order. Table 2 shows the L_1 error in the z-momentum density and the z-component of the magnetic field as a function of mesh resolution at the third order. The errors are shown at a time of 30 units, by which point the decay in the Alfven waves at second and third orders of accuracy is much larger than the amplitude of the wiggles. We see that the second and third order accurate methods do indeed achieve their design accuracies.

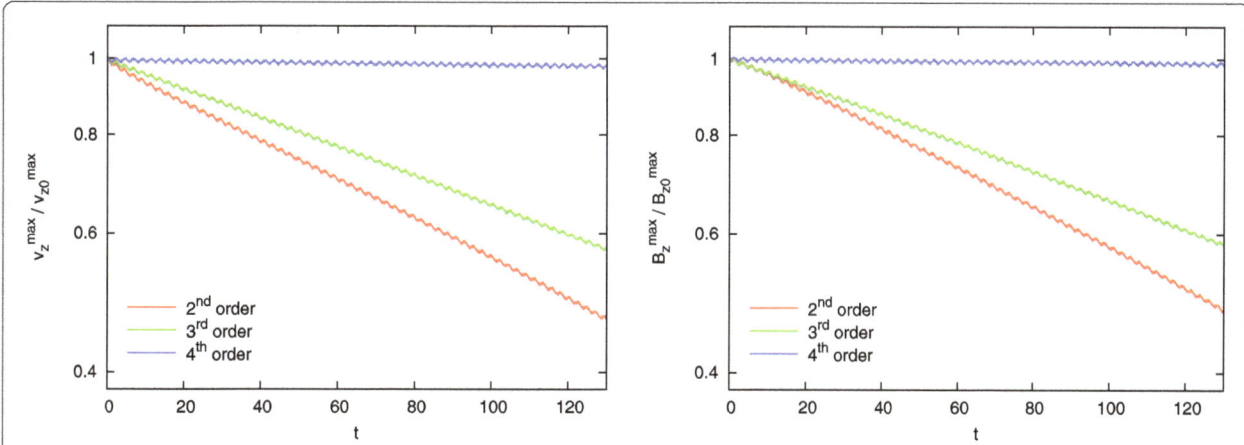

Figure 6 ADER-WENO Schemes the 2D MuSIC Riemann solver. Figure 6 shows the same Alfven wave propagation test. This time, we used different ADER-WENO schemes with increasing order of accuracy. We also used the 1D HLLI Riemann solver along with the 2D MuSIC Riemann solver with sub-structure. We see that higher order schemes produce lower dissipation.

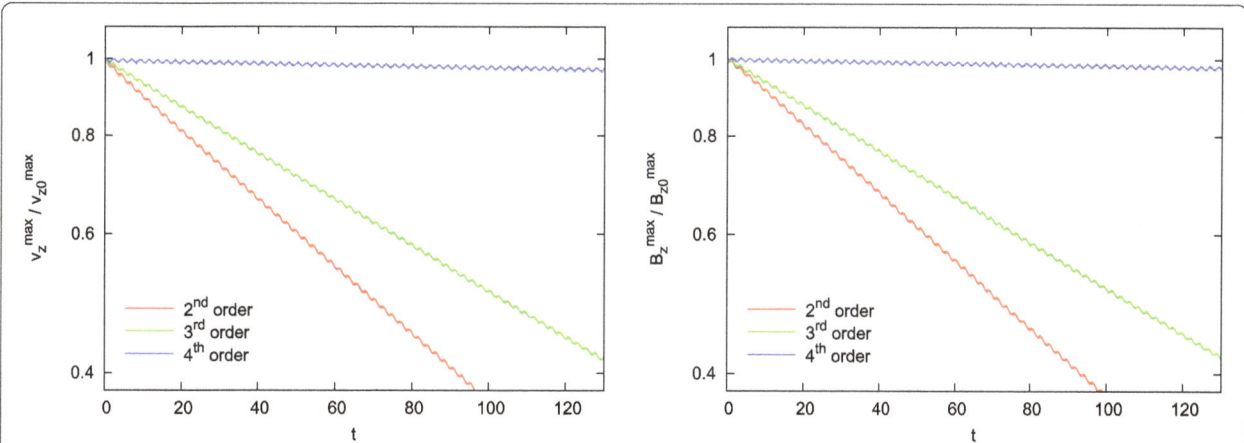

Figure 7 ADER-WENO schemes and the 2D HLL Riemann solver. Figure 7 shows the same Alfven wave propagation test as Figure 6 but when a lower quality Riemann solver is used. We used the same ADER-WENO schemes as in Figure 6. The only difference is that the simulations in this Figure were run with a 1D HLL Riemann solver and a 2D HLL Riemann solver. Comparing Figure 6 to Figure 7 enables us to appreciate the improvement that Riemann solvers with sub-structure provide in reducing dissipation.

Table 1 L_1 error in the z-momentum density and the z-component of the magnetic field as a function of mesh resolution at the second order

Zones	L_1 error z-momentum density	Accuracy	L_1 error z-magnetic field	Accuracy
60×60	2.7693×10^{-1}		1.1971×10^{-1}	
120×120	7.5660×10^{-2}	1.87	3.3246×10^{-2}	1.85
240×240	1.6083×10^{-2}	2.23	7.1989×10^{-3}	2.21

Table 2 L_1 error in the z-momentum density and the z-component of the magnetic field as a function of mesh resolution at the third order

Zones	L_1 error z-momentum density	Accuracy	L_1 error z-magnetic field	Accuracy
60×60	2.1295×10^{-1}		9.2981×10^{-2}	
120×120	3.7593×10^{-2}	2.50	1.6720×10^{-2}	2.48
240×240	5.1524×10^{-3}	2.87	2.2025×10^{-3}	2.92

5 Discussion and future prospects

In this article, we describe subtle points uniquely associated with numerical simulations of event horizon magnetospheres. All inflows must pass through an IACS, thereby rendering the interior region out of Alfven wave communication with the out-flowing wind. Thus, the IACS is a charge horizon (a one-way membrane) for the global flow since the Alfven wave is the only wave that carries a physical charge (Punsly 2004). The IACS causally excises most of the active region of space-time (the ergosphere) that would be the putative element to enforce rapid rotation of the system. The implication is that simulations of event horizon magnetospheres should be performed with numerical schemes that represent the Alfvenic properties of the system with high fidelity.

It was demonstrated that an improvement of numerical accuracy can be attained by utilizing numerical schemes based on HLLI and MuSIC Riemann solvers rather than schemes based on 1-D HLL solvers. In particular, for 1-D RMHD solvers, the HLLI Riemann solver provides the theoretical minimum dissipation of the Alfven wave and contact discontinuities that can still ensure a numerically stable scheme. In Section 4, we discussed the multidimensional extension of HLLI RMHD Riemann solver; the RMHD MuSIC Riemann solver. It is in higher dimensions that we see an even larger improvement over schemes based on 1-D HLL solvers. The MuSIC Riemann solver was shown to significantly reduce the dissipation of Alfven waves in large part to its ability to resolve the strongly interacting region that is typically ignored in schemes based on 1-D HLL Riemann solvers. We also show that very high order schemes might be free of the excessive dissipation that arises from a lower quality Riemann solver. However, that transition occurs only when schemes of fourth and higher order of accuracy are used.

Low Alfven and contact discontinuity dissipation in a numerical scheme, such as those based on MuSIC, should allow the proper propagation of Alfven wave information in the following unique circumstances endemic to event horizon magnetospheres that were discussed in the Introduction.

1. At the risk of being repetitive, the paired wind systems evolves outward and inward towards two asymptotic infinities as opposed to having a causal MHD boundary at one terminus. Thus, unlike other MHD wind problems there is a more complex critical point structure. Most specifically, all inflows pass through the inner Alfven critical surface. Thereby causally disconnecting the outflowing wind from Alfven radiation emanating from the majority of the rapidly rotating ergospheric plasma. The Alfven wave is the only wave that propagates a physical charge and thus should be involved in the establishment of the Goldreich-Julian charge density or equivalently, the field line rotation rate, Ω_F. Thus, reducing the numerical dissipation of the Alfven wave by implementing the HLLI or MuSIC Riemann solvers would seem to help in this regard.

2. The paired wind system constantly drains itself of plasma in the MHD limit. Thus, plasma injection by means of a mass floor is required. This process will dissipate MHD waves generated by the MHD

system and inject new MHD waves. The process is nontrivial and has been shown to modify Ω_F significantly in certain 3-D simulations. Since, in principle, it can modify Ω_F and mass injection perturbs the Alfven waves generated in the system, a Riemann solver lowers Alfven dissipation might shed light on the nature of the transients that occur as different plasma injection scenarios are explored.

3. Another related issue is the large numerical diffusion of plasma from the bounding accretion disk into the event horizon magnetosphere. Diffusion is substantially reduced if the Riemann solver respects the contact discontinuity. Since the source of plasma injection might be important to the establishment of Ω_F this is an important issue as well. Riemann solvers, such as the HLLI or MUSIC Riemann solvers, which treat contact discontinuities explicitly can help in this regard.

In this article, we presented both theoretical and numerical arguments that support the notion that numerical schemes based on the multi-dimensional MuSIC Riemann solver can potentially provide an improvement over existing methods of modeling 3-D event horizon magnetospheres. In particular, the MuSIC Riemann solver is both well suited for the low density environment endemic to the event horizon magnetosphere and it provides reduced dissipation of the Alfven and contact discontinuities. It is also computationally efficient which offsets the cost of improving the computational accuracy.

In Balsara and Kim (2016) we provide a subluminal scheme for RMHD as well as a discussion of the MuSIC Riemann solver. The reader who wishes to get further information can also visit the second author's website http://www.nd.edu/~dbalsara/Numerical-PDE-Course.

Competing interests
The authors declare that they have no competing interests.

Authors' contributions
BP conceived of the study and provided the contextual information on black holes. DB, JK and SG wrote most of the manuscript and provided the numerical simulations. All authors read and approved the final manuscript.

Author details
[1]1415 Granvia Altamira, Palos Verdes Estates, CA 90274, USA. [2]ICRANet, Piazza della Repubblica 10, Pescara, 65100, Italy. [3]Physics Department, University of Notre Dame du Lac, 225 Nieuwland Science Hall, Notre Dame, IN 46556, USA.

Acknowledgements
DB acknowledges support via NSF grants NSF-ACI-1307369, NSF-DMS-1361197 and NSF-ACI-1533850. DB also acknowledges support via NASA grant NASA-NNX 12A088G. Several simulations were performed on a cluster at UND that is run by the Center for Research Computing. BP thanks ICRANet for support.

References

Antón, L, Miralles, JA, Martí, JM, et al.: Relativistic magnetohydrodynamics: renormalized eigenvectors and full wave decomposition Riemann solver. Astrophys. J. Suppl. Ser. **188**, 1-31 (2010)

Balsara, D: Total variation diminishing scheme for relativistic magnetohydrodynamics. Astrophys. J. Suppl. Ser. **132**, 83-101 (2001)

Balsara, DS: Schemes for magnetohydrodynamics with divergence-free reconstruction. Astrophys. J. Suppl. Ser. **151**, 149-184 (2004)

Balsara, DS: Multidimensional extension of the HLLE Riemann solver; application to Euler and magnetohydrodynamical flows. J. Comput. Phys. **229**, 1970-1983 (2010)

Balsara, DS: A two-dimensional HLLC Riemann solver for conservation laws: application to Euler and magnetohydrodynamic flows. J. Comput. Phys. **231**, 7476-7503 (2012)

Balsara, DS: Multidimensional Riemann problem with self-similar internal structure. Part I - application to hyperbolic conservation laws on structured meshes. J. Comput. Phys. **277**, 163-200 (2014)

Balsara, DS: Three dimensional HLL Riemann solver for conservation laws on structured meshes; application to Euler and magnetohydrodynamic flows. J. Comput. Phys. **295**, 1-23 (2015)

Balsara, DS, Dumbser, M: Multidimensional Riemann problem with self-similar internal structure. Part II - application to hyperbolic conservation laws on unstructured meshes. J. Comput. Phys. **287**, 269-292 (2015)

Balsara, DS, Dumbser, M, Abgrall, R: Multidimensional HLL and HLLC Riemann solvers for unstructured meshes - with application to Euler and MHD flows. J. Comput. Phys. **261**, 172-208 (2014)

Balsara, DS, Kim, J: A subluminal relativistic magnetohydrodynamics scheme with ADER-WENO predictor and multidimensional Riemann solver-based corrector. J. Comput. Phys. **312**, 357-384 (2016)

Balsara, DS, Nkonga, B, Dumbser, M, Munz, CD: (2016b, in preparation)

Balsara, DS, Vides, J, Gurski, K, et al.: A two-dimensional Riemann solver with self-similar sub-structure - alternative formulation based on least squares projection. J. Comput. Phys. **304**, 138-161 (2016a)

Beckwith, K, Stone, JM: A second-order Godunov method for multi-dimensional relativistic magnetohydrodynamics. Astrophys. J. Suppl. Ser. **193**, 6-35 (2011)

Beskin, VS, Zheltoukhov, AA: On the structure of the magnetic field near a black hole in active galactic nuclei. Astron. Lett. **39**, 215-220 (2103)

De Villiers, J-P, Hawley, JF, Krolik, JH: Magnetically driven accretion flows in the Kerr metric. I. Models and overall structure. Astrophys. J. **599**, 1238-1253 (2003)

Del Zanna, L, Zanotti, O, Bucciantini, N, Londrillo, P: ECHO: a Eulerian conservative high-order scheme for general relativistic magnetohydrodynamics and magnetodynamics. Astron. Astrophys. **473**, 11-30 (2007)

Dumbser, M, Balsara, D: A new efficient formulation of the HLLEM Riemann solver for general conservative and non-conservative hyperbolic systems. J. Comput. Phys. **304**, 275-319 (2016)

Einfeldt, B: On Godunov-type methods for gas dynamics. SIAM J. Numer. Anal. **25**, 294-318 (1988)

Einfeldt, B, Munz, C-D, Roe, P, Sjogreen, B: On Godunov-type methods near low densities. J. Comput. Phys. **92**, 273-295 (1991)

Etienne, Z, Paschalidis, V, Haas, R, Mösta, P, Shapiro, S: IllinoisGRMHD: an open-source, user-friendly GRMHD code for dynamical spacetimes. Class. Quantum Gravity **32**, 175009 (2015)

Fragile, PC, Blaes, OM, Anninos, P, Salmonson, JD: Global general relativistic magnetohydrodynamic simulation of a tilted black hole accretion disk. Astrophys. J. **668**, 417-429 (2007)

Gammie, CF, McKinney, JC, Toth, G: A numerical scheme for general relativistic magnetohydrodynamics. Astrophys. J. **589**, 444-457 (2003)

Gardiner, TA, Stone, JM: An unsplit Godunov method for ideal MHD via constrained transport. J. Comput. Phys. **205**, 509-539 (2005)

Gardiner, TA, Stone, JM: An unsplit Godunov method for ideal MHD via constrained transport in three dimensions. J. Comput. Phys. **227**, 4123-4141 (2008)

Giacomazzo, B, Rezzolla, L: The exact solution of the Riemann problem in relativistic magnetohydrodynamics. J. Fluid Mech. **562**, 223-259 (2006)

Giacomazzo, B, Rezzolla, L: WhiskyMHD: a new numerical code for general relativistic magnetohydrodynamics. Class. Quantum Gravity **24**(12), S235-S258 (2007)

Hawley, J, Krolik, K: Magnetically driven jets in the Kerr metric. Astrophys. J. **641**, 103-116 (2006)

Honkkila, V, Janhunen, P: HLLC solver for ideal relativistic MHD. J. Comput. Phys. **223**, 643-656 (2007)

Kappeli, R, Mishra, S: Well-balanced schemes for the Euler equations with gravitation. J. Comput. Phys. **259**, 199-219 (2014)

Kappeli, R, Mishra, S: A well-balanced finite volume scheme for the Euler equations with gravitation. The exact preservation of hydrostatic equilibrium with arbitrary entropy stratification. Astron. Astrophys. **587**, A94-A110 (2016)

Kim, J, Balsara, DS: A stable HLLC Riemann solver for relativistic magnetohydrodynamics. J. Comput. Phys. **270**, 634-639 (2014)

Komissarov, S: A Godunov-type scheme for relativistic magnetohydrodynamics. Mon. Not. R. Astron. Soc. **303**, 343-366 (1999)

Komissarov, S: General relativistic magnetohydrodynamic simulations of monopole magnetospheres of black holes. Mon. Not. R. Astron. Soc. **350**, 1431-1436 (2004)

Komissarov, S: Observations of the Blandford-Znajek process and the magnetohydrodynamic Penrose process in computer simulations of black hole magnetospheres. Mon. Not. R. Astron. Soc. **359**, 801-808 (2005)

Krolik, K, Hawley, J, Hirose, S: Magnetically driven accretion flows in the Kerr metric. IV. Dynamical properties of the inner disk. Astrophys. J. **622**, 1008-1023 (2005)

McKinney, J, Blandford, R: Stability of relativistic jets from rotating, accreting black holes via fully three-dimensional magnetohydrodynamic simulations. Mon. Not. R. Astron. Soc. Lett. **394**, 126-130 (2009)

McKinney, J, Tchekhovskoy, A, Blandford, R: General relativistic magnetohydrodynamic simulations of magnetically choked accretion flows around black holes. Mon. Not. R. Astron. Soc. **423**, 3083-3117 (2012)

Mignone, A, Bodo, G: An HLLC Riemann solver for relativistic flows - II. Magnetohydrodynamics. Mon. Not. R. Astron. Soc. **368**, 1040-1054 (2006)

Mignone, A, Ugliano, M, Bodo, G: A five-wave Harten-Lax-van Leer Riemann solver for relativistic magnetohydrodynamics. Mon. Not. R. Astron. Soc. **393**, 1141-1156 (2009)

Parés, C: Numerical methods for nonconservative hyperbolic systems: a theoretical framework. SIAM J. Numer. Anal. **44**, 300-321 (2006)

Punsly, B: Fast-wave polarization charge horizons, and the time evolution of force-free magnetospheres. Astrophys. J. **612**, L41-L44 (2004)

Punsly, B: Black Hole Gravitohydromagnetics, 2nd edn. Springer, New York (2008)

White, CJ, Stone, JM: GRMHD in Athena++ using advanced Riemann solvers and staggered-mesh constrained transport (2015). arXiv:1511.00943

Zanotti, O, Dumbser, M: Efficient conservative ADER schemes based on WENO reconstruction and space-time predictor in primitive variables. Comput. Astrophys. Cosmol. **3**, 1 (2016). doi:10.1186/s40668-015-0014-x

3

PHEW: a parallel segmentation algorithm for three-dimensional AMR datasets
Application to structure detection in self-gravitating flows

Andreas Bleuler[1*], Romain Teyssier[1], Sébastien Carassou[1,2] and Davide Martizzi[1,3]

Abstract

We introduce PHEW (**P**arallel **Hi**Erarchical **W**atershed), a new segmentation algorithm to detect structures in astrophysical fluid simulations, and its implementation into the adaptive mesh refinement (AMR) code RAMSES. PHEW works on the density field defined on the adaptive mesh, and can thus be used on the gas density or the dark matter density after a projection of the particles onto the grid. The algorithm is based on a 'watershed' segmentation of the computational volume into dense regions, followed by a merging of the segmented patches based on the saddle point topology of the density field. PHEW is capable of automatically detecting connected regions above the adopted density threshold, as well as the entire set of substructures within. Our algorithm is fully parallel and uses the MPI library. We describe in great detail the parallel algorithm and perform a scaling experiment which proves the capability of PHEW to run efficiently on massively parallel systems. Future work will add a particle unbinding procedure and the calculation of halo properties onto our segmentation algorithm, thus expanding the scope of PHEW to genuine halo finding.

1 Introduction

Over the last decades, computer simulations have become an indispensable tool for studying the formation of structure on all scales in our universe. The common feature of those simulations is the clustering of matter due to self gravity. This clustering is of fractal nature in the sense that - as long as gravity is the dominant force - aggregations of matter turn out to have internal substructures, which are themselves gravitational bound, and may even contain subsubstructures. A crucial tasks in the analysis of simulations is therefore the identification of overdense regions and, ideally, their entire hierarchy of substructure.

First algorithms to perform this task have been invented in the very early days of computer simulations in Astronomy and Astrophysics. A halo finder based on spherical overdensities (SO) was described already four decades ago by Press and Schechter (1974) who used it to find structure in their simulation of 1,000 particles. Subsequently, the SO method has become one of the standard methods for halo finding. It consists in growing spherical regions around density peaks and assigning particles inside the spheres to the respective peak based on physical arguments. The also very popular friends-of-friends (FOF) method was introduced to halo finding by Davis et al. (1985). If two particles are separated by less than a user defined linking length, the particles are assigned to the same group. This results in groups of connected particles, the so-called FOF groups. On top of those two methods, a large variety of algorithms has been built over the last two decades: a recent halo finder comparison paper (Knebe et al. 2013) listed 38 different halo finders. For more detailed information about the halo finders which are on the market today, we refer to the series of papers that has emerged from the halo finding comparison project (Knebe et al. 2011; Onions et al. 2013; Knebe et al. 2013; Pujol et al. 2014).

*Correspondence: ableuler@physik.uzh.ch
[1]Institute for Computational Science, University of Zurich, Zurich, CH-8057, Switzerland
Full list of author information is available at the end of the article

On even larger scales, the identification and characterization of cosmic voids is an important task. Similar to haloes, the voids assemble in a hierarchical structure of voids and sub-voids which can be found in observational and simulation data likewise. Way et al. (2011) and Way et al. (2014) give an overview on void finding techniques and the relation to the identification of overdensities.

Automatic detection of structure is also performed at galactic scales. For example, Astronomers performing radio observations of molecular clouds entered the field when they started to identify clumps in position-position-velocity (PPV) data cubes. Stutzki and Guesten (1990) tried to fit the data by sums of triaxial Gaussian-shaped clumps and Williams et al. (1994) identified structure by contouring the dataset at evenly spaced levels without assuming an a priori shape for the clumps. More recently, Rosolowsky et al. (2008) showed how dendrograms can be used to exploit the hierarchy that naturally arises from contouring a PPV cube at multiple emission levels and used this technique to define substructures in molecular clouds.

With such a large choice of astrophysical structure finding tools at hand, one might ask the question why there needs to be yet another one. The trigger for the development of a new analysis tool was our need for 'on-the-fly' structure finding in the astrophysical simulation code (Teyssier 2002), in order to locate gas and/or dark matter clumps while the simulation is running. As pointed out in Knebe et al. (2013) there is a general trend towards 'on-the-fly' analysis for many reasons: most modern astrophysical simulations are performed on large computational infrastructure with distributed memory. The sizes of those simulations often exceed the total memory present in commonly used shared memory machines. The structure finding is therefore preferentially performed on the same machine that is running the simulation. Beyond that, the sizes of one single output of such simulations can quickly reach hundreds of GBs, up to several TBs. Storing many outputs for later post-processing is often not possible due to limited disk space, so that keeping only a catalogue of structure is the only viable solution.

Another reason for detecting structures while the simulation is advancing, is the possibility to couple the results of the halo decomposition to the simulation itself. In Bleuler and Teyssier (2014), for example, we have described a new algorithm for creation of sink particles, based on the properties of gas clumps detected 'on-the-fly'. This application requires an extremely high frequency at which structure finding must be performed. It must therefore make efficient use of the parallel infrastructure, and deliver good scaling properties for increasing numbers of MPI tasks, up to the number of CPUs the simulation is running on. Otherwise it will unacceptably slow down the simulation.

These requirements resulted in the development of PHEW (**P**arallel **H**i**E**rarchical **W**atershed), a new structure finding algorithm and its implementation into RAMSES.[a] While PHEW is not based on any pre-existing algorithm, it combines various concepts that have been used in other astrophysical structure finding tools before.

First, PHEW falls into the category of 'watershed-based' algorithms. These algorithms assign particles or cells to density peaks by following the steepest gradient, resulting in the so-called 'watershed segmentation' (see Section 2.1) of the negative density field. Other members of this category are DENMAX (Bertschinger and Gelb 1991), HOP (Eisenstein and Hut 1998), SKID (Stadel 2001), ADAPTAHOP (Aubert et al. 2004), GRASSHOPPER (Potter and Stadel, in prep). Note that in contrast to the aforementioned codes which work on the particles directly, we use a mesh to define the density field.[b] Void finding is typically performed using watershed-based algorithms too (e.g., Platen et al. 2007; Aragón-Calvo et al. 2010; Sutter et al. 2015).

Second, region merging in PHEW is based on the topological properties of saddle surfaces. This is the case as well for HOP, ADAPTAHOP and SUBFIND (Springel et al. 2001). As in the AHF halo finder (Knollmann and Knebe 2009), PHEW works on the density field deriving from particles that were previously projected onto the AMR mesh. In contrast to AHF, however, we do not use the AMR grid as a way of contouring the density field. A low density region which - for whatever reason - is refined to a high level does not compromise our results. Thus, in the landscape of existing halo finders, PHEW can be seen as filling the gap between P-HOP (Skory et al. 2010) which does not find substructures but is a MPI-parallel version of HOP, and ADAPTAHOP, a multi-threaded software that does find substructures, but has not yet been MPI-parallelized.

The aim of this paper is to present a new structure finding algorithm that: (1) can be applied to any density field defined on an adaptive grid, (2) is capable of detecting substructure, (3) is parallelized using the MPI library on distributed memory systems, and (4) is fast enough to be run at every time step of a simulation without significantly slowing down the calculation. What is not discussed in the present paper is an unbinding procedure for particles that are located inside the volume occupied by a certain halo but not gravitationally bound to it, as well as the subsequent computation of halo properties. These functionalities will be added to PHEW in the future. As briefly mentioned above, a previous version of PHEW has already been presented in Bleuler and Teyssier (2014). The algorithm described here differs from the previous one in the sense that it is now fully parallelized. This allows the algorithm to run now efficiently on thousands of CPUs and handle a complex topography with millions of density peaks and a rich hierarchy of substructures.

The article is organised as follows: in Section 2 we describe the serial version of the PHEW algorithm. In Section 3 we focus on the parallel implementation of the steps presented in Section 2. Section 4 contains scaling experiments which demonstrate the efficiency of the parallelization. Finally, we summarise and discuss our results, presenting an outlook on possible future work in Section 5.

2 The PHEW algorithm

In this section we describe the serial algorithm. As a starting point, we assume that we have a 3D density field on a AMR grid, particles have been projected onto the grid beforehand. The algorithm can be broken down in four main steps:

- Watershed segmentation;
- Saddle point search;
- Noise removal;
- Substructure merging.

In the first step, we assign every cell above a user defined density threshold to a local density maximum by ascending along the steepest gradient. This results in a primary segmentation of the computational volume into 'peak patches': regions associated to certain density peak. We establish the connectivity between the peaks by identifying the saddle points. We eliminate the peaks with a low density contrast to the background by merging them to a neighbour through their densest saddle point. The structure surviving the noise removal is considered the finest (sub)-structure. In a last step, we recursively merge the substructure to form larger and larger composite objects.

2.1 Watersheds in image processing

Before we start with a more detailed description of the algorithm, we take a quick look over the fence into the field of mathematical morphology and its application to image processing. There, watershed algorithms are a well known and extensively studied tool for image segmentation. The basic idea is that a grayscale image can be thought of as a topographic relief. A drop of water that falls somewhere onto this relief will follow the line of steepest descent until it reaches a local minimum. All points that connect to the same local minimum in that manner form a catchment basin. The watershed algorithm therefore segments the picture into catchment basins. The boundaries of the catchment basins are the actual watersheds. This technique is usually applied to the magnitude of the images gradient. In this way, the watershed lines trace regions of high gradients and segment the original image it into connected regions of small gradients. An excellent overview of the watershed techniques used in image processing is given by Roerdink and Meijster (2000).

An important difference to the watershed algorithms used for image segmentation lies in the computational cost for checking all neighbours of a cell/pixel. Working in 3D naturally increases the number of neighbours. Using an AMR grid further increases the number of possible neighbours since one has to consider possible neighbours at the same level as the original cell as well as one level above and below. Most importantly, the data structure in an AMR grid is very different from the one of a flat 2D array. The location of neighbouring cells in memory needs to be constructed before a neighbor can be checked for its density. Our main interest lies therefore in reducing the number of neighbours that have to be accessed. This aspect influences the choice of watershed algorithm for our purpose.

2.2 Watershed segmentation

In a first step, all cells above the density threshold are marked. We call those cells 'test cells'. For every test cell the densest neighbouring cell is identified and stored. If a cell has no denser neighbor, it is a local density peak. The peak obtains a peak ID which is stored as the 'peak patch label' of the corresponding cell. The test cells are sorted by decreasing density. Once sorted, every cell copies the peak patch label from its densest neighbor. The previous sorting ensures that the densest neighbour has been accessed before and has therefore already obtained its peak patch label. Thus, every cell is assigned to a peak after this one pass. All cells marked with the same peak patch label form a peak patch (see Figure 1, second panel). Note that our peak patches correspond to the catchment basins introduced in Section 2.1. Since we are working on peaks rather than minima, we introduce this new terminology to avoid the cumbersome notion of an 'inverted catchment basin'. Note that this procedure is very similar to the *hill climbing* method described in Roerdink and Meijster (2000) which was introduced by Meyer (1994).

2.3 Saddle point search

Before we can merge peak patches, we have to establish the connectivity between them. All test cells are checked for neighbouring cells that belong to a different peak patch. If such a neighbouring cell is found, the average density of the starting cell and its neighbour is considered as the density at the common surface of the two bordering peak patches. The maximum density on the connecting surface is the density of the saddle between the two peaks and stored. At the end of this step, each peak has its list of neighbouring peaks together with the corresponding saddle point densities. We denote the maximum saddle point of a peak as the 'key saddle' and the corresponding neighbour as 'key neighbour'.

2.4 Noise removal

A known problem of the watershed method is over-segmentation. The presence of a huge number of local minima - for example due to random particle noise or transient gas density fluctuations - causes segmentation into as many catchment basins as there are local minima. Generally speaking, there are two possible strategies to deal with this problem: not creating the over-segmentation in the

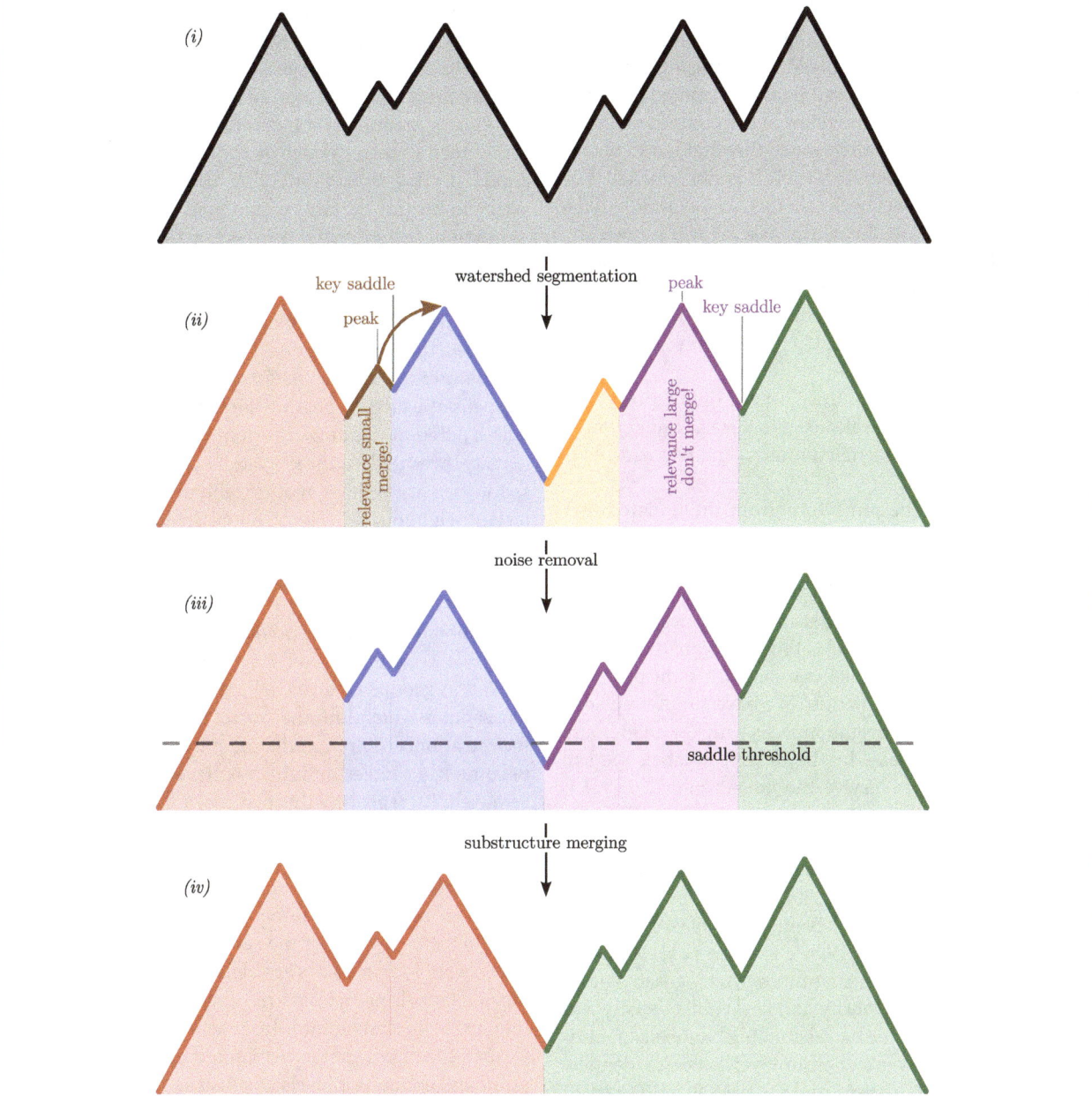

Figure 1 Working principle of PHEW. The main steps of the algorithm are visualized on a 1D density field (first panel). The segmentation into peak patches is shown in the second panel. Based on the relevance of a peak (peak-to-saddle ratio) we decide whether a peak represents 'noise' or substructure. Irrelevant peaks are merged through their highest saddle points (third panel). The surviving objects are labeled as Level 0 clumps and denote the finest level of substructure. The substructure is merged based on a saddle threshold (third panel) into parent structure (fourth panel).

first place or merging over-segmented regions. Preventing over-segmentation the can be obtained using markers to preselect allowed minima (e.g., Moga and Gabbouj 1998). This usually requires a human intervention, which in our case is not possible. Another way is to use the so-called hierarchical watershed algorithm[c] (Beucher 1994). Hierarchical watershed algorithms merge artificial catchment basins to more important ones based on some cri-

teria. What we will describe in the following turns our watershed algorithm into a hierarchical algorithm in the Beucher (1994) sense.

After having previously identified the saddle points, we classify the peaks based on their contrast to the background. We define the contrast as the ratio of the peak density to the key saddle density and name it 'relevance'. This is sketched in the second panel of Figure 1. Every peak is assigned a 'final peak' label which is initialized to the peaks own peak ID and updated whenever a peak is merged to another one. The peaks are sorted by decreasing peak density. For each peak, the key saddle is determined from the list of saddle points and the relevance is computed. Peaks with a relevance below a relevance threshold are considered noise.[d] If the peak is relevant, it is not touched. For an irrelevant peak, we check whether its key saddle links it to a denser peak. If this is the case, it will inherit the final peak label from this key peak. As in the watershed segmentation, the previous sorting makes sure that the final peak labels can propagate through long chains of connected peaks in just one loop. If a peak is both isolated and irrelevant, it is discarded.

When two peaks merge, their lists of saddle points are merged as well. If both peaks used to have a connection to the same third peak, the maximum of the two saddles is kept.

Now, we iterate the procedure: from the updated lists of saddle points, the key saddles are determined. Peaks are accessed in the order of decreasing peak density and irrelevant peaks are merged. After an iteration without any mergers, all irrelevant peaks have been merged or discarded and the noise removal is finished. Note that the described merging process follows exactly the same principle as the watershed segmentation. We have simply replaced cells by peaks, densest neighbour cells with key neighbours and the peak patch label by the final peak label. We call the structures which survive the noise removal **Level 0 clumps**. They constitute the finest structure (see Figure 1, third panel) in our hierarchy.

Using the relevance as a merging criterion results in a similar definition of a clump as it is obtained by algorithms that contour the dataset at evenly spaced levels in log-space (e.g., Williams et al. 1994). There, a peak-to-saddle ratio above a given value guarantees that a contour level will fall between peak density and key saddle density and thus the detection of the corresponding clump as an individual object. However, a contour level can coincidentally lie between the peak density and the key saddle density of an object with a very low peak-to-saddle ratio, resulting in the detection of an irrelevant density fluctuation as a clump. Merging based on the relevance removes this randomness from the analysis.

For a density field that is obtained from an underlying particle distribution, the relevance criterion can be interpreted as a signal-to-noise criterion on the basis of an individual cell. We assume a roughly constant number of particles per cell as this number is often used as a refinement criterion for dark matter simulations in RAMSES. A large relevance thus translates into a small probability that the peak density is simply drawn from a Poisson distribution with the mean being equal to the saddle point density. A *true* signal-to-noise criterion would consider the probability that the entire peak patch is consistent with being randomly drawn from the density at the saddle point. We would expect such a criterion to distinguish noise from physical structure more reliably. However, such a criterion is not compatible with our parallelization strategy of the merging procedure, as it includes quantities that are 'additive' under a merger - such as the size or the total mass of a peak patch - into the merging criterion. As we will describe in Section 2.7, this would make the outcome of the merging process depend on the exact order at which the peaks are considered for merging.

2.5 Saddle threshold merging

If desired, the remaining peaks and their associated clumps can be merged further to form composite clumps. This happens by exactly repeating the previous merging process with a different merging criterion. We have implemented a density threshold for the key saddle as a criterion. If the key saddle density is above that threshold, a peak is merged to its key neighbour (see Figure 1, fourth panel). Another possible criterion is the repeated use of the relevance threshold, this time with a higher value.

2.6 A hierarchy of saddle points

We have seen in Section 2.4 that saddle points are removed in groups or levels by merging through them. All key saddles which link their peak to a denser one are removed at once. Through the merging, other saddle points become key saddles and the next level of saddle points is removed. By repeating this process, a natural hierarchy of saddle points and clumps is produced. In Figure 2 we illustrate the construction of this hierarchy. We start with the Level 0 clumps after the noise removal (no substructure except for noise) and assume that the saddle threshold for merging is below any of the saddles depicted in Figure 2. The Level 1 saddle points are identified and used for merging. The resulting objects are Level 1 clumps as they have one level of substructure. In general, a Level n clump is formed through a merger which removes a Level n saddle point and contains n levels of substructure. This produces a very natural hierarchy of saddle points and clumps based on the levels of substructure. Note that the level of a saddle point does not reflect its density. A more traditional way of grouping substructure based on the density of the saddle that connects two substructure objects as it is for example produced by ADAPTAHOP can easily be recovered from this hierarchy.

2.7 Merging order

We will see in Section 3 that we have to drop the idea of sorting the peaks globally when we parallelize PHEW. This will alter the order in which peaks are merged in an unpredictable way. It is therefore crucial that the PHEW allows the order of mergers to change without causing different results. This not true in general. Yet, as we will show in

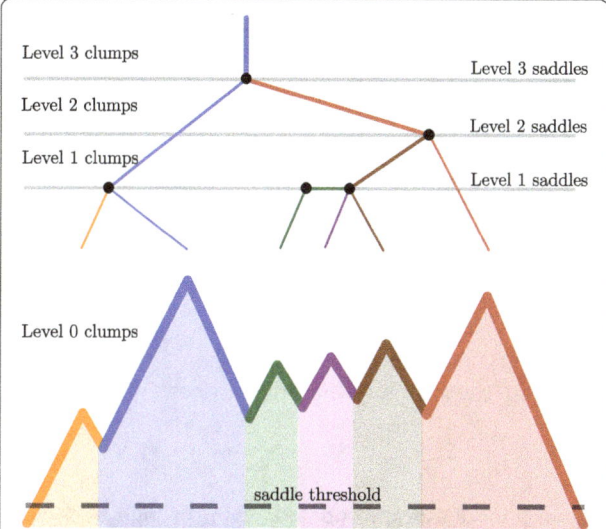

Figure 2 Hierarchy of saddle points as it is produced by our merging algorithm. Level *n* saddle points are used for merging during the *n*th round of mergers. Level *n* clumps emerge from a merger through a Level *n* saddle point and contain *n* levels of substructure.

this section, it is the case when we respect the three merging rules:

1. A peak is only merged to a denser one (upward).
2. A peak is only merged through its key saddle.
3. The density of the key saddle or the relevance are used as merging criterion.

The result of the merging procedure is uniquely determined by the set of saddle points that is used for merging. This is a subset of all saddle points. In order to affect the outcome of the merging process, changing the order of mergers therefore has to change the set of used saddle points. Let us consider a peak n connected to its key neighbour m through the key saddle s_{nm} at the very beginning of the merging process. The peak density of m is higher than that of n, $m > n$. There are three possible types of mergers related to n or m that can happen before n is considered for merging. We will show that none of them can change the fate of n.

1. A third peak might be merged into m. Due to upward merging, this cannot change the peak density of m and therefore decision if n will be merged into m is not influenced.
2. Peak m might merge into another peak m'. The saddle s_{nm} will still exist, now linking n to m'. Due to upward merging we have $m' > m > n$ which means that n is still the lower of the two peaks connected by $s_{nm'}$. The decision whether n is merged through s_{nm} is unaltered.
3. A third peak i might be merged into n. The peak density of n cannot change due to that since it

would mean that peak i had a higher density than n which contradicts the upward merging. The key saddle cannot change because this would mean that peak i had a saddle point s_{ij} higher than s_{nm}. This would imply that the saddle point s_{ni} through which i was merged into n was even higher, $s_{ni} > s_{ij}$ otherwise s_{ni} had not been the key saddle of peak i. Yet, $s_{ni} > s_{ij} > s_{nm}$ contradicts that s_{nm} is the key saddle of peak n. The peak density of n and its key saddle are thus unchanged, therefore the relevance of n is not changed either.

This shows that we can arbitrarily delay the moment when we consider a peak for merging as long as we respect the three merging rules. The mergers happening in the mean time cannot change the properties deciding if and through which saddle this peak will be merged. A possible way to prevent violation of merging rule (ii) is to consider all peaks for merging until no further mergers are possible before any new key saddle of the merged peaks is computed. This results in using the saddle points for merging on a 'level-by-level' basis. This is a key to the parallelization of PHEW since it will allow performing a big number of operations (mergers), in between each round of communication (finding new key saddles). Note that this line of argumentation breaks when we violate merging rule (iii) and use for example the clump mass as merging criterion. The mass is a property that changes with every merger. Therefore, altering the merging order does change the mass of a clump at the moment it is considered for merging and can thus change the decision whether the clumps should be merged or not.

3 Parallel implementation

We now turn to the implementation of the previously described steps in a parallel, distributed-memory framework. Where a detailed description of an algorithmic block in words would prevent readability of the paper, we refer the interested reader to a corresponding block written in pseudocode located in Algorithms 1 and 2. We assume that the computational domain has been previously decomposed into non-overlapping spatial domains, each domain containing a partition of the AMR mesh on which the density field is defined. In every MPI task, the local partition of the mesh is referred to as the 'active cells'. They are wrapped by a thin layer of cells that belong to other tasks. These ghost cells are referred to as belonging to the 'virtual boundaries'. These virtual boundaries are updated through MPI communication before PHEW is called to make sure that the densities in the virtual boundary cells are equal to the densities in the corresponding active cells hosted by other MPI tasks.

3.1 Parallel watershed

The watershed segmentation is *non-local* by nature. This can easily be understood by imagining a mountain ridge.

```
 1  for testcell ∈ {testcells} do
 2      for neighbour ∈ {neighbours} do
 3          if (PeakPatch [neighbour] ≠PeakPatch [testcell]) and (PeakPatch [neighbour] > 0) then
 4              i=GetLocalPeakIndex (PeakPatch [testcell])
 5              j=GetLocalPeakIndex (PeakPatch [neighbour])
 6              if AverageDensity (testcell,neighbour) > SaddleMatrix [i,j] then
 7                  SaddleMatrix [i,j]=AverageDensity (testcell,neighbour)
 8                  SaddleMatrix [j,i]=AverageDensity (testcell,neighbour)
 9              end
10          end
11      end
12  end
```

Algorithm 1: Pseudocode describing the construction of the local saddle point matrices

Two drops of water falling onto both sides of the ridge will initially move away into different directions. They might flow into different rivers which flow into different lakes, or they might as well end up in two rivers which join before reaching a lake. The two situations *cannot* be distinguished based on local properties. Parallelization of the watershed algorithm is therefore a non-trivial task. In the literature, one finds various approaches to parallelization for the different watershed algorithms (see e.g., Roerdink and Meijster 2000). Our technique is very close to the technique described in Moga (1997) and called 'hill climbing by locally ordered queues'.

Each task performs a loop over all its active cells, in order to identify first the test cells (cells above the density threshold). For faster access, the indices of all test cells are stored in an array. A loop over all test cells is performed where the densities of all neighbouring cells are checked. The index of the densest neighbouring cell is stored for each test cell, since it will be used several times during the algorithm. Note that the densest neighbour of a cell can lie inside the virtual boundary, while test cells are always inside the active domain.

During the first loop, all peaks (local extrema) are counted. After the loop, the number of peaks in each MPI domain are communicated between all MPI tasks, which allows each MPI task to compute a global index (ID) for its peaks. In another loop over test cells, cells which represent a peak are labeled with their global peak ID, all other test cells are initialised with a peak patch label equal to zero. The peak patch labels are updated inside the virtual boundaries using MPI communication (Figure 3, second panel). As explained in Section 2, every MPI task computes a permutation which sorts test cells in decreasing density order, using the quick sort algorithm (Press et al. 2007). Using this permutation, a sorted loop, where every cell inherits the peak patch label from its densest neighbour is performed (Figure 3, third panel). During this loop, the number of cells that have changed their peak patch label is counted. After the loop, the peak patch labels in the

virtual boundaries are updated again through MPI communications. This procedure is iterated (Figure 3, fourth panel) until no cell inside the entire computational box has changed its peak patch label during a full loop. This completes the parallel watershed segmentation.

3.2 Virtual peak boundary

As we have already described in Section 2, our peak patch merging step is analogous to the segmentation step. The patches now take the role of the cells, the peak patch label is replaced by the final peak label and the densest neighbouring cell is replaced by the key neighbouring patch. As explained before, the parallelization of the peak patch segmentation is exploiting the virtual boundaries surrounding each MPI domain. If we want to use the same strategy to parallelize the merging process, we need the analog of the virtual mesh boundary: a virtual peak boundary. In contrast to our usual virtual mesh boundary, the virtual peak boundary does not represent a fixed region in space. As the merging process advances, new connections appear and new peaks have to be introduced in the virtual peak boundary. Our virtual peak boundary is therefore more dynamic than our virtual mesh boundary.

Figure 4 shows a possible layout of peaks in memory. Note the distinction between a peaks global ID and its local index. The latter of the two is the position of the peak in local memory. The peaks that are located inside a tasks MPI domain are called active peaks. They take the first N_{active} places in memory. The active peaks are followed by the ghost peaks that belong to the virtual peak boundary. Since it is unknown at the beginning of the merging process how much space for ghost peaks will be necessary, we set

$$N_{\text{max}} = \max\left\{4 \max_{\text{tasks}}\{N_{\text{active}}\}, 1{,}000\right\}, \qquad (1)$$

as a default value that can be modified by the user. The preset N_{max} is mostly a large overestimation of the effectively

```
 1   Preparatory step - initialize two peak-based properties.
 2   for peak ∈ {active peaks} do
 3   │    alive [peak]=1
 4   │    FinalPeak [peak]=GlobalPeakID [peak]
 5   end

 6   Loop over Levels in the saddle point hierarchy.
 7   mergers=1
 8   while mergers > 0 do
 9   │    mergers=0

10   │    Propagate the final peak label through key saddle points.
11   │    LevelMergers=1
12   │    while LevelMergers > 0 do
13   │    │    LevelMergers=0
14   │    │    CommunicateSaddlepoints
15   │    │    BuildPeakCommunicator
16   │    │    ScatterCommunicate (PeakDensity,FinalPeak,alive)
17   │    │    for peak ∈ {sorted active peaks} do
18   │    │    │    if alive [peak]> 0 then
19   │    │    │    │    PSratio=PeakDensity [peak]/KeySaddle [peak]
20   │    │    │    │    if PSratio > 1.5 and PeakDensity [KeyNeighbor [peak] ]> PeakDensity [peak] then
21   │    │    │    │    │    FinalPeak [peak]=FinalPeak [KeyNeighbor [peak]]
22   │    │    │    │    │    LevelMergers=LevelMergers+1
23   │    │    │    │    end
24   │    │    │    end
25   │    │    end
26   │    │    ScatterCommunicate (FinalPeak)
27   │    │    LevelMergers=MPIsum (LevelMergers)
28   │    end

29   │    For every merger, merge the corresponding lines in the saddle point array.
30   │    for peak ∈ {all peaks} do
31   │    │    if GlobalPeakID [peak] ≠ FinalPeak [peak] then
32   │    │    │    NewIndex=GetLocalPeakIndex (FinalPeak [peak])
33   │    │    │    for column ∈ {matrix columns} do
34   │    │    │    │    if SaddleMatrix [peak,column] > SaddleMatrix [NewIndex,column] then
35   │    │    │    │    │    SaddleMatrix [NewIndex,column]=SaddleMatrix [peak,column]
36   │    │    │    │    │    SaddleMatrix [column,NewIndex]=SaddleMatrix [peak,column]
37   │    │    │    │    end
38   │    │    │    end
39   │    │    │    SaddleMatrix [NewIndex,peak]=0
40   │    │    │    SaddleMatrix [NewIndex,NewIndex]=0
41   │    │    end
42   │    end
43   │    BuildPeakCommunicator

44   │    Set alive to zero for dead peaks and count mergers.
45   │    for peak ∈ {active peaks} do
46   │    │    if GlobalPeakID [peak] ≠ FinalPeak [peak] and alive [peak]==1 then
47   │    │    │    alive [peak]=0
48   │    │    │    mergers=mergers+1
49   │    │    end
50   │    end
51   │    ScatterCommunicate (alive)
52   │    mergers=MPIsum (mergers)

53   │    Remove saddle points linking to dead peaks.
54   │    for peak ∈ {all peaks} do
55   │    │    for column ∈ {matrix columns} do
56   │    │    │    if alive [peak]==0 or alive [column]==0 then
57   │    │    │    │    SaddleMatrix [peak,column]=0
58   │    │    │    end
59   │    │    end
60   │    end
61   end
```

Algorithm 2: Pseudocode describing the parallel merger procedure

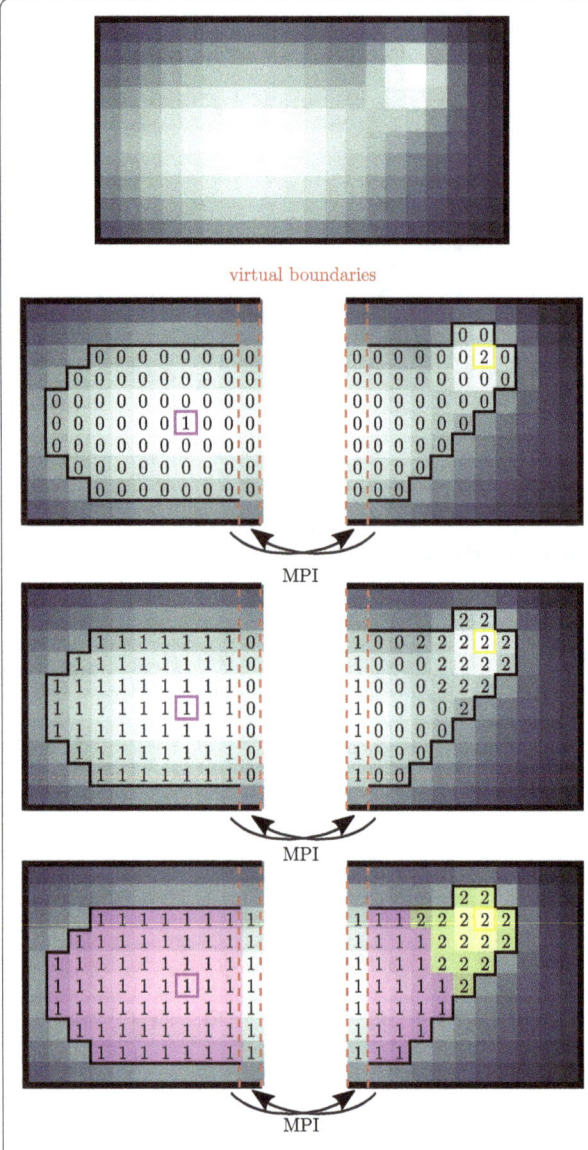

Figure 3 Parallelization of the watershed segmentation shown on a 2D field. The top panel depicts the computational box with the density field. In the second panel, the two MPI domains and the virtual boundaries are shown, the peaks have obtained their IDs and the cells are labeled. In a loop over all test cells, the peak patch labels can propagate inside the MPI domains (third panel). After the loop, the virtual boundaries are updated and the procedure is repeated (fourth panel).

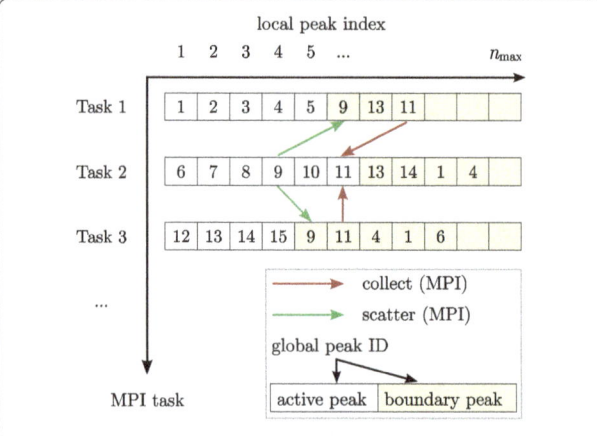

Figure 4 Example of peak layout in memory for 3 MPI tasks. The figure shows the global peak ID as a 'function' of the MPI task and the local peak ID. The local peak index for a given global peak ID is stored in a hash table.

used space in memory for peaks (see Section 4), designed to be sufficient for all setups we have tested. However, the memory consumption for peak properties is still negligible compared to the necessary space for the AMR grid.[e] All peak properties such as the peak density are allocated up to N_{max}.

Since every task is aware of its starting number of global peak IDs, switching from global peak ID to local peak index and vice versa is trivial for active peaks. To recover a

boundary peaks global ID from its local index, we simply store the global ID in memory at the position of its local index. For the opposite direction we use a hash table that contains the local peak index for a given global peak ID (hash key).[f] Whenever we introduce a new boundary peak into the virtual peak boundary, it obtains the local peak index corresponding to the first free space in memory. The global peak ID is stored and a hash key is computed. Which peaks need to be present in the virtual peak boundary depends on the connectivity of peaks. The initial state of the virtual boundary will thus be constructed while searching for saddle points that connect the peaks.

3.3 The peak communicator

By introducing a peak into the virtual peak boundary, it only obtains a local peak index. No properties except the global peak ID of a newly introduced boundary peak are present at this stage. We now describe how information is transferred from the MPI task which hosts a peak (the 'owner' of that peak) into the virtual peak boundaries of other tasks and vice versa. There are two types of communication: inward communication ('collect', red arrows in Figure 4) from all processes which have a certain peak inside their peak boundary to the owner of the peak, and outward communication ('scatter', green arrows in Figure 4) to update the peak properties in the virtual boundaries. When performing a collect communication, one has to specify whether one is computing a sum, minimum or maximum of the incoming values belonging to the same peak. When a scatter communication is performed, the peak properties of boundary peaks are overwritten with their equivalent from the peaks owner. A typical communication pattern for a peak property is therefore a collect communication followed by a scatter communication.

Before this communication can be performed, we need to build a communication structure which we refer to as the 'peak communicator'. We allocate a matrix C of size $N_{task} \times N_{task}$. The entry c_{ij} is the number of peaks inside the virtual peak boundary of task i that are owned by task j. Each task builds its line of C in a loop over the boundary peaks by looking at their global peak IDs. Through MPI communication, the lines of C are shared between all task to complete C.[g] The entries in the matrix C determine the amount of data that is sent to/received from another MPI task. This information is used to allocate send and receive buffers and to direct each entry in a tasks send buffer to the correct MPI task in a round of all-to-all communication. In order to complete the setup of the peak communicator, we use the established structure to perform a collect communication of the global peak ID. This information allows the identification of a position in the receive buffer (or in the send buffer in the case of a scatter communication) with an active peak. This completes the buildup of the communication structure. The peak communicator needs to be rebuilt whenever new peaks have potentially been added to the virtual peak boundary of any MPI task.

3.4 The saddle point matrix

To keep track of the saddle points, we establish a symmetric saddle matrix M, where the entry m_{ij} is the density of the saddle point connecting the peaks i and j. As most of the peaks patches are not touching each other, we use a sparse matrix representation of M. Note that the indices i, j are the local peak indices, which makes M a sparse matrix of virtual size $N_{max} \times N_{max}$. Since we are interested in the maximum entry of each line and the column where it is located when in comes to merging, we keep track of those two values when adding new entries into M. The maximum and its column need to be recomputed by checking each non-zero element of a line only after values have been removed from the given line in M which reduces the number necessary accesses to the sparse matrix.

The construction of the sparse matrices is performed locally the way described in Section 2.3. Whenever a connection is found to a peak that is not yet present in the virtual peak boundary, the given peak is introduced by assigning it a local index. See Algorithm 1 for the pseudocode describing the saddle point search on each task.

3.5 Communication of saddle points

We could now use a collect communication on the saddle points for every peak in the entire computational box. As a result of that, every task would have access to all saddle points of all his active peaks. The global key saddle and key neighbour could then be determined by every MPI task for his active peaks. However, this approach would introduce a lot of communication and unnecessarily fill the sparse saddle matrices. The only necessary information to perform one iteration in the merging process is the (global)

key saddle density of a peak and the corresponding key neighbour. This global maximum saddle can be found by comparing the local maxima of each MPI task. We thus minimise communication by performing a collect communication only on the local maximum of each row in the saddle point matrix. Together with the local maximum saddle density, we collect the global peak ID that denotes the local key neighbour. The owner of a peak can now compute the global key saddle for a given peak by comparing all the local maxima. The global peak ID that was received from the MPI task which hosts the global key saddle is the key neighbour of the peak. If not already present, the key neighbour is introduced into the virtual peak boundary of the owner task and the key saddle density is written into the sparse saddle matrix of the owner. Every MPI task can now perform a complete iteration in the merging process without any further communication of saddle point densities.

3.6 Merging in parallel

We are now set for the actual merging of the peaks. We introduce two new peak properties: a logical variable called `alive` which is initialised to 'true' and set to 'false' when a peak is merged into another one, and the final peak label which is initialised to the global peak ID for all active peaks. These two new properties and the peak density are updated in the virtual peak boundaries using a scatter communication. A permutation which sorts the active peaks by decreasing density is computed. Now we propagate the final peak label through the key saddles in a level-by-level fashion. On each level, we iterate until no final peak label is moved, while the virtual boundaries are updated after every iteration. This is perfectly analogous to the parallel watershed segmentation. After every level of saddle points we update the `alive` variable, the saddle point matrices and the virtual boundaries. The merger routine is described in Algorithm 2 in pseudocode. The substructure merging is performed in exactly the same way, we just replace the relevance threshold by the saddle density threshold.

4 Scaling test

We use a previously run cosmological dark matter simulation with 512^3 particles for a scaling experiment. We restart the simulation from the output corresponding to redshift $z = 0$ using various numbers of MPI tasks. Before PHEW can run, we project the particle density onto the AMR grid using the CIC (Cloud-In-Cell, Hockney and Eastwood 1981) algorithm. Once we have constructed the grid-based density field, we run PHEW with a density threshold of 80 times the cosmological critical density (noted ρ_{crit}) and a relevance threshold of 3. After merging the peak patches into Level 0 clumps (sub-haloes), we merge to form haloes by applying a saddle threshold of

Table 1 Parameters and some runtime statistics for the 1,024 task runs of the experiment

	512^3	$1,024^3$
N_{parts}	512^3	$1,024^3$
N_{tasks}	1,024	1,024
Density threshold	$80\rho_{crit}$	$80\rho_{crit}$
Relevance threshold	3	3
Saddle threshold	$200\rho_{crit}$	$200\rho_{crit}$
Number of test cells	104,360,968	835,609,288
Number of density peaks	6,714,764	53,994,995
Number of relevant clumps	1,311,208	10,612,079
Number of haloes[*]	521,185	4,234,746
Runtime	8.0 s	38.9 s
Number of iterations for…		
…watershed segmentation	7	9
…noise removal		
Level 1	7	7
Level 2	5	6
Level 3	4	4
Level 4	2	3
Level 5	1	2
Level 6	1	1
Level 7	1	1
Level 8		1
…substructure merging		
Level 1	4	3
Level 2	3	4
Level 3	3	3
Level 4	2	2
Level 5	1	2
Level 6	1	1
Level 7		1

[*]Note that we do only count the objects that contain more than 10 dark matter particles.

$200\rho_{crit}$. The first column in Table 1 summarizes parameters and runtime statistics obtained for 1,024 tasks. We see a rich hierarchy of saddle points spread over many levels. The numbers of iterations necessary show that there is structure extending over several domain boundaries at every stage of the process (peak patches, clumps, haloes). Note that PHEW finds *exactly* the same structures, independent of the number of MPI tasks that have been used. This empirically confirms what we have described in Section 2.7. It is also worth mentioning that the iteration pattern looks surprisingly similar for the other 512^3 runs in our scaling experiment. The total number of necessary iterations increases from 35 to 45 when going from 32 to 2,048 tasks while it would be only 3 when for the serial algorithm. An example of the hierarchical structure that is found by PHEW is shown in Figure 5 which depicts a halo with four levels of substructure taken from our scaling experiment.

In our numerical experiment, PHEW was run five times in a row, for five main simulation time steps following the restart. We measure the total runtime of each call to PHEW as well as the time spent on the different algorithmic steps. We find the variance of the runtimes to be negligible and

conclude that the timings are stable. Note that the preliminary construction of the density field is performed inside the watershed segmentation block. However, the CIC algorithm is quick compared to the watershed segmentation. We also measure the amount of time necessary for each MPI task to write the properties of the structure inside its domain to disk.

The runtimes for the various numbers of MPI tasks are plotted in the top two panels of Figure 6. The top panel shows satisfactory scaling of the overall algorithm up to 1,024 MPI tasks which is four times the numbers of tasks that were used to perform the original simulation. In this regime, the total runtime of PHEW is dominated by the watershed segmentation and the saddle point search. The most costly operations inside those two blocks are the construction and access of neighbouring cells. The total workload of those blocks thus scales linearly with the number of test cells per MPI task.

The second panel in Figure 6 shows that the runtime of those two blocks does actually scale over the entire range of numbers of tasks that we have tested. It also shows that the second panel in Figure 6 shows that the merging procedures scale well up to 256 tasks. The scaling of the merging process in this region is mainly controlled by two effects: with a growing number of tasks, the load imbalance of the peaks between the different MPI tasks increases. This is unavoidable as the domain decomposition is optimised for all AMR cells, not for the test cells only, and even less for the peak patches. The second reason is the growing ratio of surface to volume as the computational box is divided in smaller parts. This results in more ghost peaks per active peak which causes a higher workload per active peak. Those two effects are quantified in the first two rows of Table 2.

The solid line in the bottom panel of Figure 6 is a result of both effects mentioned above. It depicts $\max\{N_{sparse}\}$, the maximum number of used sparse matrix elements over all MPI tasks. In perfect scaling conditions, this number would decrease as $1/N_{tasks}$. We thus multiply $\max\{N_{sparse}\}$ by N_{tasks} and rescale to one at 32 tasks. We compare this to the runtime of the noise removal (also scaled). We observe that this 'worst case' number of entries in the sparse saddle point matrix does explain the scaling of the merging process up to 512 tasks. Beyond that, we believe that MPI communications become the performance bottleneck.

In Table 2 we also show the maximum ratio of ghost peaks to active peaks. For 2,048 tasks we have a value of 24%. This shows that the number N_{max} defined in Equation (1) is an overestimation of the effectively used memory for ghost peaks for this setup. In the same table, we also list the number of hast table collisions. There are very few collisions as the hash table is far from filling up and we conclude that the relatively simple hash function that we use is good enough for our purpose. Another fact worth mentioning is the relatively constant ratio of non-zero entries

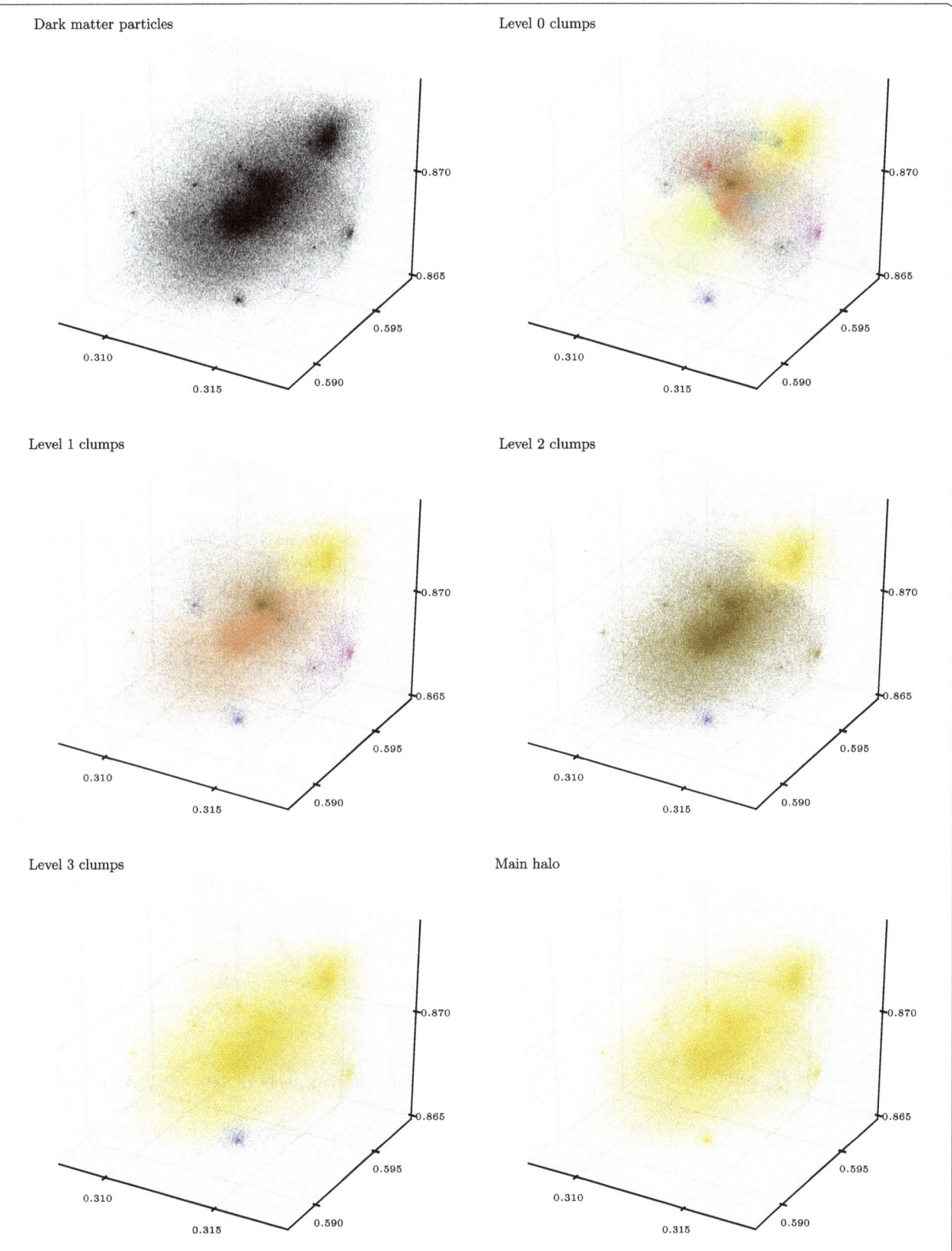

Figure 5 Visualization of PHEW applied to a dark matter halo. We show a small sub-volume of the 512^3 particle box used in our scaling experiment. The coordinates indicate the fraction of the box size. The sub-volume contains $\approx 2 \times 10^6$ particles. The objects that emerge after the noise removal (Level 0 clumps) are indicated in the second panel, where all particles belonging to the same object share a color. Every subsequent panel shows the status after a further round of merging as it is described in Section 2.6.

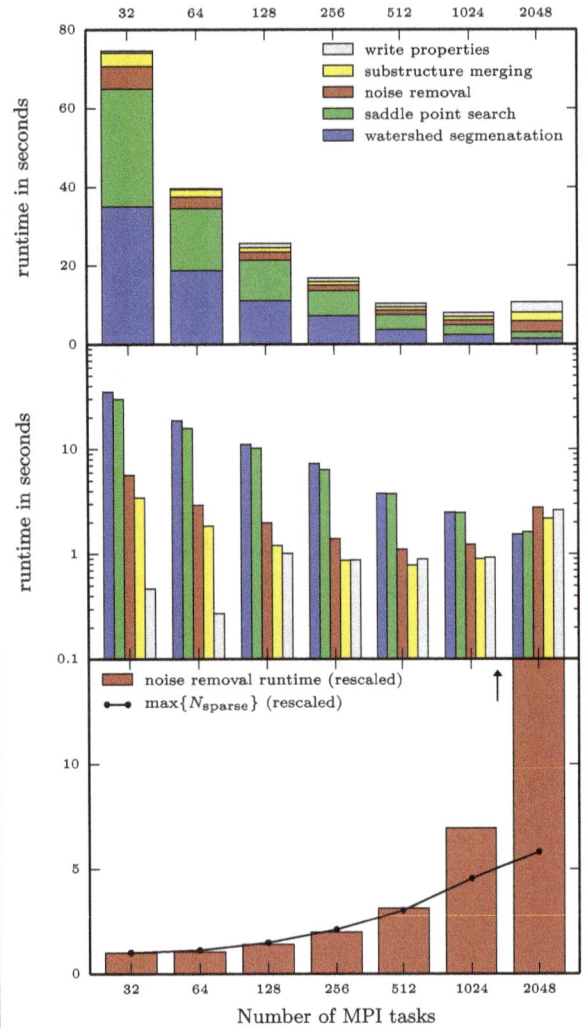

Figure 6 Scaling properties of the different parts in PHEW obtained by restarting a cosmological dark matter simulation with 512³ particles at redshift *z* = 0. The top two panels show the runtimes of the different algorithmic blocks in PHEW. The peak patch segmentation and the saddle point search exhibit excellent scaling in the entire range of MPI tasks that we have tested. The merging in our test scales well up to ~256 MPI tasks. The bottom panel shows the maximum number of sparse matrix elements over all MPI tasks compared to 1/N_{tasks} and rescaled to one at 32 MPI tasks. The increase seen in this number for of tasks is due to the growing load imbalance in terms of peaks per task and the increase in the surface to volume ratio of the domain segmentation. It explains the increase of the scaled runtime of the noise removal very well up to 512 tasks. The overall scaling of the algorithm is satisfactory up to 1,024 MPI tasks which is four times the number of CPUs the original simulation was run on.

in saddle point matrix to the number of peaks seen in the third line of Table 2. Divided by two (due to the symmetry of the saddle point matrix), this number gives a good idea of the effective number of neighbours per peak.

As a second test we perform a 'weak scaling' comparison of our 1,024 task run with another 1,024 task run but this time on a larger, 1,024³ particle box. The second column of Table 1 lists the statistics of that run. The numbers of test cells, peaks, clumps and haloes all increase by the expected factor of ≈8. We thus divide the runtimes of PHEW for this setup by 8 and compare to the runtime of the 1,024 task run on the 512³ box. This comparison is plotted in Figure 7. The figure shows that the runtime per data decreases for all parts of PHEW by increasing the size of the data. Especially the efficiency of merging routines benefit a lot from the increased size of the dataset. We thus conclude that we can enlarge the range of N_{task} where PHEW scales well, by increasing the size of the simulation.

5 Conclusions

We have presented PHEW, a new structure finding algorithm and its MPI parallel implementation into the AMR code RAMSES. PHEW finds density peaks and their associated regions in a 3D density field by performing a watershed segmentation. The merging is based on the saddle point topology. We have described a two-step approach to merging. In a first step, we merge irrelevant density fluctuations which we consider as noise. In a second step we merge the finest substructure hierarchically, into large, connected regions above the adopted density threshold. This merging process naturally results in a tree-like representation of substructure similar to the dendrograms presented by Rosolowsky et al. (2008).

The main focus of this article is on the parallel implementation of the algorithm which we have described in detail. Our implementation is truly parallel, meaning that it produces *exactly* the same results for varying numbers of MPI tasks. To test the parallelization of PHEW, we have performed a scaling experiment on a snapshot from a cosmological dark matter simulation. We have found excellent scaling in the relevant range of MPI tasks. When using the same number of MPI tasks that was used for the actual simulation, the runtime of PHEW ~10% the time it takes to advance the simulation by one time step. This allows for frequent usage of PHEW on-the-fly and thus more fine-grained information about how matter assembles in simulations.

RAMSES has recently been demonstrated to scale well up to 38,016 MPI task (Alimi et al. 2012) when used to simulate a very large cosmological volume. Even the largest haloes that PHEW will identify in such a simulation cover only a small fraction of the computational volume. This essentially turns such a setup into a weak scaling experiment for PHEW, where the scalability is determined by the domain decomposition of RAMSES. Without having applied PHEW to such a large setup, we therefore expect the algorithm to show similar scaling properties in this range as the RAMSES code. A more challenging situation for the PHEW

Table 2 Runtime diagnostics for the parallelization of PHEW when various numbers of MPI tasks are used. N_{active} and N_{ghost} are the number of active peaks and ghost peaks respectively and $N_{tot} = N_{active} + N_{ghost}$ denotes the total number of peaks per MPI task. N_{sparse} is the number of entries in the sparse saddle matrix and $N_{collisions}$ gives the number of hash table collisions. Sums, maxima and averages are taken over the all MPI tasks

N_{tasks}	32	64	128	256	512	1,024	2,048
Load imbalance ($\frac{\max\{N_{tot}\}}{\text{avg}\{N_{tot}\}}$)	1.4	1.5	1.8	2.4	2.8	3.3	3.9
Surface effect ($\frac{\sum N_{ghost}}{\sum N_{active}}$)	0.0087	0.012	0.016	0.021	0.030	0.040	0.055
Connectivity ($\frac{\sum N_{sparse}}{\sum N_{tot}}$)	9.4	9.4	9.4	9.3	9.3	9.3	9.2
$\max\{\frac{N_{ghost}}{N_{active}}\}$	0.012	0.017	0.044	0.064	0.10	0.15	0.24
$\max\{N_{tot}\}$	3.0×10^5	1.6×10^5	9.6×10^4	6.4×10^4	3.8×10^4	2.2×10^4	1.3×10^4
$\max\{N_{sparse}\}$	3.3×10^6	1.8×10^6	1.2×10^6	8.7×10^5	6.3×10^5	4.7×10^5	3.0×10^5
$\max\{N_{collisions}\}$	4	3	2	3	16	17	13

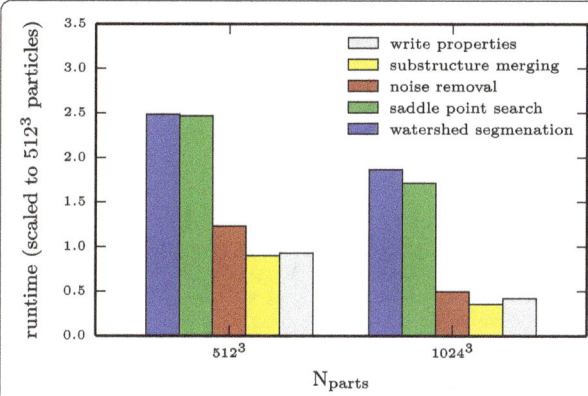

Figure 7 Weak scaling comparison of PHEW using 1,024 tasks to find structure in a 512^3 and a $1,024^3$ particle cosmological box. The PHEW runtimes for the $1,024^3$ box are divided by a factor of 8 for comparison with the runtimes for the 512^3 box. Increasing the size of the dataset improves the scaling of PHEW for large numbers of MPI tasks.

algorithm is posed by high-resolution zoom simulations of one single halo. In such a situation, the parent halo is spread over almost all MPI tasks, leading to MPI communication across the entire computational domain during the merging process and therefore slightly less favorable scaling properties.

PHEW has similarities with already existing watershed based halo finders, such as DENMAX (Bertschinger and Gelb 1991), HOP (Eisenstein and Hut 1998), SKID (Stadel 2001), ADAPTAHOP (Aubert et al. 2004), GRASSHOPPER (Potter and Stadel, in prep), but these are either not yet parallelized, do not find substructure or work only on particles. On a first sight, it looks like our approaches to defining substructure or parallelization cannot be applied to particle-based data structures since we operate on a mesh-defined density field, while the other codes work on the particle distribution directly. However, the only two concepts that we use which are naturally provided by the grid, namely a local density and the notion of a neighbour, can

be also defined for other data structures that do not rely on a grid. Once these properties are defined, the algorithm presented in this paper can be applied to particle data in the same way as we apply it to grid data.

At the current stage, our implementation of PHEW is a topological tool only, meaning that it identifies regions in space disregarding physical properties such as the kinetic or gravitational energy of the matter in that volume. For the application of PHEW as a genuine halo finder, we need to develop an unbinding procedure, which removes dark matter particles from regions they are not gravitationally bound to. We will exploit our hierarchical decomposition into substructure, to pass unbound particle to larger and larger regions, until the particles remain bound. This will unambiguously define the parent halo (or sub-halo) of the particles.

Appendix: Glossary

Clump: We use the word clump for the structure after the noise removal. It is the smallest structure that is not considered noise.

Key saddle: The highest saddle point connecting a peak to any neighbouring peak is considered the key saddle. Note that this definition slightly deviates from the one traditionally used in topography.

Key neighbour: A peaks key neighbour is the peak it is connected to through the key saddle.

Neighbouring cell: Every cell with a common face, edge or corner is considered a neighbour to a given cell.

Neighbouring peak: If a cell inside peak patch i is neighbouring a cell in peak patch j, their peaks are considered neighbouring peaks.

Noise: A peak with a small relevance (usually less than 1.5) is considered noise.

Owner: We denote the MPI task where a given peak is active as the owner of that peak.

Peak: We denote every cell hosting a local density maximum as a peak.

Peak patch: Every cell is unambiguously connected to one single density peak by recursively assigning it to the densest neighbouring cell. All cells belonging to a certain peak form the so-called peak patch. The peak patch is the equivalent to the watershed catchment basins for the negative density field.

Relevance: The relevance is defined as the ratio of a peaks density to its key saddle density or the density threshold in case of an isolated peak patch. This term is closely related to the topographical term 'prominence', which denotes the altitude difference of a peak to its highest saddle which connects the peak to a higher neighbour.

Saddle point: The density maximum on the connecting surface between two peak patches is located at the saddle point connecting the two peaks.

Test cell: Cells with a density above the adopted density threshold are called test cells. Only those are considered in our analysis.

Competing interests
The authors declare that they have no competing interests.

Authors' contributions
AB and RT are the main developers of PHEW and authors of the manuscript. SC has contributed to the application of PHEW to particle data in the course of his master thesis. DM was involved in the early development of the algorithm. All authors read and approved the final manuscript.

Author details
[1]Institute for Computational Science, University of Zurich, Zurich, CH-8057, Switzerland. [2]Institut d'Astrophysique de Paris, 98bis boulevard Arago, Paris, 75014, France. [3]Department of Astronomy, University of California, Berkeley, CA 94720-3411, USA.

Acknowledgements
The authors want to thank Stephane Colombi for his advice on substructure merging. Furthermore the authors thank Doug Potter for helpful discussions about programming techniques. The computations leading to this publication have been performed at on the zBox4 and Schroedinger Supercomputers at the University of Zurich and at the Swiss Supercomputing Centre CSCS in Lugano. This work has been supported by the Swiss National Science Foundation SNF under the project 'Computational Astrophysics' and the PASC co-design project 'Particles and Fields'.

Endnotes
[a] The RAMSES code including PHEW are publicly available and can be downloaded from http://www.bitbucket.org/rteyssie/ramses.

[b] DENMAX can be considered an in-between case since it uses a uniform grid to compute the density gradient which is then used to directly assign particles to peaks.

[c] Note that more modern approaches to region merging in image segmentation use the original image for merging while the watershed is computed on the gradient image (e.g., Peng and Zhang 2011). Using the watershed on the gradient image results in regions of similar gray values, where the densities inside our peak patches are very inhomogeneous. Approaches to region merging are thus fundamentally different in image processing than they are in our case.

[d] The relevance threshold is a user parameter that can be adapted to the setup. 1.5 is our standard choice for identifying gas clumps in RAMSES simulations. For identifying dark matter haloes, the value can be picked according to the expected number of dark matter particles per cell and the resulting Poisson noise in the density.

[e] For situations where the memory consumption due to given estimate for N_{max} becomes prohibitive, one could start with a lower number and for example double the size of the allocation on-the-fly whenever all available space for ghost peaks is occupied. However, we have not yet encountered a situation where it was necessary to use this strategy.

[f] We use a simple hash function based on the remainder of a division of the peak ID by a prime number chosen according to the maximum size of the virtual peak boundary. Collisions are dealt with by chaining in the form of a linked list (Knuth 1998). We found this to be sufficient for our purpose (see Table 2).

[g] The introduction of a N_{task}^2 sized matrix can become problematic when the number of MPI tasks is increased beyond the numbers we have tested for this publication, especially for supercomputers with relatively little available memory per core (Blue Gene architecture). In order to apply PHEW to even larger problems, one can drop the construction of the global matrix C by exploiting the fact that the n-th MPI process only needs to be aware of the n-th row and the n-th column of C, but not of the entire matrix. Considering the fact that the rows/columns of C are sparse, one can thus replace the N_{task}^2 sized matrix by a fully scalable representation of the information contained in C.

References
Alimi, J-M, et al.: First-ever full observable universe simulation. In: Proceedings of the International Conference on High Performance Computing, Networking, Storage and Analysis, p. 73. IEEE Computer Society Press, Washington (2012)

Aragón-Calvo, MA, Platen, E, van de Weygaert, R, Szalay, AS: The Spine of the Cosmic Web. Astrophys. J. **723**, 364-382 (2010). doi:10.1088/0004-637X/723/1/364

Aubert, D, Pichon, C, Colombi, S: The origin and implications of dark matter anisotropic cosmic infall on $\approx L_\star$ haloes. Mon. Not. R. Astron. Soc. **352**, 376-398 (2004). doi:10.1111/j.1365-2966.2004.07883.x

Bertschinger, E, Gelb, JM: Cosmological N-body simulations. Comput. Phys. **5**, 164-175 (1991)

Beucher, S: In: Serra, J, Soille, P (eds.) Mathematical Morphology and Its Applications to Image Processing (1994)

Bleuler, A, Teyssier, R: Towards a more realistic sink particle algorithm for the ramses code. Mon. Not. R. Astron. Soc. **445**(4), 4015-4036 (2014)

Davis, M, Efstathiou, G, Frenk, CS, White, SD: The evolution of large-scale structure in a universe dominated by cold dark matter. Astrophys. J. **292**, 371-394 (1985)

Eisenstein, DJ, Hut, P: HOP: a new group-finding algorithm for N-body simulations. Astrophys. J. **498**, 137 (1998). doi:10.1086/305535

Hockney, RW, Eastwood, JW: Computer Simulation Using Particles. McGraw-Hill, New York (1981)

Knebe, A, et al.: Haloes gone MAD: the halo-finder comparison project. Mon. Not. R. Astron. Soc. **415**, 2293-2318 (2011). doi:10.1111/j.1365-2966.2011.18858.x

Knebe, A, et al.: Structure finding in cosmological simulations: the state of affairs. Mon. Not. R. Astron. Soc. **435**, 1618-1658 (2013). doi:10.1093/mnras/stt1403

Knollmann, SR, Knebe, A: AHF: Amiga's halo finder. Astrophys. J. Suppl. Ser. **182**(2), 608 (2009)

Knuth, DE: Sorting and Searching. The Art of Computer Programming, vol. 3. Addison-Wesley, Reading (1998)

Meyer, F: Topographic distance and watershed lines. Signal Process. **38**(1), 113-125 (1994)

Moga, A: Parallel Watershed Algorithms for Image Segmentation. Tampere University of Technology, Tampere (1997)

Moga, AN, Gabbouj, M: Parallel marker-based image segmentation with watershed transformation. J. Parallel Distrib. Comput. **51**(1), 27-45 (1998)

Onions, J, et al.: Subhaloes gone Notts: spin across subhaloes and finders. Mon. Not. R. Astron. Soc. **429**, 2739-2747 (2013). doi:10.1093/mnras/sts549

Peng, B, Zhang, D: Automatic image segmentation by dynamic region merging. IEEE Trans. Image Process. **20**(12), 3592-3605 (2011)

Platen, E, van de Weygaert, R, Jones, BJT: A cosmic watershed: the WVF void detection technique. Mon. Not. R. Astron. Soc. **380**, 551-570 (2007). doi:10.1111/j.1365-2966.2007.12125.x

Potter, D, Stadel, J:. GRASSHOPPER, in prep.

Press, WH, Schechter, P: Formation of galaxies and clusters of galaxies by self-similar gravitational condensation. Astrophys. J. **187**, 425-438 (1974). doi:10.1086/152650

Press, WH, Teukolsky, SA, Vetterling, WT, Flannery, BP: Numerical Recipes 3rd Edition: The Art of Scientific Computing, 3rd edn. Cambridge University Press, New York (2007)

Pujol, A, et al.: Subhaloes gone Notts: the clustering properties of subhaloes. Mon. Not. R. Astron. Soc. **438**, 3205-3221 (2014). doi:10.1093/mnras/stt2446

Roerdink, JBTM, Meijster, A: The watershed transform: definitions, algorithms and parallelization strategies. Fundam. Inform. **41**(1-2), 187-228 (2000)

Rosolowsky, EW, Pineda, JE, Kauffmann, J, Goodman, AA: Structural analysis of molecular clouds: dendrograms. Astrophys. J. **679**(2), 1338 (2008)

Skory, S, Turk, MJ, Norman, ML, Coil, AL: Parallel hop: a scalable halo finder for massive cosmological data sets. Astrophys. J. Suppl. Ser. **191**(1), 43 (2010)

Springel, V, White, SDM, Tormen, G, Kauffmann, G: Populating a cluster of galaxies - I. Results at $z = 0$. Mon. Not. R. Astron. Soc. **328**, 726-750 (2001). doi:10.1046/j.1365-8711.2001.04912.x

Stadel, JG: Cosmological N-body simulations and their analysis. PhD thesis, University of Washington (2001)

Stutzki, J, Guesten, R: High spatial resolution isotopic CO and CS observations of M17 SW: the clumpy structure of the molecular cloud core. Astrophys. J. **356**, 513-533 (1990). doi:10.1086/168859

Sutter, PM, et al.: VIDE: The Void IDentification and Examination toolkit. Astron. Comput. **9**, 1-9 (2015)

Teyssier, R: Cosmological hydrodynamics with adaptive mesh refinement. A new high resolution code called RAMSES. Astron. Astrophys. **385**, 337-364 (2002). doi:10.1051/0004-6361:20011817

Way, MJ, Gazis, PR, Scargle, JD: Structure in the 3D Galaxy Distribution: II. Voids and Watersheds of Local Maxima and Minima (2014). arXiv:1406.6111

Way, MJ, Gazis, PR, Scargle, JD: Structure in the three-dimensional galaxy distribution. I. Methods and example results. Astrophys. J. **727**, 48 (2011). doi:10.1088/0004-637X/727/1/48

Williams, JP, de Geus, EJ, Blitz, L: Determining structure in molecular clouds. Astrophys. J. **428**, 693-712 (1994). doi:10.1086/174279

4

Implicit large eddy simulations of anisotropic weakly compressible turbulence with application to core-collapse supernovae

David Radice[*], Sean M Couch and Christian D Ott

Abstract

In the implicit large eddy simulation (ILES) paradigm, the dissipative nature of high-resolution shock-capturing schemes is exploited to provide an implicit model of turbulence. The ILES approach has been applied to different contexts, with varying degrees of success. It is the de-facto standard in many astrophysical simulations and in particular in studies of core-collapse supernovae (CCSN). Recent 3D simulations suggest that turbulence might play a crucial role in core-collapse supernova explosions, however the fidelity with which turbulence is simulated in these studies is unclear. Especially considering that the accuracy of ILES for the regime of interest in CCSN, weakly compressible and strongly anisotropic, has not been systematically assessed before. Anisotropy, in particular, could impact the dissipative properties of the flow and enhance the turbulent pressure in the radial direction, favouring the explosion. In this paper we assess the accuracy of ILES using numerical methods most commonly employed in computational astrophysics by means of a number of local simulations of driven, weakly compressible, anisotropic turbulence. Our simulations employ several different methods and span a wide range of resolutions. We report a detailed analysis of the way in which the turbulent cascade is influenced by the numerics. Our results suggest that anisotropy and compressibility in CCSN turbulence have little effect on the turbulent kinetic energy spectrum and a Kolmogorov $k^{-5/3}$ scaling is obtained in the inertial range. We find that, on the one hand, the kinetic energy dissipation rate at large scales is correctly captured even at low resolutions, suggesting that very high "effective Reynolds number" can be achieved at the largest scales of the simulation. On the other hand, the dynamics at intermediate scales appears to be completely dominated by the so-called bottleneck effect, *i.e.*, the pile up of kinetic energy close to the dissipation range due to the partial suppression of the energy cascade by numerical viscosity. An inertial range is not recovered until the point where high resolution $\sim 512^3$, which would be difficult to realize in global simulations, is reached. We discuss the consequences for CCSN simulations.

Keywords: turbulence; methods: numerical; supernovae

1 Introduction

Despite decades of studies and compelling evidence that a significant fraction (Clausen *et al.* 2015) of stars with initial masses in excess of ~ 8 solar masses explode as core-collapse supernovae (CCSN) at the end of their evolution, the exact details of the explosion mechanism are still uncertain (Woosley and Janka 2005; Janka *et al.* 2012; Burrows 2013; Foglizzo *et al.* 2015). Current state-of-the-

art 3D simulations either fail to explode or have explosion energies that fall short of the observed energies by factors of a few for most of the progenitor mass range (Janka 2012; Burrows 2013; Foglizzo *et al.* 2015).

The dynamics at the center of a star undergoing core collapse is shaped by a delicate balance between competing effects where all of the known forces: gravity, electromagnetism, weak and strong interactions, are important. The task of modeling these systems is made particularly challenging by the fact that the generation of the asymptotic explosion energies, although enormous ($\sim 10^{44}$ J), re-

[*]Correspondence: dradice@caltech.edu
TAPIR, Walter Burke Institute for Theoretical Physics, California Institute of Technology, E California Blvd, Pasadena, CA 91125, USA

quires a rather subtle, percent-level imbalance between non-linear processes over many dynamical times.

The flow of plasma in the core of a star going supernova is known to be unstable to convection (Herant 1995; Burrows *et al.* 1995; Janka and Müller 1996; Foglizzo *et al.* 2006) and/or to another large scale instability known as standing accretion shock instability (Blondin *et al.* 2003; Foglizzo *et al.* 2007). In any case, given the very large Reynolds numbers, as large as $\sim 10^{17}$ in the region of interest (Abdikamalov *et al.* 2015) (the so-called gain region, where neutrino heating dominates over neutrino cooling), it is expected that the resulting flow will be fully turbulent. It has been suggested (Murphy *et al.* 2013; Couch and Ott 2015) recently that turbulence and, in particular, turbulent pressure could tip the balance of the forces in favor of explosion. In this respect, anisotropy is of key importance, because it results in an effective radial pressure support with adiabatic index $\gamma_{turb} = 2$, much larger than that of thermal (radiation) pressure ($\gamma_{th} \simeq 4/3$). This means that turbulent kinetic energy is a much more valuable source of radial pressure support than thermal energy (see Appendix).

All of the current numerical simulations employ the implicit large eddy simulation (ILES) paradigm (Garnier *et al.* 2000; Grinstein *et al.* 2011) (also known as monotone integrated LES (MILES)) of exploiting the dissipative nature of high resolution shock capturing (HRSC) methods as an implicit turbulence model. However, the combination of the use of rather dissipative schemes and the relatively low spatial resolution that can be achieved in global simulations is such that the fidelity with which turbulence is captured is questionable (Abdikamalov *et al.* 2015).

To be useful in the context of CCSN simulations, an ILES should, at the very least, account for the right rate of decay of the kinetic energy at the largest scales while avoiding unphysical pile up of energy at smaller scales. Unfortunately, all of the current simulations seem to be strongly dominated by the so-called bottleneck effect (Abdikamalov *et al.* 2015), which corresponds to an inefficient energy transfer across intermediate scales due to the viscous suppression of non-linear interaction with smaller scales (Yakhot and Zakharov 1993; She and Jackson 1993; Falkovich 1994; Verma and Donzis 2007; Frisch *et al.* 2008). Current global simulations achieve resolutions, in the turbulent region, comparable to those of 30^3-70^3 lattices in periodic domains (Couch and O'Connor 2014; Couch and Ott 2015; Abdikamalov *et al.* 2015). At these resolutions, almost all of the dynamical range of the simulations can be expected to be directly affected by numerical viscosity (Sytine *et al.* 2000). The fidelity with which turbulence is captured in these simulations will then depend on the degree with which the numerical truncation error approximates an LES closure.

In this respect, it has been shown by Garnier *et al.* (1999) and Johnsen *et al.* (2010) that many HRSC methods can be too dissipative to yield a faithful description of turbulence at low resolutions. These studies, however, considered a different regime, decaying isotropic turbulence, while turbulence in a core-collapse supernova, as well as in many other astrophysical settings, is often strongly anisotropic (Arnett *et al.* 2009; Murphy *et al.* 2013; Couch and Ott 2015) as rotational invariance is broken by gravity. Garnier *et al.* (1999) and Johnsen *et al.* (2010) also considered different numerical schemes with respect to those used in supernova simulations. Both of these aspects can, in principle, be important. First of all, strong anisotropies could potentially influence the turbulence dynamics at the level of the energy cascade and of the dissipation (Casciola *et al.* 2007). Secondly, some of the schemes used in computational astrophysics, such as the piecewise parabolic method (PPM) (Colella and Woodward 1984) as well as some of the MUSCL (Toro 1999) schemes, have been shown, differently from some of the methods considered by Garnier *et al.* (1999) and Johnsen *et al.* (2010), to be well suited for ILES (Schmidt *et al.* 2006; Thornber *et al.* 2007).

The aim of this work is to fill the gap between existing theoretical studies and the particular applications of our interest. To this end we use a publicly available code, FLASH (Fryxell *et al.* 2000; Dubey *et al.* 2009; Lee *et al.* 2014), which is widely used in the computational astrophysics community, and perform a series of simulations of turbulence in a regime relevant for core-collapse supernovae: driven at large scale, with large anisotropies and mildly compressible. We use five different numerical setups and, for each, several resolutions in the range from 64^3 to 512^3 in a periodic domain. We study in detail the way in which the energy cascade across different scales is represented by our ILES and we discuss the use of local or lower dimensional diagnostics that can be used to assess the quality of a global simulation in a complex geometry where 3D spectra are not readily available.

The rest of this paper is organized as follows. First, in Section 2, we discuss the exact setup of our simulations and the diagnostic quantities used in our analysis. Then, in Section 3, we discuss the basic characteristics of the flow realized in our simulations. In Section 4, we present a detailed analysis of the way in which the energy cascade is captured by the different schemes at different scales. In particular, we quantify the accuracy with which different methods capture the decay rate of energy from the largest scales and the way in which energy is distributed across scales. We discuss the role of anisotropies in the context of the 4/5-law, a fundamental exact relation for isotropic and incompressible turbulence relating the statistics of velocity fluctuations with the energy dissipation rate (see Section 2.3), in Section 5. We explore the use of the 2D, transverse, energy spectrum as a diagnostic for 3D simulations in Section 6. Finally, we present a brief summary of our main findings, as well as a discussion of their implications for CCSN simulations in Section 7. Appendix contains some supplemental background material on the role of turbulence in the explosion mechanism of CCSN.

2 Methods

2.1 Numerical methods

We consider a compressible fluid with a prescribed acceleration, $\mathbf{a}$, in a unit-box with periodic boundary conditions. The code that we employ for these simulations, FLASH, solves the gas-dynamics equations in conservation form. In particular we evolve the continuity equation

$$\partial_t \rho + \nabla \cdot (\rho \mathbf{v}) = 0 \tag{1}$$

and the momentum equation

$$\partial_t(\rho \mathbf{v}) + \nabla \cdot (\rho \mathbf{v} \otimes \mathbf{v} + p\mathbf{I}) = \rho \mathbf{a}. \tag{2}$$

These equations are closed with a simple isentropic equation of state,

$$p = \rho^{4/3}, \tag{3}$$

that can be considered as a rough description of a gas dominated by radiation pressure. Since the equation of state ensures an adiabatic evolution we do not need to solve the energy equation as equations (1), (2) and (3) suffice to fully describe the flow.

Equations (1) and (2) are solved using the directionally-unsplit hydrodynamics solver of the open-source FLASH simulation framework. FLASH implements the corner transport upwind method (Colella 1990) for fully directionally-unsplit evolution of the Euler equations (Lee and Deane 2009; Lee 2013). FLASH includes several options for the order of spatial reconstruction (Lee et al. 2014), including 2nd-order TVD (Toro 1999), 3rd-order PPM (Colella and Woodward 1984), and 5th-order WENOZ (Borges et al. 2008). Fluxes are computed at 2nd-order accuracy using one of a number of approximate Riemann solvers included in FLASH, such as HLLE (Einfeldt 1988) and HLLC (Toro et al. 1994). Second-order accuracy in time is achieved via a characteristic tracing evolution of the Riemann solver input states to the time step midpoint (Colella and Woodward 1984). We remark that, in accordance with the ILES, paradigm, we do not include any additional sub-grid scale model, but relied on the implicit turbulent closure built in the numerical schemes we use for the integration of the hydrodynamics equation.

All of our simulations start with the fluid at rest $\rho = 1$, $\mathbf{v} = 0$. Turbulence is driven using the stirring module of FLASH. This module uses the Ornstein-Uhlenbeck process (Uhlenbeck and Ornstein 1930) to generate stirring modes in Fourier space. This yields an acceleration field

which smoothly decorrelates (Eswaran and Pope 1988) over a timescale T_s. The FLASH implementation permits the use of any arbitrary combination of solenoidal and compressive modes (Federrath et al. 2010). For our runs, we set $T_s = 0.5$, we use only solenoidal forcing and we restrict the accelerating field to be nonzero only in the first four Fourier modes. This forcing is designed to mimic the influence of some larger scale weakly compressible flow and, for this reason, it does not include any compressible component. This is a reasonable approximation for low Mach number convection which is well described by the anelastic approximation, e.g., Verhoeven et al. (2015). In the CCSN context, simulations show that the turbulence is highly anisotropic, being roughly twice as strong in the radial direction as either tangential direction (Murphy and Meakin 2011; Murphy et al. 2013; Handy et al. 2014; Couch and Ott 2015) since it is driven by buoyancy due to a negative radial entropy gradient. In order to emulate this behavior, the accelerating field in the x-direction (which is going to play the role of the radial direction) is scaled by a constant factor (before the solenoidal projection of the acceleration field) such that $R_{xx} \simeq 2R_{yy} \simeq 2R_{zz}$, where

$$R_{ij} = \langle \rho v_i v_j \rangle, \tag{4}$$

is the Reynolds stress tensor (to simplify the notation we considered a frame in which $\langle \rho \mathbf{v} \rangle = 0$) and $\langle \cdot \rangle$ denotes an ensemble average. Finally, the overall strength of the stirring is tuned to achieve a RMS Mach number of $\simeq 0.35$, which is typically observed in realistic CCSN simulations (Couch and Ott 2013; Müller and Janka 2015).

2.2 Energy transfer equations

In order to study the cascade of the specific kinetic energy (which we will refer to simply as "kinetic energy" or "energy" in the following), $|\mathbf{v}|^2/2$, we will consider an energy budget equation across different scales, analogous to the one commonly employed in the study of incompressible, isotropic turbulence, e.g., Frisch (1996). In particular, we consider the momentum equation (2) in non-conservation form,

$$\partial_t \mathbf{v} + (\mathbf{v} \cdot \nabla)\mathbf{v} = -\mathcal{V}\nabla p + \mathbf{a}, \tag{5}$$

where $\mathcal{V} = 1/\rho$ is the specific volume of the gas.

We can use equation (5) to derive an evolution equation for the Fourier transform of the velocity

$$\hat{\mathbf{v}}(\mathbf{k}) = \int_{\mathbb{R}^3} e^{-2\pi i \mathbf{k} \cdot \mathbf{x}} \mathbf{v}(\mathbf{x}) \, d^3 \mathbf{x}. \tag{6}$$

Transforming both sides of equation (5) we obtain

$$\partial_t \hat{\mathbf{v}} + \hat{\mathbf{v}} * 2\pi i \mathbf{k} \otimes \hat{\mathbf{v}} = -\hat{\mathcal{V}} * 2\pi i \mathbf{k} \hat{p} + \hat{\mathbf{a}}, \tag{7}$$

where $*$ denotes the convolution operator, *i.e.*,

$$[f * g](\mathbf{k}) = \int_{\mathbb{R}^3} f(\mathbf{q}) g(\mathbf{k} - \mathbf{q}) \, \mathrm{d}^3 \mathbf{q}. \tag{8}$$

If we multiply both sides of equation (7) by $\hat{\mathbf{v}}^*$ and take the real part, we obtain an equation for the 3D energy spectrum

$$\partial_t E(\mathbf{k}) = T(\mathbf{k}) + C(\mathbf{k}) + \epsilon(\mathbf{k}), \tag{9}$$

where

$$E(\mathbf{k}) = \frac{1}{2} \hat{\mathbf{v}} \cdot \hat{\mathbf{v}}^*, \tag{10}$$

$$T(\mathbf{k}) = -2\pi \Re(\hat{\mathbf{v}} * \mathbf{ik} \otimes \hat{\mathbf{v}}) \cdot \hat{\mathbf{v}}^*, \tag{11}$$

$$C(\mathbf{k}) = -2\pi \Re(\hat{\mathcal{V}} * \mathbf{i k} p) \cdot \hat{\mathbf{v}}^*, \tag{12}$$

$$\epsilon(\mathbf{k}) = \Re \hat{\mathbf{a}} \cdot \hat{\mathbf{v}}^*. \tag{13}$$

Here E is the energy spectrum (the velocity power spectral density (PSD)) and T is the same transfer term as in the classical incompressible equations and ϵ is the energy injection rate. The C term vanishes in the incompressible limit and represents the interaction between kinetic and acoustic modes. In practice, in our models, C is found to be at least one order of magnitude smaller than T at all scales and it is thus negligible. In any case, we retain C in the analysis below.

For each of the spectral quantities, S, being E, T, C or ϵ, we define the integrated spectrum, $S(k)$, as

$$S(k) = \int_{\mathbb{R}^3} S(\mathbf{k}) \delta(|\mathbf{k}| - k) \, \mathrm{d}\mathbf{k}, \tag{14}$$

$\delta(\cdot)$ being the Dirac delta function.

Integrating equation (9), we obtain the following one-dimensional energy balance equation

$$\partial_t E(k) = T(k) + C(k) + \epsilon(k). \tag{15}$$

This can also be written in terms of the energy flux across scales,

$$\Pi(k) = -\int_0^k T(\xi) \, \mathrm{d}\xi, \tag{16}$$

as

$$\partial_t E(k) + \partial_k \Pi(k) = C(k) + \epsilon(k). \tag{17}$$

Notice that *we did not assume isotropy in any of the above.*

Equation (15) is derived in the inviscid limit. In practice, our evolution method introduces dissipation in the form

of "numerical viscosity". This can be quantified in terms of the residual

$$R(k) = \partial_t E(k) - T(k) - C(k) - \epsilon(k). \tag{18}$$

This can be used to define a wave number dependent numerical viscosity:

$$\nu(k) = -\frac{1}{2} \frac{R(k)}{k^2 E(k)}. \tag{19}$$

We remark that ν does not, in general, correspond to a classical shear or bulk viscosity, but can nevertheless be interpreted as a relative measure of the dissipation acting at different wave numbers (see, *e.g.*, Fureby and Grinstein (1999); Aspden *et al.* (2009); Zhou *et al.* (2014) for alternative approaches).

In practice, since we will be working in the stationary case, after having taken the appropriate time averages, $R(k)$ reduces to

$$R(k) = -T(k) - C(k) - \epsilon(k). \tag{20}$$

Finally, since we are working in a periodic domain, which we take of size $L_x = L_y = L_z = 1$, all of the spectra are quantized and non-trivial only for k_x, k_y and k_z integers. Furthermore, all of the integrals in wave number space reduce to summations. Integrals over spherical shells are transformed to weighted sums following Eswaran and Pope (1988):

$$E(k) = \frac{4\pi k^2}{N_k} \sum_{k-1/2 < |\mathbf{k}| \le k+1/2} E(\mathbf{k}), \tag{21}$$

where N_k is the number of discrete wave-numbers in the shell $k - 1/2 < |\mathbf{k}| \le k + 1/2$.

2.3 Structure functions

The energy spectrum and its sources/fluxes give a comprehensive picture of the energy cascade and can be used to assess the level of convergence of the simulation. Unfortunately, 3D energy spectra and fluxes are not easily accessible in calculations in complex domains and/or with inhomogeneous turbulence. In these cases, local quantities in the physical domain are more easily extracted and analyzed. Hence, one of the goals of this work is to validate the use of indirect measures of convergence of ILES. Among these quantities, the structure functions of the velocity appear to be natural candidates for study.

We define the velocity increments

$$\delta v(\mathbf{x}, \mathbf{r}) = [\mathbf{v}(\mathbf{x} + \mathbf{r}) - \mathbf{v}(\mathbf{x})] \cdot \frac{\mathbf{r}}{r} \tag{22}$$

and study the quantities

$$S_p(r) = \langle \delta v^p \rangle_{j=0}, \tag{23}$$

where, $\langle \cdot \rangle_{j=0}$ denotes an ensemble average as well as a mean over all of the angles between $\mathbf{v}$ and $\mathbf{r}$ (in other words we are looking at the $j = 0$ component of the SO(3) decomposition of the structure functions (Biferale and Procaccia 2005)). In the case of homogeneous turbulence S_p does not depend on $\mathbf{x}$ and is thus a function of only the separation r.

The most important relation involving the structure functions is the so-called 4/5-law, which relates the third order structure function, $S_3(r)$, with the mean energy dissipation rate,

$$\langle \epsilon \rangle = \int_0^\infty \epsilon(k)\,dk, \tag{24}$$

and states that, for incompressible, homogeneous and isotropic turbulence (Frisch 1996):

$$S_3(r) = -\frac{4}{5}\langle \epsilon \rangle r. \tag{25}$$

Equation (25) can be derived from the Navier-Stokes equation for fully-developed, incompressible, homogeneous and isotropic turbulence (Frisch 1996). In the anisotropic or compressible case, however, equation (25) is not strictly valid and could be violated in the data. As we show in Section 5, we find equation (25) to be very well satisfied by our data, suggesting that the 3rd order structure function can be a very useful diagnostic in global simulations.

2.4 Transverse energy spectrum
Another alternative to the analysis of 3D spectra, which has been adopted by several authors in the core-collapse supernova context (Dolence *et al.* 2013; Couch and O'Connor 2014; Handy *et al.* 2014; Abdikamalov *et al.* 2015), is the use of 2D spectra computed using a spherical harmonics expansion of the velocity field tangential to one or more spherical shells in the simulation. Analogously, we emulate this by looking at quantities in the y-z plane and we define the 2D spectra

$$E_\perp(k_\perp) = \frac{1}{2}\int_{\mathbb{R}^2} \tilde{\mathbf{v}}_\perp \cdot \tilde{\mathbf{v}}_\perp^* \delta\left(\sqrt{k_y^2 + k_z^2} - k_\perp\right) dk_y\,dk_z, \tag{26}$$

where $\mathbf{v}_\perp$ is the projection of the velocity perpendicular to the x-direction and we introduced the partial Fourier

transform of $\mathbf{v}_\perp$:

$$\tilde{\mathbf{v}}_\perp(k_y, k_z) = \lim_{L_x \to +\infty} \frac{1}{L_x}\int_{-L_x/2}^{L_x/2} dx$$
$$\times \int_{\mathbb{R}^2} e^{-2\pi i(k_y y + k_z z)}\mathbf{v}_\perp(x, y, z)\,dy\,dz. \tag{27}$$

In the limit of infinite Reynolds number/resolution, the 2D spectrum is expected to have the same asymptotic behavior as the 3D spectrum, however it is a-priori unclear if $E_\perp$ is a good proxy for E at finite resolution. For this reason we find it useful to investigate this here.

As was the case for the 3D spectra, also here the spectrum is non-trivial only for integer k_y and k_z, when periodicity is taken into account. The integral in equation (26) is treated analogously to the integral in the equation (14) for the 3D case, while the average in the x-direction in equation (27) is converted to an average over the x-extent of the simulation box.

3 Basic flow properties
We employ the finite-volume HRSC (Godunov) approach in which physical states are reconstructed at inter-cell boundaries and local Riemann problems are solved to compute the physical inter cell fluxes. In particular, we perform five groups of simulations using different numerical methods. Each group is labeled using the name of the reconstruction algorithm and of the Riemann solver. For instance `TVD_HLLE`, denotes a group of simulations done using TVD reconstruction and HLLE Riemann solver. Single simulations are labeled using their resolution so that, for instance, `TVD_HLLE_N128`, denotes the `TVD_HLLE` run done using a 128^3 grid. For all of the runs the timestep is chosen to have a CFL, *i.e.*, $c\Delta t/\Delta x$, of 0.4, c being the maximum characteristics speed, with the exception of the `PPM_HLLC_CFL0.8` runs where we set the CFL to 0.8. For the TVD runs we use the monotonized central (MC) slope limiter (Toro 1999). The runs with PPM use the original flattening and artificial viscosity prescriptions from Colella and Woodward (1984). The artificial viscosity coefficient is 0.1. We remark that the use of the artificial viscosity for PPM is not really necessary in this regime (Porter and Woodward 1994), however our goal is not to perform a study of the turbulent dynamics, but to assess how each numerical method performs when used under the same condition as in a real CCSN simulation where strong shocks need to be handled in some parts of the domain.

For each group of simulations we run four resolutions: 64^3, 128^3, 256^3 and 512^3. The RMS velocity in all of the runs is $v_{\text{rms}} \simeq 0.4$, giving an eddy turnover time $\tau = 1/v_{\text{rms}} \simeq 2.5$. All of the simulations are run until time $t = 100$ ($\simeq 40$ eddy turnover times). The time evolution of a few relevant diagnostics is shown in Figure 1 for our fiducial group of runs (PPM_HLLC) at different resolutions. We can see how the flow is accelerated from rest and quickly reaches a steady, fully turbulent, state. In all cases, steady state is reached after $t \gtrsim 3$ (~1 turnover time) and the diagnostics are insensitive to the resolution. The results for the other runs (not shown) are very similar to the ones of PPM_HLLC as they all achieve very similar RMS Mach

numbers and Reynolds stresses. All of the analysis shown in the rest of the paper are performed using 380 3D snapshots (evenly spaced in time) of the data in the interval $5 \leq t \leq 100$.

A first, qualitative, comparison between the different methods can be done by looking at their visualizations. In particular, in Figure 2, we show a visualization of the magnitude of the vorticity in the x-z plane for four of the

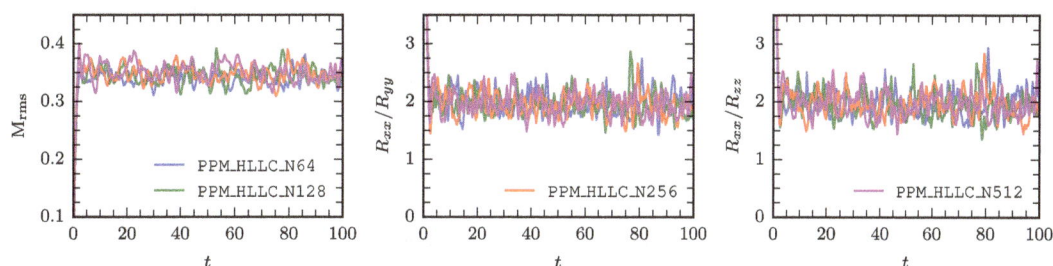

Figure 1 Time evolution of the diagnostic quantities for the fiducial set of runs PPM_HLLC with different resolutions. The left panel shows the root mean square (RMS) Mach number, while the middle and right panels show, respectively, the ratios R_{xx}/R_{yy} and R_{xx}/R_{zz}, R being the Reynolds stress tensor (equation (4)). Since the x-direction is the anisotropic direction (it would play the role of the radial direction in a CCSN) the ratios R_{xx}/R_{yy} and R_{xx}/R_{zz}, offer a global measure of the anisotropy of the flow at the largest scale. All of the quantities appear to have reached stationarity after time $t \gtrsim 3$ and oscillate around their target values. All resolutions produce the same qualitative behavior.

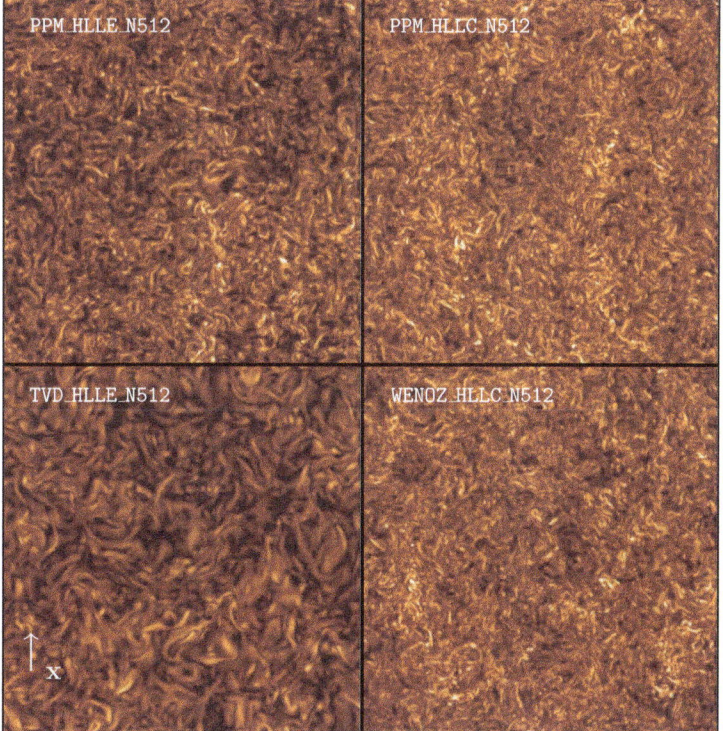

Figure 2 Square root of the magnitude of the vorticity, $\sqrt{|\nabla \times v|}$, for four of the simulations with 512^3 resolution in a slice through the middle of the x-z plane at the final time of the simulations ($t = 100$). The panels show simulations using PPM_HLLE_N512, PPM_HLLC_N512, WENOZ_HLLC_N512, and TVD_HLLC_N512 clockwise from the top left. The direction of the anisotropic driving is up in these figures. The colorcode goes linearly from 0 (no vorticity; dark colors) to 15 (light colors) and it is the same for all panels.

five schemes (excluding `PPM_HLLC_CFL0.8`) at the highest resolution (512^3). The data is taken at the final time ($t = 100$). As it can be seen from the figure, all of the simulations show the presence of thin, elongated, regions of high vorticity, as typically seen in direct numerical simulations (DNS) of homogeneous turbulent flows (Vincent and Meneguzzi 1991; Ishihara *et al.* 2007). However, the width and the intensity of the vorticity at these smaller scales depend crucially on the numerical scheme. Methods with small intrinsic numerical viscosity, such as `PPM_HLLC` and `WENOZ_HLLC`, present smaller structures and more intermittent vorticity fields with respect to more dissipative methods, such as `PPM_HLLE` and `TVD_HLLE`.

4 The energy cascade

In this section we focus our analysis on the accuracy with which the energy cascade is captured by our ILES runs. First, we focus on the largest scales of the simulation with the goal of quantifying the accuracy in the decay rate of the energy as a function of the resolution for the different methods. Next, we will look at the energy distribution at smaller scales where, in resolved simulations, the inertial range starts. Finally, we will look at the dynamics in the dissipation region and summarize.

4.1 Energy decay rate

In the limit of very large Reynolds number it is assumed, in standard turbulence phenomenology (Frisch 1996), that there exists a range of wave numbers (the inertial range) where energy injection and dissipation can be neglected in equation (17). In this range we can write (compressible effects are negligible in our simulations):

$$\partial_t E(k) + \partial_k \Pi(k) \simeq 0, \qquad (28)$$

so that stationarity requires $\Pi(k) \simeq$ const. In particular, since energy is conserved, one finds $\Pi(k) \simeq \langle \epsilon \rangle$. This means that, in the limit of large Reynolds numbers, the energy decay rate depends only on the macroscopic properties of the flow (and in particular not on the nature of the viscosity), a fact that has also been verified numerically (Kaneda *et al.* 2003). The significance of this property and its importance for the modeling of turbulence cannot be overstated.

In the context of CCSN simulations this means that the large scale kinetic energy, a crucial quantity for the dynamics of the explosion (Couch and Ott 2015), can be faithfully captured even with simulations achieving modest Reynolds numbers.

For an ILES, a basic requirement, then, is that a sufficiently high resolution should be achieved to correctly represent the energy cascade at the largest scales. What qualifies as a sufficiently high resolution is of course dependent on the details of the closure built into the scheme (and on

the accuracy required for the particular application). To quantify this, we can estimate the level of accuracy that can be reached at any given resolution, using our local simulations. In particular, we can study directly the energy flux across scales, defined by equation (16). This is shown in Figure 3 for all of the different runs.

As discussed before, we expect that $\Pi(k) \simeq \langle \epsilon \rangle$ over an extended region in Fourier space should be a direct indication that a simulation has been able to recover an inertial range. Perhaps not surprisingly, in light of previous results (Sytine *et al.* 2000), we find that regions where $\Pi \simeq \langle \epsilon \rangle$ as wide as a few wave numbers $4 \lesssim k \lesssim 10$ only appear at the highest resolutions (we will discuss the inertial range in more detail in Section 4.2). However, the amount of energy decaying from the largest scales reaches an asymptotic value much quicker than that implying that the total kinetic energy budget at the largest scales is well resolved even at modest resolutions.

We can make a more quantitative statement concerning the energy decay rate by looking at the peak of the energy flux as a function of resolution, as shown in Figure 4. We can see that at 128^3 points all of the simulations have a deviation from the asymptotic energy decay rate of less than 10%. The least dissipative methods already have an error close to the 2% level. A comparison between `PPM_HLLE` and `PPM_HLLC` reveals the profound impact that the choice of the Riemann solver has even at relatively large scale (more on the dissipative properties of the different schemes in Section 4.3).

4.2 Energy spectra

Obviously, not all of the dynamics of turbulence can be reduced to the rate at which kinetic energy decays from the injection scale. The internal dynamics of the energy cascade, far from the injection scale and far from the dissipation range, can also play an important role in many applications. To analyze this aspect we consider in Figure 5 the energy spectrum of the velocity defined by equation (10). The spectra are compensated by $k^{5/3}$ to highlight regions with Kolmogorov scaling, which might be expected in the inertial range. Since we want to focus on quantities that do not depend (or depend weakly) on the nature of the energy injection at large scale, we show all of the spectra as a function of a dimensionless wave number, $512k\Delta x$. The rationale behind this normalization is that, first of all, we assume the Kolmogorov scale η to be proportional to the grid spacing. Secondly, the 512 factor is introduced to have the dimensionless k, $512k\Delta x$ coincide with the dimensional one for the highest resolution runs. With this choice, $512k\Delta x = 512$ corresponds to a wavelength of a single grid point, $512k\Delta x = 256$ corresponds to a wavelength of two grid points and so on.

Looking at any of the groups of runs in Figure 5, one can immediately notice that the spectra obtained at different resolutions do not collapse into a single curve in the dissi-

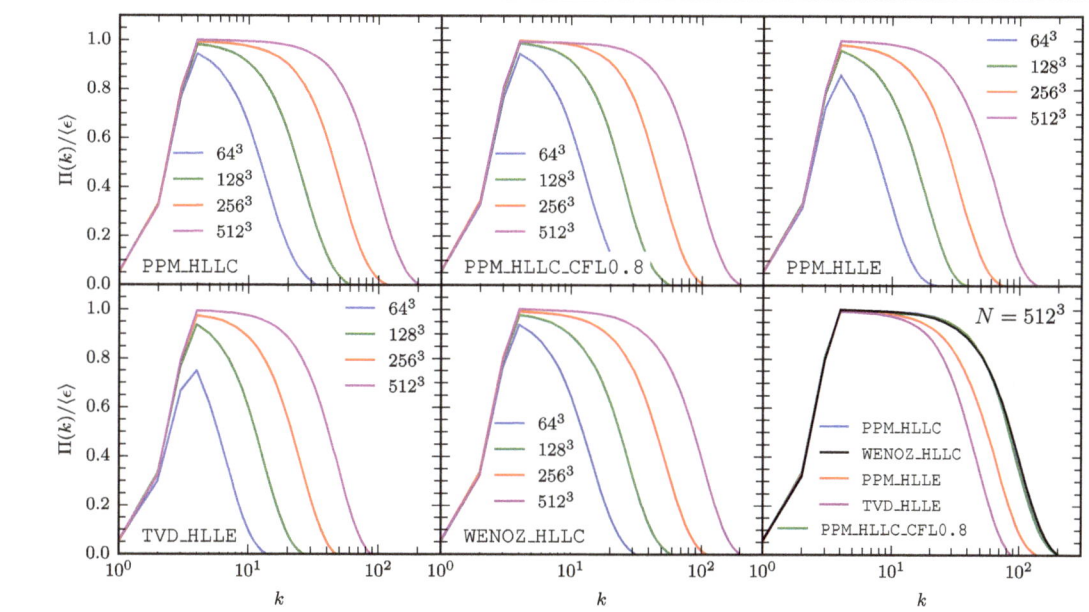

Figure 3 **Energy flux, as defined by equation (16), obtained with different numerical methods and resolutions.** The energy flux is shown normalized to the average dissipation rate given by equation (24). From left to right and from top to bottom we show the results obtained with PPM_HLLC, PPM_HLLC_CFL0.8, PPM_HLLE, TVD_HLLE and WENOZ_HLLC. The bottom right panel show a comparison of all of the methods at 512^3. All of the schemes show a good level of accuracy in the energy flux from the largest scales, with errors smaller than a few % already at low resolutions. The differences between the schemes become more marked at large wave numbers where the numerical dissipation starts to interfere with the energy cascade.

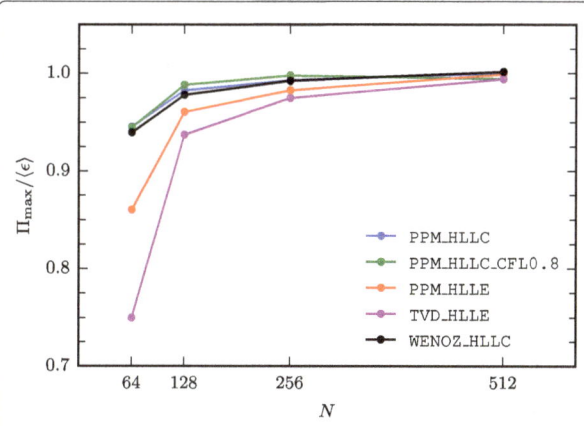

Figure 4 **Dissipation rate of the energy at the largest scales due to the turbulent cascade (not including direct dissipation by the numerical viscosity) as a function of resolution and for all of the schemes.** The dissipation rate is normalized so as to be 1 in the limit of large Reynolds numbers/resolution. At 128^3 points all of the schemes show an error of less than 10%, with the HLLC schemes already close to the 2% level.

pation region, as would be required by Kolmogorov's first similarity hypothesis (Frisch 1996) (*cf.* Gotoh *et al.* (2002)). This lack of convergence in the dissipation region could be due to the non-linear viscosity of HRSC schemes. This,

in turn, could result in an anomalous scaling of η with the grid spacing. Such scaling has been reported in the past for ILES, but it is not very well understood (Aspden *et al.* 2009). The good agreement between the three different groups of simulations employing the HLLC Riemann solver seems to support this hypothesis and suggests that the nonlinear viscosity introduced by the Riemann solver is an important ingredient in setting this scaling.

Convergence appears to be recovered at larger scales $\gtrsim 8\Delta x$ ($512k\Delta x \lesssim 64$), but the spectra appear to be dominated by the bottleneck effect. This manifests itself as a bump in the compensated spectra extending from the dissipation range until the end of the inertial range, for the simulations that show one (*e.g.*, until $512k\Delta x = 10$ for the HLLC runs), or until the energy injection scale ($512k\Delta x = 4$), for the simulations that show no or little inertial range (TVD_HLLE). The bottleneck effect is a viscous phenomenon which is also observed in direct numerical simulations. However, in the present context where viscosity is of numerical origin, it is at the very least questionable if a pronounced bottleneck is a desirable feature of the modeling. In astrophysical flows, where the Reynolds numbers are typically very large, this pile up of energy at large scales is unphysical and could affect the quantitative and qualitative outcome of a simulation (Abdikamalov *et al.* 2015). A quantification of the bottleneck effect in terms of the energy budget is discussed in Section 4.4.

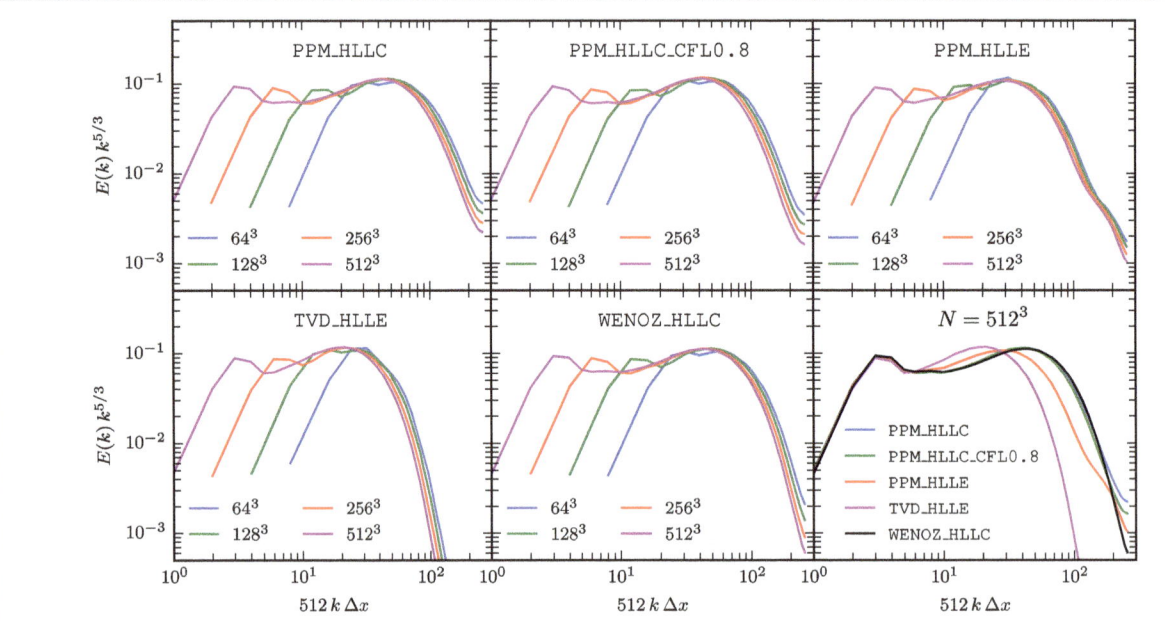

Figure 5 Energy spectra (equation (10)) obtained with different numerical methods and resolutions. The energy spectra are compensated by a $k^{5/3}$ spectrum, so that any region with Kolmogorov scaling should appear roughly flat. Furthermore, the spectra are all plotted as a function of the dimensionless wave number $512k\Delta x$ (the 512 factor is introduced to have the dimensionless wave number coincide with the dimensional one for the 512^3 runs). The first five panels show the PPM_HLLC (upper left), PPM_HLLC_CFL0.8 (upper center), PPM_HLLE (upper right), TVD_HLLE (lower left) and WENOZ_HLLC (lower center) group of runs. The last panel (lower right) shows a comparison of all of the methods at the highest resolution (512^3). An inertial range seems to be recovered only at the highest resolutions (perhaps with the exception of TVD_HLLE where no inertial range is visible). All schemes employing the HLLC Riemann solver are in very good agreement.

At even larger scales, an inertial range ($E \sim k^{-5/3}$ and $\Pi \sim$ const, see Figure 3) seems to be recovered by the least dissipative schemes (PPM and WENOZ with HLLC) in the region $4 \lesssim k \lesssim 10$. PPM_HLLE and TVD_HLLE have a more limited region, a few wave numbers at most, that could be interpreted as being an inertial range. We note that this resolution is not particularly high in comparison with state of the art DNS (Kaneda *et al.* 2003; Federrath 2013), but it would already correspond to an extremely high resolution in global CCSN simulations that typically have ~30-70 zones across the turbulent region (Abdikamalov *et al.* 2015).

The overall behavior of the spectra, as obtained by all schemes, is consistent with Kolmogorov's theory of turbulence. The anisotropic contributions to the angle-integrated spectra are too small to be detected in our data.

4.3 Numerical viscosity

At very small scales (~several grid points) the dynamics is dominated by the numerical viscosity. This can be estimated from the residual of the energy equation (17) or, equivalently, by the effective numerical viscosity $\nu(k)$ (equation (19)). The latter is shown in Figure 6 for all schemes and resolutions.

The first thing to notice is that the numerical viscosity provided by all numerical schemes is not constant, but dif-

fers by roughly an order of magnitude between low and high k. Having a wave number dependent viscosity is a desirable feature expected in any LES model (explicit or otherwise). Nevertheless, this makes the definition and calculation of the effective Reynolds number achieved in a simulation ambiguous. Meaningful ways to estimate it for ILES have been proposed (Zhou *et al.* 2014) and they can be used to ease the comparison between different simulations and assess their quality. However, one has to be very careful while using any quoted "Reynolds number" from an ILES, to estimate things like the dynamical range achieved by a simulation, because the dissipative properties of ILES differ considerably from the ones of the true Navier-Stokes equations.

Two other features can be observed in most of the numerical viscosity profiles. First, many of them exhibit a sudden reversal at high wave numbers. This is due to the fact that the numerical viscosity does not behave like a shear viscosity so that, although the numerical diffusion is strong at those scales, the numerical viscosity appears small because of a partial decoupling between vorticity and dissipation. Second, at high resolution and at the largest scales, the numerical viscosity is close to zero or even slightly negative. The reason is that the residual of equation (9) oscillates around zero and it is too small to be reliably extracted from our data: a much longer integration

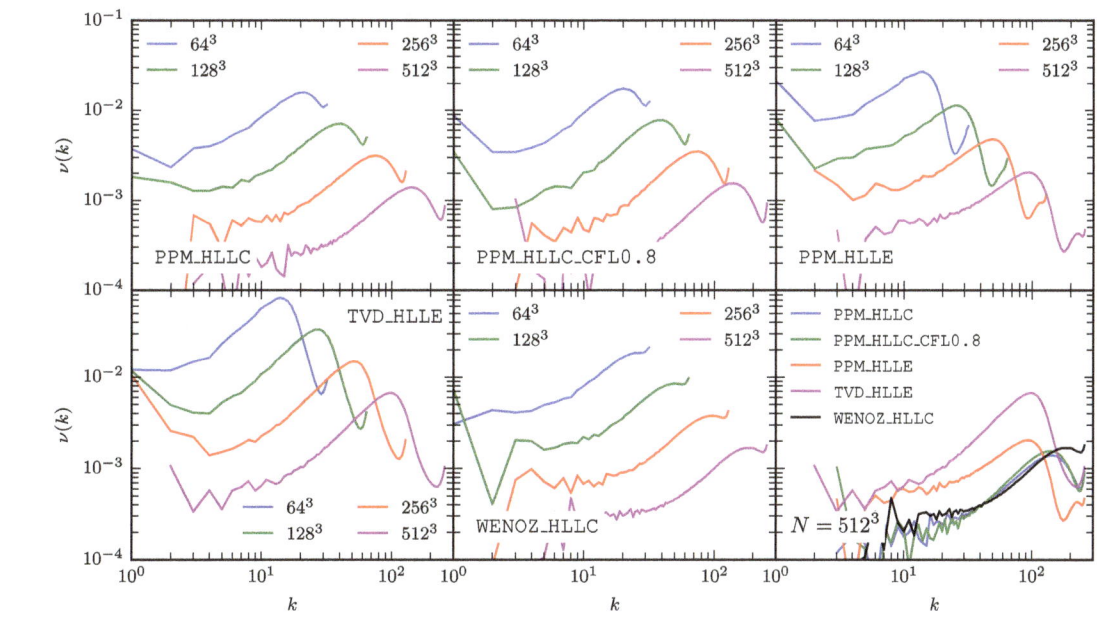

Figure 6 Numerical viscosity as a function of the wave number measured for all schemes and resolutions. The numerical viscosity is estimated using the procedure outlined in Section 2 and it is defined by equation (19). The different panels are, from left to right and from top to bottom, the results obtained with PPM_HLLC, PPM_HLLC_CFL0.8, PPM_HLLE, TVD_HLLE and WENOZ_HLLC. The bottom right panel show a comparison of all of the methods at 512^3. The numerical viscosity shows large variations across the wave number space. The choice of the Riemann solver plays a role that is at least as important as the choice of the reconstruction method in affecting the numerical viscosity throughout the entire the spectrum.

time would be needed to accumulate enough statistics for it.

Finally, a comparison between the numerical viscosity reveals two interesting effects. First, by comparing PPM_HLLC and PPM_ HLLE, we see that the choice of the Riemann solver affects the viscosity at basically all scales. Second, if we compare PPM_HLLC, PPM_HLLC_CFL0.8 and WENOZ_HLLC, we see that doubling the timestep appears to have an effect comparable to the difference between the PPM and WENOZ reconstructions at intermediate scales ($40 \lesssim k \lesssim 100$).

4.4 The energy distribution

So far we have been concerned with the energy decay rate from the largest scales, which we have shown to be well captured by the ILES (Section 4.1), and with the energy transfer in the inertial range, which we have seen to be described accurately only at much higher resolutions (Section 4.2). In a turbulent flow both of these aspects are important and a good ILES should display a distribution of energy across vortical structures at different scales that is as close as possible to the asymptotic one. Obviously, there is a limit to the accuracy that any ILES can achieve at a fixed resolution. Here, we make this statement more quantitative by considering the amount of kinetic energy that is well resolved by each simulation at a given resolution.

We introduce the cumulative energy spectrum, the integral of the energy spectrum:

$$\mathcal{E}(k) = \int_0^k E(\xi)\, d\xi. \tag{29}$$

This quantity is plotted in Figure 7, where it is normalized by

$$\frac{v_{rms}^2}{2} = \int_0^{+\infty} E(k)\, dk \tag{30}$$

to obtain the cumulative distribution function of the kinetic energy. As a reference, we also show the cumulative energy spectrum estimated from Kolmogorov's theory:

$$\mathcal{E}_{K41}(k) = \int_0^k E_{K41}(\xi)\, d\xi, \tag{31}$$

$$E_{K41}(k)$$

$$= \begin{cases} E_{PPM\_HLLC\_N512}(k), & \text{if } k \leq 4, \\ E_{PPM\_HLLC\_N512}(4)(\frac{k}{4})^{-5/3}, & k > 4. \end{cases} \tag{32}$$

We find that as the resolution increases, all schemes appear to be converging to the predictions of Kolmogorov's

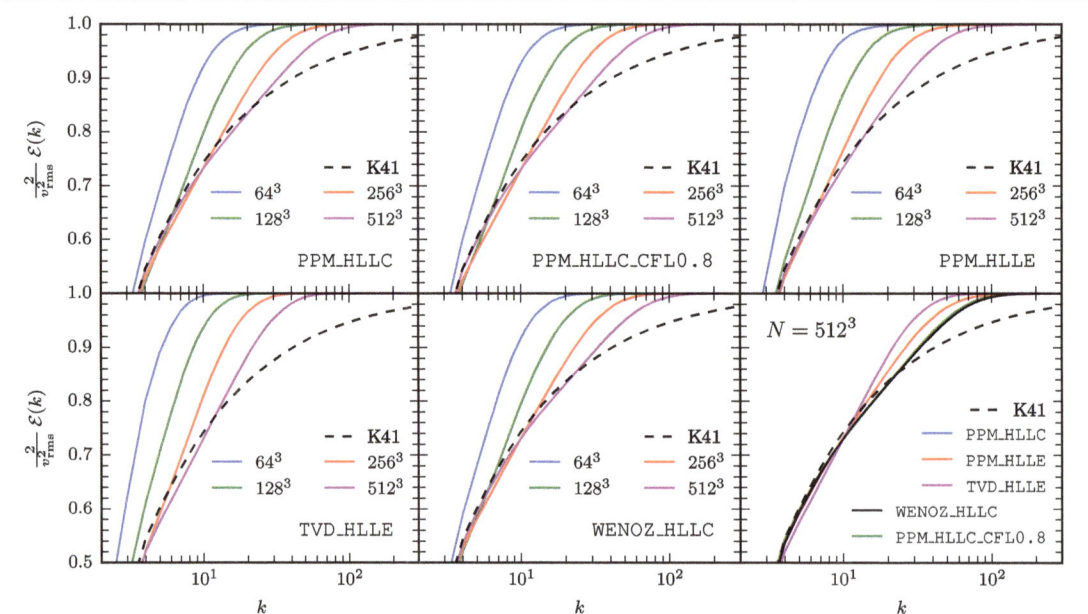

Figure 7 Cumulated energy distribution (equation (29)) for all methods and resolutions, normalized by a factor $2/v_{\rm rms}^2$ to be equal to 1 for large *k*. As a reference for comparison we also plot the asymptotic profile expected from Kolmogorov's theory (equation (31)). The different panels are, from left to right and from top to bottom, the results obtained with `PPM_HLLC`, `PPM_HLLC_CFL0.8`, `PPM_HLLE`, `TVD_HLLE` and `WENOZ_HLLC`. The bottom right panel show a comparison of all of the methods at 512^3. At low resolution all of the schemes show an excess of energy at intermediate scales, due to the bottleneck. Only at the highest resolution at least, roughly, 80% of the energy is correctly resolved.

theory. The results at finite resolution, however, are not encouraging: at 64^3 only ~50% or less of the kinetic energy is in well resolved structures, while the other ~50% have piled up at rather large scale, with a cumulative excess of ~20% at the grid scale, mostly because of the bottleneck effect. At higher resolutions, the amount of kinetic energy well captured by the ILES increases, but at 512^3 this is still only about 80% of the energy and there is still a cumulative excess of over ~5% at the grid scale ($\ell \sim \Delta x$).

5 The 4/5-law

The 4/5-law (equation (25)) is not a-priori valid in the regime of turbulence we are considering. However, the 4/5-law has been numerically verified to hold also in some situations outside the domain of validity of its derivation. For instance, for isotropic mildly compressible decaying (Porter *et al.* 2002) and driven (Benzi *et al.* 2008) turbulence. In the anisotropic case, however, anisotropic contributions cannot be excluded (Biferale *et al.* 2002), although they are known to be subdominant in some important cases (Calzavarini *et al.* 2002; Biferale *et al.* 2003; Kaneda *et al.* 2008). In this section we show that equation (25) is consistent with our data over a wide range of scales.

We compute the 3rd-order structure functions of the velocity, defined by equation (23), in a rather simple way using a random sample of 20,000 points in each of the 380 3D data dumps of our simulations. At each time, we compute the 3rd power of the velocity increments for each pair

of points and accumulate and average in time the results in bins of size Δx. The resulting structure functions are shown in Figure 8, compensated by $-\frac{5}{4} r^{-1} \langle \epsilon \rangle^{-1}$, so that the resulting quantity should be equal to one if the 4/5-law is satisfied in our data. As was the case for the energy spectra, we assume $\eta \sim \Delta x$ and plot the structure functions versus $r/\Delta x$.

The degree with which the 4/5-law is satisfied in our data is very good. We see that anisotropic contributions only play a minor role in the angle-integrated formulation of the 4/5-law. This is in agreement with the incompressible DNS of Kaneda *et al.* (2008) and has been known to be true also for Rayleigh-Bénard convection in most regimes (Lohse and Xia 2010). Our results provide an important new example where this appears to hold true. Secondly, for all of our simulations at 512^3, we find

$$\max_r \left\{ -\frac{5}{4} r^{-1} \langle \epsilon \rangle^{-1} \langle \delta v^3 \rangle_{j=0} \right\} \tag{33}$$

within 5% of 1. This level of accuracy is reached in DNS simulations achieving at least a Taylor micro-scale Reynolds number (Kaneda *et al.* 2008)

$$R_\lambda = \frac{u'\lambda}{\nu} \sim 300, \tag{34}$$

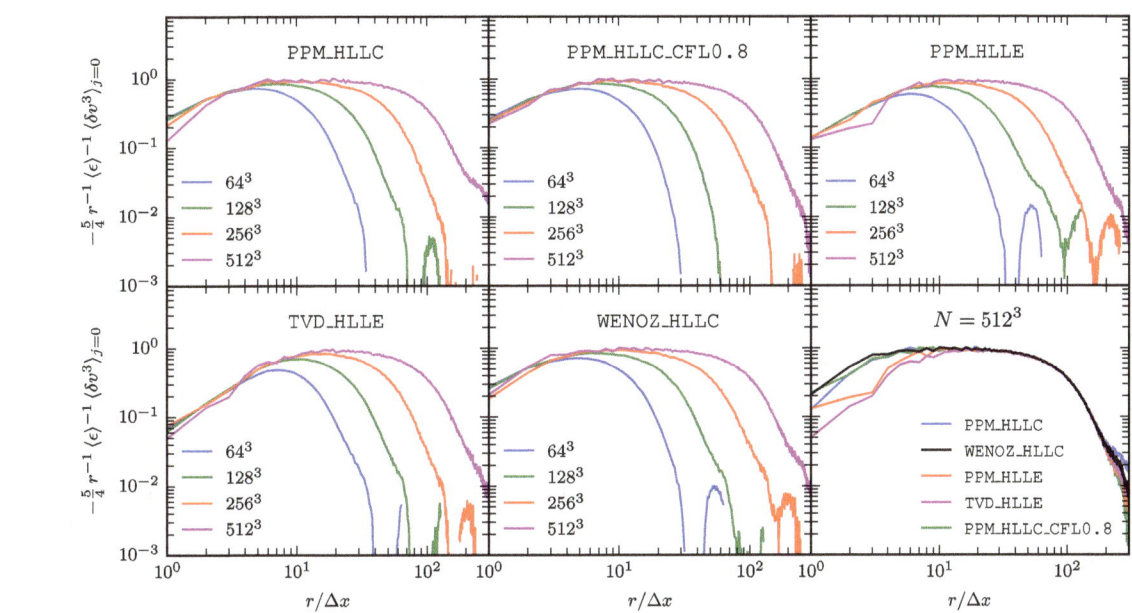

Figure 8 Compensated 3rd-order structure functions (equation (23)) for all the numerical methods and resolutions. The structure functions are compensated and scaled so that they should be close to one where the 4/5-law (equation (25)) is verified. The data is plotted as a function of the dimensionless separation $r/\Delta x$. The first five panels show the PPM_HLLC (upper left), PPM_HLLC_CFL0.8 (upper center), PPM_HLLE (upper right), TVD_HLLE (lower left) and WENOZ_HLLC (lower center) group of runs. The last panel (lower right) shows a comparison of all of the methods at the highest resolution (512^3). The 4/5-law is very well verified in our data suggesting that (1) anisotropic corrections are subdominant and (2) all of the simulations behave in a way consistent with large Reynolds numbers turbulence at the largest scales.

where ν is the kinematic viscosity, $u' = \frac{1}{\sqrt{3}} v_{\mathrm{rms}}$ and $\lambda = (15\nu u'^2/\langle\epsilon\rangle)^{1/2}$ is the Taylor micro-scale. This corresponds to a large-scale Reynold numbers $R = \frac{u'L}{\nu} \sim R_\lambda^2$, $L = 1$ being the domain size, in excess of $\sim 85,000$. Reaching these Reynolds numbers in a DNS requires resolutions between 512^3 and 1024^3 using pseudo-spectral methods (Donzis *et al.* 2008). This large-scale estimate of the Reynolds number is consistent with previous findings (Zhou *et al.* 2014), although it is several orders of magnitude larger than the one that could be naively estimated using v_{max}. For instance, for PPM_HLLC at 512^3, $v_{\mathrm{max}} \simeq 1.5 \times 10^{-3}$ and $v_{\mathrm{rms}} \simeq 0.4$ giving

$$\frac{u'L}{v_{\mathrm{max}}} \simeq 150. \qquad (35)$$

This apparent discrepancy is due to the fact that an ILES is not a DNS. As a consequence, different quantities that in a DNS depend on the Reynolds number, such as the dissipation rate or the Kolmogorov scale, behave as though the simulation had multiple values of the Reynolds number.

6 The transverse spectrum

Finally, we want to comment on the use of 2D transverse spectra in 3D simulations, a practice typically employed in the analysis of turbulence in CCSN simulations (Dolence

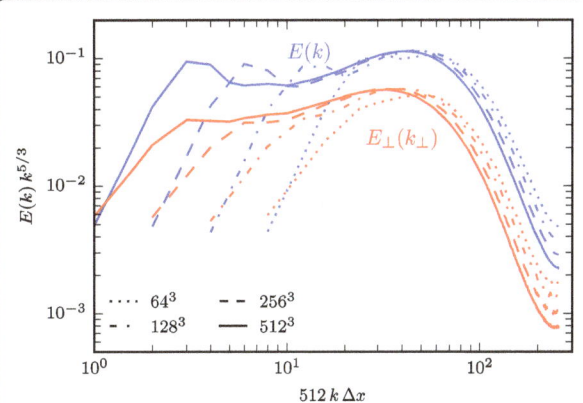

Figure 9 3D (equation (10), blue) and 2D, transverse (equation (26), red) energy spectra for the PPM_HLLC simulations. The energy spectra are compensated by $k^{5/3}$ to highlight eventual regions with Kolmogorov scaling. The spectra are plotted as a function of the dimensionless wave number $512k\Delta x$, as in Figure 5. Although $E(k)$ and $E_\perp(k_\perp)$ have similar trends, the use of the transverse spectrum can overestimate the width of the bottleneck region.

et al. 2013; Couch and O'Connor 2014; Handy *et al.* 2014; Abdikamalov *et al.* 2015).

Figure 9 shows a comparison of the 2D transverse spectrum $E_\perp(k_\perp)$ from equation (26) and the 3D energy spectrum from equation (10) for the PPM_HLLC simulations.

The other runs (not shown) have the same qualitative behavior. We can see that the transverse spectrum follows qualitatively the same trend as the 3D spectrum in terms of convergence. They are both roughly compatible with a Kolmogorov scaling, but the bottleneck appears to be more pronounced in the 2D spectrum than in the 3D spectrum. In particular, $E_\perp(k_\perp)$ only shows a very small region that suggests an inertial range, $3 \lesssim k \lesssim 5$ (as opposed to $5 \lesssim k \lesssim 10$ in $E(k)$).

Abdikamalov *et al.* (2015) concluded, also based on the analysis of 2D spectra, that turbulence in CCSN simulations is dominated by the bottleneck effect. Given the resolutions used in CCSN studies, our work supports their conclusion. However, in the light of Figure 9, we recommend that future studies supplement the analysis of 2D spectra with 3rd-order structure functions, that, as we have shown, can give a more accurate description of the energy cascade.

7 Conclusions

The details of the explosion mechanism of CCSNe have eluded our comprehension in spite of more than 50 years of studies (Woosley and Janka 2005; Janka *et al.* 2012; Burrows 2013; Foglizzo *et al.* 2015). Recent numerical advances (Murphy *et al.* 2013; Couch and Ott 2013; Couch and Ott 2015; Müller and Janka 2015) suggest that turbulence might play a fundamental role in tipping over the balance of the forces and lead to successful explosions (see also Appendix). At the same time, the level of accuracy of current simulations, which employ the ILES methodology, is unclear (Abdikamalov *et al.* 2015). Turbulence in CCSNe is mildly compressible, but strongly anisotropic (Murphy *et al.* 2013; Couch and Ott 2015). Simulations use rather dissipative numerical schemes (because they have to deal with strong shock waves and complex microphysics) and relatively low resolution, a combination (anisotropic turbulence and dissipative schemes) that has not been systematically studied before.

With the goal of assessing the reliability of ILES employed in the study of CCSNe, as well as in other areas of physics and astrophysics, we performed a series of local simulations of driven, anisotropic, weakly compressible turbulence. We compared five commonly employed numerical schemes with different reconstruction methods, Riemann solvers, and time step size. Each was run at 4 different resolutions ranging from 64^3 to 512^3. Our analysis focused on the fidelity with which the turbulent cascade is represented in each model. In particular, we performed an analysis both in Fourier space (with the velocity power-spectra and the energy flux) and in physical space (with the 3rd-order structure functions). Finally, we measured the numerical viscosity of each scheme from the residual of the specific kinetic energy equation.

We found that, on the one hand, all of the numerical setups are able to accurately capture the decay rate of kinetic

energy from the injection scale, with errors at the few % level already at 128^3 (*e.g.*, $\sim$2.5% for PPM_HLLC_N128). On the other hand, a large fraction of the energy is at unresolved scales where it piles up due to the bottleneck effect and an inertial range appears only at the highest resolutions (512^3). Even at this resolution, which would be difficult to achieve in global simulations, only roughly $\sim$80% (the exact number depends on the scheme, see Section 4.4) of the energy is resolved, the remaining $\sim$20% accumulates as excess energy at intermediate scales (the cumulative energy excess at the grid scale alone is as large as $\sim$5% of the total energy).

Current CCSN simulations have resolutions of at most of 30-70 points covering the gain region (Abdikamalov *et al.* 2015) (the energy injection scale). Based on our analysis we expect that at these resolutions even the energy decay rate from the largest scales will not be completely converged, but will show errors of up to tens of percent, depending on the numerical scheme (see Section 4.1). At smaller scales, the dynamics is going to be completely dominated by the bottleneck effect. This is in agreement with the findings of Abdikamalov *et al.* (2015), based on the use of global simulations reaching a maximum resolution of 66 grid points covering radially the extent of the gain region.

Based on our findings, we expect that, if the resolution in global simulations is increased by a factor $\sim$2 from the one of Abdikamalov *et al.* (2015), the decay rate will be converged to within a few % of the asymptotic value. This implies that the ratio between thermal and kinetic energy, a crucial quantity for the onset of the explosion, will also be converged to within a few %, at least when the energy injection rate changes slowly compared to the eddy turnover time (which is roughly $\sim$20 ms in a CCSN (Ott *et al.* 2013; Couch and O'Connor 2014)). Unfortunately, while the lead up to explosion occurs over a larger timescale of a few hundred milliseconds, the transition to explosion can happen over much shorter timescales (one turnover time or less) (Couch and Ott 2015). This means that the dynamics of the cascade over smaller time and length scales in the gain region also needs to be captured correctly since changes in the energy input rate on such short time scales will yield an inaccurate representation of the energy on large scales due to the bottleneck effect. This could require an increase of resolution by a factor $\sim$4-8 with respect to current high-resolution simulations. Additional work using semi-global or global simulations will be required to more firmly establish the resolutions requirements at the transition of the explosion.

Concerning the properties of anisotropic turbulence in our simulations, we found anisotropy contributions to the energy spectrum and to the angle-averaged formulation of the 4/5-law to be subdominant: the accuracy with which the 4/5-law is satisfied is limited only by the employed resolution and the energy spectrum appears to be consistent with Kolmogorov $k^{-5/3}$ scaling. We also found the transverse energy spectrum with respect to the direction of anisotropy, a quantity typically computed in CCSN simulations, to overestimate the bottleneck with respect to the angle-integrated 3D spectrum. For this reason, we recommend future studies of CCSN to supplement (or replace) the analysis of the transverse spectrum with the analysis of the 3rd order, angle-integrated, structure function (or, where possible, with the 3D spectrum itself).

Our results are, of course, dependent on the choice of the numerical scheme. In particular, we found significant differences in the dissipative properties of schemes employing the HLLC Riemann solver with respect to schemes using the more dissipative HLLE solver. The reconstruction order of the scheme is also important, although, while significant differences are found between TVD and PPM, the differences between PPM and WENOZ are much more minute (despite WENOZ being significantly more computationally expensive than PPM). In the end, none of the schemes we considered seems to be able to yield an accurate representation of the kinetic energy distribution across different scales at an affordable resolution for global CCSN simulations. A possible way forward would be to adopt low-dissipation numerical schemes especially designed for the use in ILES, such as the methods proposed by Hickel *et al.* (2006); Martín *et al.* (2006) or Thornber *et al.* (2008). Implementing and testing these schemes will be subject of future work.

An important limitation of the present work is that we considered a very idealized setup. On the one hand, this allows us to benchmark the behavior of ILES in a controlled environment. On the other hand, our simulations cannot fully capture all features of the turbulent convective flow in a CCSN. Unlike the situation in a CCSN, our local simulations did not include a vertical advective velocity field that is due to the accretion of the stellar mantle. However, the advective velocities are nearly constant in the regions of interest and Galilean invariance ensures that our results are unaffected. More limiting is the local nature of our simulations and the inevitable choice of boundary conditions. Moreover, our simulations could not take into account spatial variations in gravity and the large-scale radial convergence of the flow in globally spherical problems like collapsing stars. Addressing these issues will also be subject of future work.

Appendix: The role of turbulence in core-collapse supernova explosions

In this appendix we present a brief discussion of the importance of turbulent pressure in the explosion of massive stars. To set the stage, we will briefly summarize what is known of the dynamics of the most common class of CC-SNe that are relevant for our later discussion. This is done for the benefit of readers that are not supernova specialists

and it is not meant to be a complete review of the status of the field, for which we refer, instead, to the reviews of Janka *et al.* (2012) and Burrows (2013). Next, we will discuss the role of turbulence and, in particular, of turbulent pressure on the explosion mechanism, in light of some recent results (Murphy *et al.* 2013; Couch and Ott 2015).

A.1 The neutrino mechanism

Towards the end of their evolution, massive stars form massive ($\sim$1.5 solar mass) iron cores at their center. Since the iron nucleus has the largest binding energy per nucleon, no energy can be extracted from nuclear fusion beyond iron. The iron core is essentially inert and supported against gravity only by the degeneracy pressure of relativistic electrons. The mass of the iron core increases with time as more iron-group material is added by silicon shell burning. Electron capture on protons, which becomes energetically favorable at high densities, depletes the core of electrons and thus reduces the pressure supporting it against gravity. Eventually, the core becomes dynamically unstable and collapses.

During the collapse, the subsonically collapsing inner core ($\sim$0.5 solar masses) contracts until it reaches densities comparable to that in atomic nuclei ($\sim$4-5 $\times$ 10^{14} g cm^{-3}). At this point, the nuclear equation of state stiffens (due to the strong nuclear force). This halts the collapse of the inner core. It stops, bounces back and a proto-neutron star (PNS) is formed. The outer core, however, is still collapsing supersonically and a strong shock wave is launched at the interface between the inner and outer core.

It was once thought that this shock wave would travel outwards dynamically, depositing its energy in the outer layers of the star, causing the explosion. However, multiple numerical simulations performed over multiple decades have consistently shown that the initial shock fails to explode the star. Instead, it stalls due to energy losses to the dissociation of heavy nuclei into free nucleons and to the emission of neutrinos that stream away from the neutrino-semitransparent regions behind the shock (Bethe 1990). The shock generally stalls within only a few tens of milliseconds of core bounce and turns into an accretion shock standing at a radius of $\sim$100-200 km. The accretion rate through the shock is so high (a fraction of a solar mass per second) that, if nothing revitalizes the shock within $\sim$1-2 seconds, the gravitational force would overwhelm the nuclear repulsion force, collapsing the core of the supernova to a black hole, precluding explosion (*e.g.*, O'Connor and Ott (2011)).

During this time, however, the PNS will release a significant fraction of its binding energy in the form of neutrinos (of order 10^{46} J). Converting a few percent of that energy into kinetic energy would be enough to unbind the stellar envelope and power the supernova explosion. In the standard neutrino mechanism it is theorized that a small fraction neutrinos emitted from the edge of the protoneutron star is re-absorbed in the region right behind the stalled shock. The deposition of neutrino energy leads to higher thermal pressure so that the shock can eventually overcome the ram pressure of accretion and accelerates in a run-away process (Bethe 1990; Pejcha and Thompson 2012). Turbulence in the heating region behind the shock increases the time a fluid parcel spends in that region and, importantly, turbulent pressure helps in overcoming the ram pressure of accretion (see next section and Couch and O'Connor (2014)). It is, however, presently unclear if neutrino heating (even if aided by turbulence in launching the explosion) is able to provide enough energy to power the explosions to the energies inferred from astronomical observations.

A.2 Turbulent pressure and the Rankine-Hugoniot conditions

Simulations (Burrows *et al.* 1995; Murphy *et al.* 2013; Couch and Ott 2015) have shown that turbulence and, in particular, turbulent pressure behind the shock, could play an important role in aiding the explosion. To see why this is the case, we consider the Rankine-Hugoniot momentum condition for a standing accretion shock in a supernova core,

$$s\left[\rho_d v_d^r - \rho_u v_u^r\right] = \rho_d \left(v_d^r\right)^2 + p_d - \rho_u \left(v_u^r\right)^2 - p_u, \tag{36}$$

where s is the shock speed and $\cdot_d$ and $\cdot_u$ denote the downstream and upstream values respectively. For the purpose of our discussion, we can assume the upstream gas to be cold and free-falling:

$$p_u = 0 \qquad \left(v_u^r\right)^2 = \frac{GM}{r}. \tag{37}$$

For the shock to expand we must then have

$$\rho_d \left(v_d^r\right)^2 + p_d > \rho_u \frac{GM}{r}. \tag{38}$$

In the presence of turbulence, Murphy *et al.* (2013) suggested to modify equation (38) in a way akin to a Reynolds decomposition and write it as

$$\rho_d \left(\bar{v}_d^r\right)^2 + \rho_d \left(\delta v_d^r\right)^2 + p_d > \rho_u \frac{GM}{r}, \tag{39}$$

where **v** is the average velocity and $\delta \mathbf{v} = \bar{\mathbf{v}} - \mathbf{v}$ is the turbulent velocity. Although not entirely rigorous, equation (39) has been shown to be well verified in the numerical simulations if angular averages are used to compute the respective quantities (Murphy *et al.* 2013; Couch and Ott 2015).

Couch and Ott (2015) have shown that the turbulent pressure expressed in this fashion can exceed 50% of the thermal pressure, making a very significant contribution to the momentum balance in (39).

Going beyond the arguments of Murphy *et al.* (2013), we can reinterpret equation (39) as being the Rankine-Hugoniot condition for a fluid with a modified equation of state, which has two separate internal degrees of freedom: thermodynamical and turbulent. To this aim, we express δv_d^r in terms of the specific turbulent energy

$$e_{\text{turb}} = \frac{1}{2} |\delta \mathbf{v}|^2 \tag{40}$$

noting that

$$|\delta \mathbf{v}|^2 := \left(\delta v^r\right)^2 + \left(\delta v^\theta\right)^2 + \left(\delta v^\phi\right)^2, \tag{41}$$

and using the fact that

$$\left(\delta v^r\right)^2 \simeq \left(\delta v^\theta\right)^2 + \left(\delta v^\phi\right)^2 \tag{42}$$

in CCSN turbulence, to obtain

$$\left(\delta v^r\right)^2 \simeq \frac{1}{2} |\delta \mathbf{v}|^2 = e_{\text{turb}}. \tag{43}$$

Assuming the pressure varies like $p = (\gamma - 1)\rho e$, and substituting (43) into (39), we find

$$\rho_d \left(\bar{v}_d^r\right)^2 + (\gamma_{\text{th}} - 1)\rho_d e_d + (\gamma_{\text{turb}} - 1)\rho_d e_{\text{turb}}$$
$$> \rho_u \frac{GM}{r}, \tag{44}$$

where $\gamma_{\text{th}} \simeq 4/3$ is the thermodynamical adiabatic index, e_d is the downstream thermal energy and $\gamma_{\text{turb}} = 2$ is the equivalent adiabatic index associated with anisotropic CCSN turbulence. Since $\gamma_{\text{turb}} > \gamma_{\text{th}}$, we see that turbulent energy is more efficient, per unit specific internal energy, at pushing the shock than thermal energy.

We point out that, if equation (42) is dropped and turbulence is assumed to be isotropic, then $\gamma_{\text{turb}} = 5/3$, which is still larger than γ_{th}, but not as large as for the anisotropic case. This is a simple consequence of the fact that anisotropic turbulence has an anisotropic pressure, which is stronger in the radial direction.

In both cases, since the total energy is conserved, the relevant quantity is the ratio e_{turb}/e. From standard turbulent phenomenology we expect that this ratio will only depend on macroscopic parameters, such as the net heating rate, the accretion rate and so on, and not on the details of the viscosity. For this reason, we expect this ratio to be correctly captured in ILES achieving a sufficiently high resolution.

As a final remark, we point out that a similar argument has been recently proposed by Müller and Janka (2015) who formulated their equations in terms of the turbulent Mach number, as opposed to the turbulent energy.

Competing interests
The authors declare that they have no competing interests.

Authors' contributions
DR ran the simulations, performed the analysis of the data and wrote the basic draft of this paper. SMC assisted with the use of the FLASH code. CDO had the original idea that started this investigation. All of the authors took part in discussions concerning the results and contributed corrections and improvements on the early draft of the manuscript. All authors read and approved the final manuscript.

Acknowledgements
We acknowledge helpful discussions with E Abdikamalov, WD Arnett, A Burrows, R Fisher, C Meakin, P Mösta, J Murphy, M Norman, and L Roberts. This research was partially supported by the National Science Foundation under award nos. AST-1212170 and PHY-1151197 and by the Sherman Fairchild Foundation. The simulations were performed on the Caltech compute cluster Zwicky (NSF MRI-R2 award no. PHY-0960291), on the NSF XSEDE network under allocation TG-PHY100033, and on NSF/NCSA BlueWaters under NSF PRAC award no. ACI-1440083. The software used in this work was in part developed by the DOE NNSA-ASC OASCR Flash Center at the University of Chicago.

References
Abdikamalov, E, Ott, CD, Radice, D, Roberts, LF, Haas, R, Reisswig, C, Moesta, P, Klion, H, Schnetter, E: Neutrino-driven turbulent convection and standing accretion shock instability in three-dimensional core-collapse supernovae Astrophys. J. **808**, 70 (2015). doi:10.1088/0004-637X/808/1/70

Arnett, D, Meakin, C, Young, PA: Turbulent convection in stellar interiors. II. The velocity field. Astrophys. J. **690**, 1715-1729 (2009). doi:10.1088/0004-637X/690/2/1715

Aspden, A, Nikiforakis, N, Dalziel, S, Bell, J: Analysis of implicit LES methods. Commun. Appl. Math. Comput. Sci. **3**(1), 103-126 (2009). doi:10.2140/camcos.2008.3.103

Benzi, R, Biferale, L, Fisher, RT, Kadanoff, LP, Lamb, DQ, Toschi, F: Intermittency and universality in fully developed inviscid and weakly compressible turbulent flows. Phys. Rev. Lett. **100**(23), 234503 (2008). doi:10.1103/PhysRevLett.100.234503

Bethe, HA: Supernova mechanisms. Rev. Mod. Phys. **62**, 801-867 (1990)

Biferale, L, Procaccia, I: Anisotropy in turbulent flows and in turbulent transport. Phys. Rep. **414**(2-3), 43-164 (2005). doi:10.1016/j.physrep.2005.04.001

Biferale, L, Daumont, I, Lanotte, A, Toschi, F: Anomalous and dimensional scaling in anisotropic turbulence. Phys. Rev. E **66**(5), 056306 (2002). doi:10.1103/PhysRevE.66.056306

Biferale, L, Calzavarini, E, Toschi, F, Tripiccione, R: Universality of anisotropic fluctuations from numerical simulations of turbulent flows. Europhys. Lett. **64**(4), 461-467 (2003). doi:10.1209/epl/i2003-00233-9

Blondin, JM, Mezzacappa, A, DeMarino, C: Stability of standing accretion shocks, with an eye toward core-collapse supernovae. Astrophys. J. **584**, 971-980 (2003). doi:10.1086/345812

Borges, R, Carmona, M, Costa, B, Don, WS: An improved weighted essentially non-oscillatory scheme for hyperbolic conservation laws. J. Comput. Phys. **227**, 3191-3211 (2008). doi:10.1016/j.jcp.2007.11.038

Burrows, A: Colloquium: perspectives on core-collapse supernova theory. Rev. Mod. Phys. **85**, 245 (2013). doi:10.1103/RevModPhys.85.245

Burrows, A, Hayes, J, Fryxell, BA: On the nature of core-collapse supernova explosions. Astrophys. J. **450**, 830-850 (1995)

Calzavarini, E, Toschi, F, Tripiccione, R: Evidences of bolgiano-obhukhov scaling in three-dimensional Rayleigh-Bénard convection. Phys. Rev. E **66**, 016304 (2002). doi:10.1103/PhysRevE.66.016304

Casciola, CM, Gualtieri, P, Jacob, B, Piva, R: The residual anisotropy at small scales in high shear turbulence. Phys. Fluids **19**(10), 101704 (2007). doi:10.1063/1.2800043

Clausen, D, Piro, AL, Ott, CD: The black hole formation probability. Astrophys. J. **799**, 190 (2015). doi:10.1088/0004-637X/799/2/190

Colella, P: Multidimensional upwind methods for hyperbolic conservation laws. J. Comput. Phys. **87**, 171-200 (1990). doi:10.1016/0021-9991(90)90233-Q

Colella, P, Woodward, PR: The Piecewise Parabolic Method (PPM) for gas-dynamical simulations. J. Comp. Physiol. **54**, 174-201 (1984)

Couch, SM, O'Connor, EP: High-resolution three-dimensional simulations of core-collapse supernovae in multiple progenitors. Astrophys. J. **785**, 123 (2014). doi:10.1088/0004-637X/785/2/123

Couch, SM, Ott, CD: Revival of the stalled core-collapse supernova shock triggered by precollapse asphericity in the progenitor star. Astrophys. J. Lett. **778**, L7 (2013). doi:10.1088/2041-8205/778/1/L7

Couch, SM, Ott, CD: The role of turbulence in neutrino-driven core-collapse supernova explosions. Astrophys. J. **799**, 5 (2015). doi:10.1088/0004-637X/799/1/5

Dolence, JC, Burrows, A, Murphy, JW, Nordhaus, J: Dimensional dependence of the hydrodynamics of core-collapse supernovae. Astrophys. J. **765**, 110 (2013). doi:10.1088/0004-637X/765/2/110

Donzis, DA, Yeung, PK, Sreenivasan, KR: Dissipation and enstrophy in isotropic turbulence: resolution effects and scaling in direct numerical simulations. Phys. Fluids **20**(4), 045108 (2008). doi:10.1063/1.2907227

Dubey, A, Antypas, K, Ganapathy, MK, Reid, LB, Riley, K, Sheeler, D, Siegel, A, Weide, K: Extensible component-based architecture for flash, a massively parallel, multiphysics simulation code. Parallel Comput. **35**, 512-522 (2009). doi:10.1016/j.parco.2009.08.001

Einfeldt, B: On Godunov-type methods for gas dynamics. SIAM J. Numer. Anal. **25**, 294-318 (1988). doi:10.1137/0725021

Eswaran, V, Pope, S: An examination of forcing in direct numerical simulations of turbulence. Comput. Fluids **16**(3), 257-278 (1988)

Falkovich, G: Bottleneck phenomenon in developed turbulence. Phys. Fluids **6**(4), 1411-1414 (1994)

Federrath, C: On the universality of supersonic turbulence. Mon. Not. R. Astron. Soc. Lett. **436**(2), 1245-1257 (2013). doi:10.1093/mnras/stt1644

Federrath, C, Roman-Duval, J, Klessen, RS, Schmidt, W, Mac Low, M-M: Comparing the statistics of interstellar turbulence in simulations and observations. Astron. Astrophys. **512**, A81 (2010). doi:10.1051/0004-6361/200912437

Foglizzo, T, Scheck, L, Janka, H-T: Neutrino-driven convection versus advection in core-collapse supernovae. Astrophys. J. **652**, 1436-1450 (2006). doi:10.1086/508443

Foglizzo, T, Galletti, P, Scheck, L, Janka, H-T: Instability of a stalled accretion shock: evidence for the advective-acoustic cycle. Astrophys. J. **654**, 1006-1021 (2007). doi:10.1086/509612

Foglizzo, T, Kazeroni, R, Guilet, J, Masset, F, González, M, et al.: The explosion mechanism of core-collapse supernovae: progress in supernova theory and experiments. Publ. Astron. Soc. Aust. **32**, e009 (2015). doi:10.1017/pasa.2015.9

Frisch, U: Turbulence: The Legacy of A. N. Kolmogorov. Cambridge University Press, Cambridge (1996)

Frisch, U, Kurien, S, Pandit, R, Pauls, W, Ray, S, Wirth, A, Zhu, J-Z: Hyperviscosity, Galerkin truncation, and bottlenecks in turbulence. Phys. Rev. Lett. **101**(14), 144501 (2008). doi:10.1103/PhysRevLett.101.144501

Fryxell, B, Olson, K, Ricker, P, Timmes, FX, Zingale, M, Lamb, DQ, MacNeice, P, Rosner, R, Truran, JW, Tufo, H: FLASH: an adaptive mesh hydrodynamics code for modeling astrophysical thermonuclear flashes. Astrophys. J. Suppl. Ser. **131**, 273-334 (2000). doi:10.1086/317361

Fureby, C, Grinstein, FF: Monotonically integrated large eddy simulation of free shear flows. AIAA J. **37**(5), 544-556 (1999). doi:10.2514/2.772

Garnier, E, Mossi, M, Sagaut, P: On the use of shock-capturing schemes for large-eddy simulation. J. Comput. Phys. **311**, 273-311 (1999)

Garnier, E, Adams, N, Sagaut, P: Large Eddy Simulation for Compressible Flows. Springer, Berlin (2000)

Gotoh, T, Fukayama, D, Nakano, T: Velocity field statistics in homogeneous steady turbulence obtained using a high-resolution direct numerical simulation. Phys. Fluids **14**(3), 1065-1081 (2002)

Grinstein, FF, Margolin, LG, Rider, WJ: Implicit Large Eddy Simulation: Computing Turbulent Fluid Dynamics Cambridge University Press, Cambridge (2011)

Handy, T, Plewa, T, Odrzywołek, A: Toward connecting core-collapse supernova theory with observations. I. Shock revival in a 15 M$_\odot$ blue supergiant progenitor with SN 1987A energetics. Astrophys. J. **783**, 125 (2014). doi:10.1088/0004-637X/783/2/125

Herant, M: The convective engine paradigm for the supernova explosion mechanism and its consequences. Phys. Rep. **256**, 117-133 (1995). doi:10.1016/0370-1573(94)00105-C

Hickel, S, Adams, NA, Domaradzki, JA: An adaptive local deconvolution method for implicit LES. J. Comput. Phys. **213**(1), 413-436 (2006). doi:10.1016/j.jcp.2005.08.017

Ishihara, T, Kaneda, Y, Yokokawa, M, Itakura, K, Uno, A: Small-scale statistics in high-resolution direct numerical simulation of turbulence: Reynolds number dependence of one-point velocity gradient statistics. J. Fluid Mech. **592**, 335-366 (2007). doi:10.1017/S0022112007008531

Janka, H-T: Explosion mechanisms of core-collapse supernovae. Annu. Rev. Nucl. Part. Sci. **62**, 407 (2012). doi:10.1146/annurev-nucl-102711-094901

Janka, H-T, Müller, E: Neutrino heating, convection, and the mechanism of Type-II supernova explosions. Astron. Astrophys. **306**, 167-198 (1996)

Janka, H-T, Hanke, F, Hüdepohl, L, Marek, A, Müller, B, Obergaulinger, M: Core-collapse supernovae: reflections and directions. Prog. Theor. Exp. Phys. **2012**(1), 010000 (2012). doi:10.1093/ptep/pts067

Johnsen, E, Larsson, J, Bhagatwala, AV, Cabot, WH, Moin, P, Olson, BJ, Rawat, PS, Shankar, SK, Sjögreen, B, Yee, HC, Zhong, X, Lele, SK: Assessment of high-resolution methods for numerical simulations of compressible turbulence with shock waves. J. Comput. Phys. **229**(4), 1213-1237 (2010). doi:10.1016/j.jcp.2009.10.028

Kaneda, Y, Ishihara, T, Yokokawa, M, Itakura, K, Uno, A: Energy dissipation rate and energy spectrum in high resolution direct numerical simulations of turbulence in a periodic box. Phys. Fluids **15**, L21 (2003). doi:10.1063/1.1539855

Kaneda, Y, Yoshino, J, Ishihara, T: Examination of Kolmogorov's 4/5 law by high-resolution direct numerical simulation data of turbulence. J. Phys. Soc. Jpn. **77**(6), 064401 (2008). doi:10.1143/JPSJ.77.064401

Lee, D: A solution accurate, efficient and stable unsplit staggered mesh scheme for three dimensional magnetohydrodynamics. J. Comp. Physiol. **243**, 269-292 (2013). doi:10.1016/j.jcp.2013.02.049

Lee, D, Deane, AE: An unsplit staggered mesh scheme for multidimensional magnetohydrodynamics. J. Comput. Phys. **228**, 952-975 (2009). doi:10.1016/j.jcp.2008.08.026

Lee, D, Tzeferacos, P, Couch, SM, Bachan, J, Daley, C, Fatenejad, M, Flocke, N, Lamb, D, Weide, K, Dubey, A: FLASH: a multi-physics code for adaptive mesh computational fluid dynamics in astrophysics. Astrophys. J. Suppl. Ser. (2014, to be submitted)

Lohse, D, Xia, K-Q: Small-scale properties of turbulent Rayleigh-Bénard convection. Annu. Rev. Fluid Mech. **42**(1), 335-364 (2010). doi:10.1146/annurev.fluid.010908.165152

Martín, MP, Taylor, EM, Wu, M, Weirs, VG: A bandwidth-optimized WENO scheme for the effective direct numerical simulation of compressible turbulence. J. Comput. Phys. **220**(1), 270-289 (2006). doi:10.1016/j.jcp.2006.05.009

Müller, B, Janka, H-T: Non-radial instabilities and progenitor asphericities in core-collapse supernovae. Mon. Not. R. Astron. Soc. Lett. **448**, 2141-2174 (2015). doi:10.1093/mnras/stv101

Murphy, JW, Meakin, C: A global turbulence model for neutrino-driven convection in core-collapse supernovae. Astrophys. J. **742**, 74 (2011). doi:10.1088/0004-637X/742/2/74

Murphy, JW, Dolence, JC, Burrows, A: The dominance of neutrino-driven convection in core-collapse supernovae. Astrophys. J. **771**, 52 (2013). doi:10.1088/0004-637X/771/1/52

O'Connor, E, Ott, CD: Black hole formation in failing core-collapse supernovae. Astrophys. J. **730**, 70 (2011). doi:10.1088/0004-637X/730/2/70

Ott, CD, Abdikamalov, E, Mösta, P, Haas, R, Drasco, S, O'Connor, EP, Reisswig, C, Meakin, CA, Schnetter, E: General-relativistic simulations of three-dimensional core-collapse supernovae. Astrophys. J. **768**, 115 (2013). doi:10.1088/0004-637X/768/2/115

Pejcha, O, Thompson, TA: The physics of the neutrino mechanism of core-collapse supernovae. Astrophys. J. **746**, 106 (2012). doi:10.1088/0004-637X/746/1/106

Porter, D, Pouquet, A, Woodward, P: Measures of intermittency in driven supersonic flows. Phys. Rev. E **66**(2), 1-12 (2002). doi:10.1103/PhysRevE.66.026301

Porter, DH, Woodward, PR: High-resolution simulations of compressible convection using the piecewise-parabolic method. Astrophys. J. Suppl. Ser. **93**, 309-349 (1994). doi:10.1086/192057

Schmidt, W, Hillebrandt, W, Niemeyer, JC: Numerical dissipation and the bottleneck effect in simulations of compressible isotropic turbulence. Comput. Fluids **35**(4), 353-371 (2006)

She, Z, Jackson, E: On the universal form of energy spectra in fully developed turbulence. Phys. Fluids **5**(7), 1526-1528 (1993)

Sytine, IV, Porter, DH, Woodward, PR, Hodson, SW, Winkler, K-H: Convergence tests for the piecewise parabolic method and Navier-Stokes solutions for homogeneous compressible turbulence. J. Comp. Physiol. **158**, 225-238 (2000). doi:10.1006/jcph.1999.6416

Thornber, B, Mosedale, A, Drikakis, D: On the implicit large eddy simulations of homogeneous decaying turbulence. J. Comput. Phys. **226**(2), 1902-1929 (2007). doi:10.1016/j.jcp.2007.06.030

Thornber, B, Mosedale, A, Drikakis, D, Youngs, D, Williams, RJR: An improved reconstruction method for compressible flows with low Mach number features. J. Comput. Phys. **227**(10), 4873-4894 (2008). doi:10.1016/j.jcp.2008.01.036

Toro, EF: Riemann Solvers and Numerical Methods for Fluid Dynamics. Springer, Berlin (1999)

Toro, EF, Spruce, M, Speares, W: Restoration of the contact surface in the HLL-Riemann solver. Shock Waves **4**, 25-34 (1994). doi:10.1007/BF01414629

Uhlenbeck, G, Ornstein, L: On the theory of the Brownian motion. Phys. Rev. **36**, 823-841 (1930). doi:10.1103/PhysRev.36.823

Verhoeven, J, Wiesehöfer, T, Stellmach, S: Anelastic versus fully compressible turbulent Rayleigh-Bénard convection. Astrophys. J. **805**, 62 (2015). doi:10.1088/0004-637X/805/1/62

Verma, MK, Donzis, D: Energy transfer and bottleneck effect in turbulence. J. Phys. A **40**(16), 4401-4412 (2007)

Vincent, A, Meneguzzi, M: The satial structure and statistical properties of homogeneous turbulence. J. Fluid Mech. **225**, 1-20 (1991). doi:10.1017/S0022112091001957

Woosley, SE, Janka, H-T: The physics of core-collapse supernovae. Nat. Phys. **1**, 147-154 (2005). doi:10.1038/nphys172

Yakhot, V, Zakharov, V: Hidden conservation laws in hydrodynamics; energy and dissipation rate fluctuation spectra in strong turbulence. Physica D **64**(4), 379-394 (1993)

Zhou, Y, Grinstein, FF, Wachtor, AJ, Haines, BM: Estimating the effective Reynolds number in implicit large-eddy simulation. Phys. Rev. E **89**(1), 013303 (2014). doi:10.1103/PhysRevE.89.013303

5

Simulations of stripped core-collapse supernovae in close binaries

Alex Rimoldi*, Simon Portegies Zwart and Elena Maria Rossi

Abstract

We perform smoothed-particle hydrodynamical simulations of the explosion of a helium star in a close binary system, and study the effects of the explosion on the companion star as well as the effect of the presence of the companion on the supernova remnant. By simulating the mechanism of the supernova from just after core bounce until the remnant shell passes the stellar companion, we are able to separate the various phenomena leading to the final system parameters. In the final system, we measure the mass stripping and ablation from, and the additional velocity imparted to, the companion stars. Our results agree with recent work showing smaller values for these quantities compared to earlier estimates. We do find some differences, however, particularly in the velocity gained by the companion, which can be explained by the different ejecta structure that naturally results from the explosion in our simulations. These results indicate that predictions based on extrapolated Type Ia simulations should be revised. We also examine the structure of the supernova ejecta shell. The presence of the companion star produces a conical cavity in the expanding supernova remnant, and loss of material from the companion causes the supernova remnant to be more metal-rich on one side and more hydrogen-rich (from the companion material) around the cavity. Following the impact of the shell, we examine the state of the companion after being heated by the shock.

Keywords: supernovae; hydrodynamics; close binaries

1 Introduction

There is substantial evidence that most massive stars evolve in binary systems (Duquennoy and Mayor 1991; Rastegaev 2010; Sana et al. 2012). Therefore, the presence of companion star is an important consideration in the theory and observation of supernovae and supernova remnants (SNRs). In particular, while Type Ia (white-dwarf; WD) supernovae may have a companion which has deposited sufficient mass onto the WD to trigger a 'single-degenerate' explosion, many Type Ibc (stripped core-collapse) supernovae may have close companions that have been at least partly responsible for the loss of mass from the progenitor (Bersten et al. 2014; Eldridge et al. 2015; Fremling et al. 2014; Kim et al. 2015; Kuncarayakti et al. 2015).

Observational searches for supernova companions have typically focused on Type Ia explosions. Possible companions have been a subject of scrutiny in order to determine the frequency of the two main suspected (single-degenerate or double-degenerate) explosion channels (Maoz et al. 2014). Hydrogen enrichment from a companion has been searched for in Type Ia SNRs, but so far there has been no evidence of hydrogen lines (Leonard 2007; Lundqvist et al. 2015; Mattila et al. 2005). As noted in García-Senz et al. (2012), detection of H_α lines may be difficult due to confusion with Fe and Co lines due to the mostly slow ($<10^3$ km s^{-1}) hydrogen mixing with iron-peak elements.

The presence of a supernova companion is difficult to directly detect if they are low-mass stars at very large distances, and so far definitive evidence of close companions to any supernova progenitor, let alone those of Type Ibc, has been lacking. Tycho G is probably the best example of a directly imaged, suspected companion, associated with the galactic Type Ia supernova, Tycho (SN 1572; Ruiz-Lapuente et al. 2004), though recent observations put its status as a supernova companion into dispute (Kerzendorf et al. 2013; Xue and Schaefer 2015). On the other hand,

*Correspondence: rimoldi@strw.leidenuniv.nl
Leiden Observatory, Leiden University, Niels Bohrweg 2, Leiden, 2333 CA, The Netherlands

some direct searches for single-degenerate companions have ruled out giant/subgiant (evolved) stars (SN 2011fe and SNR 1006; González Hernández et al. 2012; Li et al. 2011) and even main-sequence companions (SNR 0509-67.5; Schaefer and Pagnotta 2012).

The presence of a companion due to increased emission, and therefore modification of the standard light curve, from the ejecta interacting with the companion has also been ruled out in observations of Type Ia supernovae (Olling et al. 2015). However, a recently observed supernova (iPTF14atg; Cao et al. 2015) does show evidence of interaction with a companion through the detection of an ultraviolet burst in the first several days.

Though much of the focus of previous work has been on Type Ia explosions, the phenomena of companion interactions with single-degenerate Type Ia ejecta has parallels with core-collapse supernovae in binaries, and therefore this scenario still provides a useful context. A similarity between Type Ia supernovae and Type Ibc supernovae is that the explosion energy in both is believed to by $E_{SN} \sim 10^{51}$ erg $\equiv 1$ foe (Dessart et al. 2014; Smartt 2009).[a] Also, in single-degenerate Type Ia and binary Type Ibc explosions, main-sequence companions are typically at small orbital separations. In the former, this is simply due to the requirement for Roche-lobe overflow in the companion in order to transfer mass to the WD; in the latter, this is due to binary interactions and associated dissipative processes leading to circularised close binaries (Tauris and Takens 1998, hereafter, TT98). However, while simulations of Type Ia supernovae have placed the companion at the point of Roche-lobe overflow, the orbital separations in Type Ibc supernovae can be larger than this. Therefore, simulations of the latter are needed to test the distance-dependence of results that have been extrapolated from Type Ia simulations, such as those used in Tauris (2015).

TT98 analytically investigated the consequences of a supernova in a close, circularised binary, with the goal of finding the runaway velocities of the components of a binary disrupted by a Type Ibc supernova. These predictions were based on early simulations of the effect of a supernova shell impact on a star (Fryxell and Arnett 1981) in order to determine the amount of mass lost and the change in velocity of the companion. Motivated by this problem, we perform simulations of supernovae in binary systems with properties comparable to those used in TT98.

Simulations of supernovae have been performed at many scales, ranging from hundreds of kilometres around the nascent neutron star (Janka 2012) to the impact of the ejecta shell on a companion (and the influence of the companion on the overall structure of the ejecta). The impact of Type Ia ejecta on companions has, in particular, been well studied (Liu et al. 2012; Marietta et al. 2000; Pakmor et al. 2008; Pan et al. 2012). Hirai et al. (2014) investigated the fraction of mass stripped from a giant companion star

due to a effect of a core-collapse (Type II) supernova using a two-dimensional grid-based Eulerian code. Recently, Liu et al. (2015) also presented results on the consequences of a Type Ibc supernova interacting with a binary companion using smoothed-particle hydrodynamics (SPH). These studies have often focused on the companion star without following the explosion from the moment of the supernova. As a consequence, the ejecta shell is initialised artificially via analytic prescriptions near the surface of the companion, without considering its earlier evolution. In addition, the response of the binary companion and subsequent SNR evolution is analysed in these cases from a static configuration rather than placing the binary in an orbit.

For close binary orbits it is typically assumed that the binary has circularised by this point in its evolution, so that the eccentricity of the orbit can be set to zero (TT98). We follow the same assumption in this work. Moreover, despite these close separations and, therefore, short orbital periods, in theoretical work the binary period is taken to be much shorter than the timescale over which the ejected shell impacts the companion. This can be made more explicit (as in, for example, Colgate 1970) by noting that the ejecta velocity must be larger than the escape velocity of the primary star,

$$v_{ej} > v_{esc} = \sqrt{\frac{2GM_1}{R_1}}, \tag{1}$$

where M_1 and R_1 are the mass and radius of the primary. Since the distance, a, of the companion from the primary is larger than R_1, and since the orbital velocity at that distance is

$$v_{orb} = \sqrt{\frac{GM_1}{a}}, \tag{2}$$

then it must be that $v_{ej} > v_{orb}$. In practice, the typical ejecta velocities ($10^3 \sim 10^4$ km s^{-1}) are much larger than the orbital velocities ($\sim 10^2$ km s^{-1}), hence the latter is typically ignored in analytic velocity calculations. However, matter in the ejecta in fact have a radially dependent velocity (approximately, in the homologous regime, $v_{ej}(R, t) \propto R/t$). Therefore, during the late-time interactions of the lower-density, slower (and presumably high-metallicity) ejecta with the companion, we may no longer be justified in ignoring the orbital velocity.

Also, it is likely that the companion stars in such close orbits have been synchronised with the orbital period by tidal friction (Zahn 1977). In one of the most compact binaries we consider here, a $4M_\odot$ helium star separated by $4R_\odot$ from a $1M_\odot$ companion, the orbital period is 4×10^4 s, which is still much longer than the timescale of the interaction of the supernova ejecta ($\lesssim 2 \times 10^3$ s). With synchronisation, the surface of a star at $\sim 1R_\odot$ would therefore rotate at $2\pi R_\odot/(4 \times 10^4 \text{ s}) \approx 100 \text{ km s}^{-1}$. As this is also orders of magnitude smaller than the ejecta velocity, we do not expect rotation to induce any substantial asymmetries during the shell interaction and do not consider it here.[b]

An additional important factor in the dynamics of supernovae in binaries is a possible kick imparted to the newly formed neutron star. This is likely due to a 'gravitational tugboat' effect from asymmetry in the ejecta surrounding the neutron star after the core bounce, and perhaps also high magnetic fields and the asymmetric emission of neutrinos from the proto-neutron star (Kusenko and Segrè 1996; Maruyama et al. 2011; Scheck et al. 2004, 2006; Wongwathanarat et al. 2013). For *ultra*-stripped supernova progenitors, which have very small ejecta masses, the shock expands very rapidly and the tugboat effect on the neutron star has been shown to be minimal (Suwa et al. 2015). For the range of hydrodynamic simulations in this paper we do not apply any additional kick to the neutron star.

To study this problem hydrodynamically, we simulate a supernova in an orbiting binary from just after the moment of core bounce in the supernova. To this end, we first generate stellar structure models of the binary components using a one-dimensional stellar evolution code, where we strip an evolved massive progenitor of the majority of its envelope. We then convert these stellar structures to three-dimensional stars in an SPH code, and run simulations from the moment of the supernova. We vary the mass of the primary star as well as the orbital separation independently. In particular, we are interested in investigating the dependence of the companion's removed mass and impact velocity (the radial component in velocity of the companion induced by the impact of the shell) on the initial orbital separation of the binary. We describe our numerical method in more detail in the following section.

2 Method

Throughout this work we use the Astrophysical Multipurpose Software Environment[c] (AMUSE; Pelupessy et al. 2013; Portegies Zwart et al. 2009, 2013) to perform our simulations. We first outline the technique used to generate the stellar models in our binary systems (Section 2.1). We then describe the set up of the initial hydrodynamical models from the stellar structure (Section 2.2). Finally, we describe the simulation of the supernova in the binary, with some discussion on the initial convergence tests that were performed (Section 2.3).

2.1 Stellar models

In order to generate an SPH realisation of the binary, we require a stellar evolution code that can return the internal structure of the star. Two evolution codes in AMUSE fit this criterion: MESA (Paxton et al. 2011) and EVtwin (Eggleton 1971, 2006). We chose MESA to evolve the models to their final stellar structure, motivated by difficulties in previous work in using EVtwin to obtain solutions past the carbon flash in more massive stars (de Vries et al. 2014).

Due to interactions with the binary companion (and potentially also through stellar winds), much of the mass of the primary star is lost over its lifetime, resulting the helium star progenitor of a Type Ibc supernova (for some observationally-motivated examples, see Kim et al. 2015). To obtain an estimated lifetime of the progenitor, we first use SSE, which is a fast predictor of stellar properties based on parametrised stellar evolution tracks (Hurley et al. 2000). With the intent of generating $3M_\odot$ and $4M_\odot$ Helium-star progenitors, we begin with a $12.9M_\odot$ and $16.0M_\odot$ zero-age models with metallicity $Z = 0.02$,[d] which are predicted by SSE to end with the required helium core masses.

We do not model or speculate on the specific mechanisms of the mass loss from the primary star, but instead apply a constant mass loss (removed from the outer mass shells of the MESA structure model) until the final helium star mass is reached. Because the lifetime in SSE may be an overprediction compared to the actual lifetime reached in MESA, we apply this mass loss between 80 and 90 per cent of the predicted SSE lifetime so as to not reach the end of the MESA evolution before all of the required mass is stripped. The stellar evolution is then continued until the final lifetime found in MESA.

For our helium star models, the very final stage of evolution involves a rapid expansion of the remaining, tenuous envelope. Due to interaction with the close binary companion, this small amount of material in the envelope is in fact expected to be lost from the system. For our progenitors, we use models just prior to this stage, at which the outer radius of the helium star is still compact (this corresponds to an age of 19 Myr and 14 Myr for the $3M_\odot$ and $4M_\odot$ helium stars, respectively). For consistency, we evolve our ($Z = 0.02$, $M_2 = 1M_\odot$) companion star to the same age as that used for the primary star. This means the companion is still very early on the main sequence, and therefore the effect of this small duration of stellar evolution on the structure and composition of the companion is negligible.

2.2 Hydrodynamical model set-up

We model the hydrodynamics of the supernova using the SPH code Gadget-2 (Springel 2005), running in the AMUSE framework. The SPH formalism has been shown to be effective in three-dimensional simulations of stellar phenomena (for example, Pakmor et al. 2012). One reason is that computational resources are not expended on regions of vacuum or negligible density (for example, Church et al. 2009; Lai et al. 1993), which constitute a significant fraction of the simulation volume in the current problem. Modelling a binary in a vacuum is easily handled in SPH, without the need for (low density) background fields in grid codes, which can exhibit artificial shocks from motions of other bodies within this background.

The Lagrangian nature of SPH describes advection naturally, without suffering from complications of numerical diffusion found in Eulerian codes, and we do not have to restrict the simulation to a fixed volume, which is useful in the present problem of a rapidly expanding shell of gas. A benefit to running the simulation in three dimensions is the absence of any boundary effects, which can produce on-axis artefacts (Marietta et al. 2000) or preferential wave numbers in the formation of instabilities (Warren and Blondin 2013). As with all hydrodynamical codes, the SPH method also has its drawbacks, and some of these are discussed further in the context of our convergence studies in Section 2.5.

The stellar models created in MESA are converted into SPH particles using the `star_to_sph` routine in AMUSE, in a similar method to that outlined in de Vries et al. (2014). The routine first extracts the one-dimensional hydrostatic structure of the star, represented as a function of mass coordinates, from the data generated by the stellar evolution code. The SPH particles are initialised in a homogeneous sphere constructed from a face-centred cubic lattice, and the radial positions of the particles are then adjusted so as to match the density profile from the evolution code.[e] The internal energies of the particles are then assigned from the temperature (and mean molecular weight) distribution from the stellar evolution code. We use equal-mass particles throughout these simulations (unequal-mass particles can cause additional complications such as spurious mixing; Rasio and Lombardi 1999).

The primary star is configured with a purely gravitational core particle of $1.4M_\odot$ to model the neutron star. The softening length ϵ is chosen to be equal to the smoothing length, such that, due to the compact support of the cubic spline, the smoothing kernel reaches zero at 2.8ϵ. This equality is also maintained for the SPH particles to preserve equal resolution of the gravitational and pressure forces. The zero-kinetic-energy models are relaxed over 2.5 dynamical timescales of the gas using critical damping on the velocities of the particles, where at each step the magnitude of damping is reduced so that in the final step no constraint on the velocity is imposed (for a similar approach, see de Vries et al. 2014). This is required due to effects of mapping the one-dimensional stellar structure on to the particle grid, and differences in physics between the codes, such as the value of the adiabatic exponent (which, in the SPH code, is a fixed value of $\gamma = 5/3$).

We show the one-dimensional stellar structures as well as the relaxed SPH particle densities in Figure 1. Note that for the primary stars, the density according to the SPH gas particles levels off towards the centre compared with the

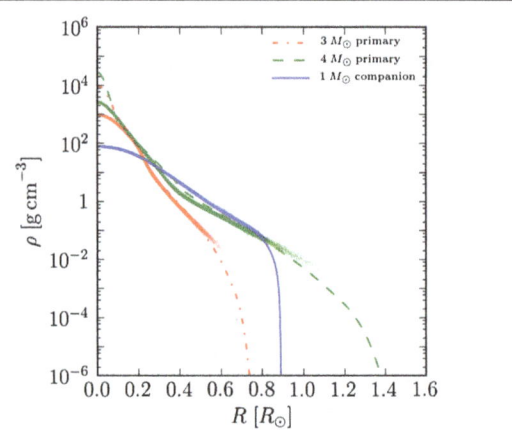

Figure 1 Comparison of the internal stellar structures of the $3M_\odot$ and $4M_\odot$ primary stars, as well as the $1M_\odot$ companion. Lines correspond to the one-dimensional models given by MESA, where the solid line is the companion star, the dot-dashed line is the $3M_\odot$ primary and the dashed line is the $4M_\odot$ primary. Overlaid in the same colour are points representing the densities calculated at the location of each SPH particle in the relaxed SPH models constructed from the MESA models (which excludes the neutron star at the origin).

MESA profiles, as the central density in this region is dominated by the single core particle (which is excluded from the densities in the plot).

We set up the binary models at different orbital separations, a, where the minimum separation is chosen to be greater than the limit of Roche-lobe overflow (RLOF) of the companion star, given by the Eggleton (1983) relation,

$$a_{\mathrm{RLOF}} = \frac{0.6q^{2/3} + \ln\left(1 + q^{1/3}\right)}{0.49q^{2/3}} R_2, \qquad (3)$$

where R_2 is the companion radius and q is the binary mass ratio M_2/M_1. Once both stars have been constructed in the SPH code, orbital velocities are determined for a circular orbit at the specified separation and applied to each star.

2.3 Simulation of the supernova explosion

The supernova is initiated using the 'thermal bomb' technique (Hirai et al. 2014; Young and Fryer 2007), which assumes the core bounce has just occurred, at which moment we inject energy into a shell of particles around the neutron star. As discussed in Young and Fryer (2007), thermal bomb approaches (along with alternative, piston-driven shocks) are not intended to embody the physical mechanism that drives the supernova. Indeed, the actual processes by which the energy gain occurs near the proto-neutron star are still not fully understood, though recent observations and insights from simulations have shed some light on the role of instabilities, asymmetries and jets in driving this process (Bruenn et al. 2013; Couch and O'Connor 2014; Couch and Ott 2015; Hanke et al.

2013; Janka 2012; Lopez et al. 2013; Milisavljevic et al. 2013).

The boundary of energy injection is specified by radius (which is equivalent to a fixed enclosed mass) instead of particle number. This allows scaling of the problem over a range of SPH particle numbers while keeping fixed the mass fraction that receives the supernova energy. The total thermal energy (a canonical 10^{51} erg $\equiv$ 1 foe) is distributed evenly amongst these N_{SN} particles, so that the specific internal energy per particle is increased by (1 foe)/($N_{SN}m_{SPH}$).

We found that a careful investigation of the effect of the injection radius was necessary. Too small a radius (and therefore N_{SN}) results in large asymmetries in the shock front that grow from intrinsic small-scale asymmetries in the initial particle distribution. On the other hand, too large a radius results in the internal energy of the supernova being distributed amongst a large number of particles, lowering the specific internal energy and therefore reducing the overall temperature in the region and weakening the shock. We found that, for the helium star models used here, injecting the supernova energy into a region $R_{SN} \lesssim 0.05 R_\odot$ generates a sufficiently spherical shock while still keeping N_{SN} sufficiently small.

To check the strength of the resulting shock, we calculated the Mach number at various stages through one of our supernovae (the $3M_\odot$ primary). After 2 s, in the high-temperature core the Mach number is ~3, but quickly grows as the shock traverses down the temperature gradient of the star (where the sound speed is lower). The Mach number exceeds 10 by shock breakout. Increasing the radius of energy injection would reduce the initial energy density and therefore the initial Mach number. However, provided the Mach number remains high (as seen in our simulations), the strong-shock conditions are upheld.

2.4 Measured parameters

With the results of these simulations it is possible to predict the final velocities (formally, at infinity) of the runaway components of supernova-disrupted binaries. For masses m relative to the neutron star mass (i.e. $m \equiv M/M_{NS}$), TT98 calculate these values in terms of the following initial parameters:

- a: the pre-supernova binary orbital separation
- v: the pre-supernova relative orbital velocities
- w: the magnitude of the kick applied to the NS
- θ and ϕ: the spherical polar angles of the NS kick vector with respect to the 'x'-axis aligned along the NS orbital velocity vector at the moment of the kick
- v_{im}: the magnitude of the radial velocity component imparted to the companion due to the impact of the supernova shell (we refer to this as the 'impact velocity')
- v_{ej}: the magnitude of the radial velocity of the ejecta shell

- m_2, m_{2f} and m_{shell}: the initial mass of the companion, the final mass of the companion after mass loss, and the mass of material in the shell, respectively (all relative to the neutron star mass)

In the original work of Wheeler et al. (1975), during the supernova shell passage over the companion star, the mass removed from the companion is parametrised by the fraction of companion radius $x = R/R_2$ as a function of the angle around the star. Above some critical fraction of the companion radius, x_{crit}, a fraction F_{strip} of the mass is stripped by the shell impact, and below it a fraction F_{ablate} of mass is ablated. The values of F_{strip} and F_{ablate} are calculated in Wheeler et al. (1975) using a polytropic star of index $n = 3$. The predictions in TT98 are based on the work of Wheeler et al. (1975) as well as mass-loss estimates from simulations of a planar slab of material hitting a star Fryxell and Arnett (1981), which have a low resolution compared with modern simulations. These results also need a corrective factor due to the shell in reality having some curvature. Higher resolution simulations, such as those presented here, provide a test of these earlier estimates, which are one of the sources of uncertainty in the results of TT98.

The magnitude of the radial impact velocity imparted to the companion by the shell, v_{im}, is theoretically determined to be

$$v_{im} = \eta v_{ej} \left(\frac{R_2}{2a} \right)^2 \frac{M_{ej}}{M_2} x_{crit}^2 \frac{1 + \ln(2v_{ej}/v_{esc})}{1 - F^*}. \tag{4}$$

Here, we use the expression from Wheeler et al. (1975) in the form adopted by Tauris (2015), which applies the substitution $(F_{strip} + F_{ablate}) = F \rightarrow F^* = (F_{strip} + F_{ablate})^\alpha$ for some α, as well as a free parameter η to account for the fact that this tends to over-predict the value of v_{im}. Effectively, η represents the final change in momentum, Δp, of the companion as a fraction of the incident momentum in the shell. As noted in Wheeler et al. (1975), corrections must be applied to this formula as it neglects the presence of a rarefaction wave back through the ejecta, geometrical effects of curvature in the shell (more important for small a), inhomogeneities in the ejecta and radiative losses behind the shock. Further phenomena can also modify the final impact velocity, such as the deformation of the companion by the shock passage (altering its cross-sectional area), the formation of a bow shock in the ejecta (during which time the flow deflects around the companion star), and shock convergence on the far side of the star (causing the asymmetric emission of material from this side of the star).

In our simulations, the two main measurements we make are, therefore, the mass loss from and impact velocity imparted to the companion, as a function of orbital separation. We measure the removed mass by calculating the specific energy for each particle,

$$e_{tot} = e_{kin} + e_{therm} + e_{pot} = \frac{1}{2}v^2 + u - \phi, \tag{5}$$

where v is the particle velocity, u is its specific internal energy and ϕ is the gravitational potential at its position. Bound particles have a negative value of e_{tot}. The amount of bound mass in the companion is time-dependent over the course of the simulation due to energy exchange between particles. The stabilisation of mass bound in the companion determines the end of our simulation, which occurs within 10 dynamical timescales of the companion star ($\sim 2 \times 10^4$ s). The mass-loss results will be discussed in Section 3.2. Measurements of the radial impact velocity that is gained by the companion will be discussed in Section 3.3.

2.5 Convergence test

One feature of SPH that requires caution is that the resolution is dependent on the local density, and therefore the method loses resolution in the lower-density, uppermost layers of the stars in our simulations. In the current problem, the mass stripped by the secondary is from these same layers. Therefore, a good test for the resolution of the simulations is to look for convergence in the quantity of mass stripped from the companion.

During the supernova, Richtmeyer-Meshkov (the impulsive form of Rayleigh-Taylor) instabilities (RMI) are expected to be present, which have been found to appear once reverse shocks form at the interfaces between discontinuities in the density gradient (Kane et al. 1999). Such discontinuities are present in Type Ibc progenitors at the interface between the carbon-oxygen boundary in the core and, if any substantial fraction of hydrogen remains in the envelope, also at the helium-hydrogen boundary. However, these discontinuities tend to be smoothed during the conversion from the one-dimensional stellar model and subsequent relaxation of the SPH particles. Proper treatment of the RMI requires a prescription of artificial conductivity that is not included in the current SPH codes in AMUSE. This instability is expected to be a significant factor in the mixing of stellar material early in the evolution of SNRs, and so any evaluation of the fate of the composition of the supernova ejecta must take this into account.

Unless the growth of RMI is explicitly seeded by some structure at the density interface, these instabilities will grow from perturbations at the numerical level of the simulation and may not, in such cases, grow substantially (Kane et al. 1999). Therefore, there is the potential for instabilities to be dependent on numerical effects such as the resolution of the simulation. Additionally, during the stripping of mass from the companion star, the initial deceleration of the shell impacting the companion may be Rayleigh-Taylor unstable, but also that the subsequent flow of the shell over the surface of the star may induce some shearing (Kelvin-Helmholtz) instabilities (KHI).

Due to the smoothing of discontinuities after relaxation of the SPH models, a lack of artificial conductivity[f] in

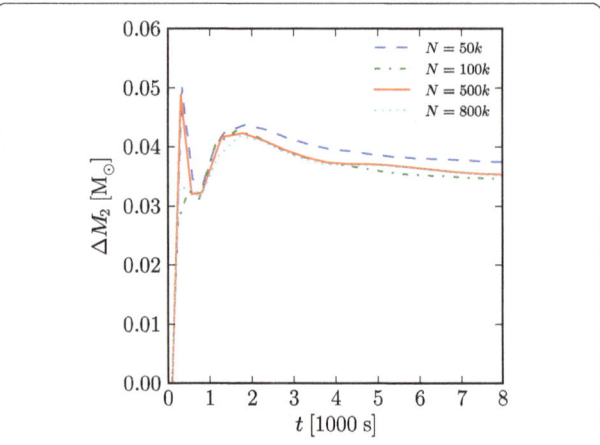

Figure 2   Results for a convergence study using the amount of mass stripped from the companion. The primary star mass was $3 M_\odot$ and the orbital separation was $4 R_\odot$ in all cases. The number of SPH particles used in each run is given in the legend.

Gadget-2 and the only perturbations being from noise in our SPH distribution, we expect that instabilities will not be fully captured in our simulations. As a result, we expect that the influence of instabilities on our results should also be reduced.

Figure 2 shows a test of varying the SPH particle number, N, based on the amount of mass lost from the companion star (evaluated using Equation 5). For low N, there is noticeable noise in the bound mass determination over time, but for $N \geq 10^5$ particles, this is no longer appreciable. As shown in Figure 2, we did not find any substantial difference in the results increasing N from 5×10^5 to 8×10^5. Accounting for this, as well as available computational resources, our simulations were run with 5×10^5 particles.

3 Results

After reviewing the initial conditions used for our simulations, we examine the early stages of the supernova (Section 3.1). We then investigate the magnitude of mass lost from the companion as a function of the orbital separation (Section 3.2), as well as the velocity imparted to the companion and the fraction of imparted momentum compared to the incident shell (Section 3.3). Next, we examine the newly formed SNR for asymmetries in morphology and metallicity (Section 3.4). Finally, we consider the subsequent evolution of a star altered by a supernova shell impact (Section 3.5).

Table 1 shows the initial conditions used in our simulations. The choice of primary and companion masses is motivated by the binary parameters used in TT98 and Tauris (2015), while the minimum orbital separations are chosen to be outside the RLOF value (Equation 3). The final two columns show the effects on the companion due to the shell impact, discussed in more detail in the remainder of this section.

Table 1 Simulation initial conditions and main results

M_1 ($M_\odot$)	M_{ej} ($M_\odot$)	a ($R_\odot$)	ΔM_2 ($M_\odot$)	v_{im} (km s^{-1})
4.0	2.6	4.5	0.021	78
4.0	2.6	5.5	0.013	57
4.0	2.6	6.5	0.0096	47
3.0	1.6	4.0	0.037	83
3.0	1.6	5.0	0.020	60
3.0	1.6	6.0	0.013	48

The first three columns indicate the initial conditions, where M_1 is the mass of the primary (helium star) and before the supernova, M_{ej} is the total ejecta mass, and a is the initial orbital separation. The last two columns are the amount of mass stripped from the companion star and the (magnitude of the) impact velocity.

3.1 Shock breakout

By approximately 20 s after the supernova is initiated, the forward shock has broken out of the surface of the helium star, during which time the fraction of SPH particles bound to the $1.4M_\odot$ neutron star drops smoothly to almost zero. We find at late times that there is some fall-back of a small amount of material, which remains bound to the neutron star. As we do not model here the complexities of the magnetic field of the new neutron star or any form of pulsar wind, it is possible that other mechanisms later expel some or all of the residual bound gas.

In Figure 3, it can be seen that there is a rapid conversion of energy from internal (thermal) energy from the moment of explosion to kinetic and potential energy as the shock passes through the star and the subsequent shell expands. By approximately 100 s following the supernova explosion, very little of the original thermal energy remains in the gas as it has been almost entirely converted into kinetic energy in the expanding shell.

Figure 4 shows the changes in density through the $3M_\odot$ helium star from shortly after core bounce until after the forward shock has reached the outer layers of the star. A lower density cavity with a very shallow gradient is seen to lag behind the expanding ejecta shell. After the shock has reached the surface (24 s), the expansion proceeds towards the companion in a self-similar manner - the variation in gradient is maintained over time, although the overall magnitude of the density drops during the expansion. We investigate the density and velocity distributions within the ejecta in more detail in Section 3.3.

3.2 Impact and mass loss from the companion

We now consider the phenomena occurring during the supernova shell impact on the companion, as well as the removal of mass during this interaction. The passing shell first strips material from the outer layers of the companion. The compression of the companion along the direction of motion of the shock causes heating of the stellar material, which results in a subsequent mass loss through ablation. This ablation of material is found to be a slower form of mass loss than the initial stripping phase.

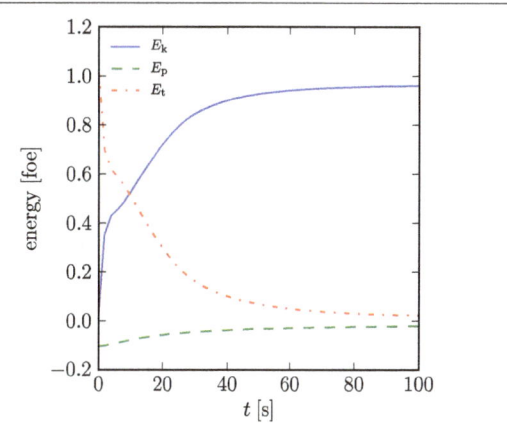

Figure 3 Distribution of total energy in the gas, in units of foe (10^{51} ergs), as a function of time following the supernova event. The energy is broken down into kinetic (E_k), potential (E_p) and internal (thermal) (E_t). This example corresponds to a primary of mass $4M_\odot$ and an orbital separation of $4.5R_\odot$.

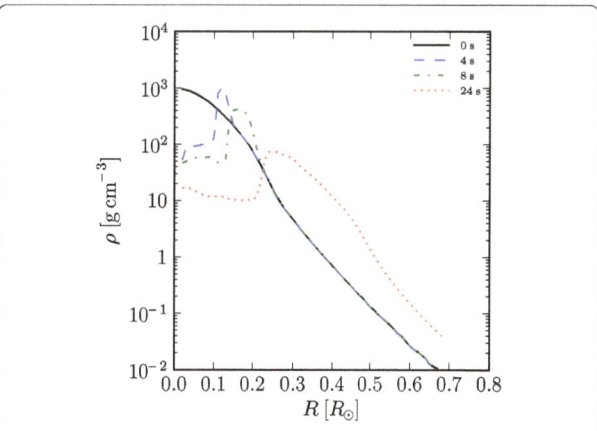

Figure 4 The radial density structure of SPH particles (excluding the core particle) at 4 s, 8 s and 24 s following a supernova in a $3.0M_\odot$ primary.

The passage of the shock through the companion can be seen in the density slices of Figure 5. The black vectors in this figure show the velocities for a random sample of all the SPH particles that were originally in the companion which have subsequently become unbound. These vectors have had the orbital velocity vector of the companion subtracted, and they are then projected onto the orbital plane. Because each SPH particle has the same mass, these vectors also indicate the relative momentum of the unbound particles.

Aside from the mass stripping from the sides of the star as predicted in Wheeler et al. (1975), there is also mass loss from the far side after the shock has passed through the star. Panel (b) of Figure 5 shows that, once the shock passes through the centre of the companion, it converges at the far

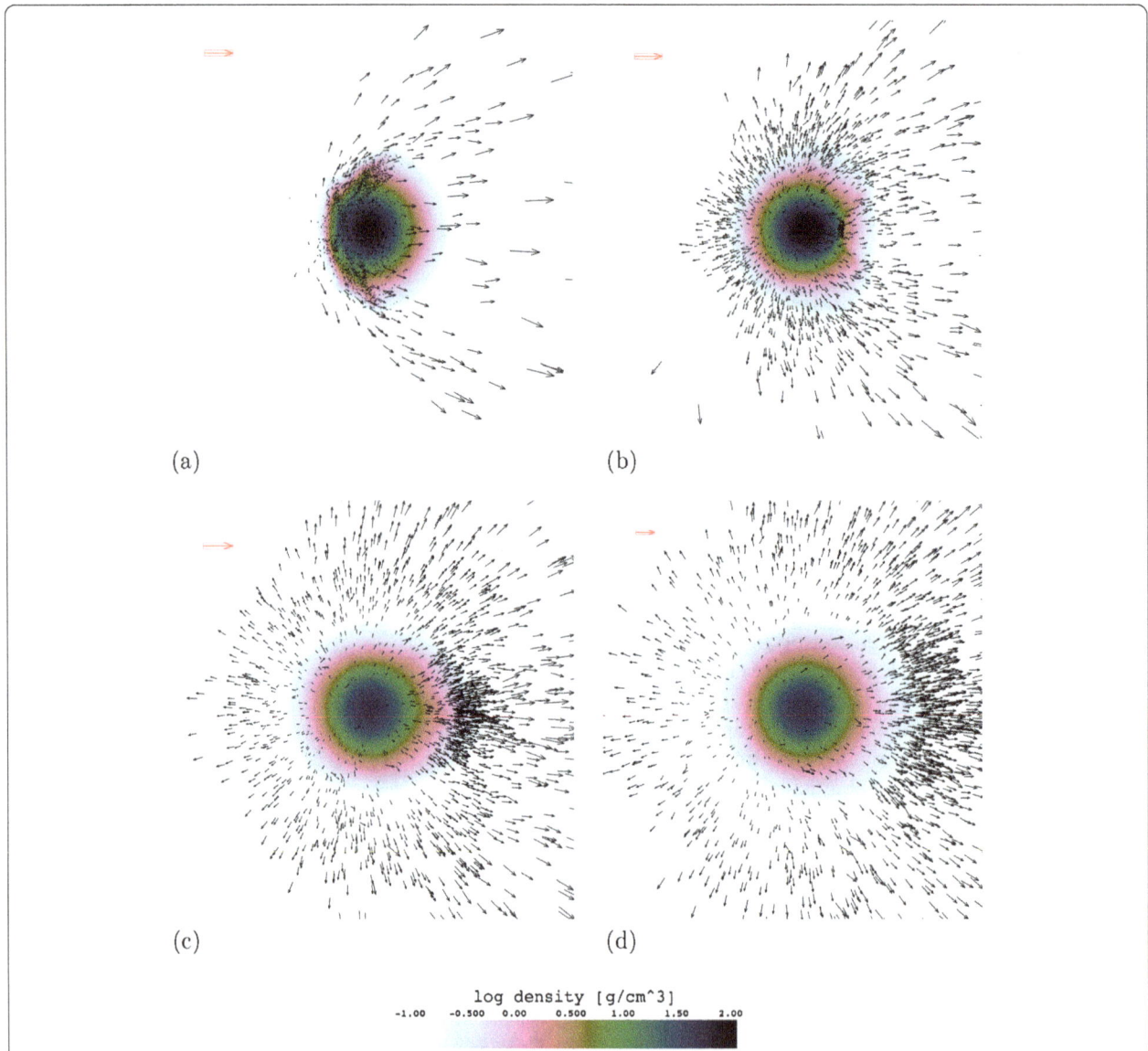

Figure 5 A slice through the *x-y* (initial orbital) plane during the passage of the supernova shock through a 1.0$M_\odot$ companion star after a 1 foe supernova in a primary star of 3.0$M_\odot$ at a distance of 4$M_\odot$. The snapshots correspond to times of (a) 433 s, (b) 1028 s, (c) 1628 s and (d) 2028 s. The shock enters the companion star from the left. The black vectors show the magnitude of the velocity projected onto the orbital plane (and with the orbital velocity of the companion subtracted) for a small random sample of all the particles removed from the companion. In each case, a reference vector (red, boxed) is given in the upper left corner; these correspond to (a) 1×10^4 km s^{-1}, (b) 3×10^3 km s^{-1}, (c) 2×10^3 km s^{-1} and (d) 1×10^3 km s^{-1}.

side of the star as it accelerates down the density gradient (similar shock convergence is seen around other spherically symmetric density gradients, such as in Rimoldi et al. 2015). This increases the local pressure on this axis, resulting in expulsion of material from the far side of the star (panels (c) and (d)) and can counter the effect of the outward kick imparted by the incident shell of material (see also Marietta et al. 2000).

In the last panel of Figure 5, the central density of the companion has dropped and it has noticeably expanded from the shock heating. During this later stage (final three panels) ablation occurs for material which has been heated to the point where the thermal energy is greater than the binding energy. Due to the shock heating, the companion becomes extended, similar to a pre-main sequence star (though its internal structure will differ from a pre-main sequence star), and its luminosity is expected to increase temporarily as it reverts to thermal equilibrium (Marietta et al. 2000; see also Section 3.5). Finally, we find a quadrupole oscillation of the companion that is induced

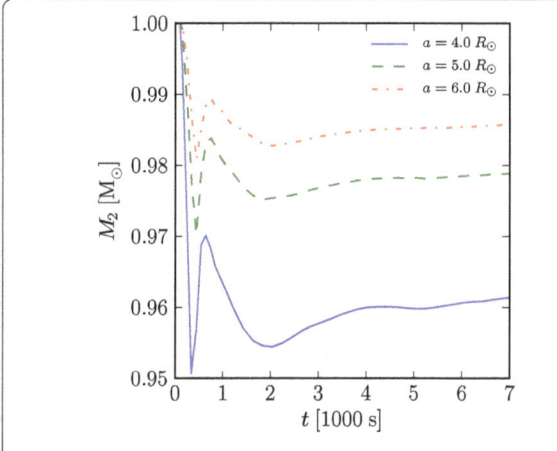

Figure 6 Mass bound to the companion using Equation 5 as a function of time since the supernova explosion for a primary helium star of $3.0M_\odot$ and a range of orbital separations.

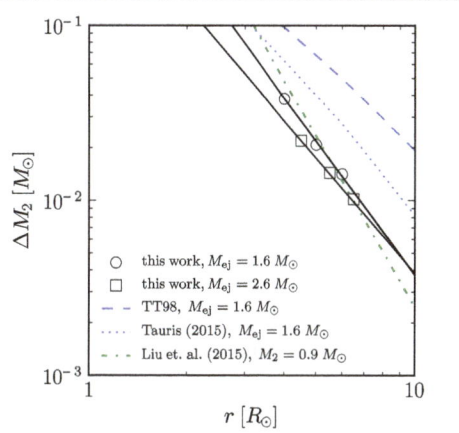

Figure 7 Final mass lost from the companion as a function of orbital separation. Circles show results with a primary star of $3M_\odot$ and squares show results for a primary star of $4M_\odot$. The secondary is $1M_\odot$ in each case. The solid lines show the best fit power-laws for each ejecta mass. The comparison curves are from the theoretical predictions of WLK75 as adapted by TT98 and Tauris (2015) (rescaled to our initial conditions), as well as the simulation results of Liu et al. (2015) for a $0.9M_\odot$ companion star. Note that the comparison with Liu et al. (2015) is not fully equivalent, as both the ejecta mass and companion mass (and therefore radius) are slightly different.

by the distortion from compression due to the shock. This ringing subsides after about one dynamical timescale of the companion star.

Figure 6 shows an example of the variation in companion mass due to the shell impact. The stripping of mass by the passing shell causes a rapid mass loss in the initial phase. There is then a brief increase in the bound mass, which has also been seen in past simulations (Liu et al. 2015; Pakmor et al. 2008; Pan et al. 2012). One possible cause of this is the formation of a reverse shock in the ejecta, which slows material with respect to the companion and increases the amount of bound mass (Pan et al. 2012). A more gradual mass loss then ensues due to the later ablation of shock-heated material. The proportion of mass lost drops rapidly even by moderate orbital separations.

In Figure 7, we show the amount of mass lost from the companion (as a fraction of its initial mass) as a function of orbital separation. The lost mass is found by subtracting the final bound mass at the last snapshot of each simulation (which occurs at 2×10^4 s) from the initial mass. We use the last time possible from the simulation as the final mass takes much of the total simulation time to reach its steady-state value. A least-squares regression gives a fit to our data of $1.3(R/R_\odot)^{-2.6}M_\odot$ for the $M_{\rm ej} = 1.6M_\odot$ data and $0.58(R/R_\odot)^{-2.2}M_\odot$ for the $M_{\rm ej} = 2.6M_\odot$ data. A variation of only 3 per cent in the values of lost mass is sufficient to obtain agreement between the fitted gradients, and therefore caution should be exercised in interpreting any difference between the two gradients. The dashed line in Figure 7 shows the prediction from TT98, the dotted line shows the fit obtained from Type Ia simulations compiled by Tauris (2015), and the dot-dashed green line is from Liu et al. (2015). The values of lost mass we find are comparable to those seen in Liu et al. (2015). Likewise, we find values of ΔM_2 less than the values that are extrapolated from simulations of Type Ia supernovae, indicating

that these values should be revised for the conditions of Type Ibc supernovae considered here.

There are some differences between our initial conditions and those from the previous work shown in Figure 7. Compared with previous Type Ia simulations, our explosion energies and ejecta masses are both slightly different. Additionally, the companion radius, R_2, shrinks slightly after relaxation of the SPH models compared with the radius from the MESA model. The predictions in TT98 and Tauris (2015) depend on these quantities in particular within the geometric parameter Ψ, used originally by Wheeler et al. (1975), defined as

$$\Psi = \left(\frac{R_2}{2a}\right)^2 \left(\frac{m_{\rm shell}}{m_2}\right)\left(\frac{v_{\rm ej}}{v_{\rm esc}} - 1\right). \tag{6}$$

This parameter is used in the determination of $x_{\rm crit}$ as well as $F_{\rm strip}$ and $F_{\rm ablate}$ in Wheeler et al. (1975) using tabulated data for an $n = 3$ polytrope. For our comparisons, we adjust these quantities (and therefore Ψ) in the TT98 and Tauris (2015) estimates to match the initial conditions of our simulations. Furthermore, the simulations in Liu et al. (2015) also use a slightly different companion mass and ejecta mass, and so their results are not completely equivalent to ours. The structure of the companion star has been shown to substantially effect the magnitude of removed mass in simulations of Type Ia supernovae (Liu et al. 2012; Meng et al. 2007; Pan et al. 2012). Slight differences in companion models can therefore responsible for some of the discrepancies.

3.3 Momentum transfer and the velocity of the companion

When the orbital separation is very small, the impact of the ejecta causes not only a significant loss of mass from the companion star but also a large change in velocity. The largest change in velocity of the companion occurs during the transfer of momentum from the shell in the initial impact. However, as the end of the shell passes over the far side of the companion, there is an overpressure acting on this side of the star when the shock converges on this axis. This causes the companion to receive a slight change in momentum in the direction opposite to the shell motion (which has been suggested in other simulations such as Fryxell and Arnett 1981; Marietta et al. 2000). In the theory of TT98, $\mathbf{v}_{\mathrm{im}}$ is defined to be an *effective* velocity that not only accounts for the momentum imparted to the companion by the passage of the shell but also the subsequent change in momentum due to (potentially asymmetric) mass loss.

We found that measuring the velocity of the companion with respect to the neutron star is complicated by the difficulty to define the baryonic centres of the binary system with the ejecta that had not yet left the binary system, the oscillatory behaviour of the companion star as a result of the shell impact, and the Brownian motion of the neutron star due to the shot-noise of the limited resolution in its vicinity.

As an alternative technique, we set up a co-rotating frame of reference that matches the original circular orbital motion. Up until the shell impact, there is no component of velocity of the companion perpendicular to this direction of motion. During and after the impact, the companion (as well as the mass unbound from it) gains a component of velocity, and therefore momentum, in the radial direction with respect to this frame. We use this to measure the momentum delivered to the companion and the material removed from the companion, as well as the radial impact velocities.

The left-hand panel of Figure 8 shows the component of momentum in the radial direction for material unbound from the companion that was not unbound in the previous time step. The first peak is due to the large amount of material initially stripped from the companion star by the shell impact. This figure also clearly indicates the burst of material out of the back of the star, seen as the second peak in the left-hand panel. The right-hand panel shows the breakdown of momenta in the radial direction for unbound and bound material originally from the companion (and their sum). This gives an alternative indication of η, where we see that although less than half of the total incident momentum in the shell is delivered to this material in total, only $\lesssim 30$ per cent of the momentum is delivered to the (bound material of the) companion star.

As the interaction of the shell with the companion is not instantaneous, we must define a point at which we measure the impact velocity. Up until the second peak seen in the left-hand panel of Figure 8, the radial velocity gained by the companion is dominated by the interaction with the passing shell. Once this interaction has ended, there is an additional, growing radial component in velocity from the eccentricity induced in the orbit. The change in gradient of p_{r} for the companion, seen in the right-hand panel of Figure 8, marks the end of the impact phase, which corresponds to just after the second peak in the left-hand panel. We define this to be the point at which we measure the impact velocity. Up to this point, the contribution to the radial velocity from any induced eccentricity is small.

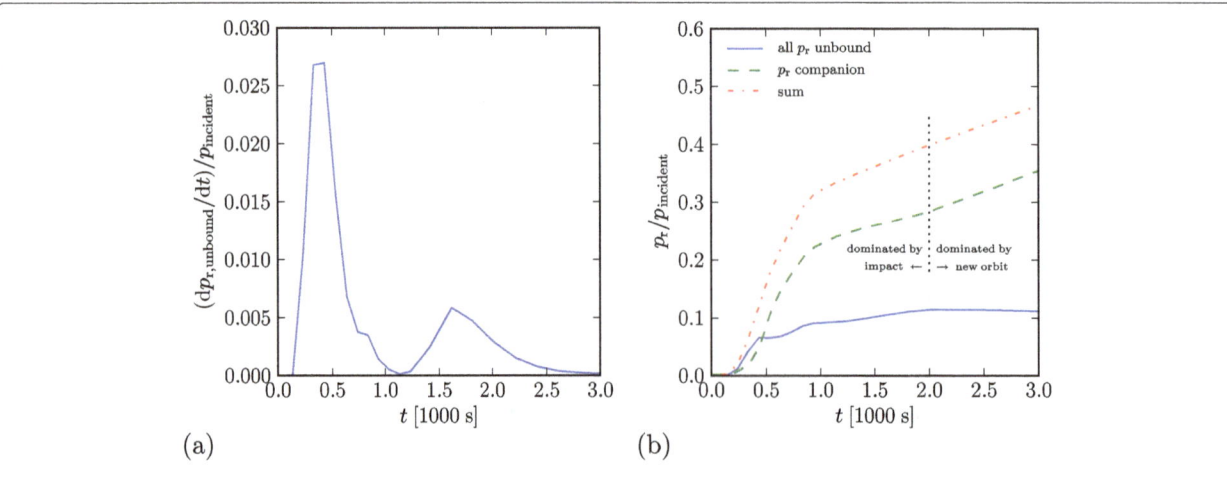

(a)   (b)

Figure 8 Components of momenta in the radial direction of (a) newly unbound material from the companion (material that was not unbound the previous snapshot) and (b) the total unbound mass from the companion, and bound mass in the companion, and the sum of these two values. Values are shown as a fraction of the total incident momentum calculated for the cross-section of shell material that impacts the companion. The example shown is for a $3M_\odot$ primary and an orbital separation of $6R_\odot$.

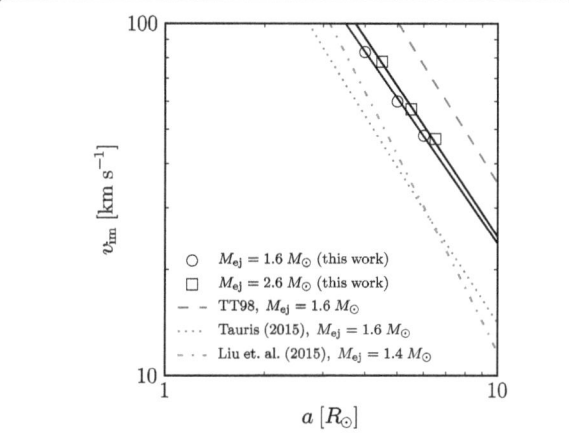

Figure 9 Magnitude of the impact velocity, v_{im}, imparted to the companion star as a function of orbital separation a. Markers and line styles correspond to those used in Figure 7. Again, as for Figure 7, the comparison with Liu et al. (2015) is not fully equivalent due to the slightly different ejecta mass and companion parameters.

Our final impact velocity magnitudes are shown in Figure 9. A least-squares regression gives a fit to these data of $556(R/R_\odot)^{-1.4}$ km s^{-1} for the $M_{ej} = 1.6M_\odot$ data and $652(R/R_\odot)^{-1.4}$ km s^{-1} for the $M_{ej} = 2.6M_\odot$ data. The velocities for both ejecta mass conditions follow a similar gradient to earlier work presented in TT98 and Tauris (2015), although it is not quite as steep as the -1.9 power-law of Liu et al. (2015). The overall scaling differs from previous work, however. The early estimate from TT98 of the impact velocities used a value of $\eta \approx 0.5$ (see Equation 4), whereas fits in Tauris (2015) and Liu et al. (2015) sit closer to $\eta \approx 0.2$. Our impact velocities lie in between these values, corresponding to $\eta \approx 0.3$. At the point of measurement of v_{im}, there will already be a small contribution in the measured v_{im} from the growing radial velocity component due to the eccentricity of the new orbit. However, even if we were to define the measured impact velocity to be earlier (before the shock convergence at the far side of the star), this still produces values of v_{im} that are larger than those seen in Liu et al. (2015). We consider a possible cause of differences in results in Section 3.3.1.

Finally, we also consider the effect of drag from the remaining material on the companion velocity, noting that there is still a non-negligible density of gas interior to the ejecta shell. For a conservative estimate of this drag force from the innermost ejecta, we neglect any outward velocity of this gas, and take a density of 10^{-3} g cm^{-3} in this material. With these values, the drag force on the companion will be

$$F_{drag} = \frac{1}{2}\rho v_2^2 C_{drag} A_2 \approx 2 \times 10^{28} \text{ N}, \qquad (7)$$

for $v_2 = 300$ km s^{-1}, where we approximate the drag coefficient of the star with a solid sphere value of $C_{drag} \approx 0.5$. For a companion mass of $M_2 = 1M_\odot$ the acceleration associated with this drag is therefore only 10^{-5} km s^{-2}. Although small, drag induced by the lower-velocity ejecta may appreciably alter the final velocity of the companion when integrated over a long timescale.

3.3.1 Ejecta profiles
We investigate in more detail our ejecta profiles as a potential cause of the discrepancy between our impact velocities and those of Liu et al. (2015). Previous work, such as that of Liu et al. (2015), has often initialised the ejecta with the assumption that it is in a homologous expansion by the time it impacts the companion, so that, for a given t, $v \propto R$. The density and velocity profiles in this ejecta are constructed from broken power-law fits to analytic treatments of the shock through the progenitor. These treatments have, in particular, been based on the polytropic envelopes (or one-dimensional structure models) of supergiant stars, and the power-law fits are to the (small and large R) asymptotic limits of a varying density gradient in the ejecta (Chevalier and Soker 1989; Matzner and McKee 1999).

In Figure 10, we show the variation of ejecta density and velocity as a function of radius from the centre of mass (by averaging the SPH particles over concentric shells) and compare with an analytic function from the equations used in Liu et al. (2015). We also show, in the bottom panel of Figure 10, the distribution of velocity over mass in the ejecta. It is clear from Figure 10 that in the ejecta from our helium star models we have a shallower density gradient through much of the outer regions compared with the power-law profiles. In this outer ejecta, the velocity and density are also higher in our models. As the impact velocity has a strong dependence on this high-velocity ejecta (Liu et al. 2015), this can explain increased impact velocities seen in our simulations.

Finally, we examined the ejecta for large-scale asymmetries by determining the shell-averaged radial profiles of density and velocity in hemispheres corresponding to the directions toward and away from the companion star. We found that the values in either direction agreed to within a few per cent, and therefore do not produce a discernible difference on the logarithmic plots in Figure 10.

3.4 Properties of the larger-scale SNR
The amount of accretion on the companion has previously been shown to decrease with increasing shell velocity (Fryxell and Arnett 1981); therefore, the high ejecta velocities in Type Ibc supernovae lead us to expect little pollution of the companion with supernova material. Indeed, we find negligible pollution of the companion star. The converse - pollution of the SNR with material from the companion - can be appreciable. A few $10^{-2}M_\odot$ of hydrogen-

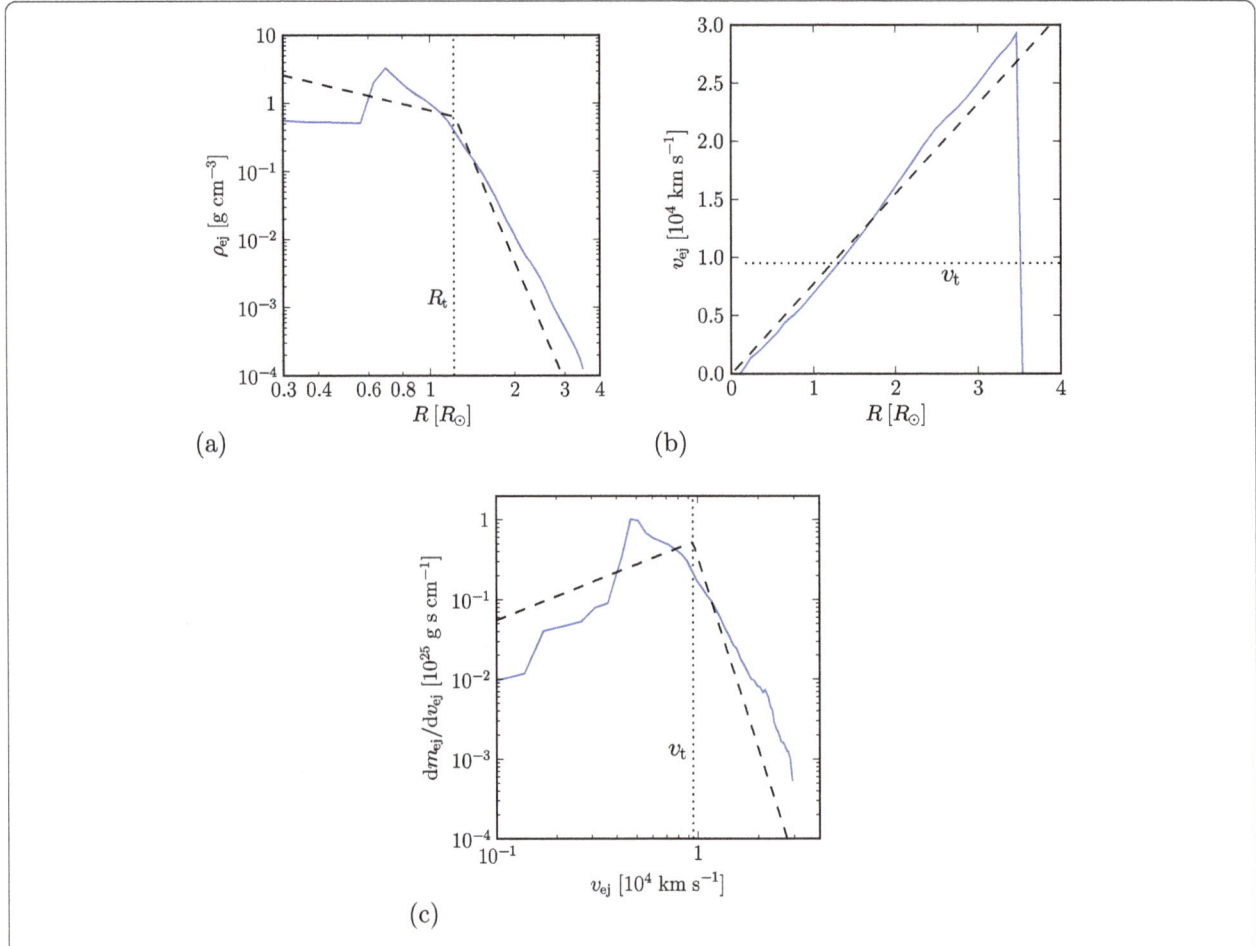

Figure 10 Blue lines show the profiles of (a) density, (b) magnitude of velocity and (c) the mass distribution of velocity within the ejecta for our simulations (binned in radial shells form the centre of mass of the ejecta). Black dashed lines show the power-law profiles used in Liu et al. (2015). Both cases were calculated for a time of 90 s after the supernova. The transition velocity (and radius at which this occurs) in the Liu et al. (2015) profiles are given as dotted lines.

rich material may be lost from the companion by the passing shell in our simulations, which may be detectable as an asymmetry in the metallicity of the SNR on the side of the companion.

Figure 11 shows a 3D rendering of the SNR and companion at 10^3 s after the moment of the supernova. At this point, a hole has been created in the passing shell due to the presence of the companion, which is seen to persist at later times. The hole in the ejecta caused by the companion is approximately 30 degrees in size for the minimum orbital separations.

Even if an ejecta hole cannot be detected morphologically, the presence of a hole in SNR ejecta may allow the inference of a companion from a burst of radiation generated during the impact with the companion, which can escape through the less optically thick region of the companion shadow cone (Kasen 2010). The hole may persist to late stages of the SNR despite some amount of refilling

due to the subsequent rarefaction wave along with hydrodynamic instabilities (García-Senz et al. 2012; Kasen 2010).

Not only do we observe a hole in the SNR due to the companion star, but we also see an increase in the density in a ring surrounding the hole, as shown most clearly in Figure 11. As shell material impacts the outer part of the companion star, where material is stripped and swept up with the ejecta, this ring of gas is also compressed in contrast with the freely expanding ejecta that do not interact with the companion. Aside from augmentation of the early supernova light curve, our results also suggest that ring-like enhancements in density of the SNR could indicate the presence of a companion star. Ring-like structures may be easier to detect than a hole in the SNR as the enhancement in density may also be associated with an increase in radiative losses in the ring.

As the full composition (the mass fraction of each species as determined by the MESA model) is recorded for

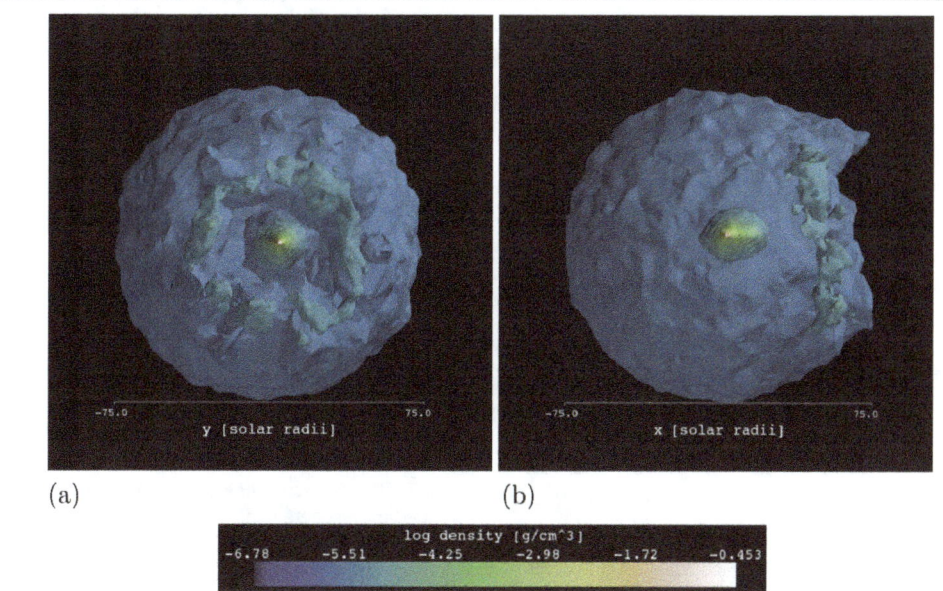

Figure 11 A 3D rendering of the gas density in the system 10^3 s after the supernova, viewed down the x-axis (a; the original axis of the binary) and y-axis (b). The companion is still distorted due to the impact, and has produced a hole in the expanding ejecta. We used the software Mayavi2 (Ramachandran and Varoquaux 2011) for the visualisation.

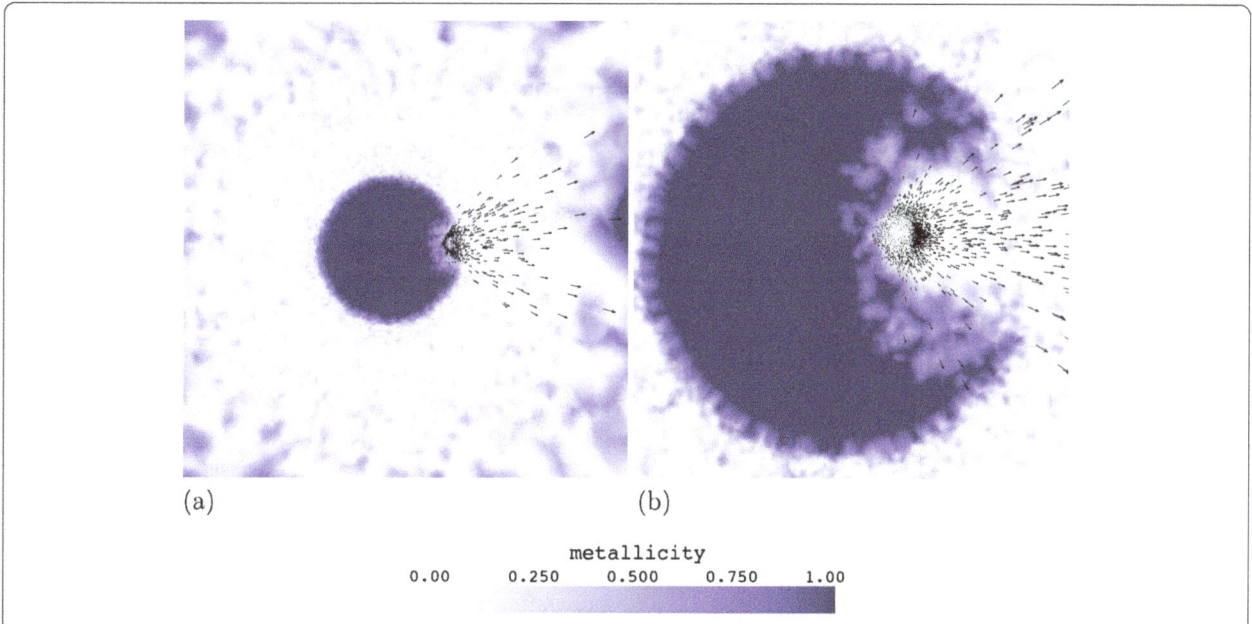

Figure 12 A $30R_\odot$ by $30R_\odot$ slice through the x-y (initial orbital) plane showing the mean metallicity ($1 - X_H - X_{He}$) within $1R_\odot$ of the plane at (a) 733 s and (b) 2028 s after a supernova in a $3M_\odot$ primary at an orbital separation of $4R_\odot$. Black vectors show samples of the momentum of the material unbound from the companion star projected onto the orbital plane, as in Figure 5.

each SPH particle, we are able to trace the dispersion of this material from the progenitor in the subsequent SNR. We do not, however, calculate the changes in composition during the supernova itself; as much of this process involves transmutation of one metal species to another in the stellar interior (where the metallicity remains $Z \sim 1$), we therefore limit the present discussion to the overall metallicity of the material.

Figure 12 illustrates that the metallicity of the SNR is highest in the innermost regions, where the ejecta represents material nearest the core of the supernova progenitor. A hole develops in this high-metallicity ejecta, at first primarily due to the shadow of the companion star (panel (a)). At later times (panel (b)), the ablation of companion material further reduces the metallicity of a large fraction of the inner part of SNR in the direction of the companion. The orbital motion of the companion star within the inner SNR during this longer period of ablation can also enlarge the region over which the gas is enriched with hydrogen.

3.5 Post-impact evolution of the companion

Following the stripping and ablation of mass from the outer layers of the companion star, we used AMUSE to investigate the stellar evolution of the companion and compare with an unperturbed stellar model evolving from the main-sequence. As we associate composition with each SPH particle from the original stellar model, we were able to convert the final SPH state of the companion back to a one-dimensional structure model by an inversion of the method to construct the SPH model outlined in Section 2.2. After the model is loaded back into MESA, we continue the stellar evolution and compared with an undisturbed companion model.

As we find negligible contamination of the companion with ejecta material, the difference in evolution is effectively due to the reduction in mass of the star. A $1M_\odot$ star with metallicity $Z = 0.02$ evolves to the through to a carbon-oxygen WD at 12.1 Gyr in MESA. On the other hand, the $1M_\odot$ model which has lost $0.04M_\odot$ of material from the supernova impact reaches this stage at a later age of 14.0 Gyr. Although the final age of the stars is noticeably different, there is little evolutionary difference between the two models on an HR diagram. It may, therefore, be difficult to distinguish a companion that has lost part of its envelope due to a supernova from $T_{\rm eff}$ and L alone. Nevertheless, the stripping and contamination in the outer layers of the star still has the potential to produce differences in chemical abundances that are spectrally distinguishable from the coeval stellar population (see also, for example, Pan et al. 2012).

More immediately, after the impact of the shell on the companion, a large amount of thermal energy is deposited in the outer layers of the star, which will dramatically affect its appearance over approximately the thermal timescale of the outer layers. To investigate this in more detail, in Figure 13 we plot the density and internal energy of a companion star subjected to a supernova in a $3M_\odot$ primary star at a separation of $4R_\odot$.

For this example, the total excess internal energy, integrated over the spherical shell $\gtrsim 0.6R_\odot$, is $\Delta U \approx 10^{47}$ erg. If this were instantaneously converted to radiation, the excess energy would be released on a diffusion timescale, which we calculate to be ~ 10 yr in these low-density outer layers of the star.[g]

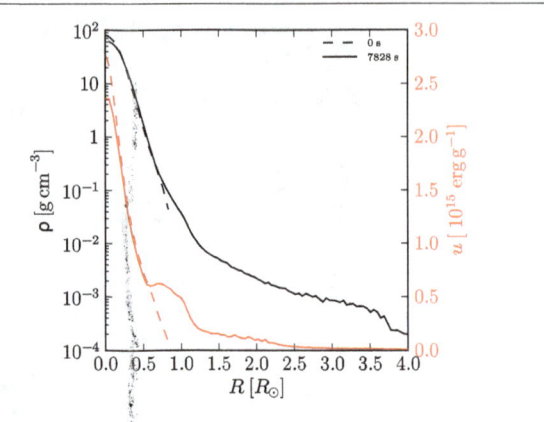

Figure 13 **Density (left-hand axis) and internal energy (right-hand axis) of the companion star before (dashed lines) and 7828 s after (solid lines) the supernova explosion.** These properties were determined by averaging across radial shells concentric with the centre of mass of the companion. All SPH particles bound to the companion were within a radius of $4.5R_\odot$.

The thermal timescale of the diffuse outer layers of such an expanded star is shorter than the canonical solar thermal timescale, and is of the order $t_{\rm therm} \approx 10^3 \sim 10^4$ yr (Marietta et al. 2000; Podsiadlowski 2003). Nevertheless, as the thermal timescale is still orders of magnitude larger than the diffusion timescale, we expect the luminosity due to this excess energy to be limited by the former.

A rough estimate of the luminosity can then be found from $\Delta U / t_{\rm therm} \approx 10^2 \sim 10^3 L_\odot$. In practice, however, the luminosity will gradually decline from a peak value to the main-sequence luminosity[h] over roughly the thermal timescale (Podsiadlowski 2003).

Finally, we note that in the example of Figure 13, the furthest extent of the SPH particles in the companion was $\sim 4R_\odot$. If the companion star were to continue expanding to a larger extent, its radius would encompass the neutron star in systems that remain bound, complicating the subsequent evolution of the system. This could increase the likelihood of accretion of material onto the neutron star, or even result in a merger between the neutron star and the companion. The interaction may also re-circularise the orbit after an eccentricity was gained from a kick to the neutron star during the supernova.

4 Discussion and conclusions

For supernovae in close binaries, the impact of the ejecta shell can have non-negligible effects on the mass and velocity of the companion star. The change in momentum of the companion is used in predictions of the final velocities of runaway stars from supernova-dissociated binaries, as seen in the recent work of Tauris (2015). These predictions are important for determining the level of contamination from these stars in searches for hypervelocity stars from other origins, such as the Hills mechanism with the super-massive black hole in the Galactic Centre (Hills 1988; Yu and Tremaine 2003).

We have performed SPH simulations of supernovae in close binaries to study the consequences of the shell impact on the companion. The overall hydrodynamic phenomena and trends we observe during these simulations are broadly consistent with previous studies of Type Ia (Liu et al. 2012; Marietta et al. 2000; Pakmor et al. 2008; Pan et al. 2012), Type Ibc (Liu et al. 2015) and Type II (Hirai et al. 2014) supernovae. In addition, we find that the gradient in the impact velocity predicted by Wheeler et al. (1975) matches our results well, with some modification of the η parameter representing the total momentum received by the companion.

While this work was in preparation, Liu et al. (2015) presented work on the effect of a Type Ibc supernova shell impacting a companion star. As with Liu et al. (2015), we find that the magnitude of mass loss and impact velocity of the companion is less than early estimates. However, the velocity induced onto the companion due to the shell impact in their work is a factor of 1.5∼2 lower than our results. Although it is not straightforward to separate the causes of such discrepancies, there are a number of differences between our calculations. One is the structure of the companion star, which is known to affect both the mass loss and impact velocity results (Liu et al. 2012, 2015; Meng et al. 2007; Pan et al. 2012). A more notable difference is that in our simulations the shell is naturally formed from the supernova explosion mechanism, as opposed to the introduction of the supernova ejecta by an analytic description. This results in a different ejecta structure, and more momentum carried in the leading edge of the ejecta, which is important in determining the final impact velocity (Liu et al. 2015).

Using the predictions from our simulations, we return to the question of runaway velocities of the components of supernova-disrupted binaries considered in TT98 and Tauris (2015). We have created a `python` code that calculates the final speeds derived by TT98 in order to investigate the analytic predictions with our simulation results. In this Monte Carlo code, an impulsive increase in velocity, $\mathbf{w}$, is imposed to the neutron star, randomly oriented from an isotropic distribution over a sphere. This is achieved by mapping from a uniform random distribution over $t \in (0,1]$ to $2\pi t$ for the angle ϕ, and from a uniform random distribution over $u \in [0,1]$ to $\cos^{-1}(2u - 1)$ for the angle θ. Figure 14 shows a comparison of the distributions of speeds with (red) and without (blue) the effect of applying $\mathbf{v}_{\mathrm{im}}$ and mass loss in the companion star.

From Figure 14, it is evident that, although adding an impact velocity to the companion (perpendicular to its orbital velocity) increases the minimum companion speed, it also in fact reduces the maximum companion speed. To clarify the discrepancies in the distributions that occur when

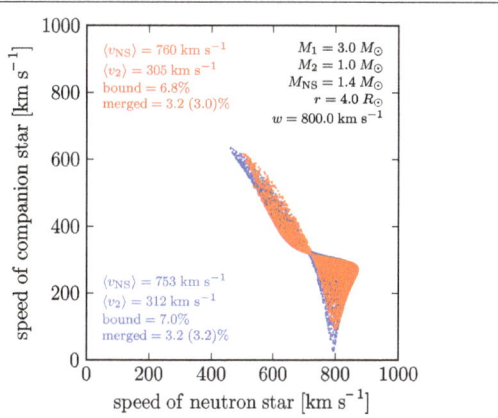

Figure 14 Comparison of Monte Carlo sampling over NS kick orientation for the case with no impact effects on the companion (blue) and with impact effects as determined from our simulations (red). Both cases show 10^4 samples of NS kick orientation with a uniform on distribution over the unit sphere. The magnitude of the NS kick velocity, w, is fixed at 800 km s^{-1} throughout. Mean values of the runaway velocities, as well as percentages of cases where the binary components remain bound and merged (and, in parentheses, merged cases that were calculated as bound). Neither bound nor merged cases appear in these distributions.

adding $\mathbf{v}_{\mathrm{im}}$, we consider the effect of NS kick angles on the final velocity of the companion star in disrupted binaries in Figure 15, analogous to Figure 4 in Tauris (2015). The white regions for high θ in each panel represent binaries that remain bound after the NS kick (and thus the runaway velocity is undefined). The grey regions represent NS kick angles for which the NS and companion star merge after the supernova. It can be seen from the lower panel that the effect of applying an impact velocity to the companion star can stabilise the systems where the NS kick is counter-aligned with the NS orbital velocity. In fact, the small region of parameter space giving large values of v_2 at $\phi = 0$ and high θ is removed after adding $\mathbf{v}_{\mathrm{im}}$ (due to these systems now remaining bound). This explains the potentially counter-intuitive result that by adding an additional velocity to the companion star in fact reduces the maximum possible velocity of runaway stars.

The main results of our work are as follows:

- We follow the supernova from just after the core bounce in a helium star generated from a stellar evolution model. Exploding a supernova in such a model produces an ejecta profile that is different from that used in previous work, which has employed an analytic function for the ejecta distribution derived from the theory of shocks travelling through a one-dimensional atmosphere. The progenitor model used here is still somewhat artificial in construction, with a constant mass loss parameter late in its evolution. As understanding of Type Ibc progenitors

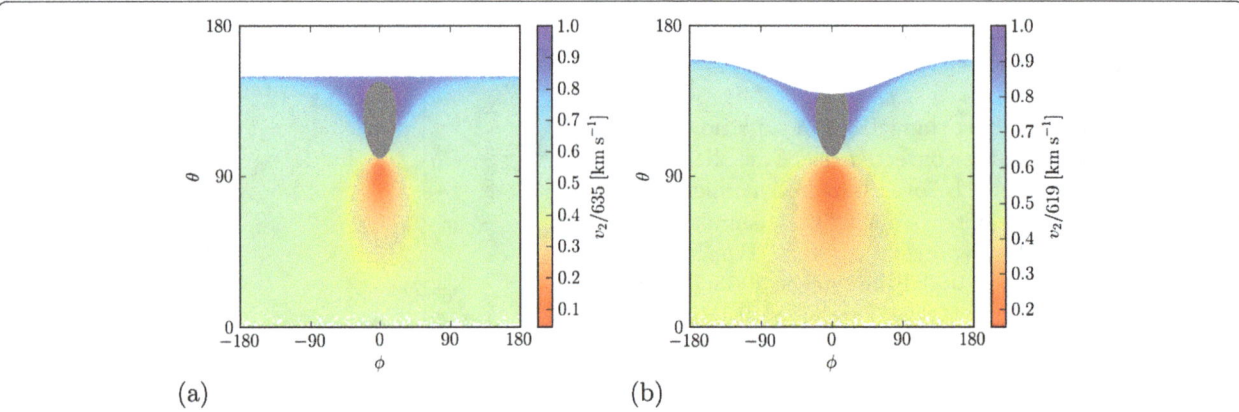

Figure 15 Distribution of NS kick angles from the *x*-axis (aligned with the pre-kick NS orbital vector) for the same parameters as the Monte Carlo runs shown in Figure 14 (but with 2 × 10⁵ samples to better fill the space in angles). The span of θ over $\phi = 0$ defines the orbital plane, where $\theta = 0$ is for a kick aligned with the NS orbital velocity vector. Panel (a) shows the case of no impact effects on the companion (blue distribution in Figure 14) and panel (b) shows the case where impact effects determined from our simulations are included (red distribution in Figure 14). Colours represent the magnitude of the companion star from disrupted binaries (as a fraction of the maximum runaway velocity). Grey shows cases where the NS and companion star merge. The white regions for large θ are cases where the binary remains bound and so there is no runaway companion.

improves (for some recent investigations, see Kim et al. 2015), future work would benefit from a more realistic progenitor model by modelling the mass loss processes in detail.

- We have investigated the mass removed from the companion in the very close binary separations seen in Type Ibc supernovae, as well as the net change in momentum of the companion star due to the shell impact and later ablation of material. We show that an extrapolation of results from Type Ia supernova simulations do not provide a good fit to the Type Ibc scenarios considered here, and we provide updated fits to the distance-dependence of these results. In agreement with Liu et al. (2015), we find lower values of the removed mass and impact velocity of the companion compared to earlier estimates; however, we find generally larger impact velocities. Discrepancies in the results are due not only to differences in companion models but also to differences in the distribution of momentum in the ejecta.

- We investigated the morphology of the SNR shortly after the shell has passed the companion, as well as the pollution of the SNR with material stripped from the companion, which, for the case of Type Ibc supernovae, may be a detectable fraction of the total mass in the ejecta (several $10^{-2} M_\odot$ out of $\sim 2 M_\odot$). The metallicity of the SNR is found to be highest in the inner regions of the SNR, and in this region the ablation of hydrogen from the outer layers of the companion star can dilute the metallicity on the side

of the SNR facing the companion, resulting in a strong asymmetry in metallicity in the orbital plane.

- The companion star is additionally found to modify the morphology of the SNR in two distinct ways: as anticipated, a hole forms in the SNR on the side of the companion; also, an increase in the SNR density is seen in a ring around the hole, which may enhance the luminosity in SNR observations.

- We have also considered subsequent state of the companion after the shell impact and removal of mass during the shell impact, and have confirmed that the luminosity of the star can be orders of magnitude larger than the main-sequence luminosity during the release of thermal energy from the shock-heated outer layers.

Competing interests
The authors declare that they have no competing interests.

Authors' contributions
AR ran the simulations, analysed the data and wrote the draft of this paper. SPZ had the idea that initiated this work and assisted with the interpretation of the results. EMR assisted with the interpretation of the impact momenta and velocity results and the runaway star Monte Carlo results. All of the authors took part in discussions concerning the results and contributed corrections and improvements on the draft of the manuscript. All authors read and approved the final manuscript.

Acknowledgements
We are grateful to Nathan de Vries for his assistance with an early version of the supernova explosion code in AMUSE, and to Arjen van Elteren and Inti Pelupessy for AMUSE code development. We also thank the two anonymous referees, whose comments helped improve this manuscript. This work was supported by the Netherlands Research Council (NWO grant numbers 612.071.305 [LGM] and 639.073.803 [VICI]), the Netherlands Research School for Astronomy (NOVA), the Interuniversity Attraction Poles Programme initiated by the Belgian Science Policy Office (IAP P7/08 CHARM), and by the European

Union's Horizon 2020 Research and Innovation programme under grant agreement No. 671564.

Endnotes

[a] More recently, the 'Bethe', B, has also been proposed as an equivalent unit in honour of Hans Bethe's work on supernovae (Weinberg 2006; Woosley and Heger 2007).

[b] As shown by Pan et al. (2012) for the Type Ia case, both orbital motion and rotation produce no substantial difference in the impact velocity gained by the companion. However, it is possible that a high rotation rate in the companion can help unbind a small additional amount of shock-heated material from the surface of the companion due to the additional rotational energy.

[c] www.amusecode.org.

[d] This 'canonical' value of the solar metallicity may be an overestimate; see Asplund et al. (2009) for a review.

[e] Randomisation of the angular orientation of the particles has the undesirable effect of the additional shot noise it generates in the initial density distribution; however, the further step of damped relaxation used here will ultimately result in a glass-like configuration.

[f] This smooths thermal energy discontinuities and is used in capturing the vortices seen in KHI. However, there has been some debate on the causes of KHI suppression in SPH; see, for example, the discussion in Gabbasov et al. (2014).

[g] To obtain this value, we integrated from $0.6R_\odot$ to the surface of the star assuming a Thomson cross-section.

[h] In fact, the luminosity from nuclear energy generation can drop lower than the main-sequence value, due to the reduction in central temperature and pressure following the initial expansion (Marietta et al. 2000).

References

Asplund, M, Grevesse, N, Sauval, AJ, Scott, P: The chemical composition of the Sun. Annu. Rev. Astron. Astrophys. 47, 481-522 (2009). doi:10.1146/annurev.astro.46.060407.145222. arXiv:0909.0948

Bersten, MC, Benvenuto, OG, Folatelli, G, Nomoto, K, Kuncarayakti, H, Srivastav, S, Anupama, GC, Quimby, R, Sahu, DK: iPTF13bvn: the first evidence of a binary progenitor for a type Ib supernova. Astron. J. 148, 68 (2014). doi:10.1088/0004-6256/148/4/68. arXiv:1403.7288

Bruenn, SW, Mezzacappa, A, Hix, WR, Lentz, EJ, Bronson Messer, OE, Lingerfelt, EJ, Blondin, JM, Endeve, E, Marronetti, P, Yakunin, KN: Axisymmetric ab initio core-collapse supernova simulations of 12-25 M_{Sun} stars. Astrophys. J. 767, L6 (2013). doi:10.1088/2041-8205/767/1/L6. arXiv:1212.1747

Cao, Y, Kulkarni, SR, Howell, DA, Gal-Yam, A, Kasliwal, MM, Valenti, S, Johansson, J, Amanullah, R, Goobar, A, Sollerman, J, Taddia, F, Horesh, A, Sagiv, I, Cenko, SB, Nugent, PE, Arcavi, I, Surace, J, Woźniak, PR, Moody, DI, Rebbapragada, UD, Bue, BD, Gehrels, N: A strong ultraviolet pulse from a newborn type Ia supernova. Nature 521, 328-331 (2015). doi:10.1038/nature14440. arXiv:1505.05158

Chevalier, RA, Soker, N: Asymmetric envelope expansion of supernova 1987A. Astrophys. J. 341, 867-882 (1989). doi:10.1086/167545

Church, RP, Dischler, J, Davies, MB, Tout, CA, Adams, T, Beer, ME: Mass transfer in eccentric binaries: the new oil-on-water smoothed particle hydrodynamics technique. Mon. Not. R. Astron. Soc. 395, 1127-1134 (2009). doi:10.1111/j.1365-2966.2009.14619.x. arXiv:0902.3509

Colgate, SA: Ejection of companion objects by supernovae. Nature 225, 247-248 (1970). doi:10.1038/225247a0

Couch, SM, O'Connor, EP: High-resolution three-dimensional simulations of core-collapse supernovae in multiple progenitors. Astrophys. J. 785, 123 (2014). doi:10.1088/0004-637X/785/2/123. arXiv:1310.5728

Couch, SM, Ott, CD: The role of turbulence in neutrino-driven core-collapse supernova explosions. Astrophys. J. 799, 5 (2015). doi:10.1088/0004-637X/799/1/5. arXiv:1408.1399

de Vries, N, Portegies Zwart, S, Figueira, J: The evolution of triples with a Roche lobe filling outer star. Mon. Not. R. Astron. Soc. 438, 1909-1921 (2014). doi:10.1093/mnras/stt1688. arXiv:1309.1475

Dessart, L, Blondin, S, Hillier, DJ, Khokhlov, A: Constraints on the explosion mechanism and progenitors of type Ia supernovae. Mon. Not. R. Astron. Soc. 441, 532-550 (2014). doi:10.1093/mnras/stu598. arXiv:1310.7747

Duquennoy, A, Mayor, M: Multiplicity among solar-type stars in the solar neighbourhood. II - Distribution of the orbital elements in an unbiased sample. Astron. Astrophys. 248, 485-524 (1991)

Eggleton, P: Evolutionary Processes in Binary and Multiple Stars. Cambridge University Press, Cambridge (2006)

Eggleton, PP: The evolution of low mass stars. Mon. Not. R. Astron. Soc. 151, 351 (1971)

Eggleton, PP: Approximations to the radii of Roche lobes. Astrophys. J. 268, 368 (1983). doi:10.1086/160960

Eldridge, JJ, Fraser, M, Maund, JR, Smartt, SJ: Possible binary progenitors for the type Ib supernova iPTF13bvn. Mon. Not. R. Astron. Soc. 446, 2689-2695 (2015). doi:10.1093/mnras/stu2197. arXiv:1408.4142

Fremling, C, Sollerman, J, Taddia, F, Ergon, M, Valenti, S, Arcavi, I, Ben-Ami, S, Cao, Y, Cenko, SB, Filippenko, AV, Gal-Yam, A, Howell, DA: The rise and fall of the type Ib supernova iPTF13bvn. Not a massive Wolf-Rayet star. Astron. Astrophys. 565, A114 (2014). doi:10.1051/0004-6361/201423884. arXiv:1403.6708

Fryxell, BA, Arnett, WD: Hydrodynamic effects of a stellar explosion on a binary companion star. Astrophys. J. 243, 994-1002 (1981). doi:10.1086/158664

Gabbasov, R, Klapp-Escribano, J, Suárez-Cansino, J, Sigalotti, L: Numerical simulations of the Kelvin-Helmholtz instability with the Gadget-2 SPH code. In: Klapp, J, Medina, A (eds.) Experimental and Computational Fluid Mechanics. Environmental Science and Engineering, pp. 291-298. Springer, Berlin (2014). ISBN 978-3-319-00115-9. doi:10.1007/978-3-319-00116-6_24

García-Senz, D, Badenes, C, Serichol, N: Is there a hidden hole in type Ia supernova remnants? Astrophys. J. 745, 75 (2012). doi:10.1088/0004-637X/745/1/75. arXiv:1110.4267

González Hernández, JI, Ruiz-Lapuente, P, Tabernero, HM, Montes, D, Canal, R, Méndez, J, Bedin, LR: No surviving evolved companions of the progenitor of SN 1006. Nature 489, 533-536 (2012). doi:10.1038/nature11447. arXiv:1210.1948

Hanke, F, Müller, B, Wongwathanarat, A, Marek, A, Janka, H-T: SASI activity in three-dimensional neutrino-hydrodynamics simulations of supernova cores. Astrophys. J. 770, 66 (2013). doi:10.1088/0004-637X/770/1/66. arXiv:1303.6269

Hills, JG: Hyper-velocity and tidal stars from binaries disrupted by a massive Galactic black hole. Nature 331, 687-689 (1988). doi:10.1038/331687a0

Hirai, R, Sawai, H, Yamada, S: The outcome of supernovae in massive binaries; removed mass, and its separation dependence. Astrophys. J. 792, 66 (2014). doi:10.1088/0004-637X/792/1/66. arXiv:1404.4297

Hurley, JR, Pols, OR, Tout, CA: Comprehensive analytic formulae for stellar evolution as a function of mass and metallicity. Mon. Not. R. Astron. Soc. 315, 543-569 (2000). doi:10.1046/j.1365-8711.2000.03426.x. arXiv:astro-ph/0001295

Janka, H-T: Explosion mechanisms of core-collapse supernovae. Annu. Rev. Nucl. Part. Sci. 62, 407-451 (2012). doi:10.1146/annurev-nucl-102711-094901. arXiv:1206.2503

Kane, J, Drake, RP, Remington, BA: An evaluation of the Richtmyer-Meshkov instability in supernova remnant formation. Astrophys. J. 511, 335-340 (1999). doi:10.1086/306685

Kasen, D: Seeing the collision of a supernova with its companion star. Astrophys. J. 708, 1025-1031 (2010). doi:10.1088/0004-637X/708/2/1025. arXiv:0909.0275

Kerzendorf, WE, Yong, D, Schmidt, BP, Simon, JD, Jeffery, CS, Anderson, J, Podsiadlowski, P, Gal-Yam, A, Silverman, JM, Filippenko, AV, Nomoto, K, Murphy, SJ, Bessell, MS, Venn, KA, Foley, RJ: A high-resolution spectroscopic search for the remaining donor for Tycho's supernova. Astrophys. J. 774, 99 (2013). doi:10.1088/0004-637X/774/2/99. arXiv:1210.2713

Kim, H-J, Yoon, S-C, Koo, B-C: Observational properties of type Ib/c supernova progenitors in binary systems. Astrophys. J. 809, 131 (2015). doi:10.1088/0004-637X/809/2/131. arXiv:1506.06354

Kuncarayakti, H, Maeda, K, Bersten, MC, Folatelli, G, Morrell, N, Hsiao, EY, González-Gaitán, S, Anderson, JP, Hamuy, M, de Jaeger, T, Gutiérrez, CP, Kawabata, KS: Nebular phase observations of the type-Ib supernova iPTF13bvn favour a binary progenitor. Astron. Astrophys. 579, A95 (2015). doi:10.1051/0004-6361/201425604. arXiv:1504.01473

Kusenko, A, Segrè, G: Pulsar velocities and neutrino oscillations. Phys. Rev. Lett. **77**, 4872-4875 (1996). doi:10.1103/PhysRevLett.77.4872. arXiv:hep-ph/9606428

Lai, D, Rasio, FA, Shapiro, SL: Collisions and close encounters between massive main-sequence stars. Astrophys. J. **412**, 593-611 (1993). doi:10.1086/172946

Leonard, DC: Constraining the type Ia supernova progenitor: the search for hydrogen in nebular spectra. Astrophys. J. **670**, 1275-1282 (2007). doi:10.1086/522367. arXiv:0710.3166

Li, W, Bloom, JS, Podsiadlowski, P, Miller, AA, Cenko, SB, Jha, SW, Sullivan, M, Howell, DA, Nugent, PE, Butler, NR, Ofek, EO, Kasliwal, MM, Richards, JW, Stockton, A, Shih, H-Y, Bildsten, L, Shara, MM, Bibby, J, Filippenko, AV, Ganeshalingam, M, Silverman, JM, Kulkarni, SR, Law, NM, Poznanski, D, Quimby, RM, McCully, C, Patel, B, Maguire, K, Shen, KJ: Exclusion of a luminous red giant as a companion star to the progenitor of supernova SN 2011fe. Nature **480**, 348-350 (2011). doi:10.1038/nature10646. arXiv:1109.1593

Liu, ZW, Pakmor, R, Röpke, FK, Edelmann, P, Wang, B, Kromer, M, Hillebrandt, W, Han, ZW: Three-dimensional simulations of the interaction between type Ia supernova ejecta and their main sequence companions. Astron. Astrophys. **548**, A2 (2012). doi:10.1051/0004-6361/201219357. arXiv:1209.4458

Liu, Z-W, Tauris, TM, Roepke, FK, Moriya, TJ, Kruckow, M, Stancliffe, RJ, Izzard, RG: The interaction of core-collapse supernova ejecta with a companion star. ArXiv e-prints, arXiv:1509.03633 (2015)

Lopez, LA, Ramirez-Ruiz, E, Castro, D, Pearson, S: The Galactic supernova remnant W49B likely originates from a jet-driven, core-collapse explosion. Astrophys. J. **764**, 50 (2013). doi:10.1088/0004-637X/764/1/50. arXiv:1301.0618

Lundqvist, P, Nyholm, A, Taddia, F, Sollerman, J, Johansson, J, Kozma, C, Lundqvist, N, Fransson, C, Garnavich, PM, Kromer, M, Shappee, BJ, Goobar, A: No trace of a single-degenerate companion in late spectra of supernovae 2011fe and 2014J. Astron. Astrophys. **577**, A39 (2015). doi:10.1051/0004-6361/201525719. arXiv:1502.00589

Maoz, D, Mannucci, F, Nelemans, G: Observational clues to the progenitors of type Ia supernovae. Annu. Rev. Astron. Astrophys. **52**, 107-170 (2014). doi:10.1146/annurev-astro-082812-141031. arXiv:1312.0628

Marietta, E, Burrows, A, Fryxell, B: Type IA supernova explosions in binary systems: the impact on the secondary star and its consequences. Astrophys. J. Suppl. Ser. **128**, 615-650 (2000). doi:10.1086/313392. arXiv:astro-ph/9908116

Maruyama, T, Kajino, T, Yasutake, N, Cheoun, M-K, Ryu, C-Y: Asymmetric neutrino emission from magnetized proto-neutron star matter including hyperons in relativistic mean field theory. Phys. Rev. D **83**(8), 081303 (2011). doi:10.1103/PhysRevD.83.081303. arXiv:1009.0976

Mattila, S, Lundqvist, P, Sollerman, J, Kozma, C, Baron, E, Fransson, C, Leibundgut, B, Nomoto, K: Early and late time VLT spectroscopy of SN 2001el - progenitor constraints for a type Ia supernova. Astron. Astrophys. **443**, 649-662 (2005). doi:10.1051/0004-6361:20052731. arXiv:astro-ph/0501433

Matzner, CD, McKee, CF: The expulsion of stellar envelopes in core-collapse supernovae. Astrophys. J. **510**, 379-403 (1999). doi:10.1086/306571. arXiv:astro-ph/9807046

Meng, X, Chen, X, Han, Z: The impact of type Ia supernova explosions on the companions in a binary system. Publ. Astron. Soc. Jpn. **59**, 835-840 (2007). doi:10.1093/pasj/59.4.835

Milisavljevic, D, Soderberg, AM, Margutti, R, Drout, MR, Howie Marion, G, Sanders, NE, Hsiao, EY, Lunnan, R, Chornock, R, Fesen, RA, Parrent, JT, Levesque, EM, Berger, E: SN 2012au: a golden link between superluminous supernovae and their lower-luminosity counterparts. Astrophys. J. Lett. **770**, L38 (2013). doi:10.1088/2041-8205/770/2/L38. arXiv:1304.0095

Olling, RP, Mushotzky, R, Shaya, EJ, Rest, A, Garnavich, PM, Tucker, BE, Kasen, D, Margheim, S, Filippenko, AV: No signature of ejecta interaction with a stellar companion in three type Ia supernovae. Nature **521**, 332-335 (2015). doi:10.1038/nature14455

Pakmor, R, Röpke, FK, Weiss, A, Hillebrandt, W: The impact of type Ia supernovae on main sequence binary companions. Astron. Astrophys. **489**, 943-951 (2008). doi:10.1051/0004-6361:200810456. arXiv:0807.3331

Pakmor, R, Edelmann, P, Röpke, FK, Hillebrandt, W: Stellar GADGET: a smoothed particle hydrodynamics code for stellar astrophysics and its application to type Ia supernovae from white dwarf mergers. Mon. Not. R. Astron. Soc. **424**, 2222-2231 (2012). doi:10.1111/j.1365-2966.2012.21383.x. arXiv:1205.5806

Pan, K-C, Ricker, PM, Taam, RE: Impact of type Ia supernova ejecta on binary companions in the single-degenerate scenario. Astrophys. J. **750**, 151 (2012). doi:10.1088/0004-637X/750/2/151. arXiv:1203.1932

Paxton, B, Bildsten, L, Dotter, A, Herwig, F, Lesaffre, P, Timmes, F: Modules for experiments in stellar astrophysics (MESA). Astrophys. J. Suppl. Ser. **192**, 3 (2011). doi:10.1088/0067-0049/192/1/3. arXiv:1009.1622

Pelupessy, FI, van Elteren, A, de Vries, N, McMillan, SLW, Drost, N, Portegies Zwart, SF: The astrophysical multipurpose software environment. Astron. Astrophys. **557**, A84 (2013). doi:10.1051/0004-6361/201321252. arXiv:1307.3016

Podsiadlowski, P: On the evolution and appearance of a surviving companion after a type Ia supernova explosion. Mon. Not. R. Astron. Soc. submitted. arXiv:astro-ph/0303660 (2003)

Portegies Zwart, S, McMillan, S, Harfst, S, Groen, D, Fujii, M, Nualláin, BÓ, Glebbeek, E, Heggie, D, Lombardi, J, Hut, P, Angelou, V, Banerjee, S, Belkus, H, Fragos, T, Fregeau, J, Gaburov, E, Izzard, R, Jurić, M, Justham, S, Sottoriva, A, Teuben, P, van Bever, J, Yaron, O, Zemp, M: A multiphysics and multiscale software environment for modeling astrophysical systems. New Astron. **14**, 369-378 (2009). doi:10.1016/j.newast.2008.10.006. arXiv:0807.1996

Portegies Zwart, S, McMillan, SLW, van Elteren, E, Pelupessy, I, de Vries, N: Multi-physics simulations using a hierarchical interchangeable software interface. Comput. Phys. Commun. **183**, 456-468 (2013). doi:10.1016/j.cpc.2012.09.024. arXiv:1204.5522

Ramachandran, P, Varoquaux, G: Mayavi: 3D visualization of scientific data. Comput. Sci. Eng. **13**(2), 40-51 (2011)

Rasio, FA, Lombardi, JC Jr.: Smoothed particle hydrodynamics calculations of stellar interactions. J. Comput. Appl. Math. **109**, 213-230 (1999). arXiv:astro-ph/9805089

Rastegaev, DA: Multiplicity and period distribution of population II field stars in solar vicinity. Astron. J. **140**, 2013-2024 (2010). doi:10.1088/0004-6256/140/6/2013. arXiv:1009.4596

Rimoldi, A, Rossi, EM, Piran, T, Portegies Zwart, S: The fate of supernova remnants near quiescent supermassive black holes. Mon. Not. R. Astron. Soc. **447**, 3096-3114 (2015). doi:10.1093/mnras/stu2630. arXiv:1501.02819

Ruiz-Lapuente, P, Comeron, F, Méndez, J, Canal, R, Smartt, SJ, Filippenko, AV, Kurucz, RL, Chornock, R, Foley, RJ, Stanishev, V, Ibata, R: The binary progenitor of Tycho Brahe's 1572 supernova. Nature **431**, 1069-1072 (2004). doi:10.1038/nature03006. arXiv:astro-ph/0410673

Sana, H, de Mink, SE, de Koter, A, Langer, N, Evans, CJ, Gieles, M, Gosset, E, Izzard, RG, Le Bouquin, J-B, Schneider, FRN: Binary interaction dominates the evolution of massive stars. Science **337**, 444 (2012). doi:10.1126/science.1223344. arXiv:1207.6397

Schaefer, BE, Pagnotta, A: An absence of ex-companion stars in the type Ia supernova remnant SNR 0509-67.5. Nature **481**, 164-166 (2012). doi:10.1038/nature10692

Scheck, L, Plewa, T, Janka, H-T, Kifonidis, K, Müller, E: Pulsar recoil by large-scale anisotropies in supernova explosions. Phys. Rev. Lett. **92**(1), 011103 (2004). doi:10.1103/PhysRevLett.92.011103. arXiv:astro-ph/0307352

Scheck, L, Kifonidis, K, Janka, H-T, Müller, E: Multidimensional supernova simulations with approximative neutrino transport. I. Neutron star kicks and the anisotropy of neutrino-driven explosions in two spatial dimensions. Astron. Astrophys. **457**, 963-986 (2006). doi:10.1051/0004-6361:20064855. arXiv:astro-ph/0601302

Smartt, SJ: Progenitors of core-collapse supernovae. Annu. Rev. Astron. Astrophys. **47**, 63-106 (2009). doi:10.1146/annurev-astro-082708-101737. arXiv:0908.0700

Springel, V: The cosmological simulation code GADGET-2. Mon. Not. R. Astron. Soc. **364**, 1105-1134 (2005). doi:10.1111/j.1365-2966.2005.09655.x. arXiv:astro-ph/0505010

Suwa, Y, Yoshida, T, Shibata, M, Umeda, H, Takahashi, K: Neutrino-driven explosions of ultra-stripped type Ic supernovae generating binary neutron stars. ArXiv e-prints, arXiv:1506.08827 (2015)

Tauris, TM: Maximum speed of hypervelocity stars ejected from binaries. Mon. Not. R. Astron. Soc. **448**, L6-L10 (2015). doi:10.1093/mnrasl/slu189. arXiv:1412.0657

Tauris, TM, Takens, RJ: Runaway velocities of stellar components originating from disrupted binaries via asymmetric supernova explosions. Astron. Astrophys. **330**, 1047-1059 (1998)

Warren, DC, Blondin, JM: Three-dimensional numerical investigations of the morphology of type Ia SNRs. Mon. Not. R. Astron. Soc. **429**, 3099-3113 (2013). doi:10.1093/mnras/sts566. arXiv:1210.7790

Weinberg, S: A Bethe unit. Phys. World **19**(2), 17 (2006)

Wheeler, JC, Lecar, M, McKee, CF: Supernovae in binary systems. Astrophys. J. **200**, 145-157 (1975). doi:10.1086/153771

Wongwathanarat, A, Janka, H-T, Müller, E: Three-dimensional neutrino-driven supernovae: neutron star kicks, spins, and asymmetric ejection of nucleosynthesis products. Astron. Astrophys. **552**, A126 (2013). doi:10.1051/0004-6361/201220636. arXiv:1210.8148

Woosley, SE, Heger, A: Nucleosynthesis and remnants in massive stars of solar metallicity. Phys. Rep. **442**(1-6), 269-283 (2007). doi:10.1016/j.physrep.2007.02.009. The Hans Bethe Centennial Volume 1906-2006

Xue, Z, Schaefer, BE: Newly determined explosion center of Tycho's supernova and the implications for proposed ex-companion stars of the progenitor. Astrophys. J. **809**, 183 (2015). doi:10.1088/0004-637X/809/2/183. arXiv:1507.06347

Young, PA, Fryer, CL: Uncertainties in supernova yields. I. One-dimensional explosions. Astrophys. J. **664**, 1033-1044 (2007). doi:10.1086/518081. arXiv:astro-ph/0612698

Yu, Q, Tremaine, S: Ejection of hypervelocity stars by the (binary) black hole in the Galactic Center. Astrophys. J. **599**, 1129-1138 (2003). doi:10.1086/379546. arXiv:astro-ph/0309084

Zahn, J-P: Tidal friction in close binary stars. Astron. Astrophys. **57**, 383-394 (1977)

6

Adaptive techniques for clustered N-body cosmological simulations

Harshitha Menon[1*], Lukasz Wesolowski[1], Gengbin Zheng[1], Pritish Jetley[1], Laxmikant Kale[1], Thomas Quinn[2] and Fabio Governato[2]

Abstract

CHANGA is an N-body cosmology simulation application implemented using CHARM++. In this paper, we present the parallel design of CHANGA and address many challenges arising due to the high dynamic ranges of clustered datasets. We propose optimizations based on adaptive techniques. We evaluate the performance of CHANGA on highly clustered datasets: a $z \sim 0$ snapshot of a 2 billion particle realization of a 25 Mpc volume, and a 52 million particle multi-resolution realization of a dwarf galaxy. For the 25 Mpc volume, we show strong scaling on up to $128K$ cores of Blue Waters. We also demonstrate scaling up to $128K$ cores of a multi-stepping run of the 2 billion particle simulation. While the scaling of the multi-stepping run is not as good as single stepping, the throughput at $128K$ cores is greater by a factor of 2. We also demonstrate strong scaling on up to $512K$ cores of Blue Waters for two large, uniform datasets with 12 and 24 billion particles.

Keywords: computational cosmology; scalability; performance analysis; dark matter

1 Introduction

Simulating the process of cosmological structure formation with enough resolution to determine galaxy morphologies requires an enormous dynamic range in space and time. Star formation (SF) is concentrated in dense gas clouds the size of just a few parsecs, while the assembly of galaxies happens over billion of years, driven by large scale structures extending over megaparsecs.

Constraints on cosmology are tightest on scales of tens of megaparsecs and larger due to observations of the Cosmic Microwave Background, giving us detailed initial conditions (Ade et al. 2013); however, our knowledge of the nonlinear evolution of the Universe and of the properties of galaxies is still imperfect, because the detailed properties of Dark Matter (Brooks 2014) and of SF (Pontzen and Governato 2014) remain only partially understood. On the other hand, simulations of large volumes of the Universe (Davis et al. 1985; Springel et al. 2005), and of individual galaxies at high resolution (Guedes et al. 2011; Hopkins et al. 2013) have been fundamental in putting the standard

hierarchical Cold Dark Matter dominated model (ΛCDM) on a robust footing (Frenk and White 2012). Further understanding requires numerical simulations of increasing dynamical range, mass and spatial resolution and physical complexity, providing a powerful incentive to develop ever more sophisticated parallel codes (Vogelsberger et al. 2012; Kim et al. 2014).

Scaling such codes to large processor count requires overcoming not only spatial resolution challenges, but also large ranges in timescales. In this paper, we compare two approaches to handling this problem. The first approach involves using different time steps for different particles in relation to their dynamical time scales, leading to an algorithm that is challenging to parallelize effectively. An alternative approach, using a single, uniformly small time step for all particles, leads to more computation, but is simpler to parallelize.

This paper presents the design of CHANGA, a parallel *n-body + SPH* cosmology code for the simulation of astrophysical systems on a wide range of spatial and time scales. Most of the physical modules of CHANGA have been imported from the well established tree + SPH code GASOLINE and we refer the readers to the existing literature (Wadsley et al. 2004, 2008, Governato et al. 2014) for more details.

*Correspondence: gplkrsh2@illinois.edu
[1]Department of Computer Science, University of Illinois at Urbana-Champaign, Urbana, USA
Full list of author information is available at the end of the article

In this paper we focus on the optimizations implemented in CHaNGa that allow it to scale to large numbers of processors, and address the challenges brought on by the high dynamic ranges of clustered datasets. We will begin with an overview of the field and place the approach taken by CHaNGa in the context of published material. We then briefly summarize some specific features of CHaNGa (some imported from GASOLINE), including force softening, smooth particle hydrodynamics, star formation, and multi-stepping. The parallel design of CHaNGa, based on over-decomposition of work, allowing a parallel run-time system to dynamically balance the load, is presented next, along with descriptions of the phases of the computation. To set the context, and a baseline, for the optimizations presented, we first describe the single-stepping performance on relatively uniform data-sets. The clustered data-sets are then introduced, and a series of performance challenges along with strategies and optimizations developed to overcome them are described. These are accompanied by detailed performance analysis using the Projections performance visualization tool (Kalé and Sinha 1993). As of Spring 2014 our performance evaluation runs demonstrate scalability to over 131,000 processor cores on NCSA's Blue Waters and up to a 3× speedup over the single-stepping algorithm.[a]

2 Current state of the art

Because of the computational challenge and the non-trivial algorithms involved, cosmological N-body simulations have been an extensively studied topic over the years. In order to frame our work in CHaNGa, we review some of the recent successes in scaling cosmological simulations on the current generations of supercomputers. However, direct comparison of the absolute performance among different codes is difficult. Different choices of accuracy criteria for the force evaluations and the time integration will have a big impact on performance, and the choices for these criteria will be determined by the various scientific goals of the simulation. For example, understanding the development of structures at very high redshift (e.g. Ishiyama et al. 2012) will present different parameter and algorithm choices than simulations that model the observations of current large scale structure (e.g. Habib et al. 2013).

2HOT (Warren 2013) is an improved version of the HOT code which has been developed over the past two decades. It uses an Oct-tree for gravity, and its gravity algorithm is very similar to that of CHaNGa. This code demonstrates near perfect strong scaling up to 262 thousand cores on Jaguar with a 128 billion particle simulation, implying 500,000 particles per core at the largest core count. The actual size of the scaling simulation (in Gigaparsecs) was not reported, but can be presumed to be a box of order 1 Gigaparsec based on the other simulations presented in

Warren (2013). 2HOT does implement a multi-step time-stepping algorithm, although it is not clear whether particles have individual time steps, and performance for the multi-step method was not presented.

The HACC (Habib et al. 2013) framework scales to millions of cores on a diverse set of architectures. It uses a modified TreePM algorithm: an FFT based particle-mesh on the large scales, a tree algorithm on intermediate scales and particle-particle on the smallest scales. HACC has been demonstrated to scale with near perfect parallel efficiency up to 16,384 nodes on Titan with 1.1 trillion particles, and up to 1.6 million cores on Sequoia with 3.6 trillion particles. These are weak scaling results, typically with millions of particles per core. They also demonstrated strong scaling up to 8,182 nodes on Titan and 16,384 cores on Sequoia.

The GreeM code (Ishiyama et al. 2012) demonstrates scaling of a trillion particle simulation to 82,944 nodes (663,522 cores) of the K computer, implying 1.5 million particles per core. This code also uses a TreePM algorithm with a hand-optimized particle force loop and a novel method to parallelize the FFT. They report that despite the new parallelization method, the FFT remains the bottleneck in their TreePM code. They also employ a multi-step method that splits the PM and particle forces, but the particles do not have individual time steps.

The GADGET-3 TreePM code (based on GADGET-2 (Springel 2005)) was used to perform a large scale structure, DM-only simulation (the 'Millenium XXL') on 12,288 cores using 303 billion particles (Angulo et al. 2012). With over 16 million particles per core, special effort was needed to optimize the memory usage of the code because the simulation was limited by memory resources.

Most of these cosmological N-body codes with published performance data scale to millions of cores with almost perfect parallel efficiency, given very large problem size (typically trillions of particles). However, it becomes even more challenging to simulate a relatively smaller problem size with higher resolution using large numbers of cores. This is due to the fact that the distribution of the particles in the simulated system tends to become more non-uniform as resolution increases, leading to load imbalance and difficult scaling. The addition of hydrodynamics and cooling only exacerbates this problem. Recent projects that coupled gravity with hydrodynamics in galaxy formation simulations and scaled past a few thousand cores include EAGLE and Illustris. The codes used (GADGET-3 and AREPO (Springel 2010)) share many of the features of CHaNGa that are necessary for galaxy formation, including individual time steps for particles, gas dynamics, and star formation/feedback prescriptions (Schaye et al. 2014; Vogelsberger et al. 2014). While some codes handle non-uniform distributions well (e.g. GADGET-3) they have not been shown yet to scale to large (100,000 or greater) core counts. Hence, to our knowledge, CHANGA is the first code to explicitly tackle both the uniform and highly clustered simulations with extremely large scaling. This is achieved by several techniques including multi-stepping and large scale dynamic load balancing described below.

3 CHANGA

The N-body/Smooth Particle Hydrodynamics (SPH) code CHANGA (Jetley et al. 2008, 2010), is an application implemented using CHARM++. CHANGA includes a number of features appropriate for the simulation of cosmological structure formation, including high force accuracy, periodic boundary conditions, evolution in comoving coordinates, adaptive time-stepping, equation of state solvers and subgrid recipes for star formation and supernovae feedback. The code is also being compared with similar codes in the AGORA comparison project (Kim et al. 2014). Cosmology research based on CHANGA includes modeling the impact of a dwarf galaxy on the Milky Way (Purcell et al. 2011), modeling the intracluster gas properties in merging galaxy clusters (Ruan et al. 2013) and distinguishing the role of Warm Dark Matter in dwarf galaxy formation and structure (Governato et al. 2014). In this section we describe the features of CHANGA, particularly as they relate to cosmological structure formation. In addition to the physics features described below, CHANGA has a number of usability features required for pushing a large simulation through a production system, such as the ability to efficiently checkpoint and restart on a different number of processors.

3.1 Gravitational force calculation

The gravitational force calculation is based on a modified version of the classic Barnes-Hut algorithm (Barnes and Hut 1986). Details of our modifications are described in Section 4, and many of our optimizations are taken from PKDGRAV (Stadel 2001), upon which our gravity calculation is based. As in PKDGRAV, we choose to expand to hexadecapole order the multipoles used for evaluating the far field due to a mass distribution within a tree node. For the force accuracies required for cosmological simulations, better than 1 percent (Power et al. 2003; Reed et al. 2003), this higher order expansion is more efficient (Quinn et al. 2013).

3.2 Force softening

When simulating dark matter and stars, the goal is to understand the evolution of a smooth distribution function that closely approaches a Boltzmann collisionless fluid. As the N-body code is sampling this distribution using particles, a more accurate representation of the underlying mass distribution is obtained if the particles are not treated as point masses, but instead have their potential softened (Dehnen 2001). Softened forces are also of practical use

since they limit the magnitude of the inter-particle force. Typically, the softening length is set at the inter-particle separation at the center of DM (Dark Matter) halos (Power et al. 2003).

Calculating the non-Newtonian forces introduced by softening adds a complication to the multipole calculation: Newtonian forces have symmetries which greatly reduce the complexity of higher order multipoles, and the number of components of the multipole moments that need to be stored. CHANGA implements softening using a cubic spline kernel, whose compact support means this complexity is not needed beyond a specified separation (convergence with Newtonian gravity is formally achieved at two softening lengths). Furthermore, rather than evaluating the more complex multipoles when softening is involved, CHANGA evaluates all forces involving softening using only the monopole moments, using a stricter opening criterion to maintain accuracy.

3.3 Periodic boundary conditions

In order to efficiently and accurately simulate a portion of an infinite Universe, we perform the calculation assuming periodic boundary conditions. Because of the long range nature of gravity, the sum over the infinite number of periodic replicas converges very slowly. CHANGA accelerates this convergence using Ewald summation (Ewald 1921), implemented similarly to Ding et al. (1992) as more fully described in Stadel (2001). This technique has the advantage that the non-periodic force calculated from the treewalk is not modified, and therefore is simple and fast. We have demonstrated in Reed et al. (2003) that, with suitable choices of the accuracy criterion, the force errors from this method do not compromise the growth of large scale structure.

3.4 Multi-stepping

In order to efficiently handle the wide range of timescales in a non-uniform cosmological simulation, CHANGA allows each particle to have its own time step. In order to amortize overheads associated with the force calculation, such as tree building, the time steps are restricted to be power-of-two subdivisions of the base time step. Details of this scheme, including how to integrate the equations of motion in coordinates that follow the expansion of the Universe, are described in Quinn et al. (1997). This scheme is also identical to that implemented in GADGET-2; see Springel (2005) for tests of its accuracy.

3.5 Smooth particle hydrodynamics

Despite being a small fraction of the energy density of the Universe, baryons play a significant role in the evolution of structure. Not only are they the means by which we can measure structure (e.g. via star light), they can also directly influence the structure of the dark matter via gravitational coupling (Pontzen and Governato 2012). Therefore, following the physics of the baryonic gas is essential

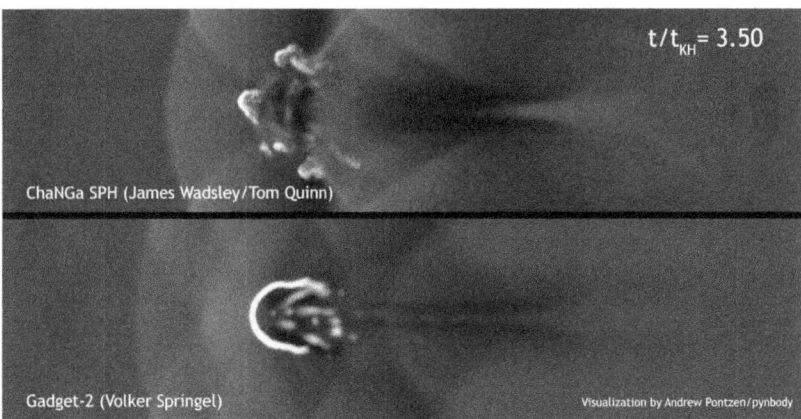

Figure 1 Updated modeling of gas physics in CHANGA. Central density slices of the time evolution of a high density cloud in pressure equilibrium in a wind. Time is in units of the Kelvin-Helmholtz growth time. CHANGA (top) vs GADGET-2 (Springel 2005) (bottom). The color density map shows how with the new SPH formulation of pressure gradients, artificial surface tension is suppressed and instabilities rapidly mix the 'blob' with the surrounding medium, while poor handling of contact discontinuities preserve the blob in the now obsolete SPH implementation of GADGET-2. We have verified that CHANGA gives results quite similar to alternative hydro codes, as the adaptive mesh refinement code ENZO (Collins et al. 2010). This figure was produced with Pynbody (Pontzen et al. 2013).

for accurate modeling of structure formation. CHANGA uses Smooth Particle Hydrodynamics (SPH) to solve the Euler equations with an implementation that closely follows (Wadsley et al. 2004). Since SPH is based on particles, implementing it is a natural extension of the algorithms to calculate gravity on a set of particles. In particular, the tree structure used for the Barnes-Hut algorithm is used to find the near neighbor particles needed for the SPH kernel sums. SPH is also relatively communication intensive compared to gravity, so we utilize the CHARM++ runtime system to adaptively overlap the communication latencies from SPH with the floating point operations needed by gravity. The current implementation of SPH in CHANGA closely follows techniques already published by independent groups and includes an updated treatment of entropy and thermal diffusion (Wadsley et al. 2008; Shen et al. 2010), pressure gradients[b] and timestepping (Durier and Dalla Vecchia 2012). This last features ensures that sudden changes in the particle internal energy, e.g. caused by feedback, are captured and propagated to neighboring particles by shortening their time step. These improvements lead to a marked improvements in the treatment of shocks (as in the Sedov-Taylor blastwave test), and cold-hot gas instabilities. A qualitative example is shown in Figure 1, where the classic 'blob' test compares CHANGA with GADGET-2.

As this paper focuses specifically on the scaling performance of CHANGA we refer to existing works (Wadsley et al. 2004; Governato et al. 2014) and Wadsley et al. (in prep.) for tests of this SPH implementation.

3.6 Star formation and feedback

Again, a necessary component of simulating structure formation is predicting the light distribution. Hence, we need a prescription for where the stars are forming. Furthermore, it is clear that star formation is a self-regulating process due to the injection of energy from supernova, ionizing radiation and stellar winds into the star-forming gas. These processes are all happening well below the resolution scale of even the highest resolution cosmological simulations, so a sub-grid model is needed to include their effects. CHANGA includes the physics of metal lines and molecular hydrogen cooling (Shen et al. 2010; Christensen et al. 2012) and feedback from supernovae (SNe). In CHANGA, we have implemented the 'blast-wave' and 'superbubbles' feedback models described in Stinson et al. (2006) and Keller et al. (2014), respectively. In both models SF occurs in high gas density regions, and the time and distance scales for energy injection into the gas is determined by physically motivated models. The 'blastwave' prescription follows an analytic model of the Sedov blast wave, and it has allowed us to successfully model a number of trends in galaxy populations including the Tully-Fisher relation (Governato et al. 2007), the mass-metallicity relation (Brooks et al. 2007), the stellar mass-halo mass relation (Munshi et al. 2013) and the formation of DM cores in dwarf galaxies (Governato et al. 2012).

4 Parallelization approach

In CHANGA, the particle distribution in space is organized in a hierarchical tree structure where each node represents a portion of the 3D space containing the particles in that volume. The root node represents the entire simulation space and the children represent sub-regions. The leaf nodes of the tree are *buckets* containing a small set of particles.

4.1 Domain decomposition

During domain decomposition, particles are divided among objects called *tree pieces* (or chares in the context of CHARM++) which are mapped onto processors by the runtime system. Typically, there are more tree pieces than the number of processors, and this over-decomposition allows the benefits of the overlapping of communication with computation and the load balancing features of CHARM++.

CHANGA supports various domain decomposition techniques, which have been evaluated previously (Sharma 2006). We used space-filling curve (SFC) decomposition for the results in this paper as that is the method currently used for most scientific studies with CHANGA.

The goal of this scheme is to identify a set of splitting points (*splitters*) along the space filling curve such that each range contains approximately equal numbers of particles. The algorithm used to identify the splitter keys is similar to the parallel histogram sort (Solomonik and Kale 2010). First, a single master object calculates a set of splitters along the SFC that partition the simulation domain into disjoint areas of roughly equal volume. It then broadcasts the splitter keys to all the tree pieces. The tree pieces evaluate the count of particles for each bin, which is reduced across all tree pieces back to the master process. The candidate keys are then adjusted based on the contributions received, and new splitters are broadcast for any bins that are not sufficiently close to an optimal partition. This process is repeated until a suitable set of splitter keys is determined such that all the bins have roughly equal numbers of particles. After the splitter keys are identified, the particles are globally distributed to tree pieces according to the splitters, where each bin corresponds to one tree piece.

4.2 Tree build

After the particles have migrated and domain decomposition is finished, each tree piece builds its tree independently. The tree build is done in a top-down manner. The algorithm starts from the root, which contains the entire simulation space, and proceeds downwards to the leaves, which are buckets containing a small number of particles, typically 8 to 12. A tree piece has information about the extent of the domain held by other tree pieces; this information is used in the tree building process. A spatial binary tree is constructed by bisecting the bounding box containing particles in the given volume. The tree building process bisects each node, starting at the root, into children, which represent sub-regions within the space, until a leaf node is constructed. If a node in the tree held by a tree piece contains particles in another tree piece, then that node becomes a boundary node.

We also take advantage of the fact that a tree piece can access other tree pieces within the same address space. All the tree pieces within the same address space are merged. After the merge, each tree piece has read-only access to the tree data structure that is constructed by merging multiple tree pieces. For additional details, we refer the reader to Jetley et al. (2008).

4.3 Tree traversal and gravity

The goal of tree traversal is to identify for each bucket of particles in the tree the list of nodes and particles whose information is needed for the gravity calculation. These *interaction lists* are constructed on a per bucket basis to amortize the overhead of the tree traversal.

Another optimization that is implemented in CHANGA to improve the performance of the gravity phase is based on the observation that nearby buckets tend to have similar interaction lists (Stadel 2001). The algorithm constructs the interaction list of a parent node before proceeding to the children, and maintains a *checklist*, passed down the tree, that reduces the number of nodes that need to be evaluated at each level.

Tree traversal requires remotely accessing nodes which are part of tree pieces on other processors. To optimize this remote data access, we have implemented a software cache, as shown in Figure 2. The *Cache Manager* serves node and particle requests made by a tree piece. If a node request is missed in the cache, then a request is sent to the corresponding tree piece. If there is already an outstanding request in the cache, no additional request is sent. When the response arrives, the requestors are informed and the walk resumes. This improves the performance by hiding the latency of remote requests and by reducing the number of messages sent and received for the remote node. To further reduce cache misses, we also perform a prefetch walk which obtains remote node information.

To effectively overlap communication and computation, we divide the tree traversal into local and remote parts. A local traversal is done on the portion of the tree which is within the local address space whereas a remote traversal is done on the remaining part of the tree and requires communication between the tree pieces. We use prioritization to give precedence to the remote traversal, which requires communication, over the computation-dominated local traversal. When the remote walk has sent out requests for the node and is waiting for the response, the local walk can be done. This enables overlap of communication with local computation and helps mask message latency. Figure 2 diagrams the gravity calculation in CHANGA with a software cache.

Sequential code in CHANGA is also well optimized. In particular, we take advantage of single-instruction, multiple-data (SIMD) parallelism inherent in the force calculation to accelerate that part of the computation using FMA or SSE vector instructions.

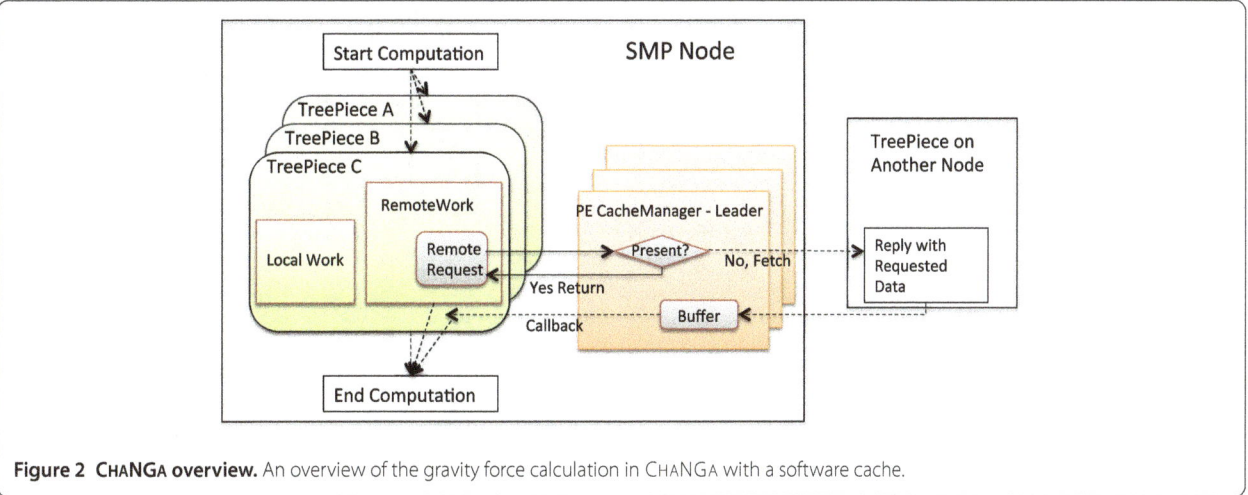

Figure 2 CHANGA **overview.** An overview of the gravity force calculation in CHANGA with a software cache.

5 Datasets and systems

We first describe the datasets used for our experiments and their characteristics. We have two large, uniform (Poisson distributed) datasets with 12 and 24 billion particles. Other than having periodic boundary conditions these two datasets are not particularly interesting for cosmology. We include them here to demonstrate the scaling of CHANGA to large core counts. *cosmo25* is a more challenging dataset: it is a 2 billion particle snapshot taken from the end (i.e. representing the current, $z \sim 0$, very clustered, structure of the Universe) of a dark matter simulation of a 25 Megaparsec cube in a ΛCDM Universe. The force softening is 340 parsecs, and the simulation represents a challenge for load balancing. This simulation was evolved using CHANGA from initial conditions at $z = 109$ based on cosmological parameters derived from the Planck data (Ade et al. 2013). The version of this simulation with gas dynamics and star formation is able to resolve the disks of spiral galaxies within this volume [Anderson et al., in preparation]. *dwarf* is our most challenging dataset: while it contains only 52 million particles spread throughout a 28.5 Megaparsec volume, most of the particles are in a single high resolution region in which a dwarf galaxy is forming. The mass resolution in this region is equivalent to having 230 billion particles in the entire volume, and the force resolution within this region is 52 parsecs. This is a high resolution version of the DWF1 simulation discussed in Governato et al. (2007). See the description of DWF1 in that paper for more details about the galactic and cosmological parameters of this simulation. While, as described above, CHANGA is capable of handling hydrodynamics and star formation, in the benchmarks below we show the performance of dark matter only simulations. We will comment on SPH performance in the discussion.

We show the performance of CHANGA on Blue Waters. Blue Waters is a hybrid Cray XE/XK system located at the National Center for Supercomputing Applications (NCSA). It contains 22,640 Cray XE6 nodes and 4,224 Cray XK7 nodes that include NVIDIA GPUs. Each dual-socket XE6 compute nodes contains two AMD Interlagos 6276 processors with a clock speed of 2.3 GHz and 64 GB of RAM.

6 Single stepping

We now describe essential optimizations required for scaling the simpler datasets that are not highly clustered, and evaluate their performance. Later sections will describe optimizations for clustered datasets.

6.1 Single stepping improvements

We observed that from-scratch domain decomposition is not required at every step, especially for datasets which are not highly clustered. After the initial domain decomposition, it needs to be performed only when there is an imbalance in the load of tree pieces. By reusing the previously determined splitters, we reduce the overhead incurred in finding the splitters as well as the number of particle migrations. We use an adaptive mechanism to determine when to perform the domain decomposition. In this approach, load statistics of the tree pieces are collected and domain decomposition is only performed if an imbalance is detected. Otherwise, only particle migration is done based on the previous splitters. We use the quiescence detection (Sinha et al. 1993) mechanism implemented in CHARM++ to determine when all the migrations are finished.

In the unoptimized version of the code, the tree build requires all tree pieces to send the information about the first and the last particle in their domain, subject to the SFC. This information is used to determine ownership of nodes in the tree but requires heavy communication. We avoid this by using the boundary information to determine a set of candidate tree pieces which may have information about

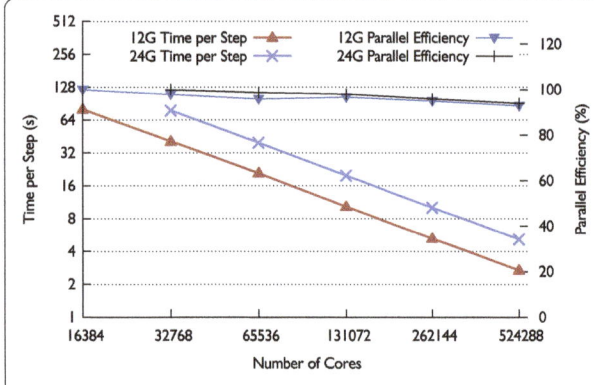

Figure 3 **Performance on Blue Waters.** Time per step and parallel efficiency for 12 and 24 billion particles on Blue Waters. Both the cases scale well achieving a parallel efficiency of 93%.

Table 1 **Breakdown of time for 1 step in seconds for 12 billion particle (top half) and 24 billion particle (bottom half) datasets run on Blue Waters with the proposed optimizations**

#cores	Gravity	DD	TB	LB	Total time
16,384	77.556	1.299	0.729	0.128	79.712
32,768	39.254	0.698	0.617	0.136	40.705
65,536	19.876	0.496	0.367	0.062	20.801
131,072	9.967	0.181	0.138	0.027	10.313
262,144	5.051	0.109	0.076	0.013	5.249
524,288	2.569	0.073	0.034	0.008	2.684
32,768	75.090	1.553	0.735	0.186	77.564
65,536	37.941	0.787	0.462	0.111	39.301
131,072	19.062	0.428	0.245	0.063	19.798
262,144	9.682	0.232	0.152	0.042	10.108
524,288	4.903	0.146	0.095	0.022	5.166

Table 2 **Breakdown of time for 1 step in seconds for 12 billion particles (top half) and 24 billion particles (bottom half) dataset run on blue waters without the proposed optimizations**

#cores	Gravity	DD	TB	LB	Total time
16,384	82.424	2.81	0.995	7.79	94.019
32,768	42.712	1.966	1.005	6.854	52.537
65,536	21.438	1.731	0.729	6.482	30.38
131,072	12.162	1.674	0.803	5.718	20.357
32,768	80.144	2.859	1.366	16.173	100.542
65,536	41.279	2.356	1.032	9.338	54.005
131,072	22.958	2.142	1.018	8.854	34.972

the required node. One of them is then queried and in case that tree piece does not have the information, it forwards it to the appropriate tree piece.

Since load balancing incurs overhead, it should be done sparingly. We use the *MetaBalancer* (Menon et al. 2012) framework in Charm++ to determine when to invoke the load balancer. *MetaBalancer* monitors the application characteristics and predicts when the load balancing should be done. *MetaBalancer* invokes the load balancer when: (1) an imbalance is detected and (2) the benefit of load balancing is more than the cost incurred due to load balancing.

6.2 Performance

Figure 3 shows strong scaling results on up to $512K$ cores on Blue Waters evolving 12 and 24 billion particles. Our application exhibits almost perfect scaling up to the maximum number of cores. Each iteration consists of domain decomposition, load balancing, tree building and the force calculation. Table 1 shows the break down of the time per step into the different phases. For the simulation evolving 12 billion particles, we achieve 93% parallel efficiency at $512K$ cores with the time per step being 2.6 seconds. For the 24 billion particle simulation, we achieve 93.8% parallel efficiency with a time per step of 5.1 seconds. The efficiency is calculated with respect to $16K$ cores and $32K$ cores for 12 and 24 billion particles, respectively.

The good scaling of the gravity phase is due to the overlap of communication and computation, the improved tree walk algorithm using an interaction list, the software request cache, prefetching, and other optimizations. The time for domain decomposition also scales with the increase in number of cores. Table 1 shows, for the 12 billion particles at $512K$ cores on Blue Waters, that domain decomposition takes on average 73 ms per step. At $128K$ cores the domain decomposition is 9 times faster in comparison to the unoptimized version. This is due to the use

of the adaptive technique to determine when to perform full domain decomposition. The tree build time also scales well and takes 34 ms at $512K$ cores. At $128K$ cores, the tree build is approximately 6 times faster than the unoptimized version. Similar trends are seen in the 24 billion particle simulation.

Table 2 contains the breakdown of the total time per step for the unoptimized version of the code. Comparing the results with Table 1, for the 12 billion particle simulation, we reduce the total time by 15 to 49%. For the 24 billion particle simulation, we reduce the total time per step by 22 to 43%. The reduction in time occurs for all phases of the application.

Figure 4 shows the time profile graph obtained using Projections (Kalé and Sinha 1993). This shows the average processor utilization over the course of one time step evolving 12 billion particles on $16K$ cores of Blue Waters. We can see that the local work, which is given a lower priority, overlaps with the communication needed for the higher-priority remote work, resulting in close to 100% processor utilization.

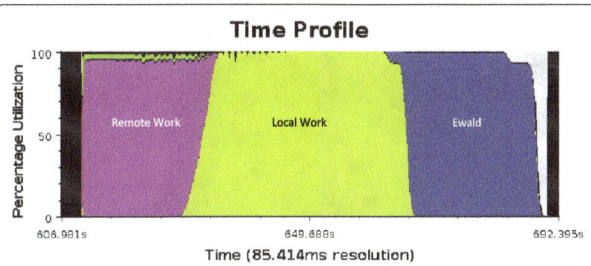

Figure 4 Processor utilization. Time profile graph which shows processor utilization over time for 16K cores on Blue Waters for 12 billion particles. This shows overlap of communication with computation to achieve high utilization.

7 Clustered dataset challenges

For datasets such as *dwarf*, the particle distribution is concentrated at the center of the simulation volume and therefore highly clustered. This creates many challenges in scaling, of which one of the most significant is communication imbalance. During the gravity phase, remote requests are sent for tree nodes that are not present in the local cache. In a clustered dataset, some tree nodes are requested many more times than others. This results in the tree pieces owning those tree nodes receiving a large volume of node request messages. Figure 5(a) shows the number of requests received by processors for the *dwarf* simulation at 8K cores on Blue Waters. We can see that a handful of processors receive as many as 30K messages. Even though there is overlap of communication with computation, this causes significant performance degradation. This is because, at this scale, there is not enough local computation to overlap seconds of delay in receiving messages. One way to mitigate this problem is to replicate the information that is being requested to prevent a few processors from being the bottleneck.

We replicate the information about the tree nodes on multiple processors ensuring that no single processor becomes overloaded. Before the gravity phase begins, tree pieces send their node information to a set of tree piece proxies on other processors. The responsibility of the tree piece proxy is to store the node information sent to it and handle requests for those nodes. When a tree piece needs to request for a remote node, it chooses randomly one of the tree piece proxies to send the request to. Figure 5(b) shows the number of messages received by the processors when four tree piece proxies are created for each tree piece. For an 8K core run on Blue Waters, replication reduces the maximum number of messages received from 32K to 4.2K, and the requests are better distributed among all processors. Figure 6 shows the time-profile graph where the x-axis is the time and y-axis is the processor utilization. Here, yellow regions constitute the local work, blue the Ewald and maroon the remote work. Note the idle time, in Figure 6(a), before the remote work begins which is due to the delay in receiving messages and the lack of local work overlap. Figure 6(b) shows the impact of replication. The remote work can start earlier due to a smaller delay in request messages. The local work overlaps with the communication until remote work is ready to start. This is a very good example that shows prioritization of remote work over local work and the overlap of communication with computation. Figure 7 shows the strong scaling performance for this dataset on core counts ranging from 1K to 16K. We compare the time for the gravity phase because the rest of the phases are the same in both cases. The gravity time is improved from 2.4 seconds to 1.7 seconds for 8K cores and from 2.1 seconds to 0.99 seconds on 16K cores. At 16K cores the parallel efficiency without replication is 48% cores whereas replication helps achieve an efficiency of 98%.

8 Multi-stepping challenges

A wide variation in mass densities can result in particles having dynamical times that vary by a large factor. In a single-stepping mode, good accuracy can only be achieved by performing the force calculation and particle position and velocity updates at the smallest timescale. However, hierarchical time stepping schemes can be used for a large

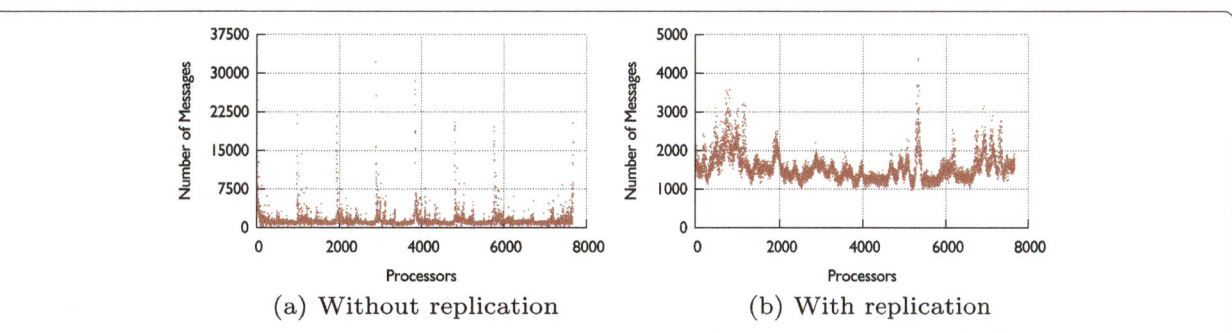

(a) Without replication   (b) With replication

Figure 5 Messages with and without replication. Number of messages received by processors for a simulation of the *dwarf* dataset on 8K cores on Blue Waters. Note that replication reduces the maximum requests received by a processor from 30K to 4.5K.

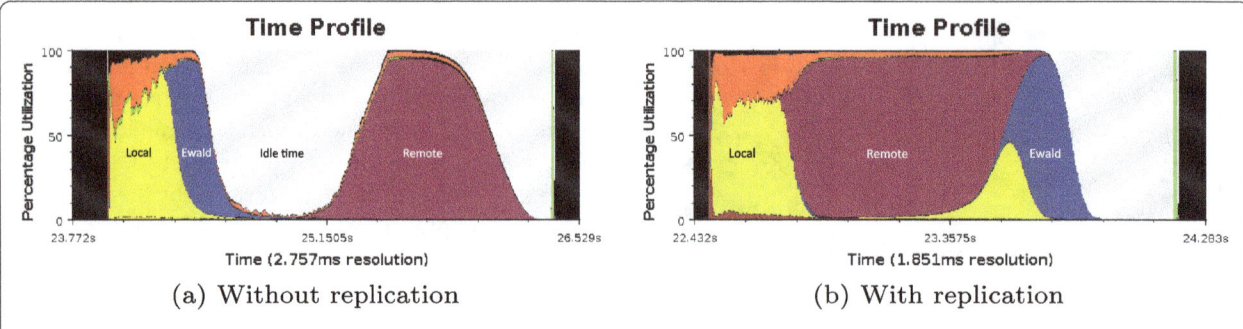

(a) Without replication (b) With replication

Figure 6 Processor utilization with and without replication. Time profile graph showing processor utilization over time for the simulation of the *dwarf* dataset on 8K cores. Note the idle time without replication which is removed by the replication and the gravity time is improved from 2.4 seconds to 1.7 seconds.

dynamic range in densities at a small additional cost. We use adaptive time scales where forces are evaluated only on relevant particles instead of evaluating forces on all the particles at the smallest time scale. In a multi-step simulation, particles are assigned to time step rungs corresponding to the shortest time scale required for an accurate simulation. Rungs corresponding to short time scales are evaluated more frequently than those for long time scales.

Using multi-stepping for clustered datasets introduces a variety of challenges. The irregular distribution of particles in the simulation space as well as the division of particles into rungs creates severe load imbalance. In general, the challenge is higher for datasets with fewer particles. We discuss various optimizations that enable CHaNGa to scale a medium-sized 2 billion particle clustered dataset, *cosmo25*, on up to 128K cores on Blue Waters. Reaching this level of performance required overcoming challenges related to load imbalance, communication overhead with a decrease in computation per processor as well as the scalability of other phases of the simulation. Strong scaling of this nature will be required to run clustered cosmological simulations on future machines with hundreds

of Petaflop/s performance, and presents a realistic proving ground for parallel strategy innovations.

8.1 Optimizations for the gravity phase

In a multiple time step simulation, the number of particles active in the fastest rung is typically only a fraction of the total number of particles being simulated. These active particles tend to be clustered, and therefore the distribution of particles among the tree pieces is highly imbalanced. One may consider performing from-scratch domain decomposition based on the active set of particles for these time steps but that results in large jumps of the domain boundaries. To prevent such sudden large variations of the boundaries, we perform from-scratch domain decomposition only when there is a significant number of particles active for that time step. But as one can imagine, this will result in tree pieces with a large variation in active particles and load. Figure 8 shows the distribution of the load on tree pieces for the fastest rung (rung 4) and the slowest rung (rung 0) of the *cosmo25* dataset. The slowest rung has tree pieces with loads distributed around the mean. But the fastest rung has only 2,405 tree pieces with active particles and some of them have a load which is 3,000 times the average load of tree pieces and 40 times the average load of the system. Even though periodic load balancing is performed to distribute the load, the maximum load of the system will be limited by the most overloaded processor which in this case is the one having the most loaded tree piece. At larger scales of 128K cores there is not enough work to be distributed among all the cores which results in significant degradation of performance. We propose two adaptive strategies to overcome this problem.

8.2 Intra-node work pushing

We use the SMP mode of CHARM++ to take advantage of the shared memory multiprocessor nodes used in HPC systems (Mei et al. 2010). The SMP mode supports multi-threading, where one CHARM++ process is assigned per

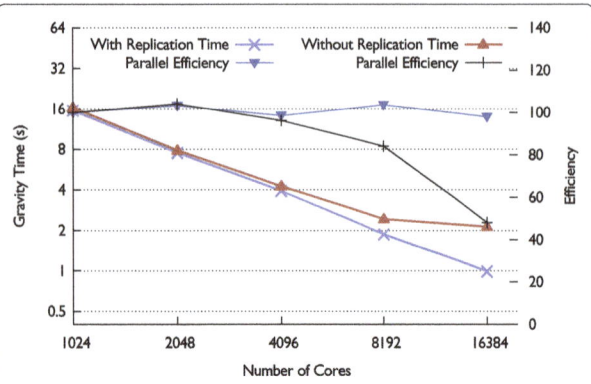

Figure 7 Performance for the *dwarf* simulation. Gravity time for the *dwarf* simulation on 8K cores on Blue Waters with and without the replication optimization.

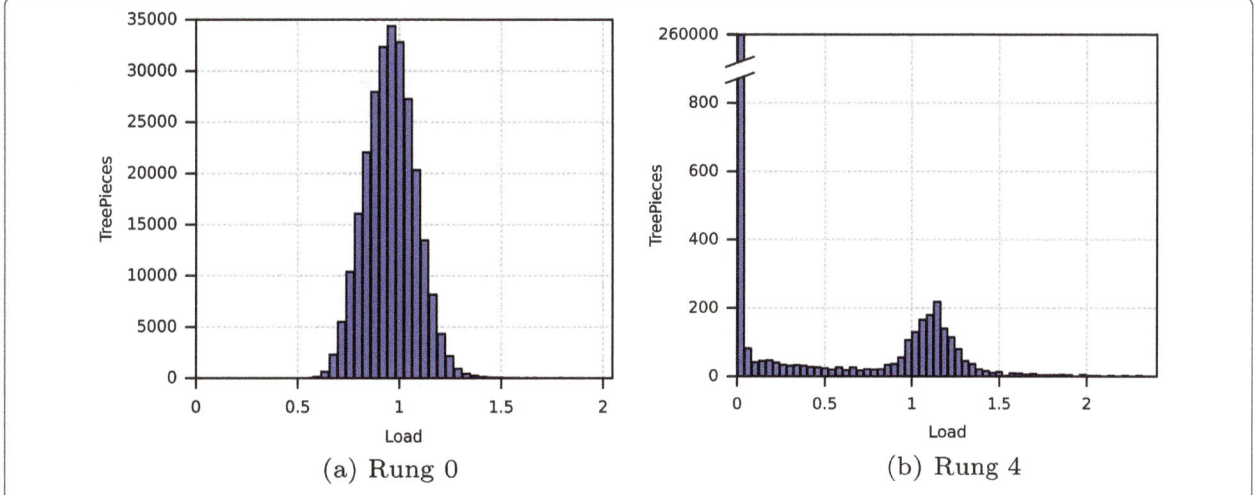

(a) Rung 0   (b) Rung 4

Figure 8 Load distribution. Distribution of tree piece load for rung 0 (slowest) and rung 4 (fastest). Rung 0 has loads distributed around the mean. Rung 4 has only 2,405 active tree pieces with a maximum load of 2.3.

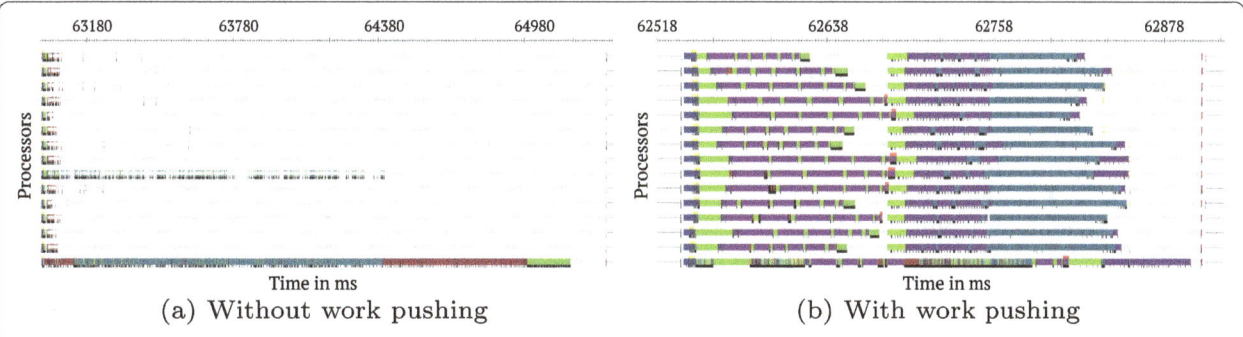

(a) Without work pushing   (b) With work pushing

Figure 9 Performance analysis with and without work pushing. Time line profile for all the PEs (rows) on a SMP process for the 16K cores run. White shows idle time and colored bars indicate busy time. Work pushing achieves better distribution of work among PEs. The total time per step reduces from 2.3 seconds to 0.3 seconds.

SMP node, with a single thread mapped to each physical core. One thread within a node is normally assigned as a communication thread responsible for internode communication, while the rest are used as worker threads that implement *processing elements (PEs)*.

Within a CHARM++ SMP process, data can be shared via pointers. The load balancing strategy works in a hierarchical fashion. Details are given in Section 8.6, but in essence it first tries to achieve load balance among the SMP processes and then balances the load among cores within the SMP process.

LBManager, which is an object present on each PE, has information about the average load of the system and the load of other PEs on the same SMP process. The *LBManager*, on identifying that a PE is overloaded, instructs overloaded tree pieces at that PE to distribute the work among other less loaded PEs within the SMP process. A tree piece is responsible for calculating forces on a set of particles in

its domain, grouped into buckets. We consider the bucket to be the smallest entity of work that can be distributed. PEs receiving a foreign bucket have access to the tree and all the data structures of the owner tree piece so that they can perform the tree traversal and gravity force calculations for the foreign bucket. Once the force calculations are done, the foreign bucket is marked as complete and the original PE is informed. Once all the foreign and local buckets are completed, the tree piece is done with the gravity calculations.

This work pushing adaptive strategy reaps the most benefit for time steps where the fastest rung is active. For the slowest rung, the forces on all the particles need to be calculated, and the load balancing is very similar to that in single stepping runs. Figure 9(a) shows the time-line view from the projections tool (Kalé and Sinha 1993) for rung 4 (the fastest rung). Here, each line corresponds to a PE and colored bars indicate busy time while white shows idle

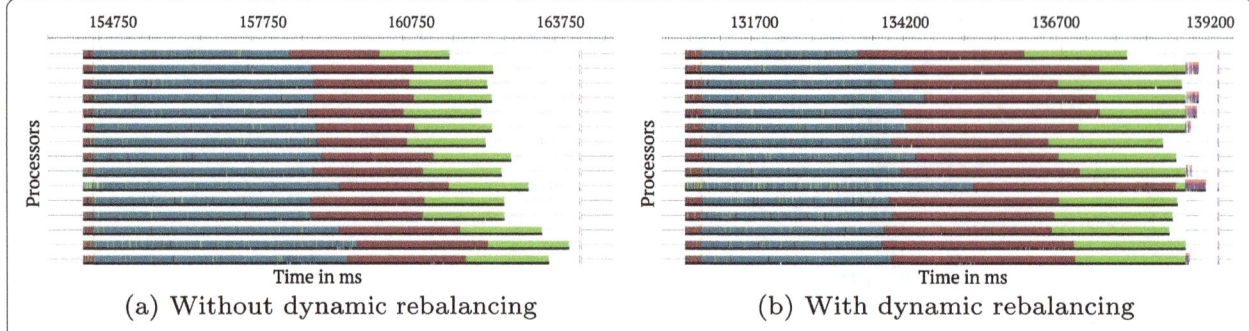

(a) Without dynamic rebalancing (b) With dynamic rebalancing

Figure 10 Performance analysis with and without dynamic rebalancing. Time line profile for all the PEs (rows) on a SMP process for the 16K cores run. White shows idle time and colored bars indicate busy time. Dynamic rebalancing eliminates trailing idle time resulting in better utilization. The total time per step reduces from 9.8 seconds to 8.5 seconds.

time. This plot is for a 32K core run on Blue Waters, and we have chosen the PE and the corresponding SMP process with the maximum load. We can see that the most loaded PE, which also contains the most loaded tree piece, is busy for about 2 seconds while other PEs are idle. Figure 9(b) shows the time line for the work pushing strategy for a set of PEs in the SMP process where one of the PEs is assigned the most loaded tree piece. With the work pushing strategy, we are able to successfully distribute the work load among other PEs within the node. This results in a reduction of the gravity time from 2.3 seconds to 0.3 seconds for the fastest rung.

8.3 Intra-node dynamic rebalancing

For clustered datasets, it is often the case at the trailing end of the gravity calculation that some of the PEs are idle while others are busy. This could be due to misprediction of load or inability of the load balancer to balance the load perfectly. Figure 10(a) shows the Projections time-line view for this scenario where the colored bars indicate busy work while the white shows idle time. We found that such slight load imbalances in the application can be mitigated by more fine-grained parallelism within the SMP process. We use an intra-node dynamic rebalancing scheme where the idle PEs within the node pick work from the busy ones. The scheme is implemented using the *CkLoop* library (Mei et al. 2010) in CHARM++, which enables fine-grained parallelism within an SMP process.

As with the work-pushing scheme, buckets are the smallest entity of work that can be reassigned.

If all the tree pieces residing on a PE have finished their work, then the PE becomes idle. At each PE, the *LBManager* maintains a PE-private variable which keeps track of its status. Since the memory address is shared among the PEs on a SMP process, the *LBManager* can access the status variable of all the PEs within the SMP process. Whenever there is a significant number of idle PEs, the dynamic rebalancing scheme kicks in. Tree pieces then create chunks of work out of the unfinished buckets and add

these to the node-level queue. The idle processors access the node-level queue and pick up work to execute. Due to the overhead associated with the node-level queue we only use the work-stealing scheme adaptively for the trailing end of the computation.

Figure 10(a) shows the time-line for the slowest rung, rung 0, of *cosmo25* dataset simulation for a 32K core run on Blue Waters. We pick a subset of PEs to show this problem. We can see that the load is almost balanced, but towards the end of the step there are some PEs which are idle while others are busy. Figure 10(b) shows the time-line with dynamic rebalancing. It is able to successfully handle small amounts of load imbalance and reduce the gravity time from 9.8 seconds to 8.5 seconds for rung 0.

8.4 SMP request cache

Data reuse can be critical in determining the performance of tree-based algorithms (Gioachin et al. 2007). Modern SMP-based supercomputers offer several levels at which data sharing can be effective. One possibility is that requests for the same remote elements from two traversals on a core can be merged. The fetched data can then be reused by all traversals on the core. Similarly, cores in the same SMP domain can share remotely fetched data. In the following we describe a two-level caching scheme that enables the data reuse across traversals on a core, as well as across cores on an SMP processor. This caching mechanism is transparent to the traversal code.

Each core on the SMP has a *private* cache, which stores pointers to remotely fetched data. There also exists one cache at the level of the SMP that is *shared* by all cores in the SMP. The shared cache contains the union of all the entries in the private caches of these PEs.

Briefly, the algorithm funnels all requests for remote data through the cache. If the data are found in the private cache, then they are immediately passed into the requesting traversal's visitor code. If the data are not found on the PE, we check whether some other piece on the PE has requested them previously. If so, a lightweight continuation is created to resume the traversal at the requested node upon its receipt. Otherwise, the more expensive, SMP-wide table lookup is performed.

We devised a scheme to manage concurrent accesses of the shared, SMP-wide cache table, where all requests for remote data generated by traversals on the SMP processor are funneled through a single core, which is termed the *fetcher* for that SMP processor. Cheap, intra-node messaging between PEs is used for efficiency.

8.5 Domain decomposition

Simulations of datasets with nonuniform distributions are characterized by extensive movement of particles across tree piece boundaries over time. When unchecked, this leads to an increasingly nonuniform distribution of particles across tree pieces and eventually precludes a good balance of load across processors. In such scenarios, it becomes useful to repeat the full domain decomposition more frequently.

The first stage of domain decomposition, as described in Section 4, involves a series of histogramming steps to determine a set of splitters that partition the simulation domain into tree pieces of roughly uniform particle count. This is implemented in terms of broadcasts from a single sorter object, which refines the splitters, to the tree pieces, and reductions of particle counts for each bin back to the sorter process. In strong scaling scenarios for highly clustered datasets, domain decomposition may become a performance bottleneck, as the number of splitters generally depends on the number of processors used in the run. Therefore, we implemented a number of optimizations aimed at improving the SFC domain decomposition performance. First, we replaced the broadcast of SFC keys from the sorter object with the broadcast of a bit vector indicating which of the bins evaluated in the previous step need further refinement. From the bit vector, the set of splitters to evaluate is determined once at each SMP node, and delivered to all tree pieces at that node for evaluation. This optimization greatly reduced the size of the buffers being broadcast. Secondly, we noticed that some histogramming steps were much more expensive than others, due to involving more splitters. This was particularly true for the first and last steps. The first histogramming step involved a full set of splitters due to none having been finalized yet. For this step, we were able to remove the broadcast of splitters by having tree pieces reuse the splitters determined the last time domain decomposition was done. We were also able to eliminate the last histogramming step in the original algorithm, in which the final set of splitters was broadcast to the tree pieces to collect a full histogram of particle counts. Instead, we modified the sorter object to preserve particle counts for all previously finalized splitters, so as to have the full set of counts at the end.

These optimizations significantly improved domain decomposition performance. For runs of the *cosmo25* dataset on Blue Waters, the time for a full domain decomposition was reduced from 3.22 s to 1.52 s on 1,024 nodes, a speedup of 2.1.

8.6 Hierarchical multistep load balancer

Even if domain decomposition assigns almost equal number of particles to tree pieces, density variations in different regions of the simulated space can result in load imbalance. We experimented with domain decomposition based on load, but the basic approach was not ideal for multistepping simulations as it led to large jumps in boundaries and significant movement of particles. Since execution time is determined by the most loaded processor, it becomes important to address the load imbalance problem without significant additional overhead.

Load balancing in CHARM++ applications like CHANGA is normally achieved by over-decomposing the problem into many more objects than processors and letting the CHARM++ dynamic load balancing framework balance the load by mapping the objects to processors (Zheng 2005). The framework can automatically instrument the computation load and communication pattern of tree pieces and other objects and store it in a distributed database. This information is then used by the load balancing strategies, which we optimized for CHANGA, to map the objects to processors. Once the decision has been made, the load balancing framework migrates the objects to newly assigned processors. Alternatively, the load of the objects and their communication pattern can be determined using a model based on *a priori* knowledge. But for CHANGA, we find that determining the load based on a heuristic called the *principle of persistence* is more accurate. Based on this heuristic we use recent history to determine the load of near-future iterations. This scheme works well for single-stepping simulations at a relatively small scale. However, multi-stepping simulations at very large scale impose several new challenges.

First, multi-stepped execution introduces some challenges in the measurement based load balancing to obtain accurate load information. Substeps within a big step in a multi-step run have selected number of active particles. Predicting the load of a tree piece based on the preceding substep will result in discrepancy between the expected load and the actual load. Therefore, we instrument and store the load of the tree pieces for different substeps/rungs separately. Whenever particles migrate from one tree piece to another, they carry a fraction of their load for the corresponding rungs for which they were active and contribute that to the new tree piece. This enables us to achieve very accurate prediction of the load of a tree piece for each substep even with migrations and multi-stepping.

Secondly, it is very challenging to collect communication pattern information in CHANGA, even at small core count, due to a very large number of messages in the simulation, which may incur significant overhead on memory when performing load balancing. Therefore, we used an alternate strategy to implicitly take communication into account during load balancing by using an ORB-based (Orthogonal Recursive Bipartitioning) strategy, which preserves the communication locality.

Lastly, in extremely large scale simulations, load balancing itself becomes a severe bottleneck. The original centralized load balancing strategies, where load balancing decision is made on one central processor, do not scale beyond a few hundred processors, which makes them unfeasible for large scale simulations. To overcome this challenge, we implemented a scalable load balancing strategy suitable for multi-stepped execution based on the hierarchical load balancing framework (Zheng 2005; Mei et al. 2011) in the CHARM++ runtime. This new load balancing strategy performs ORB to distribute the tree pieces among processors. The processors are divided into independent groups organized in a hierarchical fashion. Each group consists of 512 processors. At each level of the hierarchy, the root performs the load balancing strategy for the processors in its sub-tree. We found that two levels of the hierarchy is enough to achieve good load balance with little overhead. At higher levels of the hierarchy refinement based load balancing strategy, which minimizes the migration by considering the current assignment of tasks, is used. At the lowest level of the hierarchy we use ORB to partition the tree pieces among the processors in that subgroup. The load balancer collects the centroid information of tree pieces along with their load. Taking the centroids into account, the tree pieces are spatially partitioned into two sets along the longest dimension. Similarly, at each stage of partitioning, the processors are also partitioned. During partitioning, tree pieces are divided into two partitions such that the loads of the partitions are almost equal.

This is done recursively until one processor remains which is assigned the corresponding partition containing the tree pieces.

Another optimization to further reduce the overhead of load balancing is to combine the node level global load balancing with the intra-node load balancing strategies described in Section 8.1. We implemented such a two-level load balancing strategy, where the load is first balanced across SMP nodes, and then balanced inside each SMP node. The ORB algorithm described above is done for nodes rather than processors. Once the tree pieces are assigned to SMP nodes, they are further distributed among the PEs in the SMP node using a *greedy* strategy. This ensures that the load is equally distributed among the SMP nodes. We perform an additional step of refinement to further improve the load balance for the rare cases when the load is not evenly balanced.

8.7 Performance evaluation

We now present the scaling performance of the *cosmo25* simulation. Figure 11(a) shows the average time per iteration for this simulation with single-stepping and Figure 11(b) shows the average time per iteration with multi-stepping. In a multi-stepping run, 16 substeps constitute a big step. To compare the time for single-stepping and multi-stepping, a single big multi-step covers the same dynamical time as 16 single steps. Table 3 gives a break down of the time taken for different phases for single-stepping and multi-stepping. We can see that at $8K$ cores the single-stepping simulation takes more than 3 times the time taken by multi-stepping and at $128K$ it takes twice as long. Note that the gravity time for multi-stepping is 4.5 times faster than single stepping. Due to sufficient sequential work to overlap communication and relatively balanced tree pieces, we are able to achieve 80% efficiency for

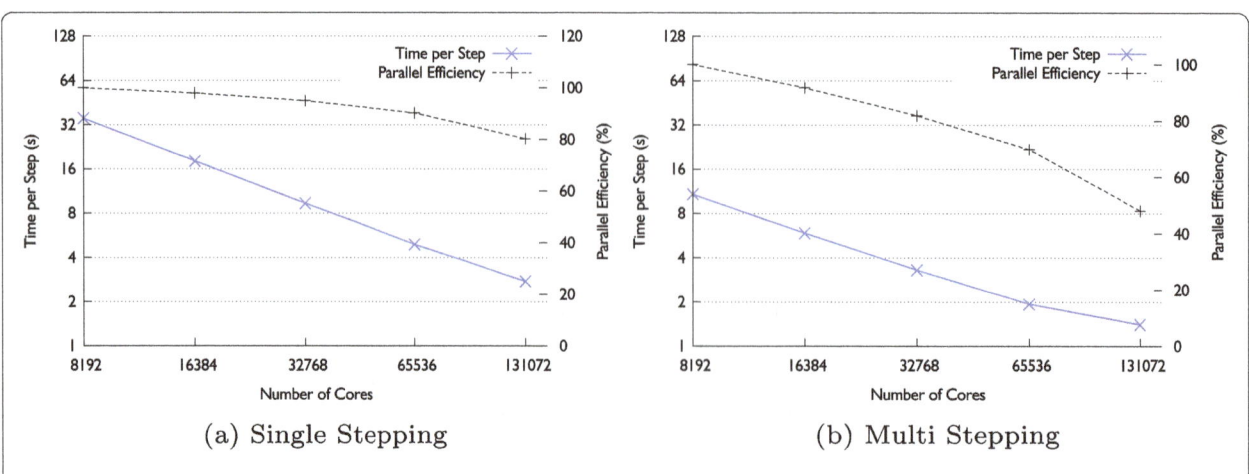

(a) Single Stepping   (b) Multi Stepping

Figure 11 Performance comparison of single stepping and multi stepping. Time per step and parallel efficiency for the *cosmo25* dataset on Blue Waters.

Table 3 Breakdown of time for 1 step in seconds for *cosmo25* dataset with single-stepping (top half) and multi-stepping (bottom half) on Blue Waters

#cores	Gravity	DD	TB	LB	Step time	16 Step time
8,192	33.433	0.441	0.292	1.423	35.589	569.424
16,384	16.952	0.210	0.148	0.851	18.161	290.576
32,768	8.643	0.132	0.091	0.496	9.362	149.792
65,536	4.395	0.163	0.073	0.295	4.926	78.816
131,072	2.353	0.134	0.066	0.216	2.769	44.304
8,192	7.45	0.83	0.47	2.1	10.85	173.6
16,384	3.73	0.79	0.32	1.07	5.91	94.56
32,768	2.1	0.46	0.2	0.55	3.31	52.96
65,536	1.1	0.35	0.12	0.37	1.94	31.04
131,072	0.77	0.24	0.07	0.33	1.41	22.56

Table 4 Breakdown of time for 1 step in seconds for *cosmo25* dataset with PKDGRAV2 on Blue Waters

#cores	Gravity	DD	TB	Step time
8,192	17.90	1.50	0.57	19.97
16,384	10.10	1.40	0.84	12.34
32,768	6.10	0.97	1.50	8.57
49,152	6.60	0.99	13.30	20.89
65,536	8.60	1.00	17.80	27.40
98,304	16.10	1.30	25.80	43.20

single-stepping at $128K$ cores with an average step time of 2.7 seconds. As described in Section 8.1 the multi-stepping run has many challenges due to irregular distribution of particles in faster rungs. Incorporating the improvements mentioned above, we are able to scale to $128K$ cores with an efficiency of 48% with respect to $8K$ cores with a time step of 1.4 seconds. Note that if we consider the gravity force calculation time, we achieve an efficiency of 60% and the gravity time is 3 times faster in multi-stepping in comparison to the single-stepping run.

8.8 Comparison with PKDGRAV

To give a sense of the absolute performance of CHANGA compared to other available N-body codes, we ran the *cosmo25* dataset with PKDGRAV2.[c] Table 4 gives the step time for PKDGRAV2 for the *cosmo25* dataset on Blue Waters. Comparing Table 4, the timings for PKDGRAV2, and Table 3, the timings for CHANGA, for the single stepping benchmark, PKDGRAV2 is faster than CHANGA on up to $32K$ cores but CHANGA continues to scale until $131K$ cores to a time per step of 2.7 s. The Multi-stepping run of CHANGA performs consistently better than PKDGRAV2.

9 Conclusion

In this paper, we have described the design and features of our highly scalable parallel gravity code CHANGA and went into the details of scaling challenges for clustered multiple time-stepping datasets. We have presented strong scaling results for uniform datasets on up to $512K$ cores

on Blue Waters evolving 12 and 24 billion particles. We also present strong scaling results for *cosmo25* and *dwarf* datasets, which are more challenging due to their highly clustered nature. We obtain good performance on up to $128K$ cores of Blue Waters and also show up to a 3 fold improvement in time with multi-stepping over single-stepping.

Many features of the CHARM++ runtime system were used to achieve these results. Starting with the standard load balancing and overlap of communication and computation enabled by the over-decomposition strategy, we employed a number of CHARM++'s features. Of particular importance were features that allowed us to replace parts of our algorithm that scaled as the number of cores, such as quiescence detection for particle movement and the hierarchical load balancer. Also of importance were features such as CkLoop, SMP Cache and node level load balancing, that exploited SMP features of almost all modern supercomputers. With these features, we can bring to bear the computational resources of many 100s of thousands of processor cores on the highly clustered, large dynamic range simulations that are necessary for understanding the formation of galaxies in the context of large scale structure.

While the focus of the work presented here was the performance of the gravity calculation, these techniques are applicable to other parts of cosmological simulations. Above, we summarized the SPH implementation in CHANGA. To give an indication of the performance of our implementation, we used the *cosmo25* dataset which actually has 23 percent of the particles labeled as 'gas'. Benchmarking this dataset with an adiabatic equation of state for the gas on $8K$ cores, we find that the SPH component of the force calculation alone takes on average 37.9 seconds compared to an average 39.6 seconds needed for the gravity calculation. However, as mentioned above, the SPH calculation is dominated by the communication, and when we overlap the SPH with the gravity calculation it adds only 15.2 seconds over a gravity calculation alone. While this result nicely demonstrates the ability of the CHARM++ runtime system to overlap communication and computation, it also indicates that there may be room for optimization of the neighbor finding algorithm. Neighbor finding is also a useful algorithm for implementing other hydrodynamic techniques. We expect that the recently developed Meshless Finite Mass and Meshless Finite Volume methods (Hopkins 2014) will scale better than SPH since they require fewer neighbors and the inter-neighbor calculations require more computation. Moving mesh methods (Springel 2010) can require the construction of a Voronoi mesh which, in turn, requires algorithms to quickly find all particles within a sphere of given radius. Again, the neighbor finding algorithm used in CHANGA can perform this task. The implementation of some of these algorithms will be the subject of future work.

Future work is also planned to improve the scaling on hybrid architectures like those of many current leadership class machines. We have had some success in getting good performance and scaling on up to 896 cores and 256 GPUs with earlier generation GPUs (Jetley et al. 2010). The CHARM++ paradigm for overlapping computing and computation also works for the overlap of data transfer from the host to the GPU, GPU gravity kernel work, and tree walk work done on the host CPUs. However, we have not addressed the balance of work, either between GPU and host CPU or among GPUs. Also, the increased performance of individual nodes enabled by GPUs or other accelerators will increase the need to optimize and hide communication costs.

Competing interests

The authors declare that they have no competing interests.

Authors' contributions

TQ is the primary researcher and supervisor of the CHANGA project. TQ, PJ, along with various contributors developed the code. TQ and FG, and others verified the code for cosmological simulations. HM, LW, TQ and LK came up with the techniques mentioned in the paper for scaling the application. H.M. developed the dynamic load balancing techniques and optimizations for the various phases of the simulation. GZ and HM developed the hierarchical load balancer. LW worked on the domain decomposition optimizations. PJ developed the SMP cache optimization. GZ optimized the performance for the Blue Waters hardware. HM performed the scaling experiments with help from TQ, GZ and LW. All the authors discussed the results and contributed extensively to the writing of the paper.

Author details

[1]Department of Computer Science, University of Illinois at Urbana-Champaign, Urbana, USA. [2]Department of Astronomy, University of Washington, Seattle, USA.

Acknowledgements

CHANGA was initially developed under NSF ITR award 0205413. Contributors to the development of the code include Graeme Lufkin, Sayantan Chakravorty, Amit Sharma, and Filippo Gioachin. This research is part of the Blue Waters sustained-petascale computing project, which is supported by the National Science Foundation (award number OCI 07-25070) and the state of Illinois. HM was supported by NSF award AST-1312913. TQ and FG where supported by NSF award AST-1311956. Use of Bluewaters was supported by NSF PRAC Award 1144357. We made use of pynbody (https://github.com/pynbody/pynbody) to create Figure 1, and we thank Andrew Pontzen for assistance in creating that figure. We also thank the referees for helpful comments that improved the manuscript.

Endnotes

[a] A public version of CHANGA is publicly available at http://hpcc.astro.washington.edu/tools/changa.html.

[b] Using a geometric density mean in the SPH force expression: $(P_i + P_j)/(\rho_i \rho_j)$ in place of $P_i/\rho_i^2 + P_j/\rho_j^2$ where P_i and ρ_i are particle pressures and densities respectively.

[c] We downloaded version 2.2.15.3.1 in June 2014 from https://hpcforge.org/projects/pkdgrav2/.

References

Planck Collaboration: Ade, PAR, Aghanim, N, Armitage-Caplan, C, Arnaud, M, Ashdown, M, Atrio-Barandela, F, Aumont, J, Baccigalupi, C, Banday, AJ, et al.: Planck 2013 results. XVI. Cosmological parameters (2013). arXiv:1303.5076

Angulo, RE, Springel, V, White, SDM, Jenkins, A, Baugh, CM, Frenk, CS: Scaling relations for galaxy clusters in the millennium-XXL simulation. Mon. Not. R. Astron. Soc. **426**, 2046-2062 (2012). doi:10.1111/j.1365-2966.2012.21830.x

Barnes, J, Hut, P: A hierarchical O(NlogN) force-calculation algorithm. Nature **324**, 446-449 (1986)

Brooks, A: Re-examining astrophysical constraints on the Dark Matter model (2014). arXiv:1407.7544

Brooks, AM, Governato, F, Booth, CM, Willman, B, Gardner, JP, Wadsley, J, Stinson, G, Quinn, T: The origin and evolution of the mass-metallicity relationship for galaxies: results from cosmological N-body simulations. Astrophys. J. Lett. **655**, 17-20 (2007). doi:10.1086/511765

Christensen, C, Quinn, T, Governato, F, Stilp, A, Shen, S, Wadsley, J: Implementing molecular hydrogen in hydrodynamic simulations of galaxy formation. Mon. Not. R. Astron. Soc. **425**, 3058-3076 (2012). doi:10.1111/j.1365-2966.2012.21628.x

Collins, DC, Xu, H, Norman, ML, Li, H, Li, S: Cosmological adaptive mesh refinement magnetohydrodynamics with Enzo. Astrophys. J. Suppl. Ser. **186**, 308-333 (2010). doi:10.1088/0067-0049/186/2/308

Davis, M, Efstathiou, G, Frenk, CS, White, SDM: The evolution of large-scale structure in a universe dominated by cold dark matter. Astrophys. J. **292**, 371-394 (1985). doi:10.1086/163168

Dehnen, W: Towards optimal softening in three-dimensional N-body codes - I. Minimizing the force error. Mon. Not. R. Astron. Soc. **324**, 273-291 (2001). doi:10.1046/j.1365-8711.2001.04237.x

Ding, H-Q, Karasawa, N, Goddard, WA III: The reduced cell multipole method for Coulomb interactions in periodic systems with million-atom unit cells. Chem. Phys. Lett. **196**, 6-10 (1992). doi:10.1016/0009-2614(92)85920-6

Durier, F, Dalla Vecchia, C: Implementation of feedback in smoothed particle hydrodynamics: towards concordance of methods. Mon. Not. R. Astron. Soc. **419**, 465-478 (2012). doi:10.1111/j.1365-2966.2011.19712.x

Ewald, PP: Die Berechnung optischer und elektrostatischer Gitterpotentiale. Ann. Phys. **369**, 253-287 (1921). doi:10.1002/andp.19213690304

Frenk, CS, White, SDM: Dark matter and cosmic structure. Ann. Phys. **524**, 507-534 (2012). doi:10.1002/andp.201200212

Gioachin, F, Sharma, A, Chakravorty, S, Mendes, C, Kale, LV, Quinn, TR: Scalable cosmology simulations on parallel machines. In: VECPAR 2006. LNCS, vol. 4395, pp. 476-489 (2007)

Governato, F, Willman, B, Mayer, L, Brooks, A, Stinson, G, Valenzuela, O, Wadsley, J, Quinn, T: Forming disc galaxies in ΛCDM simulations. Mon. Not. R. Astron. Soc. **374**, 1479-1494 (2007). doi:10.1111/j.1365-2966.2006.11266.x

Governato, F, Zolotov, A, Pontzen, A, Christensen, C, Oh, SH, Brooks, AM, Quinn, T, Shen, S, Wadsley, J: Cuspy no more: how outflows affect the central dark matter and baryon distribution in Λ cold dark matter galaxies. Mon. Not. R. Astron. Soc. **422**, 1231-1240 (2012). doi:10.1111/j.1365-2966.2012.20696.x

Governato, F, Weisz, D, Pontzen, A, Loebman, S, Reed, D, Brooks, AM, Behroozi, P, Christensen, C, Madau, P, Mayer, L, Shen, S, Walker, M, Quinn, T, Wadsley, J: Faint dwarfs as a test of DM models: WDM vs. CDM (2014). arXiv:1407.0022

Guedes, J, Callegari, S, Madau, P, Mayer, L: Forming realistic late-type spirals in a ΛCDM universe: the Eris simulation. Astrophys. J. **742**, 76 (2011). doi:10.1088/0004-637X/742/2/76

Habib, S, Morozov, V, Frontiere, N, Finkel, H, Pope, A, Heitmann, K: Hacc: extreme scaling and performance across diverse architectures. In: Proceedings of the International Conference on High Performance Computing, Networking, Storage and Analysis. SC '13, pp. 6-1610. ACM, New York (2013). doi:10.1145/2503210.2504566

Hopkins, PF: GIZMO: a new class of accurate. mesh-free hydrodynamic simulation methods (2014). arXiv:1409.7395

Hopkins, PF, Keres, D, Onorbe, J, Faucher-Giguere, C-A, Quataert, E, Murray, N, Bullock, JS: Galaxies on FIRE (Feedback In Realistic Environments): stellar feedback explains cosmologically inefficient star formation (2013). arXiv:1311.2073

Ishiyama, T, Nitadori, K, Makino, J: 4.45 Pflops astrophysical N-body simulation on K computer - the gravitational trillion-body problem (2012). arXiv:1211.4406

Jetley, P, Gioachin, F, Mendes, C, Kale, LV, Quinn, TR: Massively parallel cosmological simulations with ChaNGa. In: Proceedings of IEEE International Parallel and Distributed Processing Symposium 2008 (2008)

Jetley, P, Wesolowski, L, Gioachin, F, Kalé, LV, Quinn, TR: Scaling hierarchical N-body simulations on GPU clusters. In: Proceedings of the 2010 ACM/IEEE International Conference for High Performance Computing,

Networking, Storage and Analysis. SC '10. IEEE Computer Society, Washington (2010)

Kalé, LV, Sinha, A: Projections: a scalable performance tool. In: Parallel Systems Fair, International Parallel Processing Symposium, pp. 108-114 (1993)

Keller, BW, Wadsley, J, Benincasa, SM, Couchman, HMP: A superbubble feedback model for galaxy simulations. Mon. Not. R. Astron. Soc. **442**, 3013-3025 (2014). doi:10.1093/mnras/stu1058

Kim, J-H, Abel, T, Agertz, O, Bryan, GL, Ceverino, D, Christensen, C, Conroy, C, Dekel, A, Gnedin, NY, Goldbaum, NJ, Guedes, J, Hahn, O, Hobbs, A, Hopkins, PF, Hummels, CB, Iannuzzi, F, Keres, D, Klypin, A, Kravtsov, AV, Krumholz, MR, Kuhlen, M, Leitner, SN, Madau, P, Mayer, L, Moody, CE, Nagamine, K, Norman, ML, Onorbe, J, O'Shea, BW, Pillepich, A, Primack, JR, Quinn, T, Read, JI, Robertson, BE, Rocha, M, Rudd, DH, Shen, S, Smith, BD, Szalay, AS, Teyssier, R, Thompson, R, Todoroki, K, Turk, MJ, Wadsley, JW, Wise, JH, Zolotov, A (for the AGORA Collaboration29): The AGORA high-resolution galaxy simulations comparison project. Astrophys. J. Suppl. Ser. **210**, 14 (2014). doi:10.1088/0067-0049/210/1/14

Mei, C, Zheng, G, Gioachin, F, Kalé, LV: Optimizing a parallel runtime system for multicore clusters: a case study. In: TeraGrid'10, Pittsburgh, PA, USA (2010)

Mei, C, Sun, Y, Zheng, G, Bohm, EJ, Kalé, LV, Phillips, JC, Harrison, C: Enabling and scaling biomolecular simulations of 100 million atoms on petascale machines with a multicore-optimized message-driven runtime. In: Proceedings of the 2011 ACM/IEEE Conference on Supercomputing, Seattle, WA (2011)

Menon, H, Jain, N, Zheng, G, Kalé, LV: Automated load balancing invocation based on application characteristics. In: IEEE Cluster, Beijing, China, vol. 12 (2012)

Munshi, F, Governato, F, Brooks, AM, Christensen, C, Shen, S, Loebman, S, Moster, B, Quinn, T, Wadsley, J: Reproducing the stellar mass/halo mass relation in simulated LCDM galaxies: theory vs observational estimates. Astrophys. J. **766**, 56 (2013). doi:10.1088/0004-637X/766/1/56

Pontzen, A, Governato, F: How supernova feedback turns dark matter cusps into cores. Mon. Not. R. Astron. Soc. **421**, 3464-3471 (2012). doi:10.1111/j.1365-2966.2012.20571.x

Pontzen, A, Governato, F: Cold dark matter heats up. Nature **506**, 171-178 (2014). doi:10.1038/nature12953

Pontzen, A, Roškar, R, Stinson, GS, Woods, R: pynbody: N-Body/SPH analysis for Python. Astrophysics Source Code Library (2013). ascl:1305.002

Power, C, Navarro, JF, Jenkins, A, Frenk, CS, White, SDM, Springel, V, Stadel, J, Quinn, T: The inner structure of ΛCDM haloes - I. A numerical convergence study. Mon. Not. R. Astron. Soc. **338**, 14-34 (2003). doi:10.1046/j.1365-8711.2003.05925.x

Purcell, CW, Bullock, JS, Tollerud, EJ, Rocha, M, Chakrabarti, S: The Sagittarius impact as an architect of spirality and outer rings in the Milky Way. Nature **477**, 301-303 (2011). doi:10.1038/nature10417

Quinn, T, Katz, N, Stadel, J, Lake, G: Time stepping N-body simulations (1997). arXiv:astro-ph/9710043

Quinn, TR, Jetley, P, Kale, LV, Gioachin, F: N-body simulations with ChaNGa. In: Kale, LV, Bhatele, A (eds.) Parallel Science and Engineering Applications: The Charm++ Approach. CRC Press, Boca Raton (2013)

Reed, D, Gardner, J, Quinn, T, Stadel, J, Fardal, M, Lake, G, Governato, F: Evolution of the mass function of dark matter haloes. Mon. Not. R. Astron. Soc. **346**, 565-572 (2003). doi:10.1046/j.1365-2966.2003.07113.x

Ruan, JJ, Quinn, TR, Babul, A: The observable thermal and kinetic Sunyaev-Zel'dovich effect in merging galaxy clusters. Mon. Not. R. Astron. Soc. **432**, 3508-3519 (2013). doi:10.1093/mnras/stt701

Schaye, J, Crain, RA, Bower, RG, Furlong, M, Schaller, M, Theuns, T, Dalla Vecchia, C, Frenk, CS, McCarthy, IG, Helly, JC, Jenkins, A, Rosas-Guevara, YM, White, SDM, Baes, M, Booth, CM, Camps, P, Navarro, JF, Qu, Y, Rahmati, A, Sawala, T, Thomas, PA, Trayford, J: The EAGLE project: simulating the evolution and assembly of galaxies and their environments (2014). arXiv:1407.7040

Sharma, A: Performance evaluation of tree structures and tree traversals for parallel n-body cosmological simulations. Master's thesis, Department of Computer Science, University of Illinois at Urbana-Champaign (2006) http://charm.cs.uiuc.edu/papers/AmitMSThesis.html

Shen, S, Wadsley, J, Stinson, G: The enrichment of the intergalactic medium with adiabatic feedback - I. Metal cooling and metal diffusion. Mon. Not. R. Astron. Soc. **407**, 1581-1596 (2010). doi:10.1111/j.1365-2966.2010.17047.x

Sinha, AB, Kale, LV, Ramkumar, B: A dynamic and adaptive quiescence detection algorithm. Technical Report 93-11, Parallel Programming Laboratory, Department of Computer Science, University of Illinois, Urbana-Champaign (1993)

Solomonik, E, Kale, LV: Highly scalable parallel sorting. In: Proceedings of the 24th IEEE International Parallel and Distributed Processing Symposium (IPDPS) (2010)

Springel, V: The cosmological simulation code GADGET-2. Mon. Not. R. Astron. Soc. **364**, 1105-1134 (2005). doi:10.1111/j.1365-2966.2005.09655.x

Springel, V: E pur si muove: Galilean-invariant cosmological hydrodynamical simulations on a moving mesh. Mon. Not. R. Astron. Soc. **401**, 791-851 (2010). doi:10.1111/j.1365-2966.2009.15715.x

Springel, V, White, SDM, Jenkins, A, Frenk, CS, Yoshida, N, Gao, L, Navarro, J, Thacker, R, Croton, D, Helly, J, Peacock, JA, Cole, S, Thomas, P, Couchman, H, Evrard, A, Colberg, J, Pearce, F: Simulations of the formation, evolution and clustering of galaxies and quasars. Nature **435**, 629-636 (2005). doi:10.1038/nature03597

Stadel, JG: Cosmological N-body simulations and their analysis. PhD thesis, Department of Astronomy, University of Washington (March 2001)

Stinson, G, Seth, A, Katz, N, Wadsley, J, Governato, F, Quinn, T: Star formation and feedback in smoothed particle hydrodynamic simulations - I. Isolated galaxies. Mon. Not. R. Astron. Soc. **373**, 1074-1090 (2006). doi:10.1111/j.1365-2966.2006.11097.x

Vogelsberger, M, Sijacki, D, Kereš, D, Springel, V, Hernquist, L: Moving mesh cosmology: numerical techniques and global statistics. Mon. Not. R. Astron. Soc. **425**, 3024-3057 (2012). doi:10.1111/j.1365-2966.2012.21590.x

Vogelsberger, M, Genel, S, Springel, V, Torrey, P, Sijacki, D, Xu, D, Snyder, GF, Nelson, D, Hernquist, L: Introducing the illustris project: simulating the coevolution of dark and visible matter in the Universe (2014). arXiv:1405.2921

Wadsley, JW, Stadel, J, Quinn, T: Gasoline: a flexible, parallel implementation of TreeSPH. New Astron. **9**, 137-158 (2004)

Wadsley, JW, Veeravalli, G, Couchman, HMP: On the treatment of entropy mixing in numerical cosmology. Mon. Not. R. Astron. Soc. **387**, 427-438 (2008). doi:10.1111/j.1365-2966.2008.13260.x

Warren, MS: 2HOT: an improved parallel hashed oct-tree N-body algorithm for cosmological simulation (2013). arXiv:1310.4502

Zheng, G: Achieving high performance on extremely large parallel machines: performance prediction and load balancing. PhD thesis, Department of Computer Science, University of Illinois at Urbana-Champaign (2005)

7

The black hole accretion code

Oliver Porth[1]*[iD], Hector Olivares[1], Yosuke Mizuno[1], Ziri Younsi[1], Luciano Rezzolla[1,2],
Monika Moscibrodzka[3], Heino Falcke[3] and Michael Kramer[4]

Abstract

We present the black hole accretion code (BHAC), a new multidimensional general-relativistic magnetohydrodynamics module for the MPI-AMRVAC framework. BHAC has been designed to solve the equations of ideal general-relativistic magnetohydrodynamics in arbitrary spacetimes and exploits adaptive mesh refinement techniques with an efficient block-based approach. Several spacetimes have already been implemented and tested. We demonstrate the validity of BHAC by means of various one-, two-, and three-dimensional test problems, as well as through a close comparison with the HARM3D code in the case of a torus accreting onto a black hole. The convergence of a turbulent accretion scenario is investigated with several diagnostics and we find accretion rates and horizon-penetrating fluxes to be convergent to within a few percent when the problem is run in three dimensions. Our analysis also involves the study of the corresponding thermal synchrotron emission, which is performed by means of a new general-relativistic radiative transfer code, BHOSS. The resulting synthetic intensity maps of accretion onto black holes are found to be convergent with increasing resolution and are anticipated to play a crucial role in the interpretation of horizon-scale images resulting from upcoming radio observations of the source at the Galactic Center.

1 Introduction

Accreting black holes (BHs) are amongst the most powerful astrophysical objects in the Universe. A substantial fraction of the gravitational binding energy of the accreting gas is released within tens of gravitational radii from the BH, and this energy supplies the power for a rich phenomenology of astrophysical systems including active galactic nuclei, X-ray binaries and gamma-ray bursts. Since the radiated energy originates from the vicinity of the BH, a fully general-relativistic treatment is essential for the modelling of these objects and the flows of plasma in their vicinity.

Depending on the mass accretion rate, a given system can be found in various spectral states, with different radiation mechanisms dominating and varying degrees of coupling between radiation and gas (Fender et al. 2004; Markoff 2005). Some supermassive BHs, including the primary targets of observations by the Event-Horizon-Telescope Collaboration (EHTC[a]), i.e., Sgr A* and M87, are

accreting well below the Eddington accretion rate (Marrone et al. 2007; Ho 2009). In this regime, the accretion flow advects most of the viscously released energy into the BH rather than radiating it to infinity. Such optically thin, radiatively inefficient and geometrically thick flows are termed advection-dominated accretion flows (ADAFs, see (Narayan and Yi 1994; Narayan and Yi 1995; Abramowicz et al. 1995; Yuan and Narayan 2014)) and can be modelled without radiation feedback. Next to the ADAF, two additional radiatively inefficient accretion flows (RIAFs) exist: The advection-dominated inflow-outflow solution (ADIOS) (Blandford and Begelman 1999; Begelman 2012) and the convection-dominated accretion flow (CDAF) (Narayan et al. 2000; Quataert and Gruzinov 2000), which include respectively, the physical effects of outflows and convection. Analytical and semi-analytical approaches are reasonably successful in reproducing the main features in the spectra of ADAFs [see, e.g., Yuan et al. (2003)]. However, numerical general-relativistic magnetohydrodynamic (GRMHD) simulations are essential to gain an understanding of the detailed physical processes at play in the Galactic Centre and other low-luminosity compact objects.

*Correspondence: porth@itp.uni-frankfurt.de
[1] Institute for Theoretical Physics, Max-von-Laue-Str. 1, Frankfurt am Main, 60438, Germany
Full list of author information is available at the end of the article

Modern BH accretion-disk theory suggests that angular momentum transport is due to MHD turbulence driven by the magnetorotational instability (MRI) within a differentially rotating disk (Balbus and Hawley 1991; Balbus and Hawley 1998). Recent non-radiative GRMHD simulations of BH accretion systems in an ADAF regime have resolved these processes and reveal a flow structure that can be decomposed into a disk, a corona, a disk-wind and a highly magnetized polar funnel [see, e.g., Villiers and Hawley (2003); McKinney and Gammie (2004); McKinney (2006); McKinney and Blandford (2009)]. The simulations show complex time-dependent behaviour in the disk, corona and wind. Depending on BH spin, the polar regions of the flow contain a nearly force-free, Poynting-flux-dominated jet [see, e.g., Blandford and Znajek (1977); McKinney and Gammie (2004); Hawley and Krolik (2006); McKinney (2006)].

In addition to having to deal with highly nonlinear dynamics that spans a large range in plasma parameters, the numerical simulations also need to follow phenomena that occur across multiple physical scales. For example, in the MHD paradigm, jet acceleration is an intrinsically inefficient process that requires a few thousand gravitational radii to reach equipartition of the energy fluxes (Komissarov et al. 2007; Barkov and Komissarov 2008) (purely hydrodynamical mechanisms can however be far more efficient (Aloy and Rezzolla 2006)). Jet-environment interactions like the prominent HST-1 feature of the radio-galaxy M87 (Biretta et al. 1989; Stawarz et al. 2006; Asada and Nakamura 2012) occur on scales of $\sim 5 \times 10^5$ gravitational radii. Hence, for a self-consistent picture of accretion and ejection, jet formation and recollimation due to interaction with the environment [see, e.g., Mizuno et al. (2015)], numerical simulations must capture horizon-scale processes, as well as parsec-scale interactions with an overall spatial dynamic range of $\sim 10^5$. The computational cost of such large-scale grid-based simulations quickly becomes prohibitive. Adaptive mesh refinement (AMR) techniques promise an effective solution for problems where it is necessary to resolve small and large scale dynamics simultaneously.

Another challenging scenario is presented by radiatively efficient geometrically thin accretion disks that mandate extreme resolution in the equatorial plane in order to resolve the growth of MRI instabilities. Typically this is dealt with by means of stretched grids that concentrate resolution where needed (Avara et al. 2016; Sądowski 2016). However, when the disk is additionally tilted with respect to the spin axis of the BH (Fragile et al. 2007; McKinney et al. 2013), lack of symmetry forbids such an approach. Here an adaptive mesh that follows the warping dynamics of the disk can be of great value. The list of scenarios where AMR can have transformative qualities due to the lack of symmetries goes on, the modelling of star-disk interactions (Giannios and Sironi 2013), star-jet interactions (Barkov

et al. 2010), tidal disruption events (Tchekhovskoy et al. 2014), complex shock geometries (Nagakura and Yamada 2008; Meliani et al. 2017), and intermittency in driven-turbulence phenomena (Radice and Rezzolla 2013; Zrake and MacFadyen 2013), will benefit greatly from adaptive mesh refinement.

Over the past few years, the development of general-relativistic numerical codes employing the 3 + 1 decomposition of spacetime and conservative 'Godunov' schemes based on approximate Riemann solvers (Rezzolla and Zanotti 2013; Font 2003; Martí and Müller 2015) have led to great advances in numerical relativity. Many general-relativistic hydrodynamic (HD) and MHD codes have been developed (Hawley et al. 1984; Koide et al. 2000; De Villiers and Hawley 2003; Gammie et al. 2003; Baiotti et al. 2005; Duez et al. 2005; Anninos et al. 2005; Antón et al. 2006; Mizuno et al. 2006; Del Zanna et al. 2007; Giacomazzo and Rezzolla 2007; Radice and Rezzolla 2012; Radice et al. 2014; McKinney et al. 2014; Etienne et al. 2015; White and Stone 2015; Zanotti and Dumbser 2015; Meliani et al. 2017) and applied to study a variety of problems in high-energy astrophysics. Some of these implementations provide additional capabilities that incorporate approximate radiation transfer [see, e.g., Sądowski et al. (2013); McKinney et al. (2013); Takahashi et al. (2016)] and/or non-ideal MHD processes [see, e.g., Dionysopoulou et al. (2013); Foucart et al. (2016)]. Although these codes have been applied to many astrophysical scenarios involving compact objects and matter [for recent reviews see, e.g., Martí and Müller (2015); Baiotti and Rezzolla (2016)], full AMR is still not commonly utilised and exploited [with the exception of Anninos et al. (2005); Zanotti et al. (2015); White and Stone (2015)]. BHAC attempts to fill this gap by providing a fully-adaptive multidimensional GRMHD framework that features state-of-the-art numerical schemes.

Qualitative aspects of BH accretion simulations are code-independent [see, e.g., Villiers and Hawley (2003); Gammie et al. (2003); Anninos et al. (2005)], but quantitative variations raise questions regarding numerical convergence of the observables (Shiokawa et al. 2012; White and Stone 2015). In preparation for the upcoming EHTC observations, a large international effort, whose European contribution is represented in part by the Black-HoleCam project[b] (Goddi et al. 2016), is concerned with forward modelling of the future event horizon-scale interferometric observations of Sgr A* and M87 at submillimeter (EHTC; (Doeleman et al. 2009)) and near-infrared wavelengths (VLTI GRAVITY; (Eisenhauer et al. 2008)). To this end, GRMHD simulations have been coupled to general-relativistic radiative transfer (GRRT) calculations [see, e.g., Mościbrodzka et al. (2009); Dexter et al. (2009); Chan et al. (2015); Gold et al. (2016); Dexter et al. (2012); Mościbrodzka et al. (2016)]. In order to assess the credibility of these radiative models, it is necessary to assess the quantitative convergence of the underlying GRMHD simulations. In order to demonstrate the utility of BHAC for the EHTC science-case, we therefore validate the results obtained with BHAC against the HARM3D code (Gammie et al. 2003; Noble et al. 2009) and investigate the convergence of the GRMHD simulations and resulting observables obtained with the GRRT post-processing code BHOSS (Younsi et al. 2017).

The structure of the paper is as follows. In Section 2 we describe the governing equations and numerical methods. In Section 3 we show numerical tests in special-relativistic and general-relativistic MHD. In Section 4 the results of 2D and 3D GRMHD simulations of magnetised accreting tori are presented. In Section 5 we briefly describe the GRRT post-processing calculation and the resulting image maps from the magnetised torus simulation shown in Section 4. In Section 6 we present our conclusions and outlook.

Throughout this paper, we adopt units where the speed of light, $c = 1$, the gravitational constant, $G = 1$, and the gas mass is normalised to the central compact object mass. Greek indices run over space and time, i.e., $(0, 1, 2, 3)$, and Roman indices run over space only, i.e., $(1, 2, 3)$. We assume a $(-, +, +, +)$ signature for the spacetime metric. Self-gravity arising from the gas is neglected.

2 GRMHD formulation and numerical methods

In this section we briefly describe the covariant GRMHD equations, introduce the notation used throughout this paper, and present the numerical approach taken in our solution of the GRMHD system. The computational infrastructure underlying BHAC is the versatile open-source MPI-AMRVAC toolkit (Keppens et al. 2012; Porth et al. 2014).

In-depth derivations of the covariant fluid- and magneto-fluid dynamical equations can be found in the textbooks by (Landau and Lifshitz 2004; Weinberg 1972; Rezzolla and Zanotti 2013). We follow closely the derivation of the GRMHD equations by (Del Zanna et al. 2007). This is very similar to the 'Valencia formulation', cf. (Rezzolla and Zanotti 2013) and (Antón et al. 2006). The general considerations of the '3 + 1' split of spacetime are discussed in greater detail in (Misner et al. 1973; Gourgoulhon 2007; Alcubierre 2008).

We start from the usual set of MHD equations in covariant notation

$$
\begin{aligned}
\nabla_\mu \left(\rho u^\mu \right) &= 0, \\
\nabla_\mu T^{\mu\nu} &= 0, \\
\nabla_\mu {}^* F^{\mu\nu} &= 0,
\end{aligned}
\tag{1}
$$

which respectively constitute mass conservation, conservation of the energy-momentum tensor $T^{\mu\nu}$, and the homogeneous Faraday's law. The Faraday tensor $F^{\mu\nu}$ may be

constructed from the electric and magnetic fields E^α, B^α as measured in a generic frame U^α as

$$
F^{\mu\nu} = U^\mu E^\nu - U^\nu E^\mu - (-g)^{-1/2} \eta^{\mu\nu\lambda\delta} U_\lambda B_\delta,
\tag{2}
$$

where $\eta^{\mu\nu\lambda\delta}$ is the fully-antisymmetric symbol (see, e.g., (Rezzolla and Zanotti 2013)) and g the determinant of the spacetime four-metric. The dual Faraday tensor ${}^* F^{\mu\nu} := \frac{1}{2}(-g)^{-1/2} \eta^{\mu\nu\lambda\delta} F_{\lambda\delta}$ is then

$$
{}^* F^{\mu\nu} = U^\mu B^\nu - U^\nu B^\mu - (-g)^{-1/2} \eta^{\mu\nu\lambda\delta} U_\lambda E_\delta.
\tag{3}
$$

We are interested only in the ideal MHD limit of vanishing electric fields in the fluid frame u^μ, hence

$$
F^{\mu\nu} u_\nu = 0,
\tag{4}
$$

such that the inhomogeneous Faraday's law is not required and electric fields are dependent functions of velocities and magnetic fields. To eliminate the electric fields from the equations it is convenient to introduce vectors in the fluid frame and therefore we define the corresponding electric and magnetic field four-vectors as

$$
e^\mu := F^{\mu\nu} u_\nu, \qquad b^\mu := {}^* F^{\mu\nu} u_\nu,
\tag{5}
$$

where $e^\mu = 0$ and we obtain the constraint $u_\mu b^\mu = 0$. The Faraday tensor is then

$$
\begin{aligned}
F^{\mu\nu} &= -(-g)^{-1/2} \eta^{\mu\nu\lambda\delta} u_\lambda b_\delta, \\
{}^* F^{\mu\nu} &= b^\mu u^\nu - b^\nu u^\mu,
\end{aligned}
\tag{6}
$$

and we can write the total energy-momentum tensor in terms of the vectors u^μ and b^μ alone (Anile 1990) as

$$
T^{\mu\nu} = \rho h_{\text{tot}} u^\mu u^\nu + p_{\text{tot}} g^{\mu\nu} - b^\mu b^\nu.
\tag{7}
$$

Here the total pressure $p_{\text{tot}} = p + b^2/2$ was introduced, as well as the total specific enthalpy $h_{\text{tot}} = h + b^2/\rho$. In addition, we define the scalar $b^2 := b^\nu b_\nu$, denoting the square of the fluid frame magnetic field strength as $b^2 = B^2 - E^2$.

2.1 3 + 1 split of spacetime

We proceed to split spacetime into $3 + 1$ components by introducing a foliation into space-like hyper-surfaces Σ_t defined as iso-surfaces of a scalar time function t. This leads to the timelike unit vector normal to the slices Σ_t (Alcubierre 2008; Rezzolla and Zanotti 2013)

$$
n_\mu := -\alpha \nabla_\mu t,
\tag{8}
$$

where α is the so-called *lapse*-function. The four-velocity n^μ defines the frame of the *Eulerian observer*. If $g_{\mu\nu}$ is the metric associated with the four-dimensional manifold, we can define the metric associated with each timelike slice as

$$\gamma_{\mu\nu} := g_{\mu\nu} + n_\mu n_\nu. \tag{9}$$

This also allows us to introduce the spatial projection operator

$$\gamma_\nu^\mu := \delta_\nu^\mu + n^\mu n_\nu \tag{10}$$

such that $\gamma_\nu^\mu n_\mu = 0$ and through which we can project any four-vector V^μ (or tensor) into its temporal and spatial components.

Introducing a coordinate system adapted to the foliation Σ_t, the line element is given in $3 + 1$ form (Arnowitt et al. 2008) as

$$ds^2 = -\alpha^2\, dt^2 + \gamma_{ij}\big(dx^i + \beta^i\, dt\big)\big(dx^j + \beta^j\, dt\big), \tag{11}$$

where the spatial vector β^μ is called the *shift* vector. Written in terms of coordinates, it describes the motion of coordinate lines as seen by an Eulerian observer

$$x_{t+dt}^i = x_t^i - \beta^i\big(t, x^j\big)\, dt. \tag{12}$$

More explicitly, we write the metric $g_{\mu\nu}$ and its inverse $g^{\mu\nu}$ as

$$g_{\mu\nu} = \begin{pmatrix} -\alpha^2 + \beta_k\beta^k & \beta_i \\ \beta_j & \gamma_{ij} \end{pmatrix},$$
$$g^{\mu\nu} = \begin{pmatrix} -1/\alpha^2 & \beta^i/\alpha^2 \\ \beta^j/\alpha^2 & \gamma^{ij} - \beta^i\beta^j/\alpha^2 \end{pmatrix}. \tag{13}$$

From (13) we find the following useful relation between the determinants of the 3-metric and 4-metric

$$(-g)^{1/2} = \alpha\gamma^{1/2}. \tag{14}$$

In a coordinate system specified by (11), the four-velocity of the Eulerian observer reads

$$n_\mu = (-\alpha, 0_i), \qquad n^\mu = \big(1/\alpha, -\beta^i/\alpha\big). \tag{15}$$

It is easy to verify that this normalised vector is indeed orthogonal to any space-like vector on the foliation Σ_t. Given a fluid element with four-velocity u^μ, the Lorentz factor with respect to the Eulerian observer is[c] $\Gamma := -u^\mu n_\mu = \alpha u^0$ and we introduce the three-vectors

$$v^i := \frac{\gamma_\mu^i u^\mu}{\Gamma} = \frac{u^i}{\Gamma} + \frac{\beta^i}{\alpha}, \qquad v_i := \gamma_{ij}v^j = \frac{u_i}{\Gamma}, \tag{16}$$

which denote the fluid three-velocity.

In the following, purely spatial vectors (e.g., $v^0 = 0$) are denoted by Roman indices. Note that $\Gamma = (1 - v^2)^{-1/2}$ with $v^2 = v_i v^i$ just as in special relativity.

Further useful three-vectors are the electric and magnetic fields in the Eulerian frame

$$E^i := F^{iv}n_\nu = \alpha F^{i0}, \qquad B^i := {}^*F^{iv}n_\nu = \alpha\,{}^*F^{i0}, \tag{17}$$

which differ by a factor α from the definitions used in (Komissarov 1999; Gammie et al. 2003). Writing the general Faraday tensor (2) in terms of quantities in the Eulerian frame, the ideal MHD condition (4) leads to the well known relation

$$E^i = \gamma^{-1/2}\eta^{ijk}B_j v_k, \tag{18}$$

or put simply: $\mathbf{E} = \mathbf{B} \times \mathbf{v}$ (here η_{ijk} is the standard Levi-Civita antisymmetric symbol). Combining (6) with (17), one obtains the transformation between b^μ and B^μ as

$$b^i = \frac{B^\mu + \alpha b^0 u^i}{\Gamma}, \qquad b^0 = \frac{\Gamma(B^i v_i)}{\alpha} \tag{19}$$

which enables the dual Faraday tensor (6) to be expressed in terms of the Eulerian fields

$${}^*F^{\mu\nu} = \frac{B^\mu u^\nu - B^\nu u^\mu}{\Gamma}. \tag{20}$$

Equation (1) with the Faraday tensor in the form (20) yields the final evolution equation for B^μ. The time component of this leads to the constraint $\partial_i\sqrt{\gamma}B^i = 0$ or put more simply: $\nabla \cdot \mathbf{B} = 0$. Following (19) we obtain the scalar b^2 as

$$b^2 = \frac{B^2 + \alpha^2(b^0)^2}{\Gamma^2} = \frac{B^2}{\Gamma^2} + \big(B^i v_i\big)^2, \tag{21}$$

where $B^2 := B^i B_i$.

Using the spatial projection operator, the GRMHD Eqs. (1) can be decomposed into spatial and temporal components. We skip ahead over the involved algebra [see e.g., Del Zanna et al. (2007)] and directly state the final conservation laws

$$\partial_t(\sqrt{\gamma}\mathbf{U}) + \partial_i\big(\sqrt{\gamma}\mathbf{F}^i\big) = \sqrt{\gamma}\mathbf{S}, \tag{22}$$

with the conserved variables $\mathbf{U}$ and fluxes $\mathbf{F}^i$ defined as

$$\mathbf{U} = \begin{bmatrix} D \\ S_j \\ \tau \\ B^j \end{bmatrix}, \qquad \mathbf{F}^i = \begin{bmatrix} \mathcal{V}^i D \\ \alpha W_j^i - \beta^i S_j \\ \alpha(S^i - v^i D) - \beta^i \tau \\ \mathcal{V}^i B^j - B^i \mathcal{V}^j \end{bmatrix}, \tag{23}$$

where we define the *transport velocity* $\mathcal{V}^i := \alpha v^i - \beta^i$. Hence we solve for conservation of quantities in the Eulerian frame: the density $D := -\rho u^\nu n_\nu$, the covariant three-momentum S_j, the rescaled energy density $\tau = U - D^{\mathrm{d}}$

(where U denotes the total energy density as seen by the Eulerian observer), and the Eulerian magnetic three-fields B^j. The conserved energy density U is given by

$$U := T^{\mu\nu} n_\mu n_\nu$$

$$= \rho h \Gamma^2 - p + \frac{1}{2}\left(E^2 + B^2\right) \tag{24}$$

$$= \rho h \Gamma^2 - p + \frac{1}{2}\left[B^2\left(1 + v^2\right) - \left(B^j v_j\right)^2\right]. \tag{25}$$

The purely spatial variant of the stress-energy tensor W^{ij} was introduced for example in (23). It reads just as in special relativity

$$W^{ij} := \gamma^i_\mu \gamma^j_\nu T^{\mu\nu}$$

$$= \rho h \Gamma^2 v^i v^j - E^i E^j - B^i B^j$$

$$+ \left[p + \frac{1}{2}\left(E^2 + B^2\right)\right]\gamma^{ij} \tag{26}$$

$$= S^i v^j + p_{\text{tot}} \gamma^{ij} - \frac{B^i B^j}{\Gamma^2} - \left(B^k v_k\right)v^i B^j. \tag{27}$$

Correspondingly, the covariant three-momentum density in the Eulerian frame is

$$S_i := \gamma^\mu_i n^\alpha T_{\alpha\mu} = \rho h \Gamma^2 v_i + \eta_{ijk}\gamma^{1/2} E^j B^k \tag{28}$$

$$= \rho h \Gamma^2 v_i + B^2 v_i - \left(B^j v_j\right)B_i, \tag{29}$$

as usual. For the sources S we employ the convenient Valencia formulation without Christoffel symbols, yielding

$$S = \begin{bmatrix} 0 \\ \frac{1}{2}\alpha W^{ik}\partial_j\gamma_{ik} + S_i\partial_j\beta^i - U\partial_j\alpha \\ \frac{1}{2}W^{ik}\beta^j\partial_j\gamma_{ik} + W^j_i\partial_j\beta^i - S^j\partial_j\alpha \\ 0 \end{bmatrix} \tag{30}$$

which is valid for stationary spacetimes that are considered for the remainder of this work (Cowling approximation). Following the definitions (23) and (30), all vectors and tensors are now specified through their purely spatial variants and thus apart from the occurrence of the lapse function α and the shift vector β^i, the equations take on a form identical to the special-relativistic MHD (SRMHD) equations. This fact allows for a straightforward transformation from the SRMHD physics module of MPI-AMRVAC into a full GRMHD code.

In addition to the set of conserved variables U, knowledge of the primitive variables $P(U)$ is required for the calculation of fluxes and source terms. They are given by

$$P = \left[\rho, \Gamma v^i, p, B^i\right]. \tag{31}$$

While the transformation $U(P)$ is straightforward, the inversion $P(U)$ is a non-trivial matter which will be discussed further in Section 2.10. Note that just like in MPI-AMRVAC, we do not store the primitive variables P but extend the conserved variables by the set of *auxiliary* variables

$$A = [\Gamma, \xi], \tag{32}$$

where $\xi := \Gamma^2 \rho h$. Knowledge of A allows for quick transformation of $P(U)$. The issue of inversion then becomes a matter of finding an A consistent with both P and U.

2.2 Finite volume formulation

Since BHAC solves the equations in a finite volume formulation, we take the integral of Eq. (22) over the spatial element of each cell $\int dx^1 dx^2 dx^3$

$$\int \partial_t\left(\gamma^{1/2}U\right) dx^1 dx^2 dx^3 + \int \partial_i\left(\gamma^{1/2}F^i\right) dx^1 dx^2 dx^3$$

$$= \int \gamma^{1/2}S \, dx^1 dx^2 dx^3. \tag{33}$$

This can be written (cf. (Banyuls et al. 1997)) as

$$\partial_t(\bar{U}\Delta V) + \int_{\partial V(x^1+\Delta x^1/2)} \gamma^{1/2}F^1 \, dx^2 dx^3$$

$$- \int_{\partial V(x^1-\Delta x^1/2)} \gamma^{1/2}F^1 \, dx^2 dx^3$$

$$+ \int_{\partial V(x^2+\Delta x^2/2)} \gamma^{1/2}F^2 \, dx^1 dx^3$$

$$- \int_{\partial V(x^2-\Delta x^2/2)} \gamma^{1/2}F^2 \, dx^1 dx^3$$

$$+ \int_{\partial V(x^3+\Delta x^3/2)} \gamma^{1/2}F^3 \, dx^1 dx^2$$

$$- \int_{\partial V(x^3-\Delta x^3/2)} \gamma^{1/2}F^3 \, dx^1 dx^2$$

$$= \bar{S}\Delta V, \tag{34}$$

with the volume averages defined as

$$\bar{U} := \frac{\int \gamma^{1/2}U \, dx^1 dx^2 dx^3}{\Delta V},$$

$$\bar{S} := \frac{\int \gamma^{1/2}S \, dx^1 dx^2 dx^3}{\Delta V}, \tag{35}$$

and

$$\Delta V = \int \gamma^{1/2} dx^1 dx^2 dx^3. \tag{36}$$

We next define also the 'surfaces' ΔS^i and corresponding surface-averaged fluxes

$$\Delta S^i_{\partial V(x^i + \Delta x^i/2)} = \int_{\partial V(x^i + \Delta x^i/2)} \gamma^{1/2} \, dx^{j,j\neq i}, \tag{37}$$

and

$$\bar{F}^i_{\partial V(x^i + \Delta x^i/2)} = \frac{\int_{\partial V(x^i + \Delta x^i/2)} \gamma^{1/2} F^i \, dx^{j,j\neq i}}{\Delta S^i}. \tag{38}$$

Considering that ΔV is assumed constant in time, this leads to the evolution equation

$$\partial_t \bar{U} = -\frac{1}{\Delta V}\Big[\bar{F}^1 \Delta S^1|_{\partial V(x^1 + \Delta x^1/2)} - \bar{F}^1 \Delta S^1|_{\partial V(x^1 - \Delta x^1/2)}$$
$$+ \bar{F}^2 \Delta S^2|_{\partial V(x^2 + \Delta x^2/2)} - \bar{F}^2 \Delta S^2|_{\partial V(x^2 - \Delta x^2/2)}$$
$$+ \bar{F}^3 \Delta S^3|_{\partial V(x^3 + \Delta x^3/2)} - \bar{F}^3 \Delta S^3|_{\partial V(x^3 - \Delta x^3/2)}\Big]$$
$$+ \bar{S}. \tag{39}$$

We aim to achieve second-order accuracy and represent the interface-averaged flux, e.g., $\bar{F}^1_{\partial V(x^1 + \Delta x^1/2)}$, with the value at the midpoint, change to an intuitive index notation $F^1_{i+1/2,j,k}$, and then arrive at a semi-discrete equation for the average state in the cell (i,j,k) as

$$\frac{d\bar{U}_{i,j,k}}{dt} = -\frac{1}{\Delta V_{i,j,k}}\Big[F^1 \Delta S^1|_{i+1/2,j,k} - F^1 \Delta S^1|_{i-1/2,j,k}$$
$$+ F^2 \Delta S^2|_{i,j+1/2,k} - F^2 \Delta S^2|_{i,j-1/2,k}$$
$$+ F^3 \Delta S^3|_{i,j,k+1/2} - F^3 \Delta S^3|_{i,j,k-1/2}\Big]$$
$$+ S_{i,j,k}. \tag{40}$$

Here the source term $S_{i,j,k}$ is also evaluated at the cell barycenter to second-order accuracy (Mignone 2014). Barycenter coordinates $\bar{x}^i$ are straightforwardly defined as

$$\bar{x}^i = \frac{\int \gamma^{1/2} x^i \, dx^1 \, dx^2 \, dx^3}{\Delta V}. \tag{41}$$

This finite volume form is readily solved with the MPI–AMRVAC toolkit. For ease of implementation, we pre-compute all static integrals yielding cell volumes ΔV, Surfaces ΔS^i and barycenter coordinates. The integrations are performed numerically at the phase of initialisation using a fourth-order Simpson's rule.

For the temporal update, we interpret the semi-discrete form (40) as an ordinary differential equation in time for each cell and employ a multi-step Runge-Kutta scheme to evolve the average state in the cell $\bar{U}_{i,j,k}$, a procedure also known as 'method of lines'. At each sub-step, the point-wise interface fluxes F^i are obtained by performing a limited reconstruction operation of the cell-averaged state $\bar{U}$

to the interfaces (see Section 2.8) and employing approximate Riemann solvers, e.g., *HLL* or *TVDLF* (Section 2.9).

Several temporal update schemes are available: simple predictor-corrector, third-order Runge-Kutta (RK) RK3 (Gottlieb and Shu 1998) and the strong-stability preserving s-step, pth-order RK schemes SSPRK(s,p) schemes: SSPRK$(4,3)$, SSPRK$(5,4)$ due to (Spiteri and Ruuth 2002).[e]

2.3 Metric data-structure

The metric data-structure is built to be optimal in terms of storage while remaining convenient to use. Since the metric and its derivatives are often sparsely populated, the data is ultimately stored using index lists. For example, each element in the index list for the four-metric $g_{\mu\nu}$ holds the indices of the non-zero element together with a Fortran90 array of the corresponding metric coefficient for the grid block. A summation over indices, e.g., 'lowering' can then be cast as a loop over entries in the index-list only. For convenience, all elements can also be accessed directly over intuitive identifiers which point to the storage in the index list, e.g., m%g(mu,nu)%elem yields the grid array of the $g_{\mu\nu}$ metric coefficients as expected. Similarly, the lower-triangular indices point to the transposed indices in presence of symmetries. In addition, one block of zeros is allocated in the metric data-structure and all zero elements are set to point towards it. An overview of the available identifiers is given in Table 1.

As a consequence, only 14 grid functions are required for the Schwarzschild coordinates and 29 grid functions need to be allocated in the Kerr-Schild (KS) case. This is still less than half of the 68 grid functions which a brute-force approach would yield. The need for efficient storage management becomes apparent when we consider that the metric is required in the barycenter as well as on the interfaces, thus multiplying the required grid functions by a factor of four for three-dimensional simulations (yielding 116 grid functions in the KS case).

In order to eliminate the error-prone process of implementing complicated functions for metric derivatives, BHAC can obtain derivatives by means of an accurate complex-step numerical differentiation (Squire and Trapp 1998). This elegant method takes advantage of the Cauchy-Riemann differential equations for complex derivatives and achieves full double-precision accuracy, thereby avoiding the stepsize dilemma of common finite-differencing formulae (Martins et al. 2003). The small price to pay is that at the initialisation stage, metric elements are provided via functions of the complexified coordinates. However, the intrinsic complex arithmetic of Fortran90 allows for seamless implementation.

To promote full flexibility in the spacetime, we always calculate the inverse metric γ^{ij} using the standard LU decomposition technique (Press et al. 2007). As a result,

Table 1 Elements of the metric data-structure

Symbol	Identifier	Index list
$g_{\mu\nu}$	m%g(mu,nu)	m%nnonzero, m%nonzero(inonzero)
α	m%alpha	-
β^i	m%beta(i)	m%nnonzeroBeta, m%nonzeroBeta(inonzero)
$\sqrt{\gamma}$	m%sqrtgamma	-
γ^{ij}	m%gammainv(i,j)	-
β_i	m%betaD(i)	-
$\partial_k\gamma_{ij}$	m%dgdk(i,j,k)	m%nnonzeroDgDk, m%nonzeroDgDk(inonzero)
$\partial_j\beta^i$	m%DbetaiDj(i,j)	m%nnonzeroDbetaiDj, m%nonzeroDbetaiDj(inonzero)
$\partial_j\alpha$	m%DalphaDj(j)	m%nnonzeroDalphaDj, m%nonzeroDalphaDj(inonzero)
0	m%zero	-

GRMHD simulations on any metric can be performed after providing only the non-zero elements of the three-metric $\gamma_{ij}(x^1,x^2,x^3)$, the lapse function $\alpha(x^1,x^2,x^3)$ and the shift vector $\beta^i(x^1,x^2,x^3)$. As an additional convenience, BHAC can calculate the required elements and their derivatives entirely from the four-metric $g_{\mu\nu}(x^0,x^1,x^2,x^3)$.

2.4 Equations of state

For closure of the system (1)-(4), an equation of state (EOS) connecting the specific enthalpy h with the remaining thermodynamic variables $h(\rho,p)$ is required (Rezzolla and Zanotti 2013). The currently implemented closures are

- *Ideal gas*: $h(\rho,p) = 1 + \dfrac{\hat{\gamma}}{\hat{\gamma}-1}\dfrac{p}{\rho}$ with adiabatic index $\hat{\gamma}$.
- *Synge gas*: $h(\Theta) = \dfrac{K_3(\Theta^{-1})}{K_2(\Theta^{-1})}$, where the relativistic temperature is given by $\Theta = p/\rho$ and K_n denotes the modified Bessel function of the second kind. In fact, we use an approximation to the previous expression that does not contain Bessel functions [see Meliani et al. (2004); Keppens et al. (2012)].
- *Isentropic flow*: Assumes an ideal gas with the additional constraint $p = \kappa\rho^{\hat{\gamma}}$, where the pseudo-entropy κ may be chosen arbitrarily. This allows one to omit the energy equation entirely and only the reduced set $\boldsymbol{P} = \{\rho,v^i,B^j\}$ is solved.

As long as $h(\rho,p)$ is analytic, its implementation in BHAC is straightforward.

2.5 Divergence cleaning and augmented Faraday's law

To control the $\nabla\cdot\boldsymbol{B} = 0$ constraint on AMR grids, we have adopted a constraint dampening approach customarily used in Newtonian MHD (Dedner et al. 2002). In this approach, which is usually referred as Generalized Lagrangian Multiplier (GLM) of the Maxwell equations (but is also known as the 'divergence-cleaning' approach), we extend the usual Faraday tensor by the scalar ϕ, such that the homogeneous Maxwell equation reads

$$\nabla_\nu\left({}^*F^{\mu\nu} - \phi g^{\mu\nu}\right) = -\kappa n^\mu\phi, \tag{42}$$

and the scalar ϕ follows from contraction $\phi = ({}^*F^{\mu\nu} - \phi g^{\mu\nu})n_\mu n_\nu$. Naturally, for $\phi \to 0$, the usual set of Maxwell equations is recovered. It is straightforward to show [see, e.g., Palenzuela et al. (2009)] that (42) leads to a telegraph equation for the constraint violation parameter ϕ which becomes advected at the speed of light and decays on a timescale $1/\kappa$. With the modification (42), the time-component of Maxwell's equation now becomes an evolution equation for ϕ. After some algebra (see Appendix A), we obtain

$$\partial_t\sqrt{\gamma}\phi + \partial_i\left[\sqrt{\gamma}\left(\alpha B^i - \phi\beta^i\right)\right]$$
$$= -\sqrt{\gamma}\alpha\kappa\phi - \sqrt{\gamma}\phi\partial_i\beta^i$$
$$- \frac{1}{2}\sqrt{\gamma}\phi\gamma^{ij}\beta^k\partial_k\gamma_{ij} + \sqrt{\gamma}B^i\partial_i\alpha. \tag{43}$$

Equivalently, the modified evolution equations for B^i (see Appendix B) read

$$\partial_t\left(\sqrt{\gamma}B^j\right) + \partial_i\left(\sqrt{\gamma}\left(\mathcal{V}^iB^j - \mathcal{V}^jB^i - B^i\beta^j\right)\right)$$
$$= -\sqrt{\gamma}B^i\partial_i\beta^j - \sqrt{\gamma}\alpha\gamma^{ij}\partial_i\phi. \tag{44}$$

Now Eq. (44) replaces the usual Faraday's law and (43) is evolved alongside the modified MHD system. Due to the term $\partial_i\phi$ on the right hand side of Eq. (44), the new equation is non-hyperbolic. Hence, numerical stability can be a more involved issue than for hyperbolic equations. We find that the numerical stability of the system is enhanced when using an upwinded discretisation for $\partial_i\phi$. Note that Eqs. (43) and (44) are in agreement with (Dionysopoulou et al. 2013) when accounting for $\frac{\partial_i\sqrt{\gamma}}{\sqrt{\gamma}} = \frac{1}{2}\gamma^{lm}\partial_i\gamma_{lm}$ and taking the ideal MHD limit.

2.6 Flux-interpolated constrained transport

As an alternative to the GLM approach, the $\nabla\cdot\boldsymbol{B} = 0$ constraint can be enforced using a cell-centred version of Flux-interpolated Constrained Transport (FCT) consistent with the finite volume scheme used to evolve the hydrodynamic variables. Constrained Transport (CT) schemes aim to keep to zero at machine precision the sum of the magnetic fluxes through all surfaces bounding a cell, and therefore (in the continuous limit) the divergence of the magnetic field inside the cell. In the original version (Evans and Hawley 1988) this is achieved by evolving the magnetic flux through the cell faces and computing the circulation of the electric field along the edges bounding each face. Since each edge appears with opposite signs in the time update of two faces belonging to the same cell, the total magnetic flux leaving each cell is conserved during evolution. The magnetic field components at cell centers, necessary for performing the transformation from primitive to conserved variables and vice-versa, are then found using interpolation from the cell faces. (Toth 2000) showed that it is possible to find cell centred variants of CT schemes that go from the average field components at the cell center at a given time to those one (partial) time step ahead in a single step, without the need to compute magnetic fluxes at cell faces. The CT variant known as FCT is particularly well suited for finite volume conservative schemes as that employed by BHAC, as it calculates the electric fields necessary for the update as an average of the fluxes given by the Riemann solver. In this way, the time update for its cell centred version can be written using a form similar to (40). For example, for the update of the $\bar{B}^1$ component, we obtain

$$\frac{d\bar{B}^1_{i,j,k}}{dt} = -\frac{1}{\Delta V_{i,j,k}}\left[F^{*2}\Delta S^2|_{i,j+1/2,k} - F^{*2}\Delta S^2|_{i,j-1/2,k}\right.$$
$$\left. + F^{*3}\Delta S^3|_{i,j,k+1/2} - F^{*3}\Delta S^3|_{i,j,k-1/2}\right], \quad (45)$$

where the modified fluxes in the x^1-direction are zero and the remaining fluxes are calculated as

$$F^{*2}\Delta S^2|_{i,j-1/2,k}$$
$$= \frac{\Delta x^1_i}{8}\left(2\frac{\bar{F}^2\Delta S^2|_{i,j-1/2,k}}{\Delta x^1_i}\right.$$
$$+ \frac{\bar{F}^2\Delta S^2|_{i+1,j-1/2,k}}{\Delta x^1_{i+1}} + \frac{\bar{F}^2\Delta S^2|_{i-1,j-1/2,k}}{\Delta x^1_{i-1}}$$
$$- \frac{\bar{F}^1\Delta S^1|_{i-1/2,j,k}}{\Delta y_j} - \frac{\bar{F}^1\Delta S^1|_{i-1/2,j-1,k}}{\Delta x^2_{j-1}}$$
$$\left. - \frac{\bar{F}^1\Delta S^1|_{i+1/2,j,k}}{\Delta x^2_j} - \frac{\bar{F}^1\Delta S^1|_{i+1/2,j-1,k}}{\Delta x^2_{j-1}}\right). \quad (46)$$

The derivation of Eqs. (45) and (46) from the staggered version with magnetic fields located at cell faces is given in Appendix C. Since magnetic fields are stored at the cell center and not at the faces, the divergence conserved by the FCT method corresponds to a particular discretisation

$$\frac{1}{2}\Delta V^*(\nabla \cdot \boldsymbol{B})|_{i+1/2,j+1/2,k+1/2}$$
$$= \sum_{l_1,l_2,l_3=0,1}\left[(-1)^{1+l_1}\frac{\bar{B}^1\Delta V}{\Delta x^1} + (-1)^{1+l_2}\frac{\bar{B}^2\Delta V}{\Delta x^2}\right.$$
$$\left. + (-1)^{1+l_3}\frac{\bar{B}^3\Delta V}{\Delta x^3}\right]_{i+l_1,j+l_2,k+l_3}, \quad (47)$$

where

$$\Delta V^*|_{i+1/2,j+1/2,k+1/2} = \sum_{l_1,l_2,l_3=0,1}\Delta V|_{i+l_1,j+l_2,k+l_3}. \quad (48)$$

Equation (47) is closely related to the integral over the surface of a volume containing eight cells in 3D (see Appendix D for the derivation), and it reduces to equation (27) from (Toth 2000) in the special case of Cartesian coordinates. As mentioned before, this scheme can maintain $\nabla \cdot \boldsymbol{B} = 0$ to machine precision only if it was already zero at the initial condition. The corresponding curl operator used to setup initial conditions is derived in Appendix D.

In its current form, BHAC cannot handle both constrained transport and AMR. The reason is that special prolongation and restriction operators are required in order to avoid the creation of divergence when refining or coarsening. Due to the lack of information about the magnetic flux on cell faces, the problem of finding such divergence-preserving prolongation operators becomes underdetermined. However, storing the face-allocated (staggered) magnetic fluxes and applying the appropriate prolongation and restriction operators requires a large change in the code infrastructure on which we will report in an accompanying work.

2.7 Coordinates
Since one of the main motivations for the development of the BHAC code is to simulate BH accretion in arbitrary metric theories of gravity, the coordinates and metric datastructures have been designed to allow for maximum flexibility and can easily be extended. A list of the currently available coordinate systems is given in Table 2. In addition to the identifiers used in the code, the table lists whether numerical derivatives are used and whether the coordinates are initialised from the three-metric or the four-metric. The less well-known spacetimes and coordinates are described in the following subsection.

2.7.1 Modified Kerr-Schild coordinates
Modified KS coordinates were introduced by e.g., (McKinney and Gammie 2004) with the purpose of stretching the grid radially and being able to concentrate resolution in the equatorial region.

Table 2 Coordinates available in BHAC

Coordinates	Identifier	Num. derivatives	Init. $g_{\mu\nu}$
Cartesian	cart	No	No
Boyer-Lindquist	bl	No	No
Kerr-Schild	ks	No	No
Modified Kerr-Schild	mks	No	No
Cartesian Kerr-Schild	cks	Yes	Yes
Rezzolla & Zhidenko parametrization (Rezzolla and Zhidenko 2014)	rz	Yes	No
Horizon penetrating Rezzolla & Zhidenko coordinates	rzks	Yes	Yes
Hartle-Thorne (Hartle and Thorne 1968)	ht	Yes	Yes

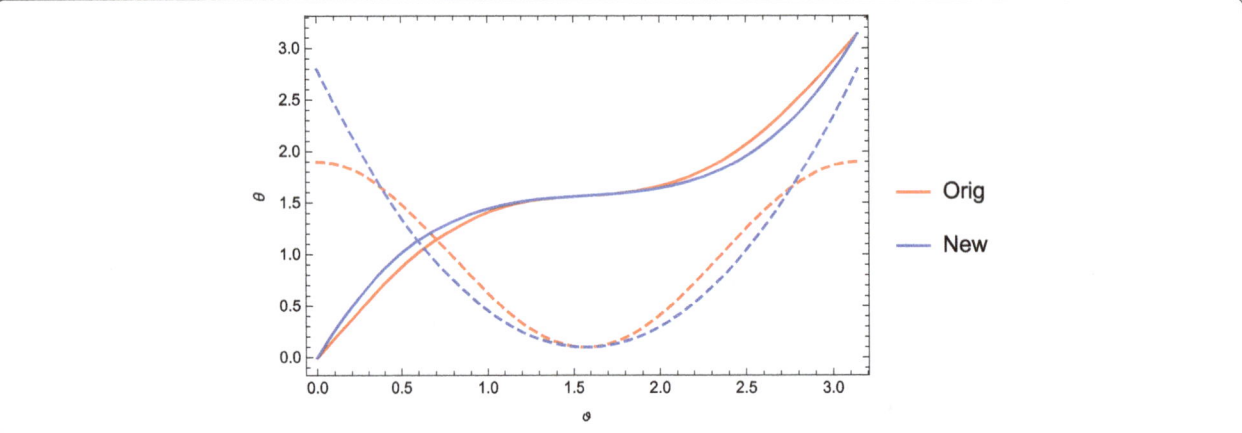

Figure 1 Modified Kerr-Schild coordinates. θ-grid stretching functions comparing the transcendental function $\vartheta(\theta_{KS})$ (solid red curves) with the cubic approach (solid blue curves) for $h = 0.9$. We also give the respective derivatives $d\theta/d\vartheta$ (dashed).

The original coordinate transformation is equivalent to:

$$r_{KS}(s) = R_0 + e^s, \tag{49}$$

$$\theta_{KS}(\vartheta) = \vartheta + \frac{h}{2}\sin(2\vartheta), \tag{50}$$

where R_0 and h are parameters which control, respectively, how much resolution is concentrated near the horizon and near the equator.

Unfortunately, the inverse of $\vartheta(\theta)$ is a transcendental equation that has to be solved numerically. To avoid this complication and still capture the functionality of the modified coordinates, we instead use the following θ-transformation

$$\theta_{KS}(\vartheta) = \vartheta + \frac{2h\vartheta}{\pi^2}(\pi - 2\vartheta)(\pi - \vartheta). \tag{51}$$

Now the solution to the cubic equation can be expressed in closed-form, and the only real root reads

$$\vartheta(\theta_{KS}) = \frac{1}{12}\pi^{2/3}\left(-\frac{2\sqrt[3]{2}(3\pi)^{2/3}(h-1)}{R(\theta_{KS})}\right.$$
$$\left. - \frac{2^{2/3}\sqrt[3]{3}R(\theta_{KS})}{h} + 6\sqrt[3]{\pi}\right), \tag{52}$$

where

$$R(\theta_{KS}) = \left[h\left(-3h\left[-108h\theta_{KS}^2 + 108\pi h\theta_{KS}\right.\right.\right.$$
$$\left.\left. + (h-4)(2\pi h + \pi)^2\right]\right)^{1/2}$$
$$\left. + 9(\pi - 2\theta_{KS})h^2\right]^{1/3}. \tag{53}$$

This is compared with the original version (50) in Figure 1 and shows a good match between the two versions of modified Kerr-Schild coordinates. The radial backtransformation follows trivially as

$$s(r_{KS}) = \ln(r_{KS} - R_0), \tag{54}$$

and the derivatives for the diagonal Jacobian are

$$\partial_s r_{KS} = e^s \tag{55}$$

$$\partial_\vartheta \theta_{KS} = 1 + 2h + 12h\left((\vartheta/\pi)^2 - \vartheta/\pi\right). \tag{56}$$

With these transformations, we obtain the new metric $g_{MKS} = J^T g_{KS} J$. Note that whenever the parameters $h = 0$ and $R_0 = 0$ are set, our MKS coordinates reduce to the standard *logarithmic* Kerr-Schild coordinates.

2.7.2 Rezzolla & Zhidenko parametrization

The Rezzolla-Zhidenko parameterisation (Rezzolla and Zhidenko 2014) has been proposed to describe spherically-symmetric BH geometries in metric theories of gravity. In particular, using a continued-fraction expansion (Padé expansion) along the radial coordinate, deviations from general relativity can be expressed using a small number of coefficients. The line element reads

$$ds^2 = -N^2(r)\,dt^2 + \frac{B^2(r)}{N^2(r)}\,dr^2$$
$$+ r^2\,d\theta^2 + r^2\sin^2\theta\,d\phi^2, \tag{57}$$

with $N(r)$ and $B(r)$ being functions of the radial coordinate r. The radial position of the event horizon is fixed at $r = r_0 > 0$ which implies that $N(r_0) = 0$. Furthermore, the radial coordinate is compactified by means of the dimensionless coordinate

$$x := 1 - \frac{r_0}{r}, \tag{58}$$

in which $x = 0$ corresponds to the position of the event horizon, while $x = 1$ corresponds to spatial infinity. Through this dimensionless coordinate, the function N can be written as

$$N^2 = xA(x), \tag{59}$$

where $A(x) > 0$ for $0 \leq x \leq 1$. Introducing additional coefficients $\epsilon, a_n,$ and b_n, the metric functions A and B are then expressed as follows

$$A(x) = 1 - \epsilon(1-x) + (a_0 - \epsilon)(1-x)^2$$
$$+ \widetilde{A}(x)(1-x)^3, \tag{60}$$
$$B(x) = 1 + b_0(1-x) + \widetilde{B}(x)(1-x)^2. \tag{61}$$

Here $\widetilde{A}$ and $\widetilde{B}$ are functions describing the metric near the event horizon and at spatial infinity. In particular, $\widetilde{A}$ and $\widetilde{B}$ have rapid convergence properties, that is by Padé approximants

$$\widetilde{A}(x) = \frac{a_1}{1 + \frac{a_2 x}{1 + \frac{a_3 x}{1+\cdots}}}, \qquad \widetilde{B}(x) = \frac{b_1}{1 + \frac{b_2 x}{1 + \frac{b_3 x}{1+\cdots}}}, \tag{62}$$

where $a_1, a_2, a_3, \ldots$ and $b_1, b_2, b_3, \ldots$ are dimensionless coefficients that can, in principle, be constrained from observations. The dimensionless parameter ϵ is fixed by the ADM mass M and the coordinate of the horizon r_0. It measures the deviation from the Schwarzschild case as

$$\epsilon = \frac{2M - r_0}{r_0} = -\left(1 - \frac{2M}{r_0}\right). \tag{63}$$

It is easy to see that at spatial infinity ($x = 1$), all coefficients contribute to (62), while at event horizon only the first two terms remain, *i.e.*

$$\widetilde{A}(0) = a_1, \qquad \widetilde{B}(0) = b_1. \tag{64}$$

Given a number of coefficients, any spherical spacetime can hence directly be simulated in BHAC. For example, the coefficients in the Rezzolla-Zhidenko parametrization for the Johannsen-Psaltis (Johannsen and Psaltis 2011) metric and for Einstein-Dilaton BHs (García et al. 1995) have already been provided in (Rezzolla and Zhidenko 2014). Typically, expansion up to a_2, b_2 yields sufficient numerical accuracy for the GRMHD simulations. The first simulations in the related horizon penetrating form of the Rezzolla-Zhidenko parametrization are discussed in (Mizuno et al. 2017).

2.8 Available reconstruction schemes

The second-order finite volume algorithm (40) requires numerical fluxes centered on the interface mid-point. As in any Godunov-type scheme [see e.g., Toro (1999), Komissarov (1999)], the fluxes are in fact computed by solving (approximate) Riemann problems at the interfaces (see Section 2.9). Hence for higher than first-order accuracy, the fluid variables need to be *reconstructed* at the interface by means of an appropriate spatial interpolation. Our reconstruction strategy is as follows. (1) Compute primitive variables $\bar{P}$ from the averages of the conserved variables $\bar{U}$ located at the cell barycenter. (2) Use the reconstruction formulae to obtain two representations for the state at the interface, one with a left-biased reconstruction stencil P^L and the other with a right-biased stencil P^R. (3) Convert the now point-wise values back to their conserved states U^L and U^R. The latter two states then serve as input for the approximate Riemann solver.

A large variety of reconstruction schemes are available, which can be grouped into standard second-order total variation diminishing (TVD) schemes like 'minmod', 'vanLeer', 'monotonized-central', 'woodward' and 'koren' [see Keppens et al. (2012), for details] and higher order methods like the third-order methods 'PPM' (Colella and Woodward 1984), 'LIMO3' (Čada and Torrilhon 2009) and the fifth-order monotonicity preserving scheme 'MP5' due to (Suresh and Huynh 1997). While the overall order of the scheme will remain second-order, the higher accuracy of the spatial discretisation usually reduces the diffusion of the scheme and improves accuracy of the solution [see, e.g., Porth et al. (2014)]. For typical GRMHD simulations with near-evacuated funnel/atmosphere regions, we find the PPM reconstruction scheme to be a good compromise between high accuracy and robustness. For simple flows, e.g., the stationary toroidal field torus discussed in Section 3.4, the compact stencil LIMO3 method is recommended.

2.9 Characteristic speed and approximate Riemann solvers

The time-update of BHAC proceeds in a dimensionally unsplit manner, thus at each Runge-Kutta substep the interface-fluxes in all directions are computed based on the previous substep. The state is then advanced to the next substep with the combined fluxes of the cell. To compute these fluxes from the reconstructed conserved variables at the interface $\boldsymbol{U}^L$ and $\boldsymbol{U}^R$, we provide two approximate Riemann solvers: (1) the Rusanov flux, also known as Total variation diminishing Lax-Friedrichs scheme (TVDLF) which is based on the largest absolute value of the characteristic waves normal to the interface c^i, and (2) the HLL solver (Harten et al. 1983), which is based on the leftmost (c^i_-) and rightmost (c^i_+) waves of the characteristic fan with respect to the interface. The HLL upwind flux function for the conserved variable $u \in \boldsymbol{U}$ is calculated as

$$
F^i(u) = \begin{cases} F^i(\boldsymbol{U}^L); & c^i_- > 0 \\ F^i(\boldsymbol{U}^R); & c^i_+ < 0 \\ \tilde{F}^i(\boldsymbol{U}^L, \boldsymbol{U}^R); & \text{otherwise} \end{cases} \tag{65}
$$

where

$$
\begin{aligned}
&\tilde{F}^i(\boldsymbol{U}^L, \boldsymbol{U}^R) \\
&:= \frac{c^i_+ F^i(\boldsymbol{U}^L) - c^i_- F^i(\boldsymbol{U}^R) + c^i_+ c^i_- (u^R - u^L)}{c^i_+ - c^i_-},
\end{aligned} \tag{66}
$$

and we set in accordance with (Davis 1988): $c^i_- = \min(\lambda^L_{i,-}, \lambda^R_{i,-})$, $c^i_+ = \max(\lambda^L_{i,+}, \lambda^R_{i,+})$.

The TVDLF flux is simply

$$
F^i(u) = \frac{1}{2} \left[F^i(\boldsymbol{U}^L) + F^i(\boldsymbol{U}^R) \right] - \frac{1}{2} c^i (u^R - u^L) \tag{67}
$$

with $c^i = \max(|c^i_-|, |c^i_+|)$.

In addition to these two standard approximate Riemann solvers, we also provide a modified TVDLF solver that preserves positivity of the conserved density D. The algorithm was first described in the context of Newtonian hydrodynamics by (Hu et al. 2013) and was successfully applied in GRHD simulations by (Radice et al. 2014). It takes advantage of the fact that the first-order Lax-Friedrichs flux $F^{i,\text{LO}}(u)$ is positivity preserving under a CFL condition CFL $\leq 1/2$. Hence the fluxes can be constructed by combining the high order flux $F^{i,\text{HO}}(u)$ (obtained e.g., by PPM reconstruction) and $F^{i,\text{LO}}(u)$ such that the updated density does not fall below a certain threshold.[f] Specifically, the modified fluxes read

$$
F^i(u) = \theta F^{i,\text{HO}}(u) + (1-\theta) F^{i,\text{LO}}(u), \tag{68}
$$

where $\theta \in [0,1]$ is chosen as a maximum value which ensures positivity of the cells adjacent to the interface (see

(Hu et al. 2013) for details of its construction). Note that although we only stipulate the density be positive, the formula (68) must be applied to all conserved variables $u \in \boldsymbol{U}$.

In relativistic MHD, the exact form of the characteristic wave speeds $\lambda_\pm$ involves solution of a quartic equation [see, e.g., Anile (1990)] which can add to the computational overhead. For simplicity, instead of calculating the exact characteristic velocities, we follow the strategy of (Gammie et al. 2003) who propose a simplified dispersion relation for the fast MHD wave $\omega^2 = a^2 k^2$. As a trade-off, the simplification can overestimate the wavespeed in the fluid frame by up to a factor of 2, yielding a slightly more diffusive behaviour. The upper bound a for the fast wavespeed is given by

$$
a^2 = c^2_s + c^2_a - c^2_s c^2_a, \tag{69}
$$

which depends on the usual sound speed and Alfvén speed

$$
c^2_s = \hat{\gamma} \frac{p}{\rho h}, \qquad c^2_a = \frac{b^2}{\rho h + b^2}, \tag{70}
$$

here given for an ideal EOS with adiabatic index $\hat{\gamma}$. As pointed out by (Del Zanna et al. 2007), the $3+1$ structure of the fluxes leads to characteristic waves of the form

$$
\lambda^i_\pm = \alpha \lambda'^i_\pm - \beta^i, \tag{71}
$$

where $\lambda'^i_\pm$ is the characteristic velocity in the corresponding special relativistic system ($\alpha \to 1$, $\beta^i \to 0$).

For the simplified isotropic dispersion relation, the characteristics can then be obtained just like in special relativistic hydrodynamics [see, e.g., Font et al. (1994), Banyuls et al. (1997), Keppens and Meliani (2008)]

$$
\begin{aligned}
\lambda'^i_\pm = &\left((1 - a^2) v^i \right. \\
&\pm \sqrt{a^2 (1 - v^2) \left[(1 - v^2 a^2) \gamma^{ii} - (1 - a^2)(v^i)^2 \right]}) \\
&\left. / (1 - v^2 a^2). \right.
\end{aligned} \tag{72}
$$

2.10 Primitive variable recovery

It is well-known that the nonlinear inversion $\boldsymbol{P}(\boldsymbol{U})$ is the Achilles heel of any relativistic (M)HD code and sophisticated schemes with multiple backup strategies have been developed over the years as a consequence (e.g., Noble et al. (2006), Faber et al. (2007), Noble et al. (2009), Etienne et al. (2012), Galeazzi et al. (2013), Hamlin and Newman (2013)). Here we briefly describe the methods used throughout this work and refer to the previously mentioned references for a more detailed discussion.

2.10.1 Primary inversions

Two primary inversion strategies are available in BHAC. The first strategy, which we denote by '1D', is a straightforward generalisation of the one-dimensional strategy described in (van der Holst et al. 2008). It involves a non-linear root finding algorithm which is implemented by means of a Newton-Raphson scheme on the auxiliary variable ξ. Once ξ is found, the velocity follows from (29)

$$v^i = \frac{S^i}{(\xi + B^2)} + \frac{B^i(B^j S_j)}{\xi(\xi + B^2)}, \tag{73}$$

and we calculate the second auxiliary variable $\Gamma = (1 - v^2)^{-1/2}$ so that $\rho = D/\Gamma$. The thermal pressure p then follows from the particular EOS in use (Section 2.4). For example, for an ideal EOS we have

$$p = \frac{\hat{\gamma}}{\hat{\gamma} - 1}\left(\frac{\xi}{\Gamma^2} - \rho\right). \tag{74}$$

For details of the consistency checks and bracketing, we refer the interested reader to (van der Holst et al. 2008).

In addition to the 1D scheme, we have implemented the '2DW' method of (Noble et al. 2006; Del Zanna et al. 2007). The 2DW inversion simultaneously solves the nonlinear Eqs. (25) and the square of the three-momentum S^2, following (29) by means of a Newton-Raphson scheme on the two variables ξ and v^2. Among all inversions tested by (Noble et al. 2006), the 2DW method was reported as the one with the smallest failure rate. We find the same trend, but also find that the lead of 2DW over 1D is rather minor in our tests.

With two distinct inversions that might fail under different circumstances, one can act as a backup strategy for the other. Typically we first attempt a 2DW inversion and switch to the 1D method when no convergence is found. The next layer of backup can be provided by the entropy method as described in the next section.

2.10.2 Entropy switch

To deal with highly magnetised regions, Noble et al. (2009); Sądowski et al. (2013) introduced the advection of entropy to provide a backup strategy for the primitive variable recovery. Similar to Noble et al. (2009), Sądowski et al. (2013), alongside the usual fluid equations, BHAC can be configured to solve an advection equation for the entropy S

$$\nabla_\mu S u^\mu = 0, \tag{75}$$

where we define

$$S := p/\rho^{\hat{\gamma}-1}, \tag{76}$$

given the adiabatic index $\hat{\gamma}$. This leads to the evolution equation

$$\partial_t \sqrt{\gamma}\Gamma S + \partial_i \sqrt{\gamma}\left(\alpha v^i - \beta^i\right)\Gamma S = 0, \tag{77}$$

for the conserved quantity ΓS. The primitive counterpart is the actual entropy $\kappa = p/\rho^{\hat{\gamma}}$, which can be recovered via $\kappa = \Gamma S/D$. In case of failure of the primary inversion scheme, using the advected entropy κ, we can attempt a recovery of primitive variables which does not depend on the conserved energy. Note that after the primitive variables are recovered from the entropy, we need to discard the conserved energy and set it to the value consistent with the entropy. On the other hand, after each successful recovery of primitive variables, the entropy is updated to $\kappa = p/\rho^{\hat{\gamma}}$, which is then advected to the next step. In addition, entropy-based inversion can be activated whenever $\beta = 2p/b^2 \leq 10^{-2}$ since the primary inversion scheme is likely to fail in these highly magnetised regions. Tests of the dynamic switching of the evolutionary equations are described in Section 3.3. In GRMHD simulations of BH accretion, the 'entropy region' is typically located in the BH magnetosphere, which is strongly magnetised and the error due to missing shock dissipation is thus expected to be small.

In the rare instances where the entropy inversion also fails to converge to a physical solution, the code is normally stopped. To force a continuation of the simulation, last resort measures that depend on the physical scenario can be employed. Often the simulation can be continued when the faulty cell is replaced with averages of the primitive variables of the neighbouring healthy cells as described in (Keppens et al. 2012). In the GRMHD accretion simulations described below, failures could happen occasionally in the highly magnetised evacuated 'funnel' region close to the outer horizon where the floors are frequently applied. We found that the best strategy is then to replace the faulty density and pressure values with the floor values and set the Eulerian velocity to zero. Note that in order to avoid generating spurious $\nabla \cdot \boldsymbol{B}$, the last resort measures should never modify the magnetic fields of the simulation.

2.11 Adaptive mesh refinement

The computational grid employed in BHAC is provided by the MPI-AMRVAC toolkit and constitutes a fully adaptive block based (oct-) tree with a fixed refinement factor of two between successive levels. That is, the domain is first split into a number of blocks with equal amount of cells (e.g., 10^3 computational cells per block). Each block can be refined into two (1D), four (2D) or eight (3D) child-blocks with an again fixed number of cells. This process of refinement can be repeated ad libitum and the data-structure can be thought of a forest (collection of trees). All operations on the grid, for example time-update, IO and problem initialisation are scheduled via a loop over a space-filling curve. We adopt the Morton Z-order curve for ease of implementation via a simple recursive algorithm.

Currently, all cells are updated with the same global time-step and hence load-balancing is achieved by cutting the space-filling curve into equal sections that are then distributed over the MPI-processes. The AMR strategy just described is applied in various astrophysical codes, for example codes employing the PARAMESH library (MacNeice et al. 2000; Fryxell et al. 2000; Zhang and MacFadyen 2006), or the recent Athena++ framework [see, e.g., White et al. (2016)]. Compared to a patch-based approach [see, e.g., Mignone et al. (2012)], the block based AMR has several advantages: (1) well-defined boundaries between neighbouring grids on different levels, (2) data is uniquely stored and updated, thus no unnecessary interpolations are performed, and (3) simple data-structure, e.g., straightforward integer arithmetic can be used to locate a particular computational block. For in-depth implementation details such as refinement/prolongation operations, indexing and ghost-cell exchange, we refer to (Keppens et al. 2012). Prolongation and restriction can be used on conservative variables or primitive variables. Typically primitive variables are chosen to avoid unphysical states which can otherwise result from the interpolations in conserved variables. The refinement criteria usually adapted is the Löhner's error estimator (Löhner 1987) on physical variables. It is a modified second derivative, normalised by the average of the gradient over one computational cell. The multidimensional generalization is given by

$$
E_{i_1 i_2 i_3} = \left(\left(\sum_p \sum_q \left(\frac{\partial^2 u}{\partial x_p \partial x_q} \Delta x_p \Delta x_q \right)^2 \right) \right.
$$

$$
\left/ \left(\sum_p \sum_q \left[\left(\left. \left| \frac{\partial u}{\partial x_p} \right| \right|_{i_p+1/2} + \left. \left| \frac{\partial u}{\partial x_p} \right| \right|_{i_p-1/2} \right) \Delta x_p \right. \right. \right.
$$

$$
\left. \left. \left. + f_{\text{wave}} \frac{\partial^2 |u|}{\partial x_p \partial x_q} \Delta x_p \Delta x_q \right]^2 \right) \right)^{1/2} . \tag{78}
$$

The indices p, q run over all dimensions $p, q = 1, \ldots, N_D$. The last term in the denominator acts as a filter to prevent refinement of small ripples, where f_{wave} is typically chosen of order 10^{-2}. This method is also used in other AMR codes such as FLASH (Calder et al. 2002), RAM (Zhang and MacFadyen 2006), PLUTO (Mignone et al. 2012) and ECHO (Zanotti et al. 2015).

3 Numerical tests
3.1 Shock tube test with gauge effect
The first code test is considered in flat spacetime and therefore no metric source terms are involved. Herein we perform one-dimensional MHD shock tube tests with gauge effects by considering gauge transformations of the spacetime. Shock tube tests are well-known tests for code validation and emphasise the nonlinear behaviour of the equations, as well as the ability to resolve discontinuities

Table 3 Shock tube with gaugeeffect setups

Case	α	β^i	γ_{11}	γ_{22}	γ_{33}
A	1	(0,0,0)	1	1	1
B	2	(0,0,0)	1	1	1
C	1	(0.4,0,0)	1	1	1
D	1	(0,0,0)	4	1	1
E	1	(0,0,0)	1	4	1
F	2	(0.4,0,0)	4	9	1

in the solutions [see, e.g., Antón et al. (2006), Del Zanna et al. (2007)].

The initial condition is given as

$$
\left(\rho, p, B^x, B^y \right) = \begin{cases} (1, 1, 0.5, 1) & x < 0, \\ (0.125, 0.1, 0.5, -1) & x > 0, \end{cases} \tag{79}
$$

and all other quantities are zero. In order to check whether the covariant fluxes are correctly implemented, we use different settings for the flat spacetime as detailed in Table 3.

In the simulations, an ideal gas EOS is employed with an adiabatic index of $\hat{\gamma} = 2$. The 1D problem is run on a uniform grid in x-direction using 1,024 cells spanning over $x \in [-1/2, 1/2]$. The simulations are terminated at $t = 0.4$. For the spatial reconstruction, we adopt the second order TVD limiter due to Koren (Koren 1993). Furthermore, RK3 timeintegration is used with Courant number set to 0.4.

Case A is the reference solution without modification of fluxes due to the three-metric, lapse or shift.[g] By means of simple transformations of flat-spacetime, all other cases can be matched with the reference solution. Case B will coincide with solution A if B is viewed at $t/2 = 0.2$. Case C will agree with case A when it is shifted in positive x-direction by $\delta x = \beta^x t = 0.16$. For case D, we rescale the domain as $x \in [-1/4, 1/4]$ and initialise the contravariant vectors as $B'^x = B^x/2$. The state at $t = 0.4$ should agree with case A when the domain is multiplied by the scale factor $h_x = 2$. For case E we initialise $B'^y = B^y/2$ and case F is initialised similarly as $B'^x = B^x/2$, $B'^y = B^y/3$.

In general, all cases agree very well with the rescaled solution. To give an example, Figure 2 shows the rescaled simulation results of case F compared to the reference solution of case A. This test demonstrates the shock-capturing ability of the MHD code and enables us to conclude that the calculation of the covariant fluxes has been implemented correctly.

3.2 Boosted loop advection
In order to test the implementation of the GLM-GRMHD system, we perform the advection of a force-free flux-tube with poloidal and toroidal components of the magnetic field in a flat spacetime.

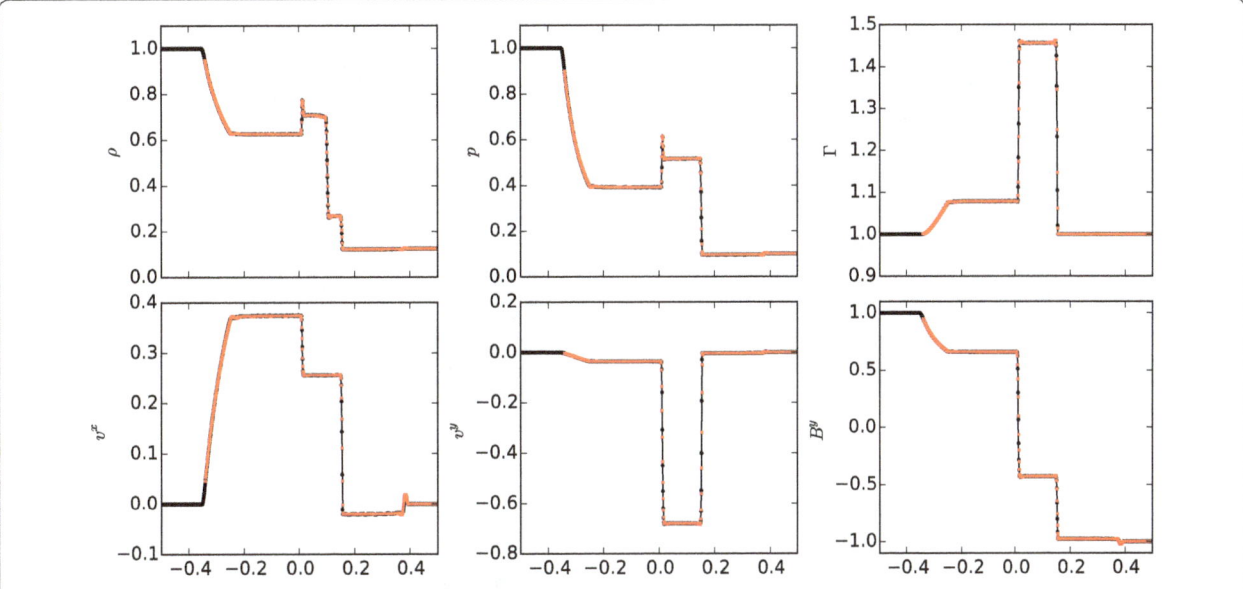

Figure 2 Shock tube with gauge effects. 1D plots of density ρ, gas pressure p, Lorentz factor Γ, velocity components v^x and v^y, and the y-component of the magnetic field for the shock tube test at $t = 0.4$. The reference solution of case A is shown as a solid black line and the rescaled solution of case F is overplotted as red squares.

The initial equilibrium configuration of a force-free flux-tube is given by a modified Lundquist tube [see e.g., Gourgouliatos et al. (2012)], where we avoid sign changes of the vertical field component B^z with the additive constant $C = 0.01$. Pressure and density are initialized as constant throughout the simulation domain. The initial pressure value is obtained from the central plasma-beta $\beta_0 = B^2(0)/2p$, where B is the magnetic field in the co-moving system. The density is set to $\rho = p/2$ yielding a relativistic hot plasma. Consequently, an adiabatic index $\gamma = 4/3$ is used. We set $\beta_0 = 0.01$, which results in a high magnetisation $\sigma_0 = B^2(0)/(\rho c^2 + 4p) \simeq 25$. The equations for the magnetic field for $r < 1$ read

$$B^\phi(r) = J_1(\alpha_t r), \tag{80}$$

$$B^z(r) = \sqrt{J_0(\alpha_t r)^2 + C}, \tag{81}$$

and

$$B^\phi(r) = 0, \tag{82}$$

$$B^z(r) = \sqrt{J_0(\alpha_t)^2 + C}, \tag{83}$$

otherwise, where J_0 and J_1 are Bessel functions of zeroth and first order respectively and the constant $\alpha_t \simeq 3.8317$ is the first root of J_0.

This configuration is then boosted to the frame moving at velocity $\boldsymbol{v} = \sqrt{2}(-v_c, -v_c, 0)$ and we test values of v_c between $0.5c$ and $0.99c$.

Standard Lorentz transformation rules result in

$$\boldsymbol{r} = \boldsymbol{r}' + (\Gamma - 1)(\boldsymbol{r}' \cdot \boldsymbol{n})\boldsymbol{n} - \Gamma t' v_c \boldsymbol{n},$$

$$\boldsymbol{B}' = \Gamma \boldsymbol{B} - \frac{\Gamma^2}{\Gamma + 1}\boldsymbol{\beta}(\boldsymbol{\beta} \cdot \boldsymbol{B}), \tag{84}$$

where t' can be set to zero and where we assumed a vanishing electric field in the co-moving system. Therefore relativistic length contraction gives the loop a squeezed appearance. A simulation domain $(x, y) \in [-1, 1]$ at a base resolution of $N_x \times N_y = 64^2$ is initialised with an additional three levels of refinement. We advect the loop for one period $(P = 2\sqrt{2}/v)$ across the domain where periodic boundary conditions are used.

The advection over the coordinates can be counteracted by setting the shift vector appropriately, i.e. $\boldsymbol{\beta} = -\boldsymbol{v}$. This is an important consistency check of the implementation. Figure 3 shows the initial and final states of the force-free magnetic flux-tube advected for one period and for the case with spacetime shifted against the advection velocity. The advected and counter-shifted cases are in good agreement, with only the truly advected case being slightly more diffused, the effect of which is reflected in the activation of more blocks on the third AMR level.

To investigate the numerical accuracy the L_1 and L_∞ norms of the out-of-plane magnetic field component B^z, as well as the divergence of magnetic field between the initial state and the simulation at a time after one advection period with different resolutions as seen in Figure 4 are checked. The error norms from analytically known solu-

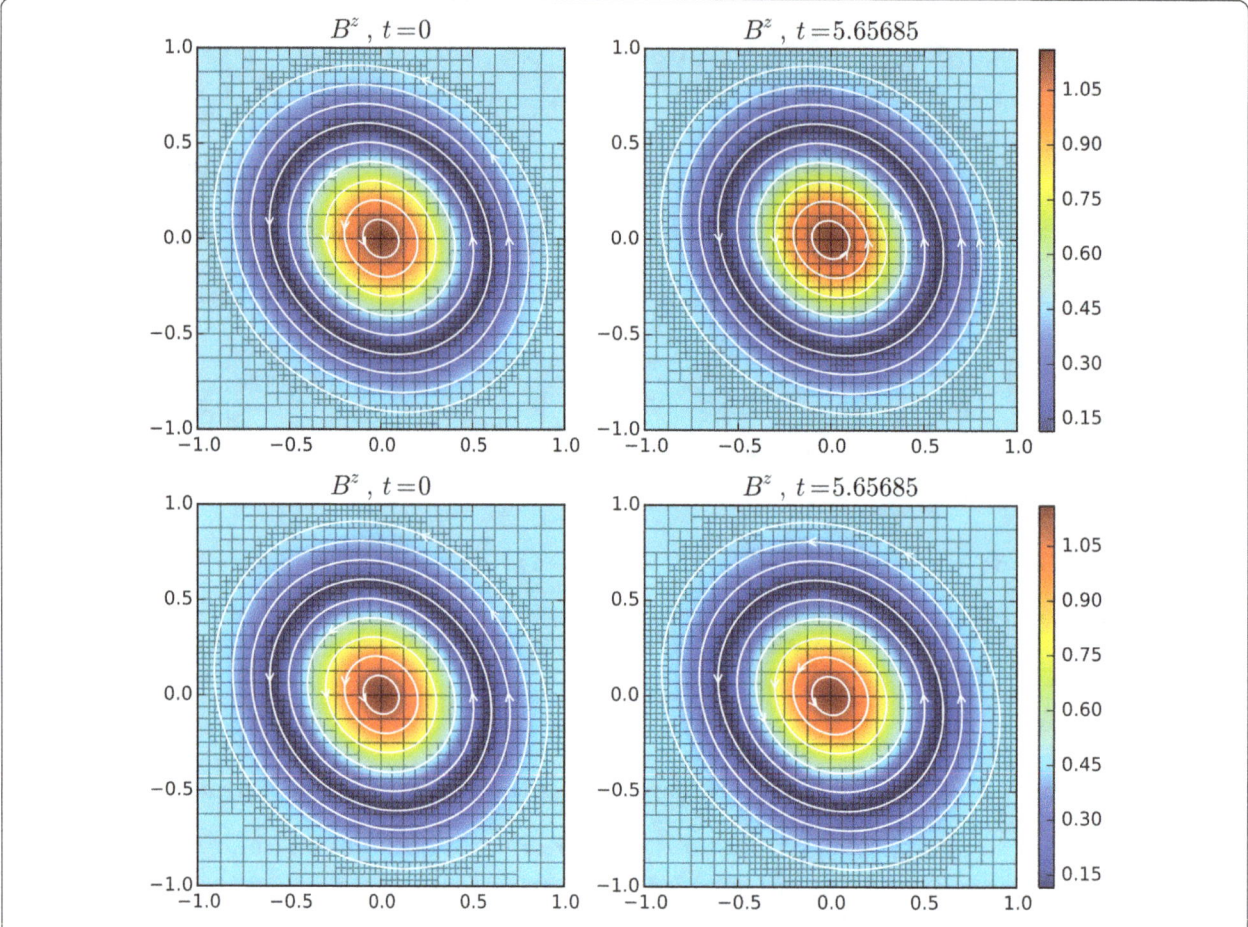

Figure 3 Boosted loop advection. Force-free magnetic loop with diagonal boost velocity $|v| = 0.5c$. *Top:* No shift, the loop is advected for one period. *Bottom:* The shift vector just opposes the (diagonal) advection velocity, $|v| = 0.5$, hence the loop remains stationary with respect to the grid. Base resolution is 64^2 cells with a total of three grid levels. The color shows the strength of the out-of-plane field component B^z and white lines are in-plane field lines of (B^x, B^z). Blocks containing 8^2 cells are indicated.

tions u^* are defined as

$$L_1(u) = \frac{1}{N_{\text{cells}}} \sum_{i,j,k} \left| \bar{u}_{i,j,k} \right.$$

$$\left. - \frac{1}{\Delta V_{i,j,k}} \int_{V_{i,j,k}} u^* \sqrt{\gamma} \, dx^1 \, dx^2 \, dx^3 \right|, \tag{85}$$

$$L_\infty(u) = \max_{i,j,k} \left| \bar{u}_{i,j,k} \right.$$

$$\left. - \frac{1}{\Delta V_{i,j,k}} \int_{V_{i,j,k}} u^* \sqrt{\gamma} \, dx^1 \, dx^2 \, dx^3 \right|, \tag{86}$$

where the summation, respectively maximum operation, includes all cells in the domain and the integrals are performed over the volume of the cell $\Delta V_{i,j,k}$. In this sense, the reported errors correspond to the mean and maximal error in the computational domain. We should note that for the test of convergence, we use a uniform grid and choose

$v = 0.5\sqrt{0.5}(1, 1, 0)$, $\beta = \sqrt{0.5}(1, 0, 0)$ resulting in an advection in direction of the upper-left diagonal. A TVD 'Koren' limiter is chosen. As expected, the convergence is second order for all cases.

3.3 Magnetised spherical accretion

A useful stress test for the conservative algorithm in a general-relativistic setting is spherical accretion onto a Schwarzschild BH with a strong radial magnetic field (Gammie et al. 2003). The steady-state solution is known as the Michel accretion solution (Michel 1972) and represents the extension to general relativity of the corresponding Newtonian solution by (Bondi 1952). The steady-state spherical accretion solution in general relativity is described in a number of works [see, e.g., Hawley et al. (1984), Rezzolla and Zanotti (2013)]. It is easy to show that the solution is not affected when a radial magnetic field of the form $B^r \propto \gamma^{-1/2}$ is added (De Villiers and Hawley 2003). This test challenges the robustness of the code and of the

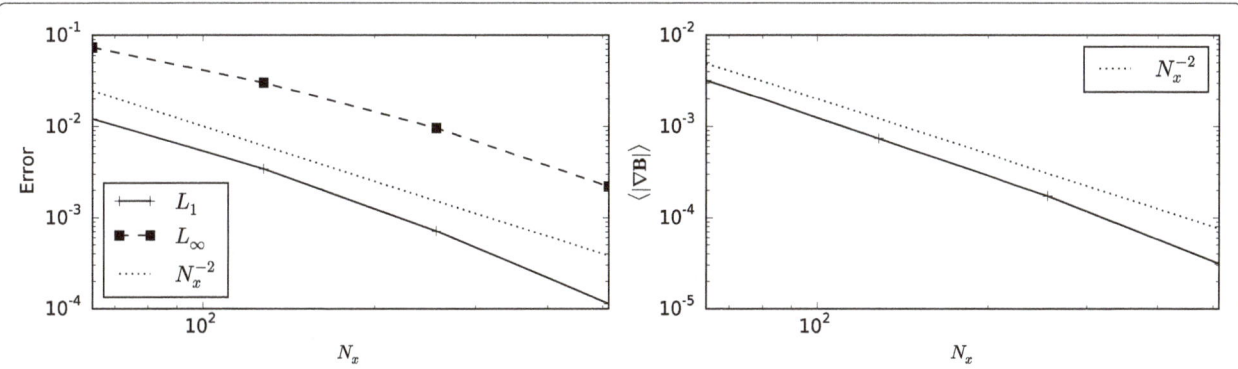

Figure 4 Error quantification: Boosted loop advection. Error of the out-of-plane magnetic field component B^z *(left)* and divergence of $\boldsymbol{B}$ *(right)*. For this test, we chose $\boldsymbol{v} = 0.5\sqrt{0.5}(1,1,0)$ and $\beta = \sqrt{0.5}(1,0,0)$, resulting in an advection in the direction of the upper-left diagonal.

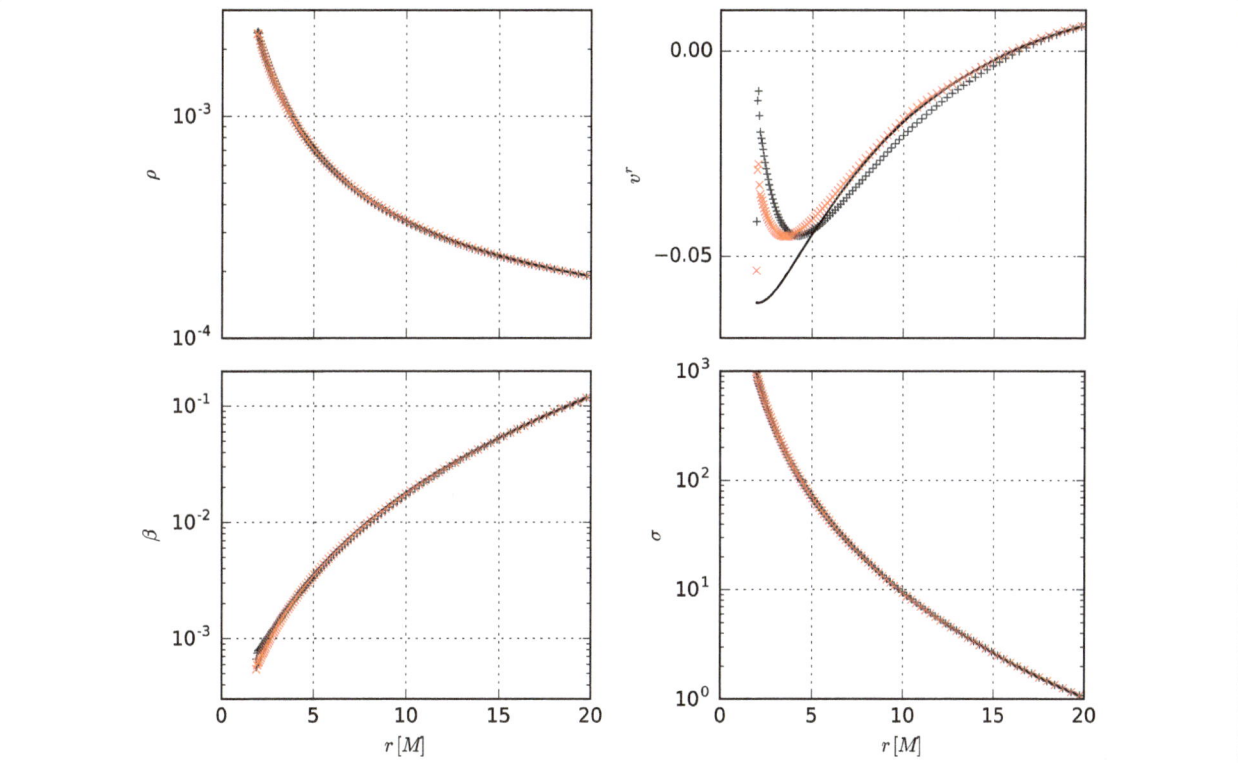

Figure 5 Profiles in the magnetised Bondi flow. Magnetized Bondi flow at $t = 100$ M in MKS coordinates with $\sigma = 10^3$ at the inner edge of the domain. The black solid curve indicates the initial exact solution. We show two realisations with resolution $N_r = 100$. Black crosses are with the standard treatment for the inversion. Red crosses switch to the entropy evolution at $\beta \leq 10^{-2}$ (here in the middle of the domain). In particular, the error in the radial three-velocity v^r decreases when switching to the entropy evolution.

inversion procedure $\boldsymbol{P}(\boldsymbol{U})$ in particular. The calculation of the initial condition follows that outlined in (Hawley et al. 1984). Here, we parametrize the field strength via $\sigma = b^2/\rho$ at the inner edge of the domain ($r = 1.9$ M). The simulation is setup in the equatorial plane using MKS coordinates corresponding to a domain of $r_{\mathrm{KS}} \in [1.9, 20]$ M. The analytic solution remains fixed at the inner and outer radial bound-

aries. We run two cases, case 1 with magnetisation up to $\sigma = 10^3$ and case 2 with a very high magnetisation reaching up to $\sigma = 10^4$. Since the problem is only 1D, the constraint $\nabla \cdot \boldsymbol{B} = 0$ has a unique solution which gets preserved via the FCT algorithm.

Figure 5 illustrates the profiles for $\sigma = 10^3$ and two inversion strategies: 2DW (black +) and 2DW with entropy

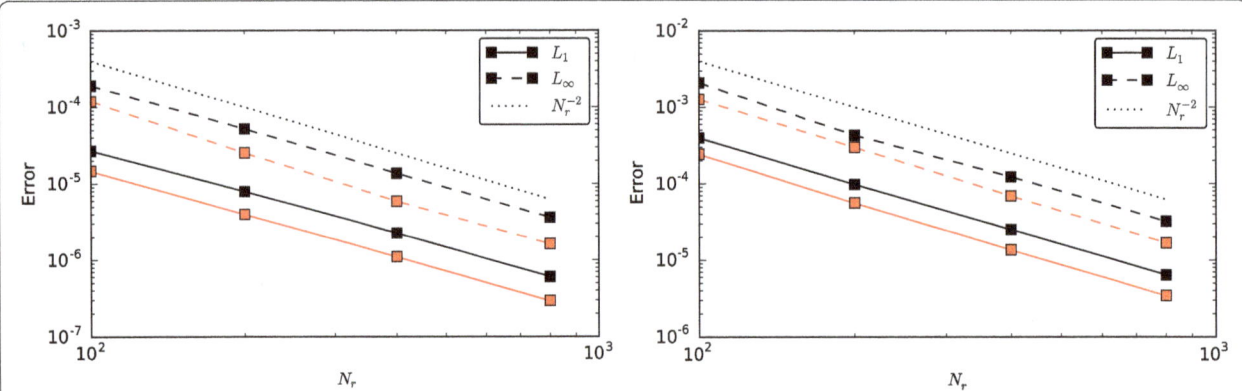

Figure 6 Error quantification: magnetised Bondi flow. Error of density ρ in the highly magnetised Bondi flow: $\sigma = 10^3$ (left) and $\sigma = 10^4$ (right). The black data points are obtained with the standard 2D inversion and the red datapoints switch to the entropy inversion at $\beta \leq 10^{-2}$. One can see that both recipes are convergent with the expected order and that the error in the entropy strategy is decreased by roughly a factor of two.

switching in regions of high magnetization $b^2/2p > 100$ (red $\times$). With the exception of the radial three-velocity near the BH horizon ($r \leq 5$ M), in both cases the simulations maintain well the steady-state solution.[h] Comparing theses results with and without entropy switching, the entropy strategy actually keeps the solution closer to the steady-state solution (solid black curves) even though the change of inversion strategy occurs in the middle of the domain, $r \simeq 10$.

The errors (L_1 and L_∞ norms) for the four cases are shown in Figure 6. Again, the second-order accuracy of the algorithm is recovered. Using the entropy strategy increases the numerical accuracy by around a factor of two and we suggest its use in the highly magnetised regime of BH magnetospheres.

3.4 Magnetised equilibrium torus

As a final validation of the code in the GRMHD regime, we perform the simulation of a magnetised equilibrium torus around a spinning BH. A generalisation of the steady-state solution of the standard hydrodynamical equilibrium torus with constant angular momentum [see, e.g., Fishbone and Moncrief (1976), Hawley et al. (1984), Font and Daigne (2002)] to MHD equilibria with toroidal magnetic fields was proposed by (Komissarov 2006). This steady-state solution is important since it constitutes a rare case of a non-trivial analytic solution in GRMHD.[i]

For the initial setup of the equilibrium torus, we adopt a particular relationship $\omega = \omega(p)$, where $\omega = \rho h$ is the fluid enthalpy and $\tilde{\omega} = \tilde{\omega}(\tilde{p}_m)$, where $\tilde{p}_m = \mathcal{L}p_m$, $\tilde{\omega} = \mathcal{L}\omega$, $p_m = b^2/2$ is the magnetic pressure, and $\mathcal{L} = g_{t\phi}g_{t\phi} - g_{tt}g_{\phi\phi}$. From these relationships, thermal and magnetic pressures are described as

$$p = K\omega^\kappa, \tag{87}$$

$$\tilde{p}_m = K_m\tilde{\omega}^\eta. \tag{88}$$

The analytical solutions can be constructed from

$$W - W_{\mathrm{in}} + \frac{\kappa}{\kappa - 1}\frac{p}{\omega} + \frac{\eta}{\eta - 1}\frac{p_m}{\omega} = 0, \tag{89}$$

for the introduced total potential W, where $W = \ln|u_t|$. The centre of the torus is located at $(r_c, \pi/2)$. At this point, we parametrize the magnetic field strength in terms of the pressure ratio

$$\beta_c = p_g(r_c, \pi/2)/p_m(r_c, \pi/2). \tag{90}$$

The gas pressure and magnetic pressure at the centre of the torus are given by

$$p_c = \omega_c(W_{\mathrm{in}} - W_c)\left(\frac{\kappa}{\kappa - 1} + \frac{\eta}{\eta - 1}\frac{1}{\beta_c}\right)^{-1},$$

$$p_{m_c} = p_c/\beta_c. \tag{91}$$

From these, the constants K and K_m for barotropic fluids are obtained.

The magnetic field distribution is given by the distribution of magnetic pressure p_m. From the consideration of a purely toroidal magnetic field one obtains

$$b^\phi = \sqrt{2p_m/\mathcal{A}},$$

$$b^t = \ell b^\phi, \tag{92}$$

where $\mathcal{A} = g_{\phi\phi} + 2lg_{t\phi} + l^2g_{tt}$ and $\ell := -u_\phi/u_t$ is the specific angular momentum.

We perform 2D simulations using logarithmic KS coordinates with $h = 0$ and $R_0 = 0$. The simulation domain is $\theta \in [0, \pi]$, $r_{\mathrm{KS}} \in [0.95r_\mathrm{h}, 50$ M], where r_h is the (outer) event horizon radius of the BH. The BH has the dimensionless spin parameter $a = 0.9$. For simplicity, we set the two indices to the same value of $\kappa = \eta = 4/3$ and also set the adiabatic index of the adopted ideal EOS to $\gamma = 4/3$. The remaining parameters are listed in the Table 4.

Initially, the velocity of the atmosphere outside of the torus is set to zero in the Eulerian frame, with density and gas pressure set to very small values of $\rho = \rho_{min} r_{BL}^{-3/2}$, $p = p_{min} r_{BL}^{-5/2}$ with $\rho_{min} = 10^{-5}$ and $p_{min} = 10^{-7}$. It is important to note that the atmosphere is free to evolve and only densities and pressures are floored according to the initial state. In the simulations we use the HLL approximate Riemann solver, third order LIMO3 reconstruction, two-step time update, and a CFL number of 0.5. We impose outflow conditions on the inner and outer boundaries of the radial direction and reflecting boundary conditions in the θ direction. As the magnetic field is purely toroidal, and will remain so during the time-evolution of this axisymmetric case, no particular $\nabla \cdot \mathbf{B} = \mathbf{0}$ treatment is used.

The top panels of Figure 7 show the density distribution at the initial state and at $t = 200$ M, as well as the plasma beta distribution at $t = 200$ M. The rotational period of the disk centre is $t_r = 68$ M. The initial torus configuration is well maintained after several rotation period. For a qualitative view of the simulations, the 1D radial and azimuthal

distributions of the density are shown in the lower two panels in Figure 7 with different grid resolutions. All but the low resolution case are visually indistinguishable from the initial condition in the bottom-left panel, showing $\rho - r$ with a linear scale. Since the atmosphere is evolved freely, small density waves propagate in the ambient medium of the torus, as seen in the $\rho - \theta$ cut. This does not adversely affect the equilibrium solution in the bulk of the torus however. Error quantification (L_1 and L_∞) is provided in Figure 8. The second-order properties of the numerical scheme are well recovered.

3.5 Differences between FCT and GLM

Having implemented two methods for divergence control, we took the opportunity to compare the results of simulations using both methods. We analysed three tests: a relativistic Orszag-Tang vortex, magnetised Michel accretion, and magnetised accretion from a Fishbone-Moncrief torus. Although much less in-depth, this comparison is in the same spirit as those performed in previous works in non-relativistic MHD (Toth 2000; Balsara and Kim 2004; Mocz et al. 2016). The well-known work by (Toth 2000) compares seven divergence-control methods, including an early non-conservative divergence-cleaning method known as the *eight-wave method* (Powell 1994), and three CT methods, finding that FCT is among the three most accurate schemes for the test problems studied. In (Balsara and Kim 2004), three divergence-cleaning schemes and one CT scheme were applied to the same test problem of supernova-induced MHD turbulence in the inter-

Table 4 Parameters for the MHD equilibrium torus test

Case	l_0	r_c	W_{in}	W_c	ω_c	β_c
A	2.8	4.62	−0.030	−0.103	1.0	0.1

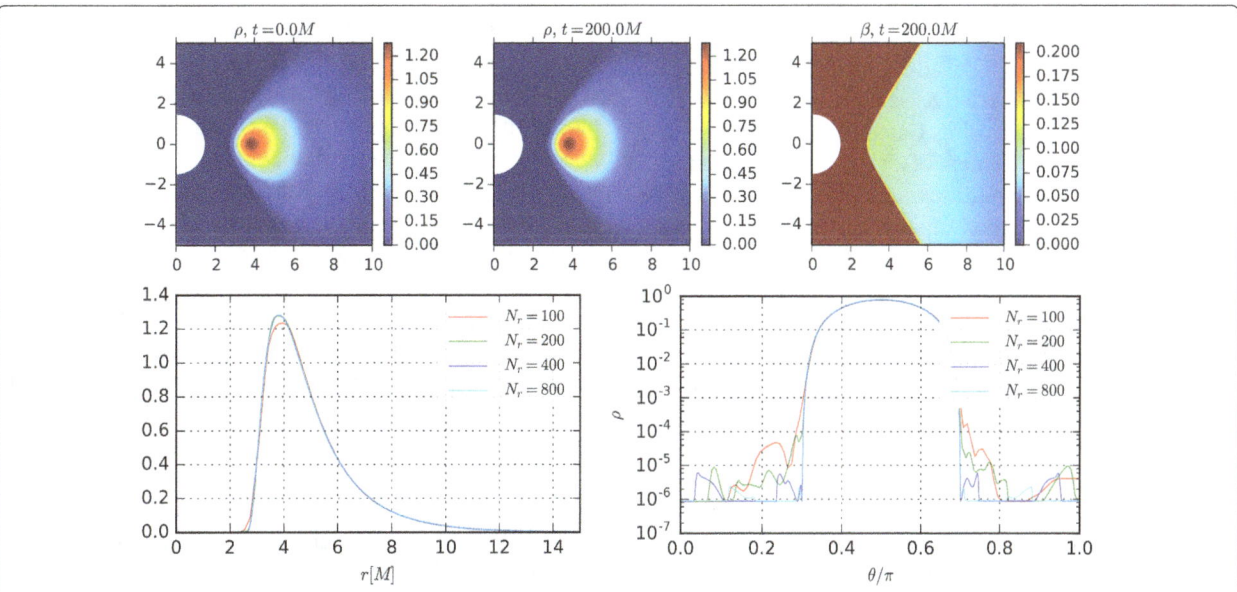

Figure 7 Evolution of the magnetised Komissarov-Torus. *Top*: qualitative view of the torus evolution at a resolution of $N_r = N_\theta = 400$. The spatial scale is given in units of M. *Left*: Initial rest-frame density distribution, *center*: density at $t = 200$ M, *right*: plasma β parameter at $t = 200$ M. *Bottom*: density slices through the torus at $t = 100$ M for constant $\theta = \pi/2$ *(left)* and $r = 5$ *(right)*.

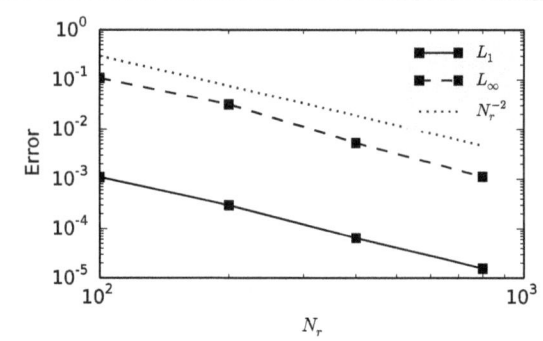

Figure 8 Error quantification: magnetised Komissarov-Torus.
Error of the density ρ in the strongly-magnetised Komissarov torus, comparing the solution at $t = 60$ M with the exact solution. The second-order behaviour of the numerical scheme is well recovered.

stellar medium. It was found that the three divergence-cleaning methods studied suffer from, among other problems, spurious oscillations in the magnetic energy, which is attributed to the non-locality introduced by the loss of hyperbolicity in the equations. Finally, in (Mocz et al. 2016), a non-staggered version of CT adapted to a moving mesh is compared to the divergence-cleaning Powell scheme (Powell et al. 1999), an improved version of the eight-wave method. They observe greater numerical stability and accuracy, and a better preservation of the magnetic field topology for the CT scheme. In their tests, the Powell scheme suffers from an artificial growth of the magnetic field. This is explained to be a result of the scheme being non-conservative.

3.5.1 Orszag-Tang vortex

The Orszag-Tang vortex (Orszag and Tang 1979) is a common problem that can be used to test MHD codes for violations of $\nabla \cdot \boldsymbol{B}$. The relativistic version presented here was performed in 2D using Cartesian coordinates in a 128×128-resolution domain of $2\pi \times 2\pi$ length units with periodic boundary conditions, and evolved for 10 time units ($c = 1$). The equation of state was chosen to be that of an ideal fluid with $\hat{\gamma} = 4/3$ and the initial conditions were set to $\rho = 1.0, p = 10.0, v^x = -0.99 \sin y, v^y = 0.99 \sin x$, $B^x = -\sin y$ and $B^y = \sin 2x$. Snapshots of the evolution are shown in Figure 9.

As can be seen in Figures 9 and 10, the general behaviour in both cases is quite similar qualitatively, with only slight differences at specific locations. For instance, when compared to GLM, FCT exhibited higher and sharper maxima for the magnitude of the magnetic field. In a similar fashion, some fine features in the Lorentz factor that can be seen in Figure 9 for FCT appear to be smeared out when using GLM, giving a false impression of symmetry under $90°$ rotations, while the actual symmetry of the problem is under $180°$ rotations. This may be an evidence of FCT being less diffusive than GLM.

3.5.2 Spherical accretion

We tested the ability of both methods to preserve a stationary solution by evolving a magnetised Michel accretion in 2D, as shown in Figure 11. We employed spherical MKS coordinates (see Section 2.7.1), $N_r \times N_\theta = 200 \times 100$ resolution, and a domain with $r \in [1.9$ M, 10 M] and $\theta \in [0, \pi]$. The fluid obeyed an ideal equation of state with $\hat{\gamma} = 5/3$ and the sonic radius was located at $r_c = 8$, and the magnetic field was normalised so that the maximum magnetisation was $\sigma = 10^3$. We repeated the numerical experiment of section Section 3.3, now in 2D. As shown in Figure 11, numerical artefacts start to become noticeable at these later times. For instance, with these extreme magnetisations, for GLM we observe spurious features near the poles at $\theta = 0$ and $\theta = \pi$, as well as deviations in the velocity field near the outer boundary $r = 10$ M. The polar region is of special interest for jet simulations, where the divergence-control method must be robust enough to interplay with the axial boundary conditions. The bottom of Figure 11 shows the profiles of several quantities at $\theta = \pi/2$. Both divergence-control methods produce an excellent agreement between the solution at different times in the equatorial region. The rightmost column in the bottom of Figure 11 shows the relative errors in the radial component of the magnetic field for each method. The errors for FCT are not only one order of magnitude lower than for GLM, but also behave differently, remaining at the same level near the more-magnetised inner region instead of growing as seen for GLM.

3.5.3 Accreting torus

To compare both methods in a setting closer to our intended scientific applications, we simulated accretion from a magnetised perturbed Fishbone-Moncrief torus around a Kerr BH with $M = 1$ and $a = 0.9375$. We employed modified spherical MKS coordinates as described in Section 2.7.1 and a domain where $r \in [1.29, 2500]$ and $\theta \in [0, \pi]$ with a resolution of $N_r \times N_\theta = 512 \times 256$, and evolved the system until $t = 2,000$ M. At the radial boundaries, we imposed *noinflow* boundary conditions while at the boundaries with the polar axis we imposed symmetric boundary conditions for the scalar variables and the radial vector components and antisymmetric boundary conditions for the azimuthal and polar components. In the BHAC code, *noinflow* boundary conditions are implemented via continuous extrapolation of the primitive variables and by replacing the three-velocity with zero in case inflowing velocities are present in the solution. The fluid obeyed an ideal equation of state with $\hat{\gamma} = 4/3$. The inner edge of the torus was located at $r_{in} = 6.0$ and the maximum density was located at $r_{max} = 12.0$. The initial magnetic field configuration consisted of a single loop with $A_\phi \propto (\rho/\rho_{max} - \rho_{cut})$ and zero for $\rho < \rho_{cut} = 0.2$. To simulate vacuum, the region outside the torus was filled with a

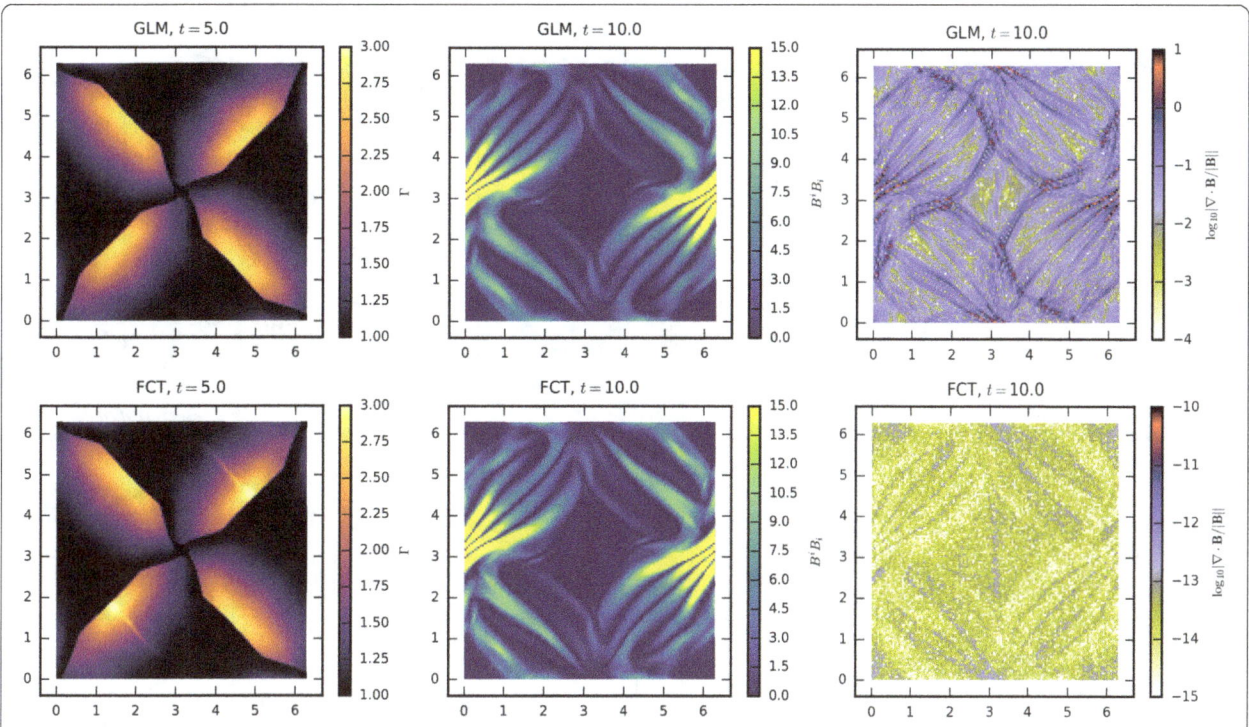

Figure 9 Relativistic Orszag-Tang vortex: in-plane quantities. Relativistic Orszag-Tang vortex. *Left column*: small differences can be observed in this snapshot of the Lorentz factor at $t = 5.0$. Some features that appear when using CT are flattened when using GLM, possibly due to a greater diffusivity of the latter. *Middle column*: final snapshot of B^lB_l. Good agreement between the two methods can be seen, except at some extreme points. *Right column*: violation of $\nabla \cdot \boldsymbol{B} = 0$.

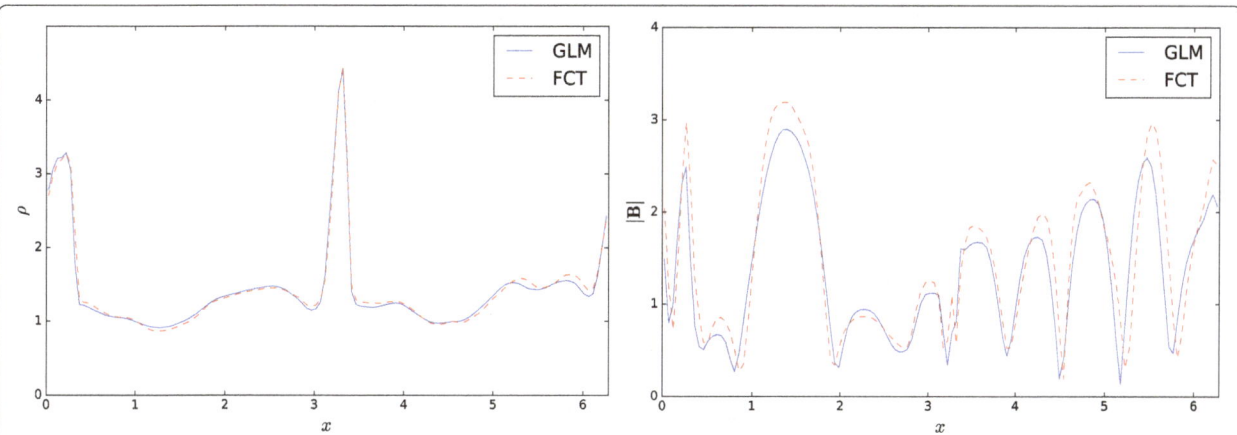

Figure 10 Relativistic Orszag-Tang vortex: horizontal cuts. Relativistic Orszag-Tang vortex: cuts at $y = \pi/2$ and $t = 10.0$ of the density ρ (left) and the magnitude of the magnetic field $|B|$ (right). While for ρ there is in general a good agreement, FCT tends to produce higher maxima for the magnetic field.

tenuous atmosphere as is customarily done in these types of simulation. In this case, the prescription for the atmosphere was $\rho_{atm} = \rho_{min}r^{-3/2}$ and $p_{atm} = p_{min}r^{-5/2}$, where $\rho_{min} = 10^{-5}$ and $p_{min} = 1/3 \times 10^{-7}$. A qualitative difference can be seen even at early times of the simulation. The two upper panels of Figure 12 show a snapshot of the simu-

lation at $t = 20$ M using both GLM and FCT. For GLM some of the magnetic field has diffused out of the original torus, magnetising the atmosphere. This artefact is visible for GLM from almost the beginning of the simulation ($t \approx 20$ M), while for FCT it is minimal. Even though this particular artefact is not of crucial importance for the

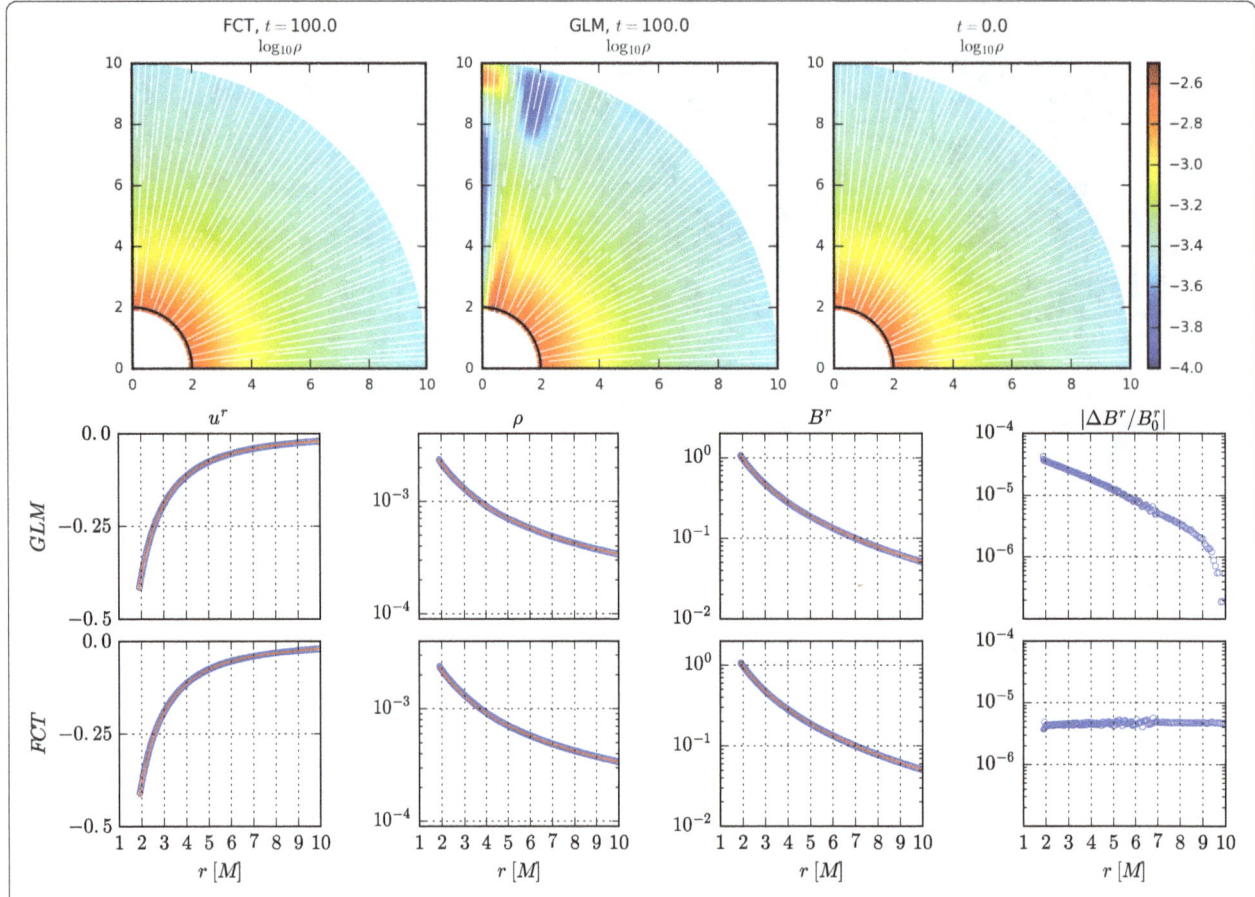

Figure 11 Magnetised Bondi flow: comparing FCT and GLM. *Top*: logarithmic density and streamlines in 2D magnetised Michel accretion at times $t = 0$ M (left) and $t = 100$ M using GLM (centre) and FCT (right). The horizon is marked by the black line at $r = 2$. *Bottom*: profiles at $\theta = \pi/2$ of, from left to right, radial 3-velocity, density and radial magnetic field at $t = 0$ M (blue circles) and $t = 100$ M (red line) using GLM (upper) and FCT (lower). The last column shows the relative difference between the magnetic field at $t = 100$ M and at the initial condition.

subsequent dynamics of the simulation, this points to a higher inclination of GLM to produce spurious magnetic field structures. At later times (bottom panels of Figure 12), the most noticeable difference is the smaller amount of turbulent magnetic structures and the bigger plasma magnetisation inside the funnel in FCT, as compared to GLM. This latter difference indicates that the choice of technique to control $\nabla \cdot \boldsymbol{B}$ may have an effect on the possibility of jet formation in GRMHD simulations, although this specific effect was not extensively studied.

To summarise this small section on the comparison between both divergence-control techniques, we found from the three tests performed that FCT seems to be less diffusive than GLM, is able to preserve for a longer time a stationary solution, and seems to create less spurious structures in the magnetic field. However, it still has the inconvenient property that it is not possible to implement a cell-entered version of it whilst fully incorporating AMR. As mentioned previously, we are currently working on a staggered implementation adapted to AMR, and this will be described in a separate work.

4 Torus simulations

4.1 Initial conditions

We consider a hydrodynamic equilibrium torus threaded by a weak magnetic field loop. The particular equilibrium torus solution with constant angular momentum was first presented by Fishbone and Moncrief (1976) and Kozlowski et al. (1978) and is now a standard test for GRMHD simulations [see, e.g., Font and Daigne (2002), Zanotti et al. (2003), Antón et al. (2006), Rezzolla and Zanotti (2013), White et al. (2016)]. To facilitate cross-comparison, we set the initial conditions in the torus close to those adopted by Shiokawa et al. (2012), White et al. (2016). Hence the spacetime is a Kerr BH with dimensionless spin parameter $a = 0.9375$. The inner radius of the torus is set to $r_{\text{in}} = 6$ M and the density maximum is located at $r_{\text{max}} = 12$ M, where radial and azimuthal positions refer to Boyer-Lindquist coordinates. With these choices, the orbital period of the

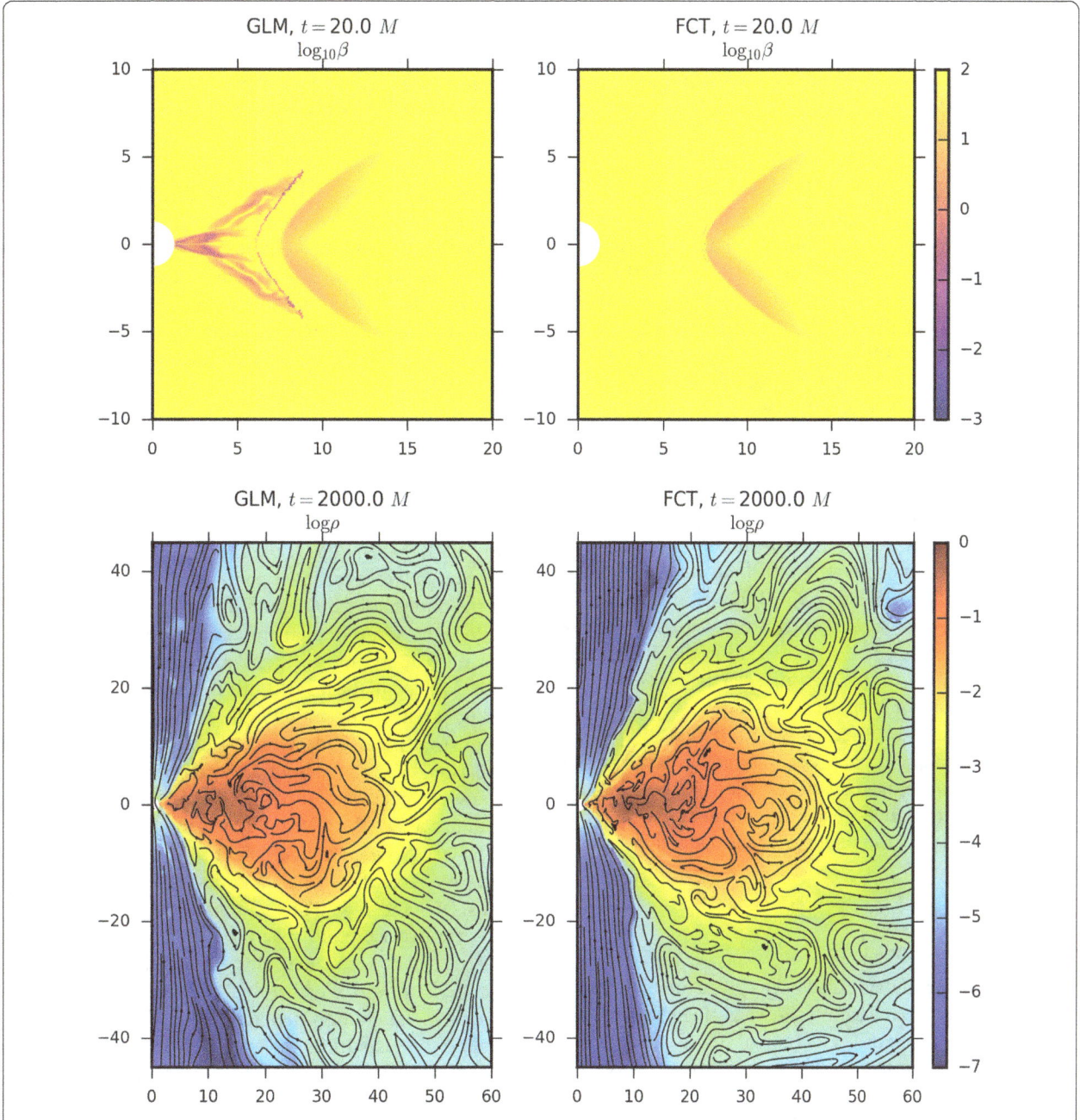

Figure 12 Magnetised torus: comparing FCT and GLM. Magnetised torus: plasma β at $t = 20$ M (top) and density and magnetic field lines $t = 2,000$ M (bottom) using GLM (left) and FCT (right).

torus at the density maximum becomes $T = 247$ M. We adopt an ideal gas EOS with an adiabatic index of $\hat{\gamma} = 4/3$. A weak single magnetic field loop defined by the vector potential

$$A_\phi \propto \max(\rho/\rho_{max} - 0.2, 0), \tag{93}$$

is added to the stationary solution. The field strength is set such that $2p_{max}/b_{max}^2 = 100$, where global maxima of pressure p_{max} and magnetic field strength b_{max}^2 do not necessarily coincide. In order to excite the MRI inside the torus, the thermal pressure is perturbed by 4% white noise.

As with any fluid code, vacuum regions must be avoided and hence we apply floor values for the rest-mass density ($\rho_{fl} = 10^{-5} r^{-3/2}$) and the gas pressure ($p_{fl} = 1/3 \times 10^{-7} r^{-5/2}$). In practice, for all cells which satisfy $\rho \leq \rho_{fl}$ we set $\rho = \rho_{fl}$, in addition if $p \leq p_{fl}$, we set $p = p_{fl}$.

The simulations are performed using horizon penetrating logarithmic KS coordinates (corresponding to our set of modified KS coordinates with $h = 0$ and $R_0 = 0$). In the 2D cases, the simulation domain covers $r_{KS} \in [0.96 r_h, 2{,}500 \text{ M}]$ and $\theta \in [0, \pi]$, where $r_h \simeq 1.35$ M. In the 3D cases, we slightly excise the axial region $\theta \in [0.02\pi, 0.98\pi]$ and adopt $\phi \in [0, 2\pi]$. We set the boundary conditions in the horizon and at $r = 2{,}500$ M to zero gradient in primitive variables. The θ-boundary is handled as follows: when the domain extends all the way to the poles (as in our 2D cases), we adopt 'hard' boundary conditions, thus setting the flux through the pole manually to zero. For the excised cone in the 3D cases, we use reflecting 'soft' boundary conditions on primitive variables.

The time-update is performed with a two-step predictor corrector based on the TVDLF fluxes and PPM reconstruction. Furthermore, we set the CFL number to 0.4 and use the FCT algorithm. Typically, the runs are stopped after an evolution for $t = 5{,}000$ M, ensuring that no signal from the outflow boundaries can disturb the inner regions. To check convergence, we adopt the following resolutions: $N_r \times N_\theta \in \{256 \times 128, 512 \times 256, 1{,}024 \times 512\}$ in the 2D cases and $N_r \times N_\theta \times N_\phi \in \{128 \times 64 \times 64, 192 \times 96 \times 96, 256 \times 128 \times 128, 384 \times 192 \times 192\}$ in the 3D runs. In the following, the runs are identified via their resolution in θ-direction. For the purpose of validation, we ran the 2D cases also with the HARM3D code (Noble et al. 2009).[j]

To facilitate a quantitative comparison, we report radial profiles of disk-averaged quantities similar to Shiokawa et al. (2012), White et al. (2016), Beckwith et al. (2008). For a quantity $q(r, \theta, \phi, t)$, the shell average is defined as

$$\langle q(r,t) \rangle := \frac{\int_0^{2\pi} \int_{\theta\min}^{\theta\max} q(r,\theta,\phi,t) \sqrt{-g} \, d\phi \, d\theta}{\int_0^{2\pi} \int_{\theta\min}^{\theta\max} \sqrt{-g} \, d\phi \, d\theta}, \qquad (94)$$

which is then further averaged over a given time interval to yield $\langle q(r) \rangle$ (note that we omit the weighting with the density as done by Shiokawa et al. (2012), White et al. (2016)). The limits $\theta_{\min} = \pi/3$, $\theta_{\max} = 2\pi/3$ ensure that atmosphere material is not taken into account in the averaging. The time-evolution is monitored with the accretion rate $\dot{M}$ and the magnetic flux threading the horizon ϕ_B

$$\dot{M} := \int_0^{2\pi} \int_0^{\pi} \rho u^r \sqrt{-g} \, d\theta \, d\phi, \qquad (95)$$

$$\phi_B := \frac{1}{2} \int_0^{2\pi} \int_0^{\pi} |B^r| \sqrt{-g} \, d\theta \, d\phi, \qquad (96)$$

where both quantities are evaluated at the outer horizon r_h.

4.2 2D results

Figure 13 illustrates the qualitative time evolution of the torus by means of the rest-frame density ρ, plasma-β and

the magnetisation $\sigma = b^2/\rho$. After $t \simeq 300$ M, the MRI-driven turbulence leads to accretion onto the central BH. The accretion rate and magnetic flux threading the BH then quickly saturate into a quasi-stationary state (see also Figure 14). The accreted magnetic flux fills the polar regions and gives rise to a strongly magnetised funnel with densities and pressures dropping to their floor values. For the adopted floor values we hence obtain values of plasma-β as low as 10^{-5} and magnetisations peaking at $\sigma \approx 10^3$ in the inner BH magnetosphere. These extreme values pose a stringent test for the robustness of the code and, consequently, the funnel region must be handled with the auxiliary inversion based on the entropy switch (see Section 2.10.2).

4.2.1 Comparison to HARM3D

For validation purposes we simulated the same initial conditions with the HARM3D code. Wherever possible, we have made identical choices for the algorithm used in both codes, that is: PPM reconstruction, TVDLF Riemann solver and a two step time update. It is important to note that the outer radial boundary differs in both codes: while the HARM3D setup implements outflow boundary conditions at $r = 50$ M, in the BHAC runs the domain and radial grid is doubled in the logarithmic Kerr-Schild coordinates, yielding identical resolution in the region of interest. This ensures that no boundary effects compromise the BHAC simulation. Next to the boundary conditions, also the initial random perturbation varies in both codes which can amount to a slightly different dynamical evolution.

After verifying good agreement in the qualitative evolution, we quantify with both codes $\dot{M}$ and ϕ_B according to Eqs. (95) and (96). The results are shown in Figure 14. Onset-time of accretion, magnitude and overall behaviour are in excellent agreement, despite the chaotic nature of the turbulent flow. We also find the same trend with respect to the resolution-dependence of the results: upon doubling the resolution, the accretion rate $\langle \dot{M} \rangle$, averaged over $t \in [1{,}000, 2{,}000]$, increases significantly by a factor of 1.908 and 1.843 for BHAC and HARM, respectively. For $\langle \phi_B \rangle$, the factors are 1.437 and 1.484. At a given resolution, the values for $\langle \dot{M} \rangle$ and $\langle \phi_B \rangle$ agree between the two codes within their standard deviations. Furthermore, we have verified that these same resolution variations are within the run-to-run deviations due to a different random number seed for the initial perturbation.

Further validation is provided in Figure 15 where disk-averaged profiles for the two highest resolution 2D runs are shown according to Eq. (94). The quantities of interest are the rest-frame density ρ, the dimensionless temperature $\Theta := p/\rho c^2$, the magnitude of the fluid-frame magnetic field $|B| = \sqrt{b^2}$, thermal and magnetic pressures P_{gas}, P_{mag} and the plasma-β. Again we set the averaging time $t \in [1{,}000, 2{,}000]$ M with both codes. The agreement can

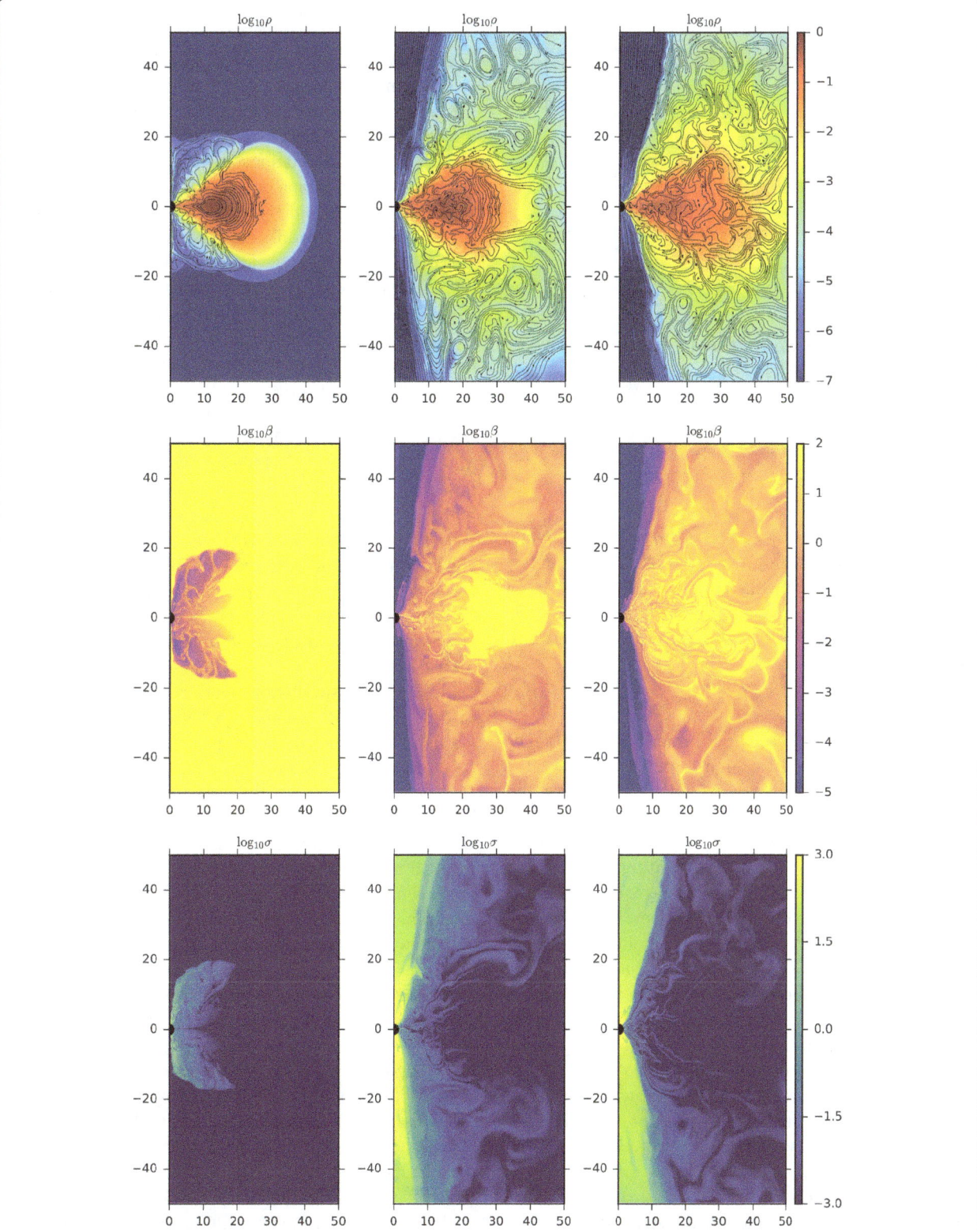

Figure 13 2D magnetised torus evolution. Evolution of the 2D magnetised torus with resolution 1,024 × 512 for times $t/M \in \{300, 1,000, 2,000\}$. We show logarithmic rest-frame density (top), logarithmic plasma β (middle) and the logarithm of the magnetisation parameter $\sigma = b^2/\rho$ (bottom). Magnetic field lines are traced out in the first panel using black contour lines. One can clearly make out the development of the MRI and evacuation of a strongly magnetised funnel reaching values of $\beta < 10^{-5}$ and $\sigma \approx 10^3$.

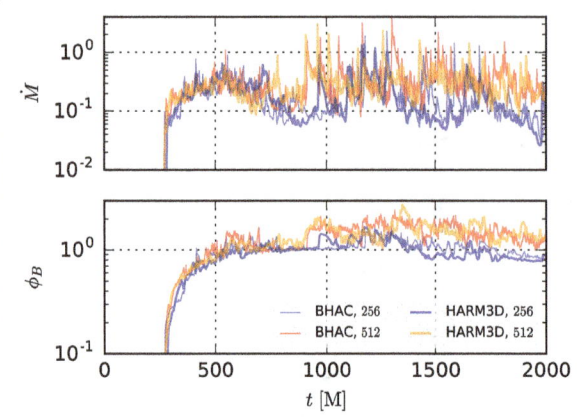

Figure 14 Accretion rates in the 2D magnetised torus. Accretion rates and horizon-penetrating magnetic flux in the 2D validation runs. We show two resolutions with each code: BHAC (blue, red) and HARM3D (dark blue, orange). Despite the chaotic nature of the turbulent accretion both codes show very good qualitative and quantitative agreement.

be considered as very good, that is apart from a slightly higher magnetisation in HARM for $r \in [20, 30]$, the differences of which are well within the 1σ standard deviation over the averaging time. Small systematic departures at the outer edge of the HARM domain are likely attributable to boundary effects.

4.3 3D results

We now turn to the 3D runs performed with BHAC. The qualitative evolution of the high resolution run is illustrated in Figure 16 showing rest-frame density and b^2 on the two slices $z = 0$ and $y = 0$. Overall, the evolution progresses in a similar manner to the 2D cases: MRI-driven accretion starts at $t \approx 300$ M and enters saturation at around $t \simeq 1,000$ M. Similar values for the magnetisation in the funnel region are also obtained. However, since the MRI cannot be sustained in axisymmetry as poloidal field cannot be re-generated via the ideal MHD induction equation (Cowling 1933), we expect to see qualitative differences between the 2D and 3D cases at late times.

Four different numerical resolutions were run which allows a first convergence analysis of the magnetised torus accretion scenario. Based on the convergence study, we can estimate which numerical resolutions are required for meaningful observational predictions derived from GRMHD simulations of this type.

Since we attempt to solve the set of dissipation-free ideal MHD equations, convergence in the strict sense cannot be achieved in the presence of a turbulent cascade [see also the discussion in Sorathia et al. (2012), Hawley et al. (2013)].[k] Instead, given sufficient scale separation, one might hope to find convergence in quantities of interest like the disk averages and accretion rates. The convergence

of various indicators in similar GRMHD torus simulations was addressed for example by (Shiokawa et al. 2012). The authors found signs for convergence in most quantifications when adopting a resolution of $192 \times 192 \times 128$, however no convergence was found in the correlation length of the magnetic field. Hence the question as to whether GRMHD torus simulations can be converged with the available computational power is still an open one.

From Figures 17 and 18, it is clear that the resolution of the $N_\theta = 64$ run is insufficient: a peculiar mini-torus is apparent in the disk-averaged density which diminishes with increasing resolution. Also the onset-time of accretion and the saturation values differ significantly between the $N_\theta = 64$ run and its high-resolution counterparts. These differences diminish between the high-resolution runs and we can see signs of convergence in the accretion rate: increasing resolution from $N_\theta = 128$ to $N_\theta = 192$ appears to not have a strong effect on $\dot{M}$. Also the evolution of ϕ_B agrees quite well between $N_\theta = 128$ and $N_\theta = 192$. Hence the systematic resolution dependence of $\dot{M}$ and ϕ_B in the (even higher resolution) 2D simulations appears to be an artefact of the axisymmetry. It is also noteworthy that the variability amplitude of the accretion rate is reduced in the 3D cases. It appears that the superposition of uncorrelated accretion events distributed over the ϕ-coordinate tends to smear out the sharp variability that results in the axisymmetric case.

Although the simulations were run until $t = 5,000$ M, in order to enable comparison with the 2D simulations, we deliberately set the averaging time to $t \in [1,000$ M, $2,000$ M]. Figure 18 shows that as the resolution is increased, the disk-averaged 3D data approaches the much higher resolution 2D results shown in Figure 15, indicating that the dynamics are dominated by the axisymmetric MRI modes at early times. When the resolution is increased from $N_\theta = 128$ to $N_\theta = 192$, the disk-averaged profiles generally agree within their standard deviations, although we observe a continuing trend towards higher gas pressures and magnetic pressures in the outer regions $r \in [30$ M$, 50$ M$]$. The overall computational cost quickly becomes significant: for the $N_\theta = 128$ simulation we spent 100 K CPU hours on the Iboga cluster equipped with Intel(R) Xeon(R) E5-2640 v4 processors. As the runtime scales with resolution according to N_θ^4, doubling resolution would cost a considerable 1.6 M CPU hours.

4.4 Effect of AMR

In order to investigate the effect of the AMR treatment, we have performed a 2D AMR-GRMHD simulation of the torus setup. It is clear that whether a simulation can benefit from adaptive mesh refinement is very much dependent on the physical scenario under investigation. For example, in the hydrodynamic simulations of recoiling BHs due to (Meliani et al. 2017), refinement on the spiral shock was

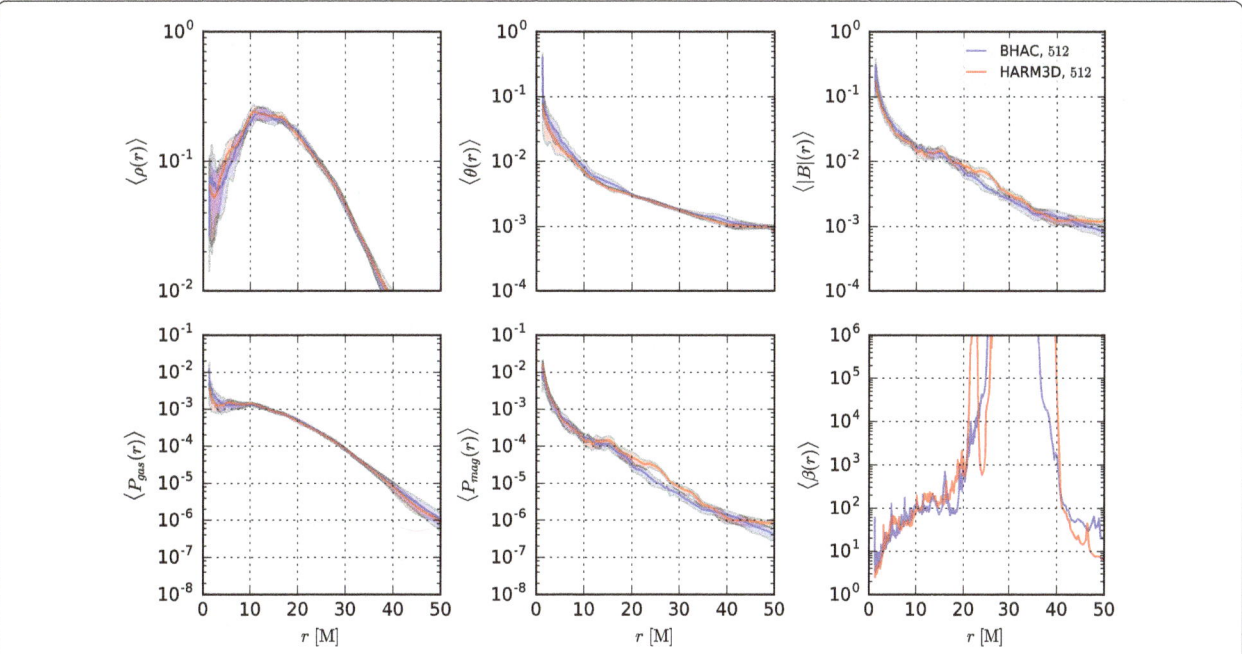

Figure 15 Disk-averaged quantities in the 2D validation runs. Disk-averaged quantities in the 2D validation runs. The blue curves are obtained with BHAC and the red curves with HARM3D in a two-dimensional setting. The shaded regions mark the 1σ standard deviation of the spatially-averaged snapshots (omitted for the highly fluctuating $\langle\beta\rangle$). Apart from a slightly higher magnetisation in HARM for $r \in [20, 30]$, we find excellent agreement between both codes.

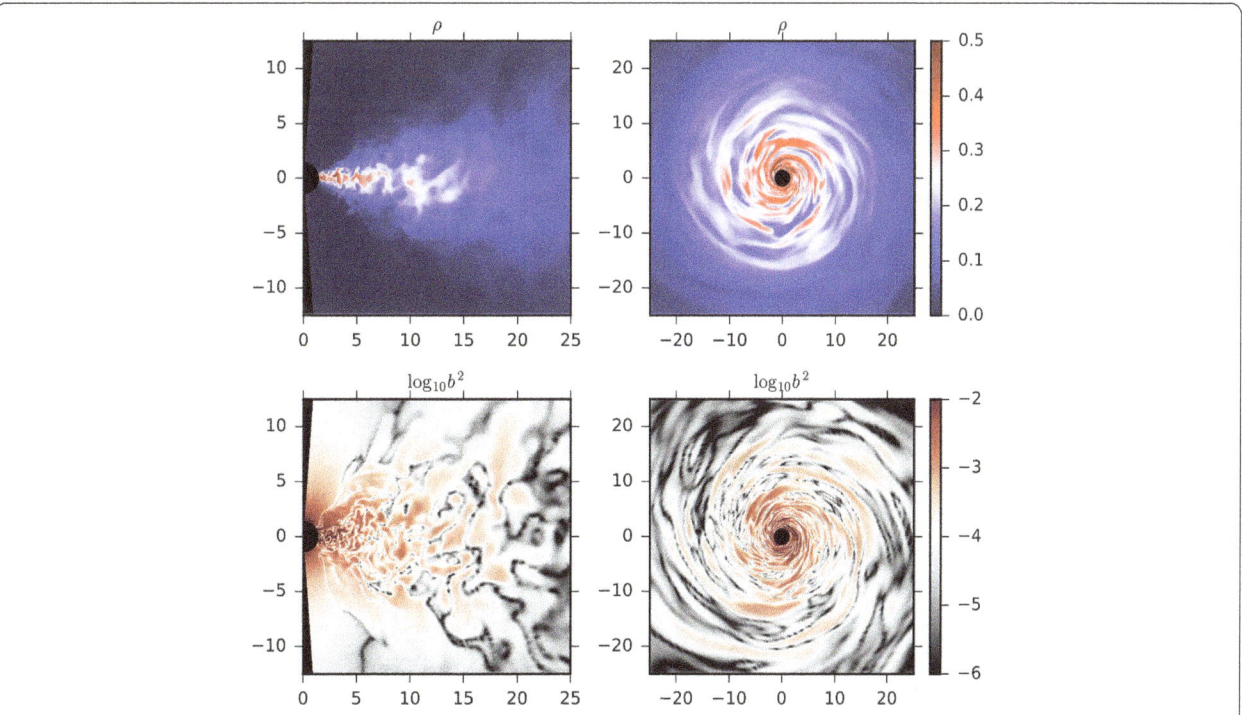

Figure 16 3D torus evolution. Fluid-frame density (top) and $\log_{10} b^2$ (bottom) for $t = 3{,}000$ M on the $y = 0$ plane (left) and the $z = 0$ plane (right) in the 3D magnetised torus run with resolution $384 \times 192 \times 192$.

demonstrated to yield significant speedups at a comparable quality of solution. This is understandable as the numerical error is dominated by the shock hypersurface. In the turbulent accretion problem, whether automated mesh refinement yields any benefits is not clear.

The initial conditions for this test are the same as those used in Section 4.1. However, due to the limitation of current AMR treatment, we resort to the GLM divergence cleaning method. Three refinement levels are used and re-

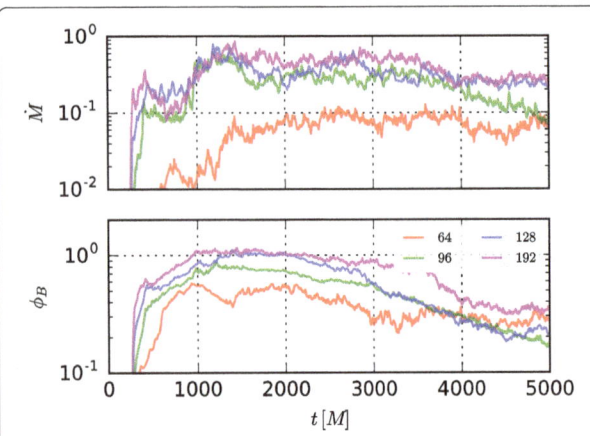

Figure 17 Accretion rates in the 2D magnetised torus. Accretion rates and horizon-penetrating magnetic flux in the 3D runs for varying numerical resolution. We show results from four different resolutions labeled according to the number of cells in θ-direction.

finement is triggered by the error estimator due to (Löhner 1987) with the tolerance set to $\epsilon_t = 0.1$ (see Section 2.11). The numerical resolution in the base level is set to $N_r \times N_\theta = 128 \times 128$. To test the validity and efficiency, we also perform the same simulation in a uniform grid with resolution of $N_r \times N_\theta = 512 \times 512$ which corresponds to the resolution on the highest AMR level.

Figure 19 shows the densities at $t = 2,000$ M as well as the time-averaged density and plasma beta for the AMR and uniform cases. The averaged quantities are calculated in the time interval of $t \in [1,000 \, \text{M}, 2,000 \, \text{M}]$. The overall behaviour is quite similar in both cases. Naturally, differences are seen in the turbulent structure in the torus and wind region for a single snapshot. However, in terms of averaged quantities, the difference becomes marginal. In order to better quantify the difference between the AMR and uniform runs, the mass accretion rate and horizon penetrating magnetic flux are shown in Figure 20. These quantities exhibit a similar behaviour in both cases. In particular, the difference between the AMR run and the uniform run is smaller than the one from different resolution uniform runs and compatible with the run-to-run variation due to a different random number seed (cf. Section 4.2). This is unsurprising since the error estimator triggers refinement of the innermost torus region to the highest level of AMR during most of the simulation time. The development of small scale turbulence by the MRI is clearly captured and it leads to similar mass accretion onto the BH.

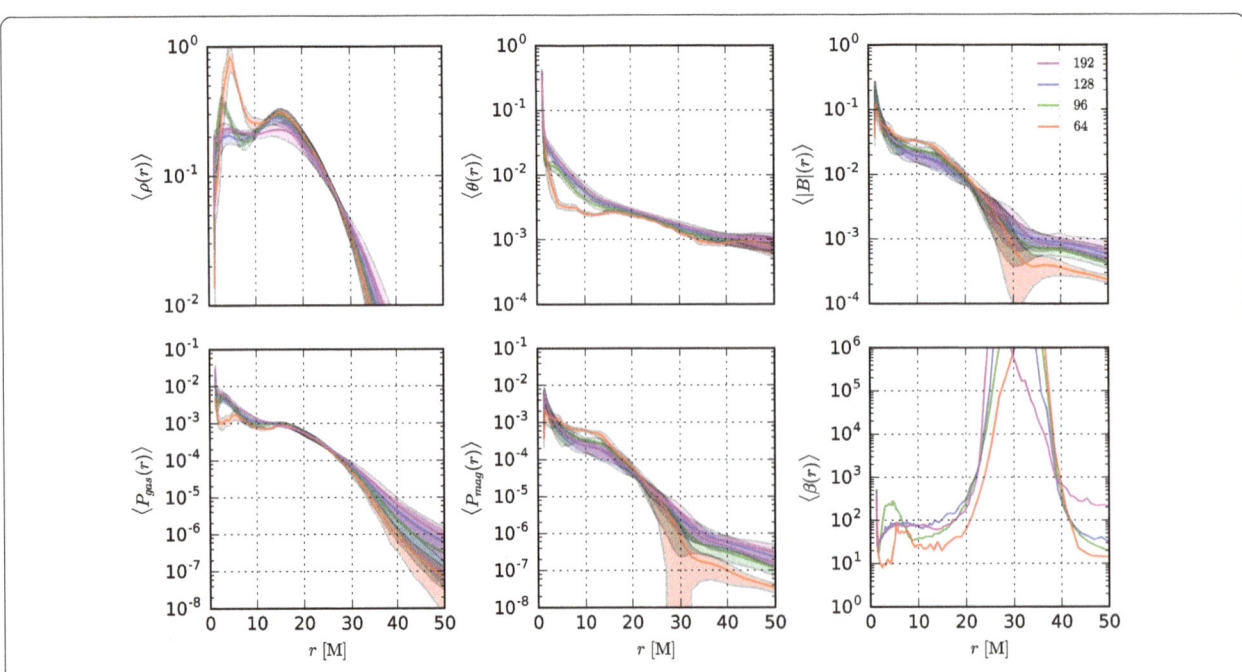

Figure 18 Disk-averaged quantities in the 3D torus runs. Disk-averaged quantities in the 3D runs for varying numerical resolution. The shaded regions mark the 1σ standard deviation of the spatially-averaged snapshots as in Figure 15.

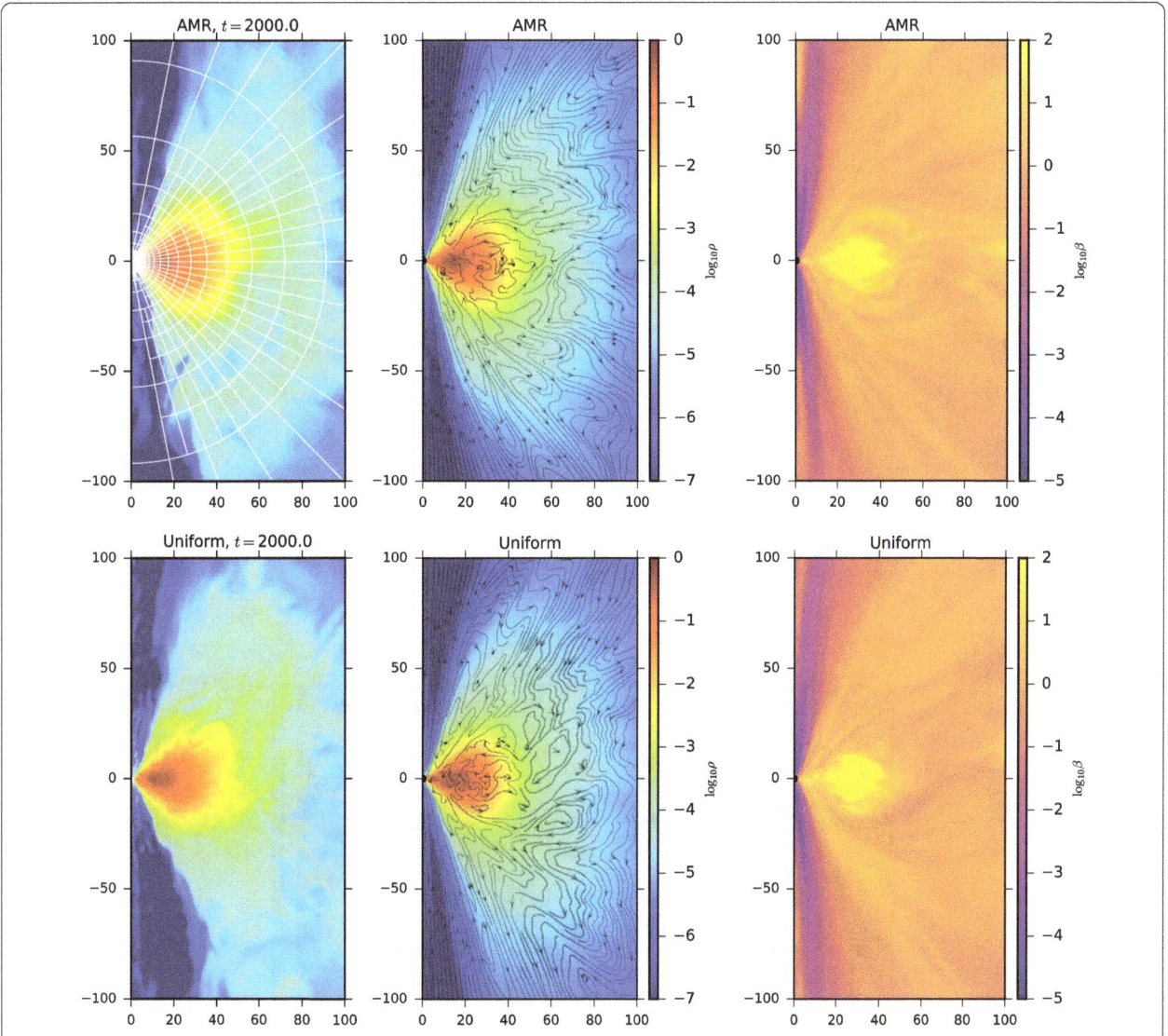

Figure 19 Evolution of the 2D torus runs with AMR. 2D logarithmic density at $t = 2,000$ M (left), averaged density (middle), and averaged plasma beta (right) of the 2D magnetised torus with three-levels AMR (top panels) and uniform resolution 512×512 (bottom panels). Magnetic field lines are traced out in the middle panels using black contour lines. The averaged quantities are calculated in the time interval $t \in [1,000$ M, $2,000$ M]. AMR blocks containing 16^2 cells are indicated in the upper left panel.

One of the important merits of using AMR is the possibility to resolve small and large scale dynamics simultaneously with lower computational cost than uniform grids. Figure 21 shows the large scale structure of the averaged magnetisation after 10,000 M of simulation time. The averaged quantities are calculated in the time interval $t \in [6,000$ M, $10,000$ M]. In order to allow the large-scale magnetic field structure to settle down, we average over a later simulation time compared to the previous non-AMR cases. From the figure the collimation angle and magnetisation of the highly magnetised funnel in the

AMR case are slightly wider than those in uniform case but the large-scale global structure is very similar in both cases.

A comparison of the computational time for a uniform resolution with 512^2 and the equivalent AMR run (three-level AMR) is shown in Table 5. It is encouraging that even in the naive three-level AMR simulation we obtain qualitatively similar results comparable to the high resolution uniform run, but with having spent only 64% of the computational time of the uniform run.[1] Figure 22 shows the evolution of the total number of cells during

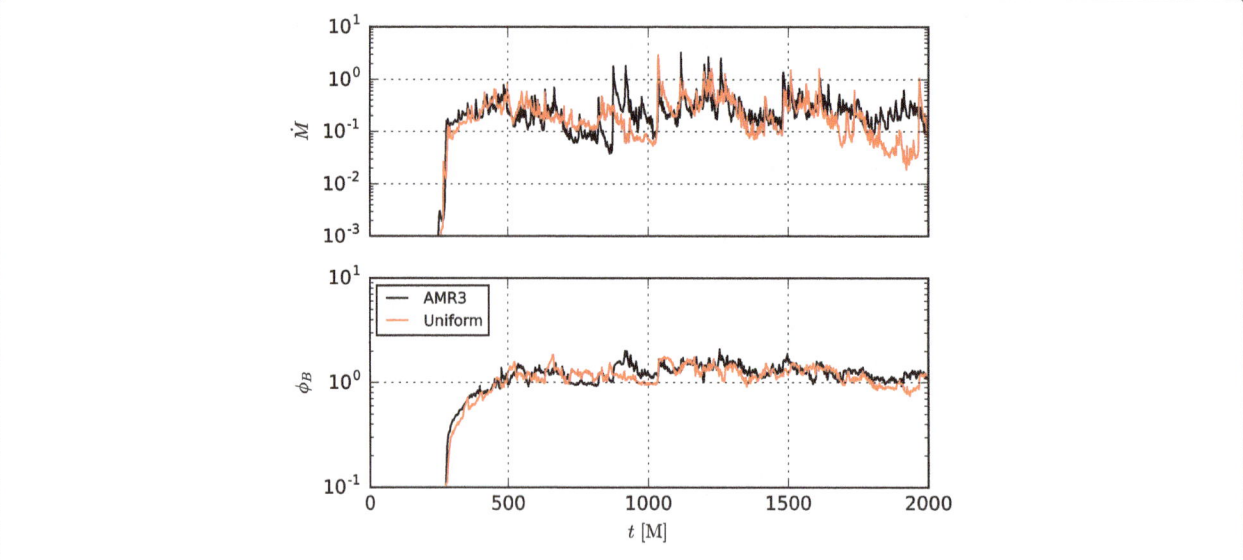

Figure 20 Accretion rates comparing AMR and uniform resolution. Accretion rates and horizon penetrating magnetic flux of the 2D magnetised torus with three levels of AMR (black) and uniform resolution 512×512 (red).

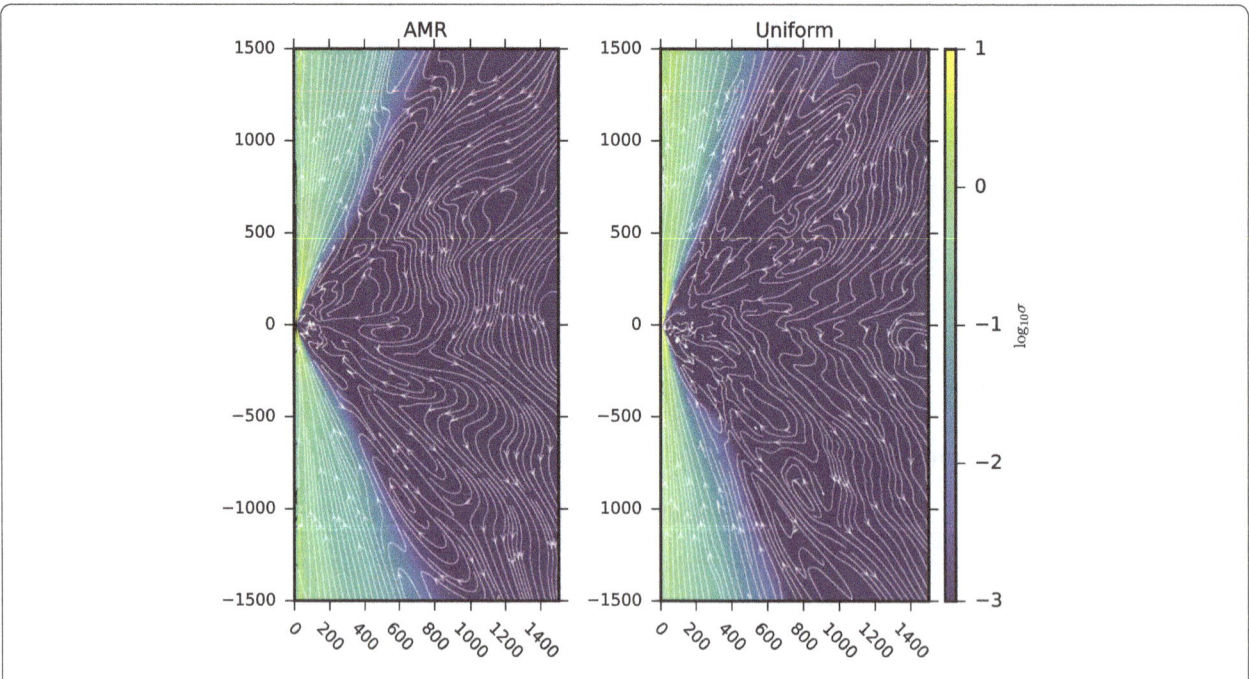

Figure 21 Magnetisation on large scales comparing AMR and uniform resolution. 2D logarithmic averaged magnetisation of the magnetised torus with three levels of AMR (left) and uniform resolution 512×512 (right). Magnetic field lines are traced out by white contour-lines. The averaged quantities are calculated in the time interval of $t \in [6,000\,M, 10,000\,M]$.

the simulations of AMR cases. Initially less than 2^{16} cells are used even when we use three AMR levels, which is a similar number of cells as the uniform grid case with 256×256. When the simulation starts, the total cell number increases rapidly due to development of turbulence in the torus which is triggering higher refinement. We note that the total number of cells is still half of the total number of cells in the corresponding high-resolution uniform grid simulation (512×512), thus resulting in a direct reduction of computational cost. With increasing dynamic range, we expect the advantages of AMR to increase significantly, rendering it a useful tool for simulations involving structures spanning multiple scales. We leave a more detailed discussion on the effect of the AMR refinement strategy and various divergence-control methods to a future paper.

5 Radiation post-processing

In order to compute synthetic observable images of the BH shadow and surrounding accretion flow it is necessary to perform general-relativistic ray-tracing and GRRT post-processing [see, e.g., Fuerst and Wu (2004), Vincent et al. (2011), Younsi et al. (2012), Younsi and Wu (2015), Chan et al. (2015), Dexter (2016), Pu et al. (2016), Younsi et al. (2016)]. In this article the GRRT code BHOSS (Black Hole Observations in Stationary Spacetimes) (Younsi et al. 2017) is used to perform these calculations. From BHAC, GRMHD simulation data are produced which are subsequently used as input for BHOSS. Although BHAC has full AMR capabilities, for the GRRT it is most expedient to output GRMHD data that has been re-gridded to a uniform grid.

Since these calculations are performed in post-processing, the effects of radiation forces acting on the plasma during its magnetohydrodynamical evolution are not included. Additionally, the fast-light approximation has also been adopted in this study, i.e., it is assumed that the light-crossing timescale is shorter than the dynamical timescale of the GRMHD simulation and the dynamical evolution of the GRMHD simulation as light rays propagate through it is not considered. Such calculations are considered in an upcoming article (Younsi et al. 2017).

Several different coordinate representations of the Kerr metric are implemented in BHOSS, including Boyer-Lindquist (BL), Logarithmic BL, Cartesian BL, Kerr-Schild (KS), Logarithmic KS, Modified KS and Cartesian KS. All GRMHD simulation data used in this study are specified in Logarithmic KS coordinates. Although BHOSS can switch between all coordinate systems on the fly, it is most straightforward to perform the GRRT calculations in the same coordinate system as the GRMHD data, only adaptively switching to e.g., Cartesian KS when near the polar region. This avoids the need to transform between coordinate systems at every point along every ray in the GRMHD data interpolation, saving computational time.

5.1 Radiative transfer equation

Electromagnetic radiation is described by null geodesics of the background spacetime (in this case Kerr), and these are calculated in BHOSS using a Runge-Kutta-Fehlberg integrator with fourth order adaptive step sizing and 5th order error control. Any spacetime metric may be considered in BHOSS, as long as the contravariant or covariant

Table 5 CPU hours (CPUH) spent by the simulations of the 2D magnetised torus at uniform resolution and fraction of that time spent by the equivalent AMR runs up to $t = 2{,}000$ M

Grid size ($N_r \times N_\theta$)	CPU time uniform [CPUH]	Equiv. AMR time fraction [$\epsilon_t = 0.1$]
512×512	674.0	0.643

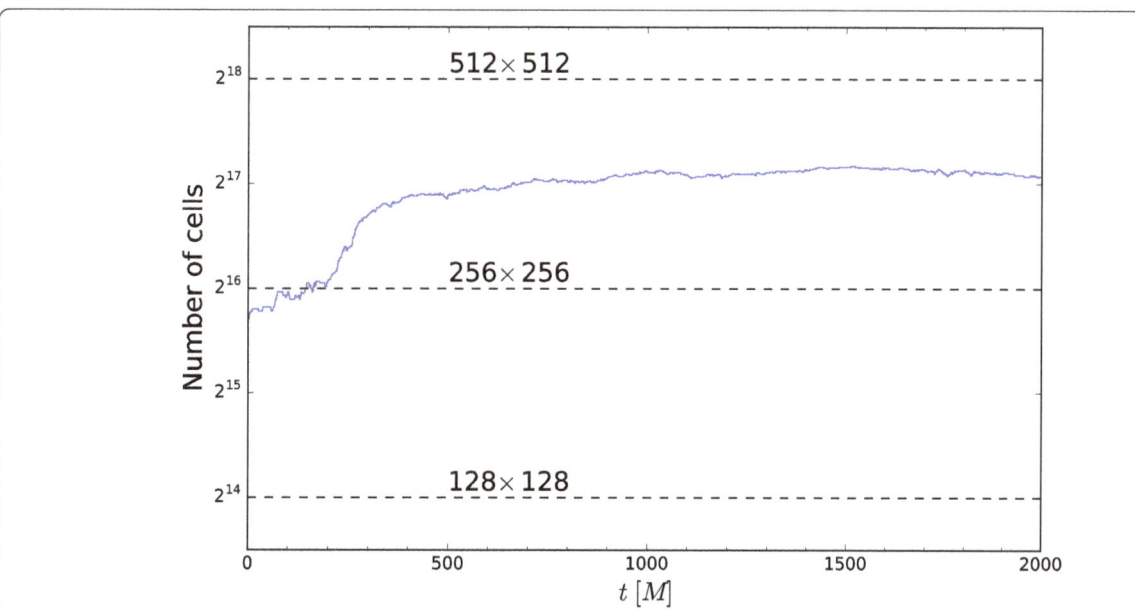

Figure 22 Number of cells as a function of time for the AMR simulation. Number of cells as a function of time for the AMR simulation. The dotted lines show the resolution of uniform grids equivalent to each of the three AMR levels.

metric tensor components are specified, even if they are only tabulated on a grid. For the calculations presented in this article the Kerr spacetime is written algebraically and in closed-form.

The observer is calculated by constructing a local orthonormal tetrad using trial basis vectors. These basis vectors are then orthonormalized using the metric tensor through a modified Gram-Schmidt procedure. The initial conditions of each ray for the coordinate system under consideration are then calculated and the geodesics are integrated backwards in time from the observer, until they either: (i) escape to infinity (exit the computational domain), (ii) are captured by the BH, or (iii) are effectively absorbed by the accretion flow.

In order to perform these calculations the GRRT equation is integrated in parallel with the geodesic equations of motion of each ray. Written in covariant form, the (unpolarized) GRRT equation, in the absence of scattering, may be written Younsi et al. (2012) as

$$\frac{d\mathcal{I}}{d\lambda} = -k_\mu u^\mu \left(-\alpha_{\nu,0}\mathcal{I} + \frac{j_{\nu,0}}{\nu_0^3} \right), \tag{97}$$

where $\mathcal{I} := I_\nu/\nu^3$ is the Lorentz-invariant intensity, I_ν is the specific intensity, ν is the frequency of radiation, $\alpha_{\nu,0}$ is the specific absorption coefficient and $j_{\nu,0}$ is the specific emission coefficient. The subscript 'ν' denotes evaluation of a quantity at a specific frequency, ν, and a subscript '0' denotes evaluation of a quantity in the local fluid rest frame. The terms k_μ and u^μ are the photon 4-momentum and the fluid 4-velocity of the emitting medium, respectively. The former is calculated from the geodesic integration and the latter is determined from the GRMHD simulation data. The affine parameter is denoted by λ.

By introducing the optical depth along the ray

$$\tau_\nu(\lambda) = -\int_{\lambda_0}^{\lambda} \alpha_{\nu,0}(\lambda')k_\mu u^\mu \, d\lambda', \tag{98}$$

together with the Lorentz-invariant emission coefficient ($\eta = j_\nu/\nu^2$) and Lorentz-invariant absorption coefficient ($\chi = \nu\alpha_\nu$), the GRRT Eq. (97) may be rewritten as

$$\frac{d\mathcal{I}}{d\tau_\nu} = -\mathcal{I} + \frac{\eta}{\chi}. \tag{99}$$

Following (Younsi et al. 2012), Eq. (99) may be reduced to two differential equations

$$\gamma \frac{d\tau_\nu}{d\lambda} = \alpha_{\nu,0}, \tag{100}$$

$$\gamma \frac{d\mathcal{I}}{d\lambda} = \frac{j_{\nu,0}}{\nu_0^3}\exp(-\tau_\nu), \tag{101}$$

where

$$\gamma = \frac{\nu}{\nu_0} = \frac{(k_\alpha u^\alpha)_{\text{obs}}}{(k_\beta u^\beta)_0}, \tag{102}$$

is the relative energy shift between the observer ('obs') and the emitting fluid element. Integrating the GRRT equation in terms of the optical depth in the manner presented provides two major advantages. Firstly, the calculation of the geodesic and of the radiative transfer equation may be performed simultaneously, rather than having to calculate the entire geodesic, store it in memory, and then perform the radiative transfer afterwards. Secondly, by integrating in terms of the optical depth we may specify a threshold value (typically of order unity) whereby the geodesic integration is terminated when encountering optically thick media exceeding this threshold. The combination of these two methods saves significant computational time and expense.

5.2 BHOSS-simulated emission from Sgr A*

Having in mind the upcoming radio observations of the BH candidate Sgr A* at the Galactic Centre, the following discussion presents synthetic images of Sgr A*. The GRMHD simulations evolve a single fluid (of ions) and are scale-free in length and mass. Consequently a scaling must be applied before performing GRRT calculations. Within BHOSS this means specifying the BH mass, which sets the length and time scales, and specifying either the mass accretion rate or an electron density scale, which scales the gas density, temperature and magnetic field strength to that of a radiating electron.

Since the GRMHD simulation is of a single fluid, it is necessary to adopt a prescription for the local electron temperature and rest-mass density. Several such prescriptions exist, some which scale using the mass accretion rate [see, e.g., Mościbrodzka et al. (2009), Mościbrodzka et al. (2014), Dexter et al. (2010)], scale using density to determine the electron number density and physical accretion rate [see, e.g., Chan et al. (2015), Chan et al. (2015)], and some by employing a time-dependent smoothing model of the mass accretion rate [see, e.g., Shiokawa et al. (2012)].

The dimensionless proton temperature, Θ_p, is defined as

$$\Theta_p := \frac{k_B T_p}{m_p c^2}, \tag{103}$$

where k_B is the Boltzmann constant, T_p is the geometrical (i.e., in physical units) proton temperature and m_p is the proton mass. This is then calculated from the GRMHD simulation density (ρ) and pressure (p) as

$$\Theta_p = \frac{3p}{\rho}, \tag{104}$$

where the fact that the equation of state is ideal and that $\hat{\gamma} = 4/3$ has been assumed. The magnetic field strength in geometrical units, B_{geo}, is readily obtained from the code magnetic field strength $B = \sqrt{b_\mu b^\mu}$ as

$$B_{\text{geo}} = c\left(\frac{\rho_{\text{geo}}}{\rho}\right)^{1/2} B. \tag{105}$$

What remains is to specify T_e (or $\Theta_e := k_B T_e/m_e c^2$) and ρ_{geo}. For simplicity we adopt the prescription of (Mości-brodzka et al. 2009), wherein T_p/T_e is assumed to be a fixed ratio. Whilst such an approximation is rather crude, to zeroth order the protons and electrons may be assumed to be coupled in this way. To scale the electron number density we adopt the method of (Chan et al. 2015), assuming a density scale typically of order 10^7cm^{-3}. A somewhat more sophisticated approach is to employ a thresholding of the fluid plasma beta where, when the local plasma beta exceeds some threshold the electrons and protons are coupled as previously mentioned (disk region), but when not exceeded (typically in the funnel region) the electron temperature is assumed to be constant (Mości-brodzka et al. 2014; Mościbrodzka et al. 2016; Chan et al. 2015). Since plasma beta is found to decrease with resolution (Shiokawa et al. 2012) and in this paper we seek only to demonstrate the convergence of our simulated shadow images obtained from the GRMHD data in regions where the density is non-negligible, we adopt the former model.

For the plasma emissivity we use the approximate formula for thermal magnetobremsstrahlung (Leung et al. 2011), which is given by

$$j_\nu = \left(\frac{\sqrt{2}\pi e^2}{3c}\right) n_e \frac{\nu_s}{K_2(\Theta_e^{-1})}\left(X^{1/2} + 2^{11/12}X^{1/6}\right)^2$$
$$\times \exp(-X^{1/3}), \tag{106}$$

where e is the electron charge, n_e the electron number density, and

$$X := \frac{\nu}{\nu_s}, \tag{107}$$

$$\nu_s = \left(\frac{e}{9\pi m_e c}\right)B\Theta_e^2 \sin\vartheta, \tag{108}$$

and ϑ is the pitch angle of the photon with respect to the magnetic field. The absorption coefficient is readily obtained from Kirchoff's law.

Each image is generated using a uniform grid of $1{,}000 \times 1{,}000$ rays, sampling 60 uniformly logarithmically spaced frequency bins between 10^9 Hz and 10^{15} Hz. All panels depict the observed image as seen at an observational fre-

quency of 230 GHz, i.e. the frequency at which the EHT will image Sgr A*. This resolution is chosen because the integrated flux over the entire ray-traced image is convergent: doubling the resolution from 500×500 to $1{,}000 \times 1{,}000$ yields an increase of 0.17%, and from $1{,}000 \times 1{,}000$ to $2{,}000 \times 2{,}000$ an increase of 0.09%.

In practical GRRT calculations only simulation data which has already reached a quasi-steady state, typically $t > 2{,}000$ M, is used. In this study we focus on the observational appearance of the accretion flow and BH shadow image. The detailed discussion of the spectrum, variability and plasma models warrants a separate study.

5.3 Comparison of images

A natural and important question arises from GRRT calculations of BH shadows: do ray-traced images of GRMHD simulation data converge as the resolution of the GRMHD simulation is increased? The existence of an optimal resolution, beyond which differences in images are small, implies that one can save additional computational time and expense by running the simulation at this optimal resolution. It would also imply that the GRMHD data satisfactorily capture the small-scale structure, turbulence and variations of the accretion flow. As such, it is informative to investigate the convergence of BH shadow images obtained from GRMHD simulation data of differing resolutions, both quantitatively and qualitatively.

To address this question we first generate a series of four snapshot images at $t = 2{,}500$ M of the the accretion flow and BH shadow from GRMHD simulation data. The resolution of these data are $2\mathcal{N} \times \mathcal{N} \times \mathcal{N}$ in (r, θ, ϕ), i.e., twice as much resolution in the radial direction compared to the zenith and azimuthal directions. The images depicted in Figure 23 correspond to $\mathcal{N} = 64, 96, 128$ and 192 respectively. Here, the proton to electron temperature ratio was chosen as $T_p/T_e = 3$ (similar to (Mościbrodzka et al. 2009; Mościbrodzka et al. 2014)), the electron number density scaling as 5×10^7 cm^{-3}, the BH mass is set to $4.5 \times 10^6 M_\odot$, the source distance is 8.4×10^3 pc, the dimensionless BH spin parameter 0.9375 and the observer inclination angle with respect to the BH spin axis is $60°$.

A direct consequence of increasing the resolution of the GRMHD data is resolving the fine-scale turbulent structure of the accretion flow. The characteristic dark shadow delineating the BH shadow can be clearly seen in all images. As the resolution of the GRMHD data is increased, the images become less diffuse. It is difficult with the naked eye to draw firm physical conclusions, and so in the following we perform a quantitative pixel-by-pixel analysis.

With these snapshot images we may perform a quantitative measure of the difference between any two images through introducing the (normalised) cross-correlation. For two given two-dimensional arrays $f(x, y)$ and $g(x, y)$

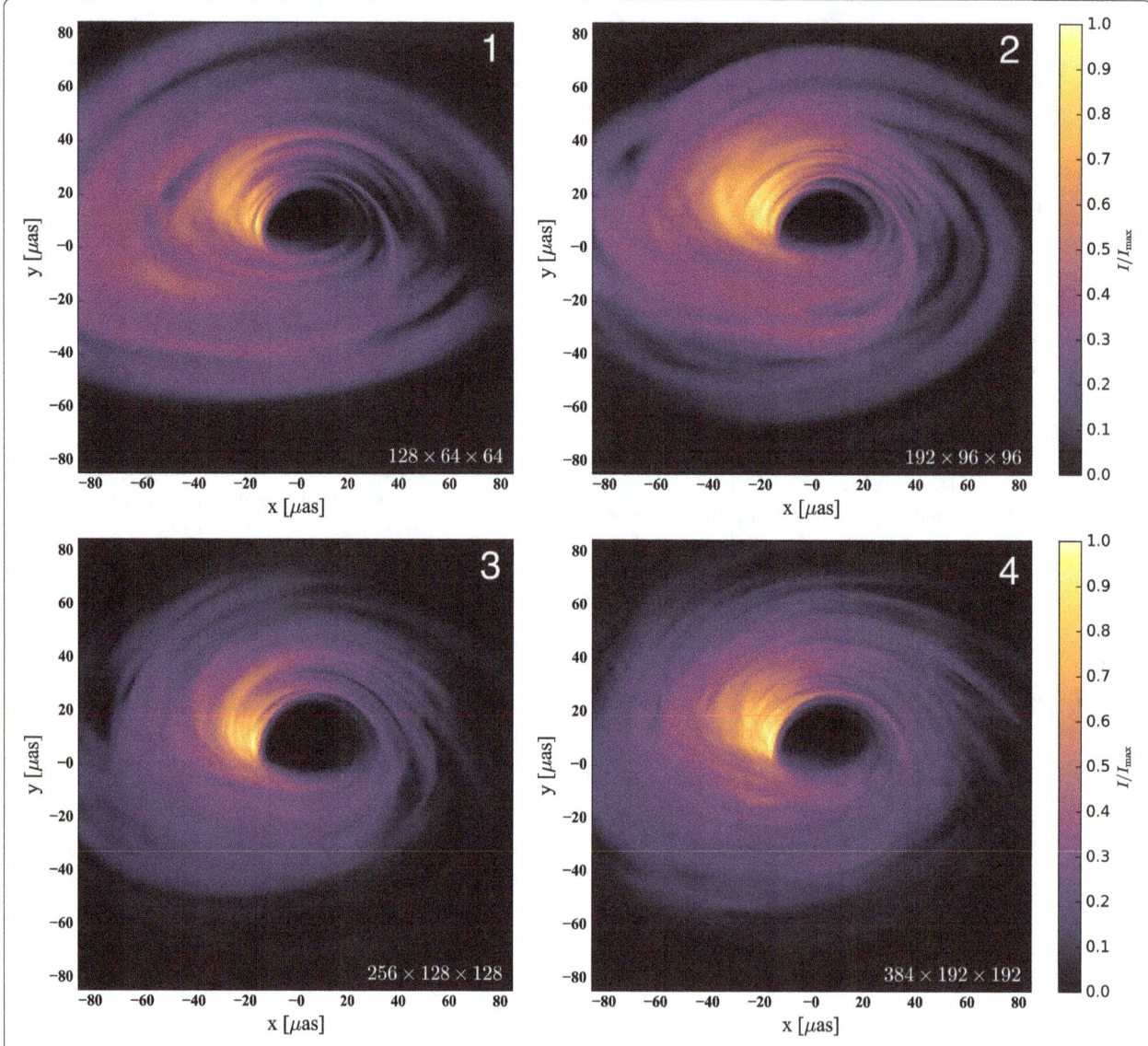

Figure 23 GRRT of 3D torus for increasing resolution. Snapshot images of 3D GRMHD simulation data with parameters chosen to mimic the emission from Sgr A*. The resolution of the simulation data is indicated in the bottom-right corner of each panel and discussed in the text.

(i.e., 2D images), a measure of similarity or difference may be calculated through the cross-correlation $\mathcal{C}$, where $\mathcal{C} \in [-1, 1]$. The normalised cross-correlation is defined as

$$\mathcal{C} := \mathcal{C}_{i,j}$$

$$:= \frac{1}{N\sigma_f\sigma_g} \sum_{x,y} \left\{ [f(x,y) - \mu_f][g(x,y) - \mu_g] \right\}, \qquad (109)$$

where μ_f, σ_f and μ_g, σ_g correspond to the mean and standard deviation of f and g respectively, and N is equal to the size of either f or g. In the examples considered here the images are all of equal size and dimension, so $N = N_f = N_g$.

Equation (109) may be interpreted as the inner product between two data arrays, with the value of $\mathcal{C}$ expressing the degree to which the data are aligned with respect to each other. When $\mathcal{C} = 1$ the data are identical, save for a multiplicative constant, when $\mathcal{C} = 0$ the data are completely uncorrelated, and when $\mathcal{C} < 0$ the data are negatively correlated.

Each image pixel has an intensity value represented as a single greyscale value between zero and one. Given the relative intensity data of two different images, Eq. (109) is then employed to determine the normalised cross-correlation between the two images. This procedure applied to the panels in Fig. (23) yields the following symmetric matrix of cross-correlation values between the images

$$
\mathcal{C}_{i,j} = \begin{pmatrix} 1 & 0.839495 & 0.809205 & 0.856958 \\ - & 1 & 0.867578 & 0.917560 \\ - & - & 1 & 0.948544 \\ - & - & - & 1 \end{pmatrix}. \quad (110)
$$

A visual representation of the pixel-by-pixel differences is given in Figure 24. Indices i and j, where $(i,j) = (1,4)$, denote the images being cross-correlated.

The rightmost column of Eq. (110) denotes the cross-correlation values, $\mathcal{C}_{i,4}$, in descending order between images, i.e., the cross-correlation of image 4 with images 1, 2, 3 and 4 respectively. Since $\mathcal{C}_{i+1,4} > \mathcal{C}_{i,4}$ it is clear that the similarity between images increases as the resolution of the GRMHD simulation is increased. Similarly, for image 3 it is found that $\mathcal{C}_{i+1,3} > \mathcal{C}_{i,3}$. Finally, it also follows that $\mathcal{C}_{3,4} > \mathcal{C}_{2,3} > \mathcal{C}_{1,2}$, i.e., the correlation between successive pairs of images increases with increasing resolution, demonstrating the convergence of the GRMHD simulations with increasing grid resolution. Whilst the lowest resolution of $128 \times 64 \times 64$ is certainly insufficient, both difference images and cross-correlation measures indicate that a resolution of $256 \times 128 \times 128$ is sufficient and represents a good compromise.

6 Conclusions and outlook

We have described the capabilities of BHAC, a new multidimensional general-relativistic magnetohydrodynamics code developed to perform hydrodynamical and MHD simulations of accretion flows onto compact objects in arbitrary stationary spacetimes exploiting the numerous advantages of AMR techniques. The code has been tested with several one-, two- and three- dimensional scenarios in special-relativistic and general-relativistic MHD regimes. For validation, GRMHD simulations of MRI unstable tori have been compared with another well-known and tested GRMHD code, the HARM3D code. BHAC shows very good agreement with the HARM3D results, both qualitatively and quantitatively. As a first demonstration of the AMR capabilities in multi-scale simulations, we performed the magnetized-torus accretion test with and without AMR. Despite the latter intrinsically implies an overhead of $\sim 10\%$, the AMR runtime amounted to 65% of that relative to the uniform grid, simply as a result of the more economical use of grid cells in the block based AMR. At the same time, the AMR results agree very well with the more expensive uniform-grid results. With increasing dynamic range, we expect the advantages of AMR to increase even more significantly, rendering it a useful tool for simulations involving structures of multiple physical scales.

Currently, two methods controlling the divergence of the magnetic field are available in BHAC and we compared them in three test problems. Although solutions obtained with the cell-centered flux-interpolated constrained transport (FCT) algorithm and the divergence cleaning scheme (GLM) yield the same (correct) physical behaviour in the case of weak magnetic fields, FCT performs considerably better in the presence of strong magnetic fields. In particular, FCT is less diffusive than GLM, is able to preserve a stationary solution, and it creates less spurious structures in the magnetic field. For example, the use of GLM in the case of accretion scenarios with strong magnetic fields leads to worrisome artefacts in the highly magnetised funnel region. The development of a constrained transport scheme compatible with AMR is ongoing and will be presented in a separate work (Olivares et al. 2017).

The EHTC and its European contribution, the BlackHoleCam project (Goddi et al. 2016), aim at obtaining horizon-scale radio images of the BH candidate at the Galactic Center. In anticipation of these results, we have used the 3D GRMHD simulations as input for GRRT calculations with the newly developed BHOSS code (Younsi et al. 2017). We found that the intensity maps resulting from different resolution GRMHD simulations agree very well, even when comparing snapshot data that was not time averaged. In particular, the normalised cross-correlation between images achieves up to 94.8% similarity between the two highest resolution runs. Furthermore, the agreement between two images converges as the resolution of the GRMHD simulation is increased. Based on this comparison, we find that moderate grid resolutions of $256 \times 128 \times 128$ (corresponding to physical resolutions of $\Delta r_{KS} \times \Delta \theta_{KS} \times \Delta \phi_{KS} = 0.04 \, \mathrm{M} \times 0.024 \, \mathrm{rad} \times 0.05 \, \mathrm{rad}$ at the horizon) yield sufficiently converged intensity maps. Given the large and likely degenerate parameter space and the uncertainty in modelling of the electron distribution, this result is encouraging, as it demonstrates that the predicted synthetic image is quite robust against the ever-present time variability, but also against the impact that the grid resolution of the GRMHD simulations might have. In addition, independent information on the spatial orientation and magnitude of the spin, such as the one that could be deduced from the dynamics of a pulsar near Sgr A* (Psaltis et al. 2016), would greatly reduce the space of degenerate solutions and further increase the robustness of the predictions that BHAC will provide in terms of synthetic images.

Finally, we have demonstrated the excellent flexibility of BHAC with a variety of different astrophysical scenarios that are ongoing and will be published shortly. These include: oscillating hydrodynamical equilibrium tori for the modelling of quasi-periodic oscillations (de Avellar et al. 2017), episodic jet formation (Porth et al. 2016) and magnetised tori orbiting non-rotating dilaton BHs (Mizuno et al. 2017).

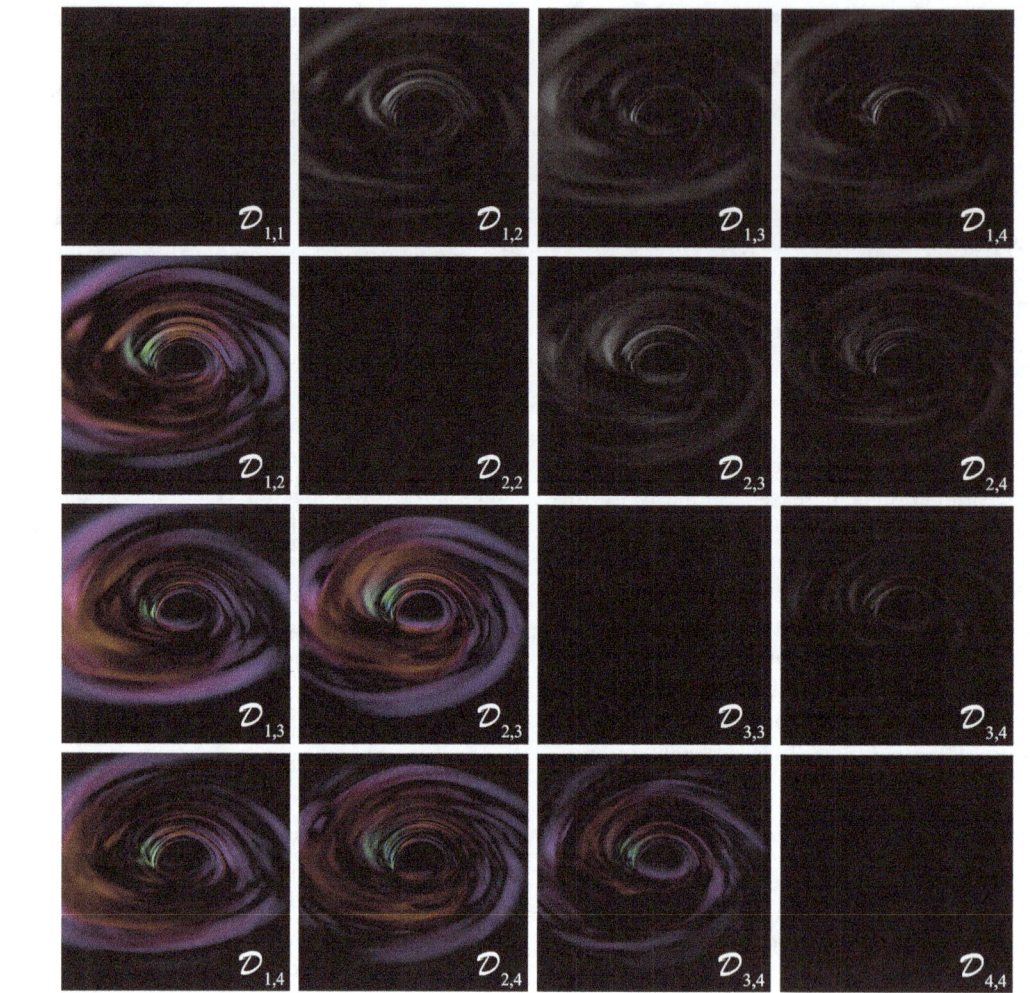

Figure 24 Matrix of image differences $\mathcal{D}_{i,j}$. Matrix of image differences $\mathcal{D}_{i,j}$ of the four panels in Figure 23. Upper diagonal panels are greyscale differences. Lower diagonal panels are identical to corresponding upper diagonal panels but with differences illustrated with RGB pixel values. Black panels correspond to $\mathcal{D}_{i,j}$, i.e., trivially the difference between an image and itself.

Appendix A: Evolution of the scalar ϕ

To obtain the evolution equation for ϕ in the augmented Faraday's law, we project (42) onto the Eulerian observer by contracting with $-n_\mu$ as

$$-\nabla_\nu\left({}^*F^{\mu\nu}n_\mu - \phi n^\nu\right) = -\kappa\phi - \left({}^*F^{\mu\nu} - \phi g^{\mu\nu}\right)\nabla_\nu n_\mu \quad (111)$$

$$\Rightarrow \quad \nabla_\nu B^\nu + \nabla_\nu \phi n^\nu$$

$$= -\kappa\phi - \left({}^*F^{\mu\nu} - \phi g^{\mu\nu}\right)\nabla_\nu n_\mu \quad (112)$$

$$\Rightarrow \quad (-g)^{-1/2}\partial_i\left[\gamma^{1/2}\alpha B^i\right] + \nabla_\nu \phi n^\nu$$

$$= -\kappa\phi - \left({}^*F^{\mu\nu} - \phi g^{\mu\nu}\right)\nabla_\nu n_\mu, \quad (113)$$

where we used $B^\nu = -n_\mu{}^*F^{\mu\nu}$. Using the definition of extrinsic curvature $K_{\mu\nu} := -\gamma_\mu^\lambda\nabla_\lambda n_\nu$, we can write [Eq. (7.62)

in (Rezzolla and Zanotti 2013)]

$$\nabla_\nu n_\mu = -n_\nu a_\mu - K_{\mu\nu}, \quad (114)$$

where we used the 'acceleration' of the Eulerian observer $a_\mu := n^\lambda\nabla_\lambda n_\mu$ which satisfies $n^\mu a_\mu = 0$. With the identity $a_i = \alpha^{-1}\partial_i\alpha$ (York 1979) and exploiting the symmetries of ${}^*F^{\mu\nu}$ and $K_{\mu\nu}$, is straightforward to show that

$$(\gamma)^{-1/2}\partial_i\left[(\gamma)^{1/2}\alpha B^i\right] + \alpha F^{*\mu\nu}\nabla_\nu n_\mu$$

$$= (\gamma)^{-1/2}\partial_i\left[(\gamma)^{1/2}\alpha B^i\right] - B^i\partial_i\alpha. \quad (115)$$

Hence it follows that

$$\partial_t(\sqrt{\gamma}\phi) + \partial_i\left[\sqrt{\gamma}\left(\alpha B^i - \phi\beta^i\right)\right]$$

$$= -\alpha\gamma^{1/2}\kappa\phi + \alpha\gamma^{1/2}\phi g^{\mu\nu}\nabla_\nu n_\mu + \gamma^{1/2}B^i\partial_i\alpha. \quad (116)$$

Using again Eq. (114), the source term $\mathcal{S} := \sqrt{\gamma}\alpha\phi g^{\mu\nu}\nabla_\nu n_\mu$ can be rewritten as

$$\mathcal{S} = -\sqrt{\gamma}\alpha\phi g^{\mu\nu}K_{\mu\nu}, \tag{117}$$

where the first term drops out due to the orthogonality $n^\mu a_\mu = 0$. For a symmetric tensor $S^{\mu\nu}$, we have

$$\alpha S^{\mu\nu}K_{\mu\nu} = \alpha S^{ij}K_{ij}$$
$$= S^i_j\partial_i\beta^j + \frac{1}{2}S^{ij}\beta^k\partial_k\gamma_{ij}. \tag{118}$$

This follows from the relation $\Gamma^0_{ij} = -K_{ij}\alpha^{-1}$ where Γ^0_{ij} are elements of the 4-Christoffel symbols [see e.g., (B.9) of (Alcubierre 2008)]. Thus

$$\mathcal{S} = -\sqrt{\gamma}\phi\left(\partial_i\beta^i + \frac{1}{2}\gamma^{ij}\beta^k\partial_k\gamma_{ij}\right) \tag{119}$$

$$= -\sqrt{\gamma}\phi\partial_i\beta^i - \sqrt{\gamma}\phi\frac{1}{2}\gamma^{ij}\beta^k\partial_k\gamma_{ij}. \tag{120}$$

Appendix B: Modified Faraday's law

The augmented Faraday's law follows from the j-component of (42) as

$$\nabla_\nu\left({}^*F^{j\nu} - \phi g^{j\nu}\right) = \kappa\phi\beta^j/\alpha \tag{121}$$

$$\Rightarrow \quad (-g)^{-1/2}\left\{\partial_t\left[\sqrt{\gamma}(-B^j)\right]\right.$$
$$\left. + \partial_i\left[\sqrt{\gamma}(\mathcal{V}^jB^i - \mathcal{V}^iB^j)\right]\right\}$$
$$+ g^{j\lambda}\partial_\lambda(-\phi) = \kappa\phi\beta^j/\alpha \tag{122}$$

$$\Rightarrow \quad \partial_t\left(\sqrt{\gamma}B^j\right) + \partial_i\left[\sqrt{\gamma}(\mathcal{V}^iB^j - \mathcal{V}^jB^i)\right]$$
$$+ \sqrt{\gamma}\alpha g^{j\lambda}\partial_\lambda\phi = -\kappa\phi\sqrt{\gamma}\beta^j \tag{123}$$

$$\Rightarrow \quad \partial_t\left(\sqrt{\gamma}B^j\right) + \partial_i\left[\sqrt{\gamma}(\mathcal{V}^iB^j - \mathcal{V}^jB^i)\right]$$
$$+ \frac{\beta^j}{\alpha}\partial_t(\sqrt{\gamma}\phi) + \sqrt{\gamma}\alpha\gamma^{ji}\partial_i\phi$$
$$- \sqrt{\gamma}\frac{\beta^i\beta^j}{\alpha}\partial_i\phi = -\sqrt{\gamma}\kappa\phi\beta^j. \tag{124}$$

We see that apart from the gradient ϕ-term, we obtain another term that involves the time-derivative of $(\sqrt{\gamma}\phi)$. Hence we need to plug in Eq. (43). We rewrite the term $\beta^j\partial_t(\sqrt{\gamma}\phi)/\alpha$ simplifying the lengthy expression

$$\frac{\beta^j}{\alpha}\partial_t(\sqrt{\gamma}\phi) = -\frac{\beta^j}{\alpha}\partial_i\left[\sqrt{\gamma}(\alpha B^i - \phi\beta^i)\right]$$
$$- \sqrt{\gamma}\kappa\phi\beta^j - \frac{\beta^j}{\alpha}\sqrt{\gamma}\phi\partial_i\beta^i$$
$$- \frac{1}{2}\frac{\beta^j}{\alpha}\sqrt{\gamma}\phi\gamma^{il}\beta^k\partial_k\gamma_{il}$$

$$+ \frac{\beta^j}{\alpha}\sqrt{\gamma}B^i\partial_i\alpha \tag{125}$$

$$= -\partial_i\left[\sqrt{\gamma}B^i\beta^j\right] + \sqrt{\gamma}B^i\partial_i\beta^j$$
$$+ \sqrt{\gamma}\frac{\beta^i\beta^j}{\alpha}\partial_i\phi - \sqrt{\gamma}\kappa\phi\beta^j. \tag{126}$$

Substituting this into (124) yields the modified Faraday's law (44).

Appendix C: Derivation of cell centred formulas for FCT

In the 3 + 1 decomposition, for the case of a stationary spacetime the induction equation can be written in component form as

$$\partial_t\sqrt{\gamma}B^a + \partial_b\left(-\eta^{abc}E_c\right) = 0. \tag{127}$$

Integrating this on each of the surfaces bounding a cell with vertexes at $x^1_{i+l_1}, x^2_{i+l_2}, x^3_{i+l_3}$ with $l = \pm 1/2$, and using the Stokes theorem, we obtain the evolution equations for the magnetic flux in CT, for instance

$$\frac{d\Phi_{i+1/2,j,k}}{dt} = G_{i+1/2,j+1/2,k} - G_{i+1/2,j-1/2,k}$$
$$- G_{i+1/2,j,k+1/2} + G_{i+1/2,j,k-1/2}, \tag{128}$$

where

$$\Phi_{i+1/2,j,k} = \int_{\partial V(x^1_{i+1/2})} \gamma^{1/2}B^1\,dx^2\,dx^3, \tag{129}$$

with each G representing a line integral of the form

$$G_{i+1/2,j+1/2,k} = -\int_{x^3_{k-1/2}}^{x^3_{k+1/2}} E_3|_{x^1_{i+1/2},x^2_{j+1/2}}\,dx^3. \tag{130}$$

The fact that each of these integrals appear in the evolution equation of two magnetic fluxes guarantees the conservation of divergence, as will be explained in the next Section.

On the other hand, the numerical fluxes corresponding to the magnetic field components that are returned by the Riemann solver are surface integrals of the electric field, for example, the flux in the x^2-direction for B^1 is

$$\Delta S^2\bar{F}^2|_{i,j+1/2,k} = \int_{x^1_{i-1/2}}^{x^1_{i+1/2}}\int_{x^3_{k-1/2}}^{x^3_{k+1/2}} E_{x^3}|_{j+1/2}\,dx^3\,dx^1. \tag{131}$$

The innermost integral is the same as that of Eq. (130), so the average flux can be interpreted as

$$\Delta S^2\bar{F}^2|_{i,j+1/2,k} = -\Delta x_i\tilde{G}_{i,j+1/2,k}, \tag{132}$$

where $\tilde{G}_{i,j+1/2,k}$ is the mean value of the integral from Eq. (130). To second-order accuracy, this integral takes the value $\tilde{G}_{i,j+1/2,k}$ at the middle of the cell, therefore $G_{i+1/2,j+1/2,k}$ can be found by interpolating the averaged fluxes from the four adjacent cell faces as

$$
\begin{aligned}
G_{i+1/2,j+1/2,k} \\
= \frac{1}{4}\left(\frac{\Delta S^2 \bar{F}^2|_{i,j+1/2,k}}{\Delta x} + \frac{\Delta S^2 \bar{F}^2|_{i+1,j+1/2,k}}{\Delta x_{i+1}} \right. \\
\left. - \frac{\Delta S^1 \bar{F}^1|_{i+1/2,j,k}}{\Delta y_j} - \frac{\Delta S^1 \bar{F}^1|_{i+1/2,j+1,k}}{\Delta y_{j+1}} \right).
\end{aligned} \tag{133}
$$

Since we implemented a cell-centred version of FCT, we are interested in the evolution of the average magnetic field at the cell centres. To second order accuracy, the rate of change of the average value of the x^1-component of the magnetic field is

$$
\begin{aligned}
\int_{x_{i-1/2}}^{x_{x+1/2}} \frac{d\Phi}{dt} dx &= \Delta V_{ijk} \frac{d\bar{B}^x}{dt} \\
&\approx \frac{\Delta x_i}{2}\left(\left.\frac{d\Phi}{dt}\right|_{x_{i+1/2}} + \left.\frac{d\Phi}{dt}\right|_{x_{i-1/2}} \right).
\end{aligned} \tag{134}
$$

Now we substitute Eq. (133) into Eq. (128) and Eq. (128) into Eq. (134). After some algebra, we finally obtain Eqs. (45) and (46).

Appendix D: Discretisation of $\nabla \cdot B$ and zero-divergence initial conditions

CT schemes aim to maintain to zero at machine precision the discretisation of the divergence given by

$$
\begin{aligned}
(\nabla \cdot B)_{i,j,k} = \frac{1}{\Delta V_{i,j,k}}(&\Phi_{i+1/2,j,k} - \Phi_{i-1/2,j,k} + \Phi_{i,j+1/2,k} \\
&- \Phi_{i,j-1/2,k} + \Phi_{i,j,k+1/2} - \Phi_{i,j,k-1/2}),
\end{aligned} \tag{135}
$$

which can be thought of as the volume average of the quantity $\partial_a(\gamma^{1/2}B^a)$ in the given cell.

When calculating the evolution equation for $(\nabla \cdot B)_{i,j,k}$, each of the integrals G appear with opposite signs in the expression for $d\Phi/dt$ (128) and cancel to machine precision. Therefore, if this discretisation of the divergence was originally zero, it will be zero to machine precision during the rest of the simulation.

However, in the cell-centred version of FCT employed here, we lack information concerning the magnetic flux at cell faces, so Eq. (135) cannot be used to monitor the creation of divergence. We will therefore find a derived quantity that we can monitor based on the other available quantities.

We calculate the average value of the divergence of eight cells sharing a vertex as

$$
\begin{aligned}
(\nabla \cdot B)_{i+1/2,j+1/2,k+1/2} \\
= \frac{1}{\Delta V^*}\sum_{l_1,l_2,l_3=0,1} \Delta V(\nabla \cdot B)|_{i+l_1,j+l_2,k+l_3}.
\end{aligned} \tag{136}
$$

When substituting Eq. (135), the right hand side of Eq. (136) consists of a sum of terms of the form

$$
\begin{aligned}
\sum_{l_2,l_3=0,1}(&\Phi_{i+3/2,j+l_2,k+l_3} - \Phi_{i+1/2,j+l_2,k+l_3} \\
&+ \Phi_{i+1/2,j+l_2,k+l_3} - \Phi_{i-1/2,j+l_2,k+l_3}),
\end{aligned}
$$

for each direction. Using the same second-order approximation as for the time-update,

$$
\Delta V_{i,j,k}\bar{B}^x_{i,j,k} \approx \frac{\Delta x_i}{2}(\Phi_{i+1/2,j,k} + \Phi_{i-1/2,j,k}), \tag{137}
$$

this becomes

$$
\sum_{l_1,l_2,l_3=0,1}\left[(-1)^{1+l_1}\frac{\bar{B}^1 \Delta V}{\Delta x^1}\right]_{i+l_1,j+l_2,k+l_3}.
$$

Finally, summing over the three directions, we recover Eq. (47). Since the same second-order approximation is used both for the definition and for the time update of $\bar{B}^a$, the definition of divergence given by Eq. (47) is conserved to machine precision during each evolution step.

To obtain a divergence-free initial condition, we calculate the magnetic field as the curl of a vector potential. First, we calculate the magnetic flux at each of the cell faces as

$$
\begin{aligned}
\Phi_{i+1/2,j,k} = \mathcal{A}_{i+1/2,j+1/2,k} - \mathcal{A}_{i+1/2,j-1/2,k} \\
- \mathcal{A}_{i+1/2,j,k+1/2} + \mathcal{A}_{i+1/2,j,k-1/2},
\end{aligned} \tag{138}
$$

where $\mathcal{A}$ are line integrals of the vector potential along the cell edges

$$
\mathcal{A}_{i+1/2,j+1/2,k} = \int_{x^3_{k-1/2}}^{x^3_{k+1/2}} A_3|_{x^1_{i+1/2},x^2_{j+1/2}} dx^3. \tag{139}
$$

Then we use again the second order approximation from Eq. (137) to find the average magnetic field components at the cell center. By construction, in this way we obtain a divergence-free initial condition using either of the discretization of divergence in Eqs. (135) or (47).

Competing interests
The authors declare that they have no competing interests.

Funding

This research is supported by the ERC synergy grant 'BlackHoleCam: Imaging the Event Horizon of Black Holes' (Grant No. 610058), by 'NewCompStar', COST Action MP1304, by the LOEWE-Program in HIC for FAIR, and by the European Union's Horizon 2020 Research and Innovation Programme (Grant 671698) (call FETHPC-1-2014, project ExaHyPE). ZY acknowledges support from an Alexander von Humboldt Fellowship. HO is supported in part by a CONACYT-DAAD scholarship.

Authors' contributions

The implementation of the GRMHD equations was performed by OP. The FCT algorithm was implemented and tested by HO. YM contributed with code tests and ZY performed the radiative-transfer calculations. MM provided HARM3D validation data. The project was initiated by LR, HF and MK and was closely supervised by LR. All authors read and approved the final manuscript.

Author details

[1]Institute for Theoretical Physics, Max-von-Laue-Str. 1, Frankfurt am Main, 60438, Germany. [2]Frankfurt Institute for Advanced Studies, Ruth-Moufang-Straße 1, Frankfurt am Main, D-60438, Germany. [3]Department of Astrophysics/IMAPP, Radboud University Nijmegen, P.O. Box 9010, Nijmegen, 65008, The Netherlands. [4]Max-Planck-Institut für Radioastronomie, Auf dem Hügel 69, Bonn, D-53121, Germany.

Acknowledgements

It is a pleasure to thank Christian Fromm, Mariafelicia de Laurentis, Thomas Bronzwaer, Jordy Davelaar, Elias Most and Federico Guercilena for discussions. We are grateful to Scott Noble for the ability to use the HARM3D code for comparison and to Zakaria Meliani for input on the construction of BHAC. The initial setup for the toroidal-field equilibrium torus was kindly provided by Chris Fragile. The simulations were performed on LOEWE at the CSC-Frankfurt and Iboga at ITP Frankfurt. We acknowledge technical support from Thomas Coelho.

Endnotes

[a] http://www.eventhorizontelescope.org

[b] http://www.blackholecam.org

[c] This quantity is often indicated as W (Antón et al. 2006; Rezzolla and Zanotti 2013).

[d] Using $\tau = U - D$ instead of U improves accuracy in regions of low energy and enables one to consistently recover the Newtonian limit.

[e] For implementation details, see Porth et al. (2014).

[f] In the general-relativistic hydrodynamic WhiskyTHC code (Radice and Rezzolla 2012; Radice et al. 2014), this desirable property allows to set floors on density close to the limit of floating point precision $\sim 10^{-16} \rho_{\mathrm{ref}}$.

[g] We note that for the reference solution we have relied here on the extensive set of tests performed in flat spacetime within the MPI-AMRVAC framework; however, we could also have employed as reference solution the 'exact' solution as derived in Ref. (Giacomazzo and Rezzolla 2006).

[h] Note that the discrepancy in v^r appears less dramatic when viewed in terms of the four-velocity u^r.

[i] We thank Chris Fragile for providing subroutines for this test case.

[j] The results were kindly provided by Monika Moscibrodzka.

[k] Even when the dissipation length is well resolved, high-Reynolds number flows show indications for positive Lyapunov exponents and thus non-convergent chaotic behaviour see, e.g., Lecoanet et al. (2016).

[l] Since we use the same Courant limited timestep for all grid-levels, the speedup is entirely due to saving in computational cells. The additional speedup that would be gained from (Berger and Oliger 1984)-type hierarchical timesteps can be estimated from the level population of the simulation: the expected additional gain is only $\sim 8\%$ for this setup.

References

Abramowicz, MA, Chen, X, Kato, S, Lasota, J-P, Regev, O: Thermal equilibria of accretion disks. Astrophys. J. Lett. **438**, 37-39 (1995). doi:10.1086/187709 arXiv:astro-ph/9409018

Alcubierre, M: Introduction to 3 + 1 Numerical Relativity. Oxford University Press, Oxford (2008). doi:10.1093/acprof:oso/9780199205677.001.0001

Aloy, MA, Rezzolla, L: A powerful hydrodynamic booster for relativistic jets. Astrophys. J. **640**, 115-118 (2006)

Anile, AM: Relativistic Fluids and Magneto-Fluids. Cambridge University Press, Cambridge (1990)

Anninos, P, Fragile, PC, Salmonson, JD: Cosmos++: relativistic magnetohydrodynamics on unstructured grids with local adaptive refinement. Astrophys. J. **635**, 723 (2005). doi:10.1086/497294

Antón, L, Zanotti, O, Miralles, JA, Martí, JM, Ibáñez, JM, Font, JA, Pons, JA: Numerical 3 + 1 general relativistic magnetohydrodynamics: a local characteristic approach. Astrophys. J. **637**, 296 (2006) arXiv:astro-ph/0506063

Arnowitt, R, Deser, S, Misner, CW: Republication of: the dynamics of general relativity. Gen. Relativ. Gravit. **40**, 1997-2027 (2008). doi:10.1007/s10714-008-0661-1 arXiv:gr-qc/0405109

Asada, K, Nakamura, M: The structure of the M87 Jet: a transition from parabolic to conical streamlines. Astrophys. J. Lett. **745**, 28 (2012). doi:10.1088/2041-8205/745/2/L28 arXiv:1110.1793

Avara, MJ, McKinney, JC, Reynolds, CS: Efficiency of thin magnetically arrested discs around black holes. Mon. Not. R. Astron. Soc. **462**, 636-648 (2016). doi:10.1093/mnras/stw1643 arXiv:1508.05323

Baiotti, L, Hawke, I, Montero, PJ, Löffler, F, Rezzolla, L, Stergioulas, N, Font, JA, Seidel, E: Three-dimensional relativistic simulations of rotating neutron-star collapse to a Kerr black hole. Phys. Rev. D **71**(2), 024035 (2005). doi:10.1103/PhysRevD.71.024035 arXiv:gr-qc/0403029

Baiotti, L, Rezzolla, L: Binary neutron-star mergers: a review of Einstein's richest laboratory (2016). arXiv:1607.03540

Balbus, SA, Hawley, JF: A powerful local shear instability in weakly magnetized disks. I - linear analysis. II - nonlinear evolution. Astrophys. J. **376**, 214-233 (1991). doi:10.1086/170270

Balbus, SA, Hawley, JF: Instability, turbulence, and enhanced transport in accretion disks. Rev. Mod. Phys. **70**, 1-53 (1998). doi:10.1103/RevModPhys.70.1

Balsara, DS, Kim, J: A comparison between divergence-cleaning and staggered-mesh formulations for numerical magnetohydrodynamics. Astrophys. J. **602**, 1079-1090 (2004). doi:10.1086/381051 arXiv:astro-ph/0310728

Banyuls, F, Font, JA, Ibáñez, JM, Martí, JM, Miralles, JA: Numerical 3 + 1 general-relativistic hydrodynamics: a local characteristic approach. Astrophys. J. **476**, 221 (1997). doi:10.1086/303604

Barkov, MV, Aharonian, FA, Bosch-Ramon, V: Gamma-ray flares from red giant/jet interactions in active galactic nuclei. Astrophys. J. **724**, 1517-1523 (2010). doi:10.1088/0004-637X/724/2/1517 arXiv:1005.5252

Barkov, MV, Komissarov, SS: Stellar explosions powered by the Blandford-Znajek mechanism. Mon. Not. R. Astron. Soc. **385**, 28-32 (2008). doi:10.1111/j.1745-3933.2008.00427.x

Beckwith, K, Hawley, JF, Krolik, JH: The influence of magnetic field geometry on the evolution of black hole accretion flows: similar disks, drastically different jets. Astrophys. J. **678**, 1180-1199 (2008). doi:10.1086/533492

Begelman, MC: Radiatively inefficient accretion: breezes, winds and hyperaccretion. Mon. Not. R. Astron. Soc. **420**, 2912-2923 (2012). doi:10.1111/j.1365-2966.2011.20071.x arXiv:1110.5356

Berger, MJ, Oliger, J: Adaptive mesh refinement for hyperbolic partial differential equations. J. Comput. Phys. **53**, 484-512 (1984)

Biretta, JA, Owen, FN, Cornwell, TJ: A search for motion and flux variations in the M87 jet. Astrophys. J. **342**, 128-134 (1989). doi:10.1086/167581

Blandford, RD, Begelman, MC: On the fate of gas accreting at a low rate on to a black hole. Mon. Not. R. Astron. Soc. **303**, 1-5 (1999). doi:10.1046/j.1365-8711.1999.02358.x arXiv:astro-ph/9809083

Blandford, RD, Znajek, RL: Electromagnetic extraction of energy from Kerr black holes. Mon. Not. R. Astron. Soc. **179**, 433-456 (1977)

Bondi, H: On spherically symmetric accretion. Mon. Not. R. Astron. Soc. **112**, 195 (1952)

Čada, M, Torrilhon, M: Compact third-order limiter functions for finite volume methods. J. Comput. Phys. **228**, 4118-4145 (2009). doi:10.1016/j.jcp.2009.02.020

Calder, AC, Fryxell, B, Plewa, T, Rosner, R, Dursi, LJ, Weirs, VG, Dupont, T, Robey, HF, Kane, JO, Remington, BA, Drake, RP, Dimonte, G, Zingale, M, Timmes, FX, Olson, K, Ricker, P, MacNeice, P, Tufo, HM: On validating an astrophysical simulation code. Astrophys. J. Suppl. Ser. **143**, 201-229 (2002). doi:10.1086/342267 arXiv:astro-ph/0206251

Chan, C-K, Psaltis, D, Özel, F, Medeiros, L, Marrone, D, Sądowski, A, Narayan, R: Fast variability and millimeter/IR flares in GRMHD models of Sgr A* from strong-field gravitational lensing. Astrophys. J. **812**, 103 (2015). doi:10.1088/0004-637X/812/2/103 arXiv:1505.01500

Chan, C-K, Psaltis, D, Özel, F, Narayan, R, Sądowski, A: The power of imaging: constraining the plasma properties of GRMHD simulations using EHT observations of Sgr A*. Astrophys. J. **799**, 1 (2015). doi:10.1088/0004-637X/799/1/1 arXiv:1410.3492

Colella, P, Woodward, PR: The piecewise parabolic method (ppm) for gas-dynamical simulations. J. Comput. Phys. **54**(1), 174-201 (1984). doi:10.1016/0021-9991(84)90143-8

Cowling, TG: The magnetic field of sunspots. Mon. Not. R. Astron. Soc. **94**, 39-48 (1933). doi:10.1093/mnras/94.1.39

Davis, SF: Simplified second-order Godunov-type methods. SIAM J. Sci. Stat. Comput. **9**, 445-473 (1988)

de Avellar, M, et al.: (2017) in prepration

De Villiers, J-P, Hawley, JF: A numerical method for general relativistic magnetohydrodynamics. Astrophys. J. **589**, 458-480 (2003). doi:10.1086/373949 arXiv:astro-ph/0210518

Dedner, A, Kemm, F, Kröner, D, Munz, CD, Schnitzer, T, Wesenberg, M: Hyperbolic divergence cleaning for the MHD equations. J. Comput. Phys. **175**, 645-673 (2002). doi:10.1006/jcph.2001.6961

Del Zanna, L, Zanotti, O, Bucciantini, N, Londrillo, P: ECHO: a Eulerian conservative high-order scheme for general relativistic magnetohydrodynamics and magnetodynamics. Astron. Astrophys. **473**, 11-30 (2007). doi:10.1051/0004-6361:20077093 arXiv:0704.3206

Dexter, J: A public code for general relativistic, polarised radiative transfer around spinning black holes. Mon. Not. R. Astron. Soc. **462**, 115-136 (2016). doi:10.1093/mnras/stw1526 arXiv:1602.03184

Dexter, J, Agol, E, Fragile, PC: Millimeter flares and VLBI visibilities from relativistic simulations of magnetized accretion onto the galactic center black hole. Astrophys. J. Lett. **703**, 142-146 (2009). doi:10.1088/0004-637X/703/2/L142 arXiv:0909.0267

Dexter, J, Agol, E, Fragile, PC, McKinney, JC: The submillimeter bump in Sgr A* from relativistic MHD simulations. Astrophys. J. **717**, 1092-1104 (2010). doi:10.1088/0004-637X/717/2/1092 arXiv:1005.4062

Dexter, J, McKinney, JC, Agol, E: The size of the jet launching region in M87. Mon. Not. R. Astron. Soc. **421**, 1517-1528 (2012). doi:10.1111/j.1365-2966.2012.20409.x arXiv:1109.6011

Dionysopoulou, K, Alic, D, Palenzuela, C, Rezzolla, L, Giacomazzo, B: General-relativistic resistive magnetohydrodynamics in three dimensions: formulation and tests. Phys. Rev. D **88**(4), 044020 (2013). doi:10.1103/PhysRevD.88.044020 arXiv:1208.3487

Doeleman, S, Agol, E, Backer, D, Baganoff, F, Bower, GC, Broderick, A, Fabian, A, Fish, V, Gammie, C, Ho, P, Honman, M, Krichbaum, T, Loeb, A, Marrone, D, Reid, M, Rogers, A, Shapiro, I, Strittmatter, P, Tilanus, R, Weintroub, J, Whitney, A, Wright, M, Ziurys, L: Imaging an event horizon: submm-VLBI of a super massive black hole. In: Astro2010: The Astronomy and Astrophysics Decadal Survey. ArXiv Astrophysics e-Prints, vol. 2010 (2009) arXiv:0906.3899

Duez, MD, Liu, YT, Shapiro, SL, Stephens, BC: Relativistic magnetohydrodynamics in dynamical spacetimes: numerical methods and tests. Phys. Rev. D **72**(2), 024028 (2005). doi:10.1103/PhysRevD.72.024028 arXiv:astro-ph/0503420

Eisenhauer, F, Perrin, G, Brandner, W, Straubmeier, C, Richichi, A, Gillessen, S, Berger, JP, Hippler, S, Eckart, A, Schöller, M, Rabien, S, Cassaing, F, Lenzen, R, Thiel, M, Clénet, Y, Ramos, JR, Kellner, S, Fédou, P, Baumeister, H, Hofmann, R, Gendron, E, Boehm, A, Bartko, H, Haubois, X, Klein, R, Dodds-Eden, K, Houairi, K, Hormuth, F, Gräter, A, Jocou, L, Naranjo, V, Genzel, R, Kervella, P, Henning, T, Hamaus, N, Lacour, S, Neumann, U, Haug, M, Malbet, F, Laun, W, Kolmeder, J, Paumard, T, Rohloff, R-R, Pfuhl, O, Perraut, K, Ziegleder, J, Rouan, D, Rousset, G: GRAVITY: getting to the event horizon of Sgr A*. In: Optical and Infrared Interferometry. Proc. SPIE, vol. 7013, p. 70132 (2008). doi:10.1117/12.788407 arXiv:0808.0063

Etienne, ZB, Liu, YT, Paschalidis, V, Shapiro, SL: General relativistic simulations of black-hole-neutron-star mergers: effects of magnetic fields. Phys. Rev. D **85**(6), 064029 (2012). doi:10.1103/PhysRevD.85.064029 arXiv:1112.0568

Etienne, ZB, Paschalidis, V, Haas, R, Mösta, P, Shapiro, SL: IllinoisGRMHD: an open-source, user-friendly GRMHD code for dynamical spacetimes. Class. Quantum Gravity **32**(17), 175009 (2015). doi:10.1088/0264-9381/32/17/175009 arXiv:1501.07276

Evans, CR, Hawley, JF: Simulation of magnetohydrodynamic flows - a constrained transport method. Astrophys. J. **332**, 659-677 (1988). doi:10.1086/166684

Faber, JA, Baumgarte, TW, Etienne, ZB, Shapiro, SL, Taniguchi, K: Relativistic hydrodynamics in the presence of puncture black holes. Phys. Rev. D **76**(10), 104021 (2007). doi:10.1103/PhysRevD.76.104021 arXiv:0708.2436

Fender, RP, Belloni, TM, Gallo, E: Towards a unified model for black hole X-ray binary jets. Mon. Not. R. Astron. Soc. **355**, 1105-1118 (2004). doi:10.1111/j.1365-2966.2004.08384.x arXiv:astro-ph/0409360

Fishbone, LG, Moncrief, V: Relativistic fluid disks in orbit around Kerr black holes. Astrophys. J. **207**, 962-976 (1976)

Font, JA: Numerical hydrodynamics in general relativity. Living Rev. Relativ. **6**, 4 (2003)

Font, JA, Daigne, F: The runaway instability of thick discs around black holes - I. The constant angular momentum case. Mon. Not. R. Astron. Soc. **334**, 383-400 (2002)

Font, JA, Ibanez, JM, Marquina, A, Marti, JM: Multidimensional relativistic hydrodynamics: characteristic fields and modern high-resolution shock-capturing schemes. Astron. Astrophys. **282**, 304-314 (1994)

Foucart, F, Chandra, M, Gammie, CF, Quataert, E: Evolution of accretion discs around a Kerr black hole using extended magnetohydrodynamics. Mon. Not. R. Astron. Soc. **456**, 1332-1345 (2016). doi:10.1093/mnras/stv2687 arXiv:1511.04445

Fragile, PC, Blaes, OM, Anninos, P, Salmonson, JD: Global general relativistic magnetohydrodynamic simulation of a tilted black hole accretion disk. Astrophys. J. **668**, 417-429 (2007). doi:10.1086/521092 arXiv:0706.4303

Fryxell, B, Olson, K, Ricker, P, Timmes, FX, Zingale, M, Lamb, DQ, MacNeice, P, Rosner, R, Truran, JW, Tufo, H: FLASH: an adaptive mesh hydrodynamics code for modeling astrophysical thermonuclear flashes. Astrophys. J. Suppl. Ser. **131**, 273-334 (2000). doi:10.1086/317361

Fuerst, SV, Wu, K: Radiation transfer of emission lines in curved space-time. Astron. Astrophys. **424**, 733-746 (2004). doi:10.1051/0004-6361:20035814 arXiv:astro-ph/0406401

Galeazzi, F, Kastaun, W, Rezzolla, L, Font, JA: Implementation of a simplified approach to radiative transfer in general relativity. Phys. Rev. D **88**(6), 064009 (2013). doi:10.1103/PhysRevD.88.064009 arXiv:1306.4953

Gammie, CF, McKinney, JC, Tóth, G: Harm: a numerical scheme for general relativistic magnetohydrodynamics. Astrophys. J. **589**, 458 (2003) arXiv:astro-ph/0301509

García, A, Galtsov, D, Kechkin, O: Class of stationary axisymmetric solutions of the Einstein-Maxwell-Dilaton-Axion field equations. Phys. Rev. Lett. **74**, 1276-1279 (1995). doi:10.1103/PhysRevLett.74.1276

Giacomazzo, B, Rezzolla, L: The exact solution of the Riemann problem in relativistic MHD. J. Fluid Mech. **562**, 223-259 (2006) arXiv:gr-qc/0507102

Giacomazzo, B, Rezzolla, L: WhiskyMHD: a new numerical code for general relativistic magnetohydrodynamics. Class. Quantum Gravity **24**, 235 (2007). doi:10.1088/0264-9381/24/12/S16 arXiv:gr-qc/0701109

Giannios, D, Sironi, L: The S2 star as a probe of the accretion disc of Sgr A*. Mon. Not. R. Astron. Soc. **433**, 25-29 (2013). doi:10.1093/mnrasl/slt051 arXiv:1303.2115

Goddi, C, Falcke, H, Kramer, M, Rezzolla, L, Brinkerink, C, Bronzwaer, T, Deane, R, De Laurentis, M, Desvignes, G, Davelaar, JRJ, Eisenhauer, F, Eatough, R, Fraga-Encinas, R, Fromm, CM, Gillessen, S, Grenzebach, A, Issaoun, S, Janßen, M, Konoplya, R, Krichbaum, TP, Laing, R, Liu, K, Lu, R-S, Mizuno, Y, Moscibrodzka, M, Müller, C, Olivares, H, Porth, O, Pfuhl, O, Ros, E, Roelofs, F, Schuster, K, Tilanus, R, Torne, P, van Bemmel, I, van Langevelde, HJ, Wex, N, Younsi, Z, Zhidenko, A: BlackHoleCam: fundamental physics of the Galactic center. Int. J. Mod. Phys. D (2016) submitted. arXiv:1606.08879

Gold, R, McKinney, JC, Johnson, MD, Doeleman, SS: Probing the magnetic field structure in Sgr A* on black hole horizon scales with polarized radiative transfer simulations. ArXiv e-prints (2016). arXiv:1601.05550

Gottlieb, S, Shu, C: Total Variation Diminishing Runge-Kutta schemes. Math. Comput. **67**, 73-85 (1998)

Gourgoulhon, E: 3 + 1 Formalism and bases of numerical relativity. ArXiv General Relativity and Quantum Cosmology e-prints (2007). arXiv:gr-qc/0703035

Gourgouliatos, KN, Fendt, C, Clausen-Brown, E, Lyutikov, M: Magnetic field structure of relativistic jets without current sheets. Mon. Not. R. Astron.

Soc. **419**, 3048-3059 (2012). doi:10.1111/j.1365-2966.2011.19946.x arXiv:1110.0838

Hamlin, ND, Newman, WI: Role of the Kelvin-Helmholtz instability in the evolution of magnetized relativistic sheared plasma flows. Phys. Rev. **87**(4), 043101 (2013). doi:10.1103/PhysRevE.87.043101

Harten, A, Lax, PD, van Leer, B: On upstream differencing and Godunov-type schemes for hyperbolic conservation laws. SIAM Rev. **25**, 35 (1983). doi:10.1137/1025002

Hartle, JB, Thorne, KS: Slowly rotating relativistic stars. II. Models for neutron stars and supermassive stars. Astrophys. J. **153**, 807 (1968). doi:10.1086/149707

Hawley, JF, Krolik, JH: Magnetically driven jets in the Kerr metric. Astrophys. J. **641**, 103-116 (2006). doi:10.1086/500385 arXiv:astro-ph/0512227

Hawley, JF, Richers, SA, Guan, X, Krolik, JH: Testing convergence for global accretion disks. Astrophys. J. **772**, 102 (2013). doi:10.1088/0004-637X/772/2/102 arXiv:1306.0243

Hawley, JF, Smarr, LL, Wilson, JR: A numerical study of nonspherical black hole accretion. I equations and test problems. Astrophys. J. **277**, 296-311 (1984). doi:10.1086/161696

Ho, LC: Radiatively inefficient accretion in nearby galaxies. Astrophys. J. **699**, 626-637 (2009). doi:10.1088/0004-637X/699/1/626 arXiv:0906.4104

Hu, XY, Adams, NA, Shu, C-W: Positivity-preserving method for high-order conservative schemes solving compressible Euler equations. J. Comput. Phys. **242**, 169-180 (2013). doi:10.1016/j.jcp.2013.01.024 arXiv:1203.1540

Johannsen, T, Psaltis, D: Metric for rapidly spinning black holes suitable for strong-field tests of the no-hair theorem. Phys. Rev. D **83**(12), 124015 (2011). doi:10.1103/PhysRevD.83.124015 arXiv:1105.3191

Keppens, R, Meliani, Z: Linear wave propagation in relativistic magnetohydrodynamics. Phys. Plasmas **15**(10), 102103 (2008). doi:10.1063/1.2991408 arXiv:0810.2416

Keppens, R, Meliani, Z, van Marle, AJ, Delmont, P, Vlasis, A, van der Holst, B: Parallel, grid-adaptive approaches for relativistic hydro and magnetohydrodynamics. J. Comput. Phys. **231**, 718-744 (2012). doi:10.1016/j.jcp.2011.01.020

Koide, S, Meier, DL, Shibata, K, Kudoh, T: General relativistic simulation of early jet formation in a rapidly rotating black hole magnetosphere. Astrophys. J. **536**, 668-674 (2000)

Komissarov, SS: A Godunov-type scheme for relativistic magnetohydrodynamics. Mon. Not. R. Astron. Soc. **303**, 343-366 (1999). doi:10.1046/j.1365-8711.1999.02244.x

Komissarov, SS: Magnetized tori around Kerr black holes: analytic solutions with a toroidal magnetic field. Mon. Not. R. Astron. Soc. **368**, 993-1000 (2006). doi:10.1111/j.1365-2966.2006.10183.x arXiv:astro-ph/0601678

Komissarov, SS, Barkov, MV, Vlahakis, N, Königl, A: Magnetic acceleration of relativistic active galactic nucleus jets. Mon. Not. R. Astron. Soc. **380**, 51-70 (2007). doi:10.1111/j.1365-2966.2007.12050.x arXiv:astro-ph/0703146

Koren, B: Numerical Methods for Advection-Diffusion Problems. Notes on Numerical Fluid Mechanics, vol. 45. Vieweg, Braunschweig (1993)

Kozlowski, M, Jaroszynski, M, Abramowicz, MA: The analytic theory of fluid disks orbiting the Kerr black hole. Astron. Astrophys. **63**, 209-220 (1978)

Landau, LD, Lifshitz, EM: The Classical Theory of Fields. Course of Theoretical Physics, vol. 2. Elsevier, Oxford (2004)

Lecoanet, D, McCourt, M, Quataert, E, Burns, KJ, Vasil, GM, Oishi, JS, Brown, BP, Stone, JM, O'Leary, RM: A validated non-linear Kelvin-Helmholtz benchmark for numerical hydrodynamics. Mon. Not. R. Astron. Soc. **455**, 4274-4288 (2016). doi:10.1093/mnras/stv2564 arXiv:1509.03630

Leung, PK, Gammie, CF, Noble, SC: Numerical calculation of magnetobremsstrahlung emission and absorption coefficients. Astrophys. J. **737**, 21 (2011). doi:10.1088/0004-637X/737/1/21

Löhner, R: An adaptive finite element scheme for transient problems in CFD. Comput. Methods Appl. Mech. Eng. **61**, 323-338 (1987). doi:10.1016/0045-7825(87)90098-3

MacNeice, P, Olson, KM, Mobarry, C, de Fainchtein, R, Packer, C: Paramesh: A parallel adaptive mesh refinement community toolkit. Comput. Phys. Commun. **126**(3), 330-354 (2000). doi:10.1016/S0010-4655(99)00501-9

Markoff, S: Sagittarius A* in context: daily flares as a probe of the fundamental X-ray emission process in accreting black holes. Astrophys. J. Lett. **618**, 103-106 (2005). doi:10.1086/427841 arXiv:astro-ph/0412140

Marrone, DP, Moran, JM, Zhao, J-H, Rao, R: An unambiguous detection of Faraday rotation in Sagittarius A*. Astrophys. J. Lett. **654**, 57-60 (2007). doi:10.1086/510850 arXiv:astro-ph/0611791

Martí, JM, Müller, E: Grid-based methods in relativistic hydrodynamics and magnetohydrodynamics. Living Rev. Comput. Astrophys. **1** (2015). doi:10.1007/lrca-2015-3

Martins, JR, Sturdza, P, Alonso, JJ: The complex-step derivative approximation. ACM Trans. Math. Softw. **29**(3), 245-262 (2003). doi:10.1145/838250.838251

McKinney, JC: General relativistic magnetohydrodynamic simulations of the jet formation and large-scale propagation from black hole accretion systems. Mon. Not. R. Astron. Soc. **368**, 1561-1582 (2006). doi:10.1111/j.1365-2966.2006.10256.x

McKinney, JC, Blandford, RD: Stability of relativistic jets from rotating, accreting black holes via fully three-dimensional magnetohydrodynamic simulations. Mon. Not. R. Astron. Soc. **394**, 126-130 (2009). doi:10.1111/j.1745-3933.2009.00625.x

McKinney, JC, Gammie, CF: A measurement of the electromagnetic luminosity of a Kerr black hole. Astrophys. J. **611**, 977-995 (2004). doi:10.1086/422244 arXiv:astro-ph/0404512

McKinney, JC, Tchekhovskoy, A, Blandford, RD: Alignment of magnetized accretion disks and relativistic jets with spinning black holes. Science **339**, 49 (2013). doi:10.1126/science.1230811 arXiv:1211.3651

McKinney, JC, Tchekhovskoy, A, Sadowski, A, Narayan, R: Three-dimensional general relativistic radiation magnetohydrodynamical simulation of super-Eddington accretion, using a new code HARMRAD with M1 closure. ArXiv e-prints (2013). arXiv:1312.6127

McKinney, JC, Tchekhovskoy, A, Sadowski, A, Narayan, R: Three-dimensional general relativistic radiation magnetohydrodynamical simulation of super-Eddington accretion, using a new code HARMRAD with M1 closure. Mon. Not. R. Astron. Soc. **441**, 3177-3208 (2014). doi:10.1093/mnras/stu762 arXiv:1312.6127

Meliani, Z, Mizuno, Y, Olivares, H, Porth, O, Rezzolla, L, Younsi, Z: Simulations of recoiling black holes: adaptive mesh refinement and radiative transfer. Astron. Astrophys. **598**, 38 (2017). doi:10.1051/0004-6361/201629191 arXiv:1606.08192

Meliani, Z, Sauty, C, Tsinganos, K, Vlahakis, N: Relativistic Parker winds with variable effective polytropic index. Astron. Astrophys. **425**, 773-781 (2004). doi:10.1051/0004-6361:20035653 arXiv:astro-ph/0407100

Michel, FC: Accretion of matter by condensed objects. Astrophys. Space Sci. **15**, 153 (1972)

Mignone, A: High-order conservative reconstruction schemes for finite volume methods in cylindrical and spherical coordinates. J. Comput. Phys. **270**, 784-814 (2014). doi:10.1016/j.jcp.2014.04.001 arXiv:1404.0537

Mignone, A, Zanni, C, Tzeferacos, P, van Straalen, B, Colella, P, Bodo, G: The PLUTO code for adaptive mesh computations in astrophysical fluid dynamics. Astrophys. J. Suppl. Ser. **198**, 7 (2012). doi:10.1088/0067-0049/198/1/7 arXiv:1110.0740

Misner, CW, Thorne, KS, Wheeler, JA: Gravitation. W. H. Freeman, San Francisco (1973)

Mizuno, Y, et al.: General relativistic magnetohydrodynamic simulations of an accretion torus in a non-rotating dilaton black hole (2017) in prepratation

Mizuno, Y, Gómez, JL, Nishikawa, K-I, Meli, A, Hardee, PE, Rezzolla, L: Recollimation shocks in magnetized relativistic jets. Astrophys. J. **809**, 38 (2015). doi:10.1088/0004-637X/809/1/38 arXiv:1505.00933

Mizuno, Y, Nishikawa, K-I, Koide, S, Hardee, P, Fishman, GJ: RAISHIN: A high-resolution three-dimensional general relativistic magnetohydrodynamics code. ArXiv Astrophysics e-prints (2006). arXiv:astro-ph/0609004

Mocz, P, Pakmor, R, Springel, V, Vogelsberger, M, Marinacci, F, Hernquist, L: A moving mesh unstaggered constrained transport scheme for magnetohydrodynamics. Mon. Not. R. Astron. Soc. **463**, 477-488 (2016). doi:10.1093/mnras/stw2004 arXiv:1606.02310

Mościbrodzka, M, Falcke, H, Shiokawa, H: General relativistic magnetohydrodynamical simulations of the jet in M 87. Astron. Astrophys. **586**, 38 (2016). doi:10.1051/0004-6361/201526630 arXiv:1510.07243

Mościbrodzka, M, Falcke, H, Shiokawa, H, Gammie, CF: Observational appearance of inefficient accretion flows and jets in 3D GRMHD simulations: application to Sagittarius A*. Astron. Astrophys. **570**, 7 (2014). doi:10.1051/0004-6361/201424358 arXiv:1408.4743

Mościbrodzka, M, Gammie, CF, Dolence, JC, Shiokawa, H, Leung, PK: Radiative models of SGR A* from GRMHD simulations. Astrophys. J. **706**, 497-507 (2009). doi:10.1088/0004-637X/706/1/497 arXiv:0909.5431

Nagakura, H, Yamada, S: General relativistic hydrodynamic simulations and linear analysis of the standing accretion shock instability around a black hole. Astrophys. J. **689**, 391-406 (2008). doi:10.1086/590325 arXiv:0808.4141

Narayan, R, Igumenshchev, IV, Abramowicz, MA: Self-similar accretion flows with convection. Astrophys. J. **539**, 798-808 (2000). doi:10.1086/309268 arXiv:astro-ph/9912449

Narayan, R, Yi, I: Advection-dominated accretion: a self-similar solution. Astrophys. J. Lett. **428**, 13-16 (1994). doi:10.1086/187381 arXiv:astro-ph/9403052

Narayan, R, Yi, I: Advection-dominated accretion: underfed black holes and neutron stars. Astrophys. J. **452**, 710 (1995). doi:10.1086/176343 arXiv:astro-ph/9411059

Noble, SC, Gammie, CF, McKinney, JC, Del Zanna, L: Primitive variable solvers for conservative general relativistic magnetohydrodynamics. Astrophys. J. **641**, 626-637 (2006). doi:10.1086/500349 arXiv:astro-ph/0512420

Noble, SC, Krolik, JH, Hawley, JF: Direct calculation of the radiative efficiency of an accretion disk around a black hole. Astrophys. J. **692**, 411-421 (2009). doi:10.1088/0004-637X/692/1/411 arXiv:0808.3140

Olivares, H, et al.: (2017) in prepratation

Orszag, SA, Tang, C-M: Small-scale structure of two-dimensional magnetohydrodynamic turbulence. J. Fluid Mech. **90**, 129-143 (1979). doi:10.1017/S002211207900210X

Palenzuela, C, Lehner, L, Reula, O, Rezzolla, L: Beyond ideal MHD: towards a more realistic modelling of relativistic astrophysical plasmas. Mon. Not. R. Astron. Soc. **394**, 1727-1740 (2009). doi:10.1111/j.1365-2966.2009.14454.x arXiv:0810.1838

Porth, O, Olivares, H, Mizuno, Y, Younsi, Z, Rezzolla, L, Moscibrodzka, M, Falcke, H, Kramer, M: The black hole accretion code. ArXiv e-prints (2016). arXiv:1611.09720

Porth, O, Xia, C, Hendrix, T, Moschou, SP, Keppens, R: MPI-AMRVAC for solar and astrophysics. Astrophys. J. Suppl. Ser. **214**, 4 (2014). doi:10.1088/0067-0049/214/1/4 arXiv:1407.2052

Powell, KG: Approximate Riemann solver for magnetohydrodynamics (that works in more than one dimension). Computer Applications in Science and Engineering (ICASE). Technical report (1994)

Powell, KG, Roe, PL, Linde, TJ, Gombosi, TI, De Zeeuw, DL: A solution-adaptive upwind scheme for ideal magnetohydrodynamics. J. Comput. Phys. **154**, 284-309 (1999). doi:10.1006/jcph.1999.6299

Press, WH, Teukolsky, SA, Vetterling, WT, Flannery, BP: Numerical Recipes 3rd Edition: The Art of Scientific Computing Cambridge University Press, Cambridge (2007)

Psaltis, D, Wex, N, Kramer, M: A quantitative test of the no-hair theorem with Sgr A* using stars, pulsars, and the event horizon telescope. Astrophys. J. **818**, 121 (2016). doi:10.3847/0004-637X/818/2/121 arXiv:1510.00394

Pu, H-Y, Yun, K, Younsi, Z, Yoon, S-J: Odyssey: a public GPU-based code for general relativistic radiative transfer in Kerr spacetime. Astrophys. J. **820**, 105 (2016). doi:10.3847/0004-637X/820/2/105 arXiv:1601.02063

Quataert, E, Gruzinov, A: Convection-dominated accretion flows. Astrophys. J. **539**, 809-814 (2000). doi:10.1086/309267 arXiv:astro-ph/9912440

Radice, D, Rezzolla, L: THC: a new high-order finite-difference high-resolution shock-capturing code for special-relativistic hydrodynamics. Astron. Astrophys. **547**, 26 (2012). doi:10.1051/0004-6361/201219735 arXiv:1206.6502

Radice, D, Rezzolla, L: Universality and intermittency in relativistic turbulent flows of a hot plasma. Astrophys. J. **766**, 10 (2013). doi:10.1088/2041-8205/766/1/L10 arXiv:1209.2936

Radice, D, Rezzolla, L, Galeazzi, F: Beyond second-order convergence in simulations of binary neutron stars in full general-relativity. Mon. Not. R. Astron. Soc. Lett. **437**, 46-50 (2014). doi:10.1093/mnrasl/slt137 arXiv:1306.6052

Radice, D, Rezzolla, L, Galeazzi, F: High-order fully general-relativistic hydrodynamics: new approaches and tests. Class. Quantum Gravity **31**(7), 075012 (2014). doi:10.1088/0264-9381/31/7/075012 arXiv:1312.5004

Rezzolla, L, Zanotti, O: Relativistic Hydrodynamics. Oxford University Press, Oxford (2013). doi:10.1093/acprof:oso/9780198528906.001.0001

Rezzolla, L, Zhidenko, A: New parametrization for spherically symmetric black holes in metric theories of gravity. Phys. Rev. D **90**(8), 084009 (2014). doi:10.1103/PhysRevD.90.084009 arXiv:1407.3086

Sądowski, A: Thin accretion discs are stabilized by a strong magnetic field. Mon. Not. R. Astron. Soc. **459**, 4397-4407 (2016). doi:10.1093/mnras/stw913

Sądowski, A, Narayan, R, Tchekhovskoy, A, Zhu, Y: Semi-implicit scheme for treating radiation under M1 closure in general relativistic conservative fluid dynamics codes. Mon. Not. R. Astron. Soc. **429**, 3533-3550 (2013). doi:10.1093/mnras/sts632 arXiv:1212.5050

Shiokawa, H, Dolence, JC, Gammie, CF, Noble, SC: Global general relativistic magnetohydrodynamic simulations of black hole accretion flows: a convergence study. Astrophys. J. **744**, 187 (2012). doi:10.1088/0004-637X/744/2/187 arXiv:1111.0396

Sorathia, KA, Reynolds, CS, Stone, JM, Beckwith, K: Global simulations of accretion disks. I. Convergence and comparisons with local models. Astrophys. J. **749**, 189 (2012). doi:10.1088/0004-637X/749/2/189 arXiv:1106.4019

Spiteri, RJ, Ruuth, SJ: A new class of optimal high-order strong-stability-preserving time discretization methods. SIAM J. Numer. Anal. **40**(2), 469-491 (2002)

Squire, W, Trapp, G: Using complex variables to estimate derivatives of real functions. SIAM Rev. **40**, 110-112 (1998). doi:10.1137/S003614459631241X

Stawarz, Ł, Aharonian, F, Kataoka, J, Ostrowski, M, Siemiginowska, A, Sikora, M: Dynamics and high-energy emission of the flaring HST-1 knot in the M 87 jet. Mon. Not. R. Astron. Soc. **370**, 981-992 (2006). doi:10.1111/j.1365-2966.2006.10525.x arXiv:astro-ph/0602220

Suresh, A, Huynh, HT: Accurate monotonicity-preserving schemes with Runge-Kutta time stepping. J. Comput. Phys. **136**(1), 83-99 (1997). doi:10.1006/jcph.1997.5745

Takahashi, HR, Ohsuga, K, Kawashima, T, Sekiguchi, Y: Formation of overheated regions and truncated disks around black holes: three-dimensional general relativistic radiation-magnetohydrodynamics simulations. Astrophys. J. **826**, 23 (2016). doi:10.3847/0004-637X/826/1/23 arXiv:1605.04992

Tchekhovskoy, A, Metzger, BD, Giannios, D, Kelley, LZ: Swift J1644+57 gone MAD: the case for dynamically important magnetic flux threading the black hole in a jetted tidal disruption event. Mon. Not. R. Astron. Soc. **437**, 2744-2760 (2014). doi:10.1093/mnras/stt2085 arXiv:1301.1982

Toro, EF: Riemann Solvers and Numerical Methods for Fluid Dynamics. Springer, Berlin (1999)

Toth, G: The div b=0 constraint in shock-capturing magnetohydrodynamics codes. J. Comput. Phys. **161**, 605-652 (2000). doi:10.1006/jcph.2000.6519

van der Holst, B, Keppens, R, Meliani, Z: A multidimensional grid-adaptive relativistic magnetofluid code. Comput. Phys. Commun. **179**, 617-627 (2008). doi:10.1016/j.cpc.2008.05.005 arXiv:0807.0713

Villiers, JPD, Hawley, JF: Global general relativistic magnetohydrodynamic simulations of accretion tori. Astrophys. J. **592**, 1060 (2003). arXiv:astro-ph/0210518

Vincent, FH, Paumard, T, Gourgoulhon, E, Perrin, G: GYOTO: a new general relativistic ray-tracing code. Class. Quantum Gravity **28**(22), 225011 (2011). doi:10.1088/0264-9381/28/22/225011 arXiv:1109.4769

Weinberg, S: Gravitation and Cosmology: Principles and Applications of the General Theory of Relativity. Wiley, New York (1972)

White, CJ, Stone, JM: GRMHD in Athena++ using advanced Riemann slvers and staggered-mesh constrained transport. ArXiv e-prints (2015). arXiv:1511.00943

White, CJ, Stone, JM, Gammie, CF: An extension of the Athena++ code framework for GRMHD based on advanced Riemann solvers and staggered-mesh constrained transport. Astrophys. J. Suppl. Ser. **225**, 22 (2016). doi:10.3847/0067-0049/225/2/22 arXiv:1511.00943

York, JW: Kinematics and dynamics of general relativity. In: Smarr, LL (ed.) Sources of Gravitational Radiation, pp. 83-126. Cambridge University Press, Cambridge (1979)

Younsi, Z, et al.: (2017) in prepratation

Younsi, Z, Wu, K: Variations in emission from episodic plasmoid ejecta around black holes. Mon. Not. R. Astron. Soc. **454**, 3283-3298 (2015). doi:10.1093/mnras/stv2203 arXiv:1510.01700

Younsi, Z, Wu, K, Fuerst, SV: General relativistic radiative transfer: formulation and emission from structured tori around black holes. Astron. Astrophys. **545**, 13 (2012). doi:10.1051/0004-6361/201219599 arXiv:1207.4234

Younsi, Z, Zhidenko, A, Rezzolla, L, Konoplya, R, Mizuno, Y: New method for shadow calculations: application to parametrized axisymmetric black holes. Phys. Rev. D **94**(8), 084025 (2016). doi:10.1103/PhysRevD.94.084025 arXiv:1607.05767

Yuan, F, Narayan, R: Hot accretion flows around black holes. Annu. Rev. Astron. Astrophys. **52**, 529-588 (2014). doi:10.1146/annurev-astro-082812-141003. arXiv:1401.0586

Yuan, F, Quataert, E, Narayan, R: Nonthermal electrons in radiatively inefficient accretion flow models of Sagittarius A*. Astrophys. J. **598**, 301-312 (2003). doi:10.1086/378716 arXiv:astro-ph/0304125

Zanotti, O, Dumbser, M: A high order special relativistic hydrodynamic and magnetohydrodynamic code with space-time adaptive mesh refinement. Comput. Phys. Commun. **188**, 110-127 (2015). doi:10.1016/j.cpc.2014.11.015 arXiv:1312.7784

Zanotti, O, Fambri, F, Dumbser, M: Solving the relativistic magnetohydrodynamics equations with ADER discontinuous Galerkin methods, a posteriori subcell limiting and adaptive mesh refinement. Mon. Not. R. Astron. Soc. **452**, 3010-3029 (2015). doi:10.1093/mnras/stv1510 arXiv:1504.07458

Zanotti, O, Rezzolla, L, Font, JA: Quasi-periodic accretion and gravitational waves from oscillating 'toroidal neutron stars' around a Schwarzschild black hole. Mon. Not. R. Astron. Soc. **341**, 832-848 (2003)

Zhang, W, MacFadyen, AIR: A relativistic adaptive mesh refinement hydrodynamics code. Astrophys. J. Suppl. Ser. **164**, 255 (2006)

Zrake, J, MacFadyen, AI: Spectral and intermittency properties of relativistic turbulence. Astrophys. J. **763**, 12 (2013). doi:10.1088/2041-8205/763/1/L12 arXiv:1210.4066

8

Efficient conservative ADER schemes based on WENO reconstruction and space-time predictor in primitive variables

Olindo Zanotti[*] and Michael Dumbser

Abstract

We present a new version of conservative ADER-WENO finite volume schemes, in which both the high order spatial reconstruction as well as the time evolution of the reconstruction polynomials in the local space-time predictor stage are performed in *primitive* variables, rather than in conserved ones. To obtain a conservative method, the underlying finite volume scheme is still written in terms of the cell averages of the conserved quantities. Therefore, our new approach performs the spatial WENO reconstruction *twice*: the *first* WENO reconstruction is carried out on the known *cell averages* of the conservative variables. The WENO polynomials are then used at the cell centers to compute *point values* of the *conserved variables*, which are subsequently converted into *point values* of the *primitive variables*. This is the only place where the conversion from conservative to primitive variables is needed in the new scheme. Then, a *second* WENO reconstruction is performed on the point values of the primitive variables to obtain piecewise high order reconstruction polynomials of the primitive variables. The reconstruction polynomials are subsequently evolved in time with a *novel* space-time finite element predictor that is directly applied to the governing PDE written in *primitive form*. The resulting space-time polynomials of the primitive variables can then be directly used as input for the numerical fluxes at the cell boundaries in the underlying *conservative* finite volume scheme. Hence, the number of necessary conversions from the conserved to the primitive variables is reduced to just *one single conversion* at each cell center. We have verified the validity of the new approach over a wide range of hyperbolic systems, including the classical Euler equations of gas dynamics, the special relativistic hydrodynamics (RHD) and ideal magnetohydrodynamics (RMHD) equations, as well as the Baer-Nunziato model for compressible two-phase flows. In all cases we have noticed that the new ADER schemes provide *less oscillatory solutions* when compared to ADER finite volume schemes based on the reconstruction in conserved variables, especially for the RMHD and the Baer-Nunziato equations. For the RHD and RMHD equations, the overall accuracy is improved and the CPU time is reduced by about 25 %. Because of its increased accuracy and due to the reduced computational cost, we recommend to use this version of ADER as the standard one in the relativistic framework. At the end of the paper, the new approach has also been extended to ADER-DG schemes on space-time adaptive grids (AMR).

Keywords: high order WENO reconstruction in primitive variables; ADER-WENO finite volume schemes; ADER discontinuous Galerkin schemes; AMR; hyperbolic conservation laws; relativistic hydrodynamics and magnetohydrodynamics; Baer-Nunziato model

[*]Correspondence: olindo.zanotti@unitn.it
Laboratory of Applied Mathematics, Department of Civil, Environmental and Mechanical Engineering, University of Trento, Via Mesiano 77, Trento, 38123, Italy

1 Introduction

Since their introduction by Toro and Titarev (Toro et al. 2001; Titarev and Toro 2002; Toro and Titarev 2002; Titarev and Toro 2005; Toro and Titarev 2006), ADER (arbitrary high order derivatives) schemes for hyperbolic partial differential equations (PDE) have been improved and developed along different directions. A key feature of these methods is their ability to achieve uniformly high order of accuracy in space and time in a single step, without the need of intermediate Runge-Kutta stages (Pareschi et al. 2005; Pidatella et al. 2015), by exploiting the approximate solution of a Generalized Riemann Problem (GRP) at cell boundaries. ADER schemes have been first conceived within the finite volume (FV) framework, but they were soon extended also to the discontinuous Galerkin (DG) finite element framework (Dumbser and Munz 2006; Taube et al. 2007) and to a unified formulation of FV and DG schemes, namely the so-called $\mathbb{P}_N\mathbb{P}_M$ approach (Dumbser et al. 2008a). In the original ADER approach by Toro and Titarev, the approximate solution of the GRP is obtained through the solution of a conventional Riemann problem between the boundary-extrapolated values, and a sequence of linearized Riemann problems for the spatial derivatives. The required time derivatives in the GRP are obtained via the so-called Cauchy-Kowalevski procedure, which consists in replacing the time derivatives of the Taylor expansion at each interface with spatial derivatives of appropriate order, by resorting to the strong differential form of the PDE. Such an approach, though formally elegant, becomes prohibitive or even impossible as the complexity of the equations increases, especially for multidimensional problems and for relativistic hydrodynamics and magneto-hydrodynamics. On the contrary, in the modern reformulation of ADER (Dumbser et al. 2008b; Dumbser et al. 2008a; Balsara et al. 2013), the approximate solution of the GRP is achieved by first evolving the data locally inside each cell through a *local space-time discontinuous Galerkin predictor* (LSDG) step that is based on a weak form of the PDE, and, second, by solving a sequence of classical Riemann problems along the time axis at each element interface. This approach has the additional benefit that it can successfully cope with stiff source terms in the equations, a fact which is often encountered in physical applications. For these reasons, ADER schemes have been applied to real physical problems mostly in their modern version. Notable examples of applications include the study of Navier-Stokes equations, with or without chemical reactions (Hidalgo and Dumbser 2011; Dumbser 2010), geophysical flows (Dumbser et al. 2009), complex three-dimensional free surface flows (Dumbser 2013), relativistic magnetic reconnection (Dumbser and Zanotti 2009; Zanotti and Dumbser 2011), and the study of the Richtmyer-Meshkov instability in the relativistic regime (Zanotti and Dumbser 2015). In the last few years, ADER schemes have been enriched with several additional properties, reaching a high level of flexibility. First of all, ADER schemes have been soon extended to deal with non-conservative systems of hyperbolic PDE (Toro and Hidalgo 2009; Dumbser et al. 2009; Dumbser et al. 2014), by resorting to path-conservative methods (Parés and Castro 2004; Pares 2006). ADER schemes

have also been extended to the Lagrangian framework, in which they are currently applied to the solution of multidimensional problems on unstructured meshes for various systems of equations, (Boscheri and Dumbser 2013; Dumbser and Boscheri 2013; Boscheri et al. 2014a; Boscheri et al. 2014b; Boscheri and Dumbser 2014). On another side, ADER schemes have been combined with Adaptive Mesh Refinement (AMR) techniques (Dumbser et al. 2013; Zanotti and Dumbser 2015), exploiting the local properties of the discontinuous Galerkin predictor step, which is applied cell-by-cell irrespective of the level of refinement of the neighbour cells. Moreover, ADER schemes have also been used in combination with discontinuous Galerkin methods, even in the presence of shock waves and other discontinuities within the flow, thanks to a novel *a posteriori* sub-cell finite volume limiter technique based on the MOOD approach (Clain et al. 2011; Diot et al. 2012), that is designed to stabilize the discrete solution wherever the DG approach fails and produces spurious oscillations or negative densities and pressures (Dumbser et al. 2014; Zanotti et al. 2015a; Zanotti et al. 2015b).

The various implementations of ADER schemes mentioned so far differ under several aspects, but they all share the following common features: they apply the local space-time discontinuous Galerkin predictor to the conserved variables, which in turn implies that, if a WENO finite volume scheme is used, the spatial WENO reconstruction is also performed in terms of the conserved variables. Although this may be regarded as a reasonable choice, it has two fundamental drawbacks. The first one has to do with the fact that, as shown by Munz (1986), the reconstruction in conserved variables provides the worst shock capturing fidelity when compared to the reconstruction performed either in primitive or in characteristic variables. The second drawback is instead related to computational performance. Since the computation of the numerical fluxes requires the calculation of integrals via Gaussian quadrature, the physical fluxes must necessarily be computed at each space-time Gauss-Legendre quadrature point. However, there are systems of equations (e.g. the relativistic hydrodynamics or magnetohydrodynamics equations) for which the physical fluxes can only be written in terms of the primitive variables. As a result, a conversion from the conserved to the primitive variables is necessary for the calculations of the fluxes, and this operation, which is never analytic for such systems of equations, is rather expensive. For these reasons it would be very desirable to have an ADER scheme in which both the reconstruction and the subsequent local space-time discontinuous Galerkin predictor are performed in primitive variables. It is the aim of the present paper to explore this possibility. It is also worth stressing that in the context of high order finite difference Godunov methods, based on traditional Runge-Kutta discretization in time, the reconstruction in primitive variables has been proved to be very successful by Del Zanna et al. (2007) in their ECHO general relativistic code (see also Bucciantini and Del Zanna 2011; Zanotti et al. 2011). In spite of the obvious differences among the numerical schemes adopted, the approach that we propose here and the ECHO-approach share the common feature of requiring a single (per cell) conversion from the conserved to the primitive variables.

The plan of the paper is the following: in Section 2 we describe the numerical method, with particular emphasis on Section 2.3 and on Section 2.4, where the spatial reconstruction strategy and the local space-time discontinuous Galerkin predictor in primitive variable are described. The results of our new approach are presented in Section 3 for a set of four different systems of equations. In Section 4 we show that the new strategy can also be extended to pure discontinuous Galerkin schemes, even in the presence of space-time adaptive meshes (AMR). Finally, Section 5 is devoted to the conclusions of the work.

2 Numerical method

We present our new approach for purely regular Cartesian meshes, although there is no conceptual reason preventing the extension to general curvilinear or unstructured meshes, which may be considered in future studies.

2.1 Formulation of the equations

We consider hyperbolic systems of balance laws that contain both conservative and non-conservative terms, i.e.

$$\frac{\partial \mathbf{Q}}{\partial t} + \nabla \cdot \mathbf{F}(\mathbf{Q}) + \mathbf{B}(\mathbf{Q}) \cdot \nabla \mathbf{Q} = S(\mathbf{Q}), \tag{1}$$

where $\mathbf{Q} \in \Omega_\mathbf{Q} \subset \mathbb{R}^\nu$ is the state vector of the ν *conserved variables*, which, for the typical gas dynamics equations, are related to the conservation of mass, momentum and energy. $\mathbf{F}(\mathbf{Q}) = [\mathbf{f}^x(\mathbf{Q}), \mathbf{f}^y(\mathbf{Q}), \mathbf{f}^z(\mathbf{Q})]$ is the flux tensor[a] for the conservative part of the PDE system, while $\mathbf{B}(\mathbf{Q}) = [\mathbf{B}_x(\mathbf{Q}), \mathbf{B}_y(\mathbf{Q}), \mathbf{B}_z(\mathbf{Q})]$ represents the non-conservative part of it. Finally, $\mathbf{S}(\mathbf{Q})$ is the vector of the source terms, which may or may not be present. In the follow up of our discussion it is convenient to recast the system (1) in quasilinear form as

$$\frac{\partial \mathbf{Q}}{\partial t} + \mathbf{A}(\mathbf{Q}) \cdot \nabla \mathbf{Q} = \mathbf{S}(\mathbf{Q}), \tag{2}$$

where $\mathbf{A}(\mathbf{Q}) = [\mathbf{A}_x, \mathbf{A}_y, \mathbf{A}_z] = \partial \mathbf{F}(\mathbf{Q})/\partial \mathbf{Q} + \mathbf{B}(\mathbf{Q})$ accounts for both the conservative and the non-conservative contributions. As we shall see below, a proper discretization of Eq. (2) can provide the time evolution of the conserved variables $\mathbf{Q}$, but when the *primitive variables* $\mathbf{V}$ are adopted instead, Eq. (2) translates into

$$\frac{\partial \mathbf{V}}{\partial t} + \mathbf{C}(\mathbf{Q}) \cdot \nabla \mathbf{V} = \left(\frac{\partial \mathbf{Q}}{\partial \mathbf{V}}\right)^{-1} \mathbf{S}(\mathbf{Q}),$$

$$\text{with } \mathbf{C}(\mathbf{Q}) = \left(\frac{\partial \mathbf{Q}}{\partial \mathbf{V}}\right)^{-1} \mathbf{A}(\mathbf{Q}) \left(\frac{\partial \mathbf{Q}}{\partial \mathbf{V}}\right). \tag{3}$$

In the following we suppose that the conserved variables $\mathbf{Q}$ can always be written *analytically* in terms of the primitive variables $\mathbf{V}$, i.e. the functions

$$\mathbf{Q} = \mathbf{Q}(\mathbf{V}) \tag{4}$$

are supposed to be analytic for all PDE systems under consideration. On the contrary, the conversion from the conserved to the primitive variables, henceforth the *cons-to-prim conversion*, is not always available in closed form, i.e. the functions

$$\mathbf{V} = \mathbf{V}(\mathbf{Q}) \tag{5}$$

may *not* be analytic (e.g. for relativistic hydrodynamics and magnetohydrodynamics to be discussed in Section 3.2), thus requiring an approximate numerical solution. As a result, the matrix $(\frac{\partial \mathbf{Q}}{\partial \mathbf{V}})^{-1}$, which in principle could be simply computed as

$$\left(\frac{\partial \mathbf{Q}}{\partial \mathbf{V}}\right)^{-1} = \left(\frac{\partial \mathbf{V}}{\partial \mathbf{Q}}\right), \tag{6}$$

in practice it cannot be obtained in this manner, but it must be computed as

$$\left(\frac{\partial \mathbf{Q}}{\partial \mathbf{V}}\right)^{-1} = \mathbf{M}^{-1}, \tag{7}$$

where we have introduced the notation

$$\mathbf{M} = \left(\frac{\partial \mathbf{Q}}{\partial \mathbf{V}}\right), \tag{8}$$

which will be used repeatedly below. Since $\mathbf{Q}(\mathbf{V})$ is supposed to be analytic, the matrix $\mathbf{M}$ can be easily computed. Equation (1) will serve us as the master equation to evolve the cell averages of the conserved variables $\mathbf{Q}$ via a standard finite volume scheme. However, both the spatial WENO reconstruction and the subsequent LSDG predictor will act on the primitive variables $\mathbf{V}$, hence relying on the alternative formulation given by Eq. (3). The necessary steps to obtain such a scheme are described in the Sections 2.2-2.4 below.

2.2 The finite volume scheme

In Cartesian coordinates, we discretize the computational domain Ω through space-time control volumes $\mathcal{I}_{ijk} = I_{ijk} \times$

$[t^n, t^n + \Delta t] = [x_{i-\frac{1}{2}}, x_{i+\frac{1}{2}}] \times [y_{j-\frac{1}{2}}, y_{j+\frac{1}{2}}] \times [z_{k-\frac{1}{2}}, z_{k+\frac{1}{2}}] \times [t^n, t^n + \Delta t]$, with $\Delta x_i = x_{i+\frac{1}{2}} - x_{i-\frac{1}{2}}$, $\Delta y_j = y_{j+\frac{1}{2}} - y_{j-\frac{1}{2}}$, $\Delta z_k = z_{k+\frac{1}{2}} - z_{k-\frac{1}{2}}$ and $\Delta t = t^{n+1} - t^n$. Integration of Eq. (1) over $\mathcal{I}_{ijk}$ yields the usual finite volume discretization

$$
\begin{aligned}
\bar{\mathbf{Q}}_{ijk}^{n+1} = \bar{\mathbf{Q}}_{ijk}^{n} - \frac{\Delta t}{\Delta x_i} & \left[\left(\mathbf{f}_{i+\frac{1}{2},j,k}^{x} - \mathbf{f}_{i-\frac{1}{2},j,k}^{x} \right) \right. \\
& \left. + \frac{1}{2} \left(D_{i+\frac{1}{2},j,k}^{x} + D_{i-\frac{1}{2},j,k}^{x} \right) \right] \\
- \frac{\Delta t}{\Delta y_j} & \left[\left(\mathbf{f}_{i,j+\frac{1}{2},k}^{y} - \mathbf{f}_{i,j-\frac{1}{2},k}^{y} \right) \right. \\
& \left. + \frac{1}{2} \left(D_{i,j+\frac{1}{2},k}^{y} + D_{i,j-\frac{1}{2},k}^{y} \right) \right] \\
- \frac{\Delta t}{\Delta z_k} & \left[\left(\mathbf{f}_{i,j,k+\frac{1}{2}}^{z} - \mathbf{f}_{i,j,k-\frac{1}{2}}^{z} \right) \right. \\
& \left. + \frac{1}{2} \left(D_{i,j,k+\frac{1}{2}}^{z} + D_{i,j,k-\frac{1}{2}}^{z} \right) \right] \\
& + \Delta t (\bar{\mathbf{S}}_{ijk} - \bar{\mathbf{P}}_{ijk}),
\end{aligned}
\tag{9}
$$

where the cell average

$$
\begin{aligned}
\bar{\mathbf{Q}}_{ijk}^{n} = & \frac{1}{\Delta x_i} \frac{1}{\Delta y_j} \frac{1}{\Delta z_k} \\
& \times \int_{x_{i-\frac{1}{2}}}^{x_{i+\frac{1}{2}}} \int_{y_{j-\frac{1}{2}}}^{y_{j+\frac{1}{2}}} \int_{z_{k-\frac{1}{2}}}^{z_{k+\frac{1}{2}}} \mathbf{Q}(x,y,z,t^n) \, dz \, dy \, dx
\end{aligned}
\tag{10}
$$

is the spatial average of the vector of conserved quantities at time t^n. In Eq. (9) we recognize two different sets of terms, namely those due to the conservative part of the system (1), and those coming from the non-conservative part of it. In the former set we include the three time-averaged fluxes

$$
\begin{aligned}
\mathbf{f}_{i+\frac{1}{2},jk}^{x} = & \frac{1}{\Delta t} \frac{1}{\Delta y_j} \frac{1}{\Delta z_k} \\
& \times \int_{t^n}^{t^{n+1}} \int_{y_{j-\frac{1}{2}}}^{y_{j+\frac{1}{2}}} \int_{z_{k-\frac{1}{2}}}^{z_{k+\frac{1}{2}}} \tilde{\mathbf{f}}^x \left(\mathbf{v}_h^-(x_{i+\frac{1}{2}}, y, z, t), \right. \\
& \left. \mathbf{v}_h^+(x_{i+\frac{1}{2}}, y, z, t) \right) dz \, dy \, dt,
\end{aligned}
\tag{11}
$$

$$
\begin{aligned}
\mathbf{f}_{i,j+\frac{1}{2},k}^{y} = & \frac{1}{\Delta t} \frac{1}{\Delta x_i} \frac{1}{\Delta z_k} \\
& \times \int_{t^n}^{t^{n+1}} \int_{x_{i-\frac{1}{2}}}^{x_{i+\frac{1}{2}}} \int_{z_{k-\frac{1}{2}}}^{z_{k+\frac{1}{2}}} \tilde{\mathbf{f}}^y \left(\mathbf{v}_h^-(x, y_{j+\frac{1}{2}}, z, t), \right. \\
& \left. \mathbf{v}_h^+(x, y_{j+\frac{1}{2}}, z, t) \right) dz \, dx \, dt,
\end{aligned}
\tag{12}
$$

$$
\begin{aligned}
\mathbf{f}_{i,j,k+\frac{1}{2}}^{z} = & \frac{1}{\Delta t} \frac{1}{\Delta x_i} \frac{1}{\Delta y_j} \\
& \times \int_{t^n}^{t^{n+1}} \int_{x_{i-\frac{1}{2}}}^{x_{i+\frac{1}{2}}} \int_{y_{j-\frac{1}{2}}}^{y_{j+\frac{1}{2}}} \tilde{\mathbf{f}}^z \left(\mathbf{v}_h^-(x, y, z_{k+\frac{1}{2}}, t), \right. \\
& \left. \mathbf{v}_h^+(x, y, z_{k+\frac{1}{2}}, t) \right) dy \, dx \, dt
\end{aligned}
\tag{13}
$$

and the space-time averaged source term

$$
\begin{aligned}
\bar{\mathbf{S}}_{ijk} = & \frac{1}{\Delta t} \frac{1}{\Delta x_i} \frac{1}{\Delta y_j} \frac{1}{\Delta z_k} \\
& \times \int_{t^n}^{t^{n+1}} \int_{x_{i-\frac{1}{2}}}^{x_{i+\frac{1}{2}}} \int_{y_{j-\frac{1}{2}}}^{y_{j+\frac{1}{2}}} \int_{z_{k-\frac{1}{2}}}^{z_{k+\frac{1}{2}}} \mathbf{S}(\mathbf{v}_h(x, y, \\
& z, t)) \, dz \, dy \, dx \, dt.
\end{aligned}
\tag{14}
$$

We emphasize that the terms $\mathbf{v}_h$ in Eqs. (11)-(14), as well as in the few equations below, are piecewise space-time polynomials of degree M in *primitive variables*, computed according to a suitable LSDG predictor based on the formulation (3), as we will discuss in Section 2.4. This marks a striking difference with respect to traditional ADER schemes, in which such polynomials are instead computed in conserved variables and are denoted as $\mathbf{q}_h$ (see, e.g. Hidalgo and Dumbser 2011). The integrals over the smooth part of the non-conservative terms in Eq. (9) yield the following contribution,

$$
\begin{aligned}
\bar{\mathbf{P}}_{ijk} = & \frac{1}{\Delta t} \frac{1}{\Delta x_i} \frac{1}{\Delta y_j} \frac{1}{\Delta z_k} \\
& \times \int_{t^n}^{t^{n+1}} \int_{x_{i-\frac{1}{2}}}^{x_{i+\frac{1}{2}}} \int_{y_{j-\frac{1}{2}}}^{y_{j+\frac{1}{2}}} \int_{z_{k-\frac{1}{2}}}^{z_{k+\frac{1}{2}}} \mathbf{B}(\mathbf{v}_h) \\
& \times \mathbf{M} \nabla \mathbf{v}_h \, dz \, dy \, dx \, dt,
\end{aligned}
\tag{15}
$$

while the *jumps* across the element boundaries are treated within the framework of path-conservative schemes (Parés and Castro 2004; Pares 2006; Muñoz and Parés 2007; Castro et al. 2006; Castro et al. 2008a; Castro et al. 2008b) based on the Dal Maso-Le Floch-Murat theory (Dal Maso et al. 1995) as

$$
\begin{aligned}
D_{i+\frac{1}{2},j,k}^{x} = & \frac{1}{\Delta t} \frac{1}{\Delta y_j} \frac{1}{\Delta z_k} \\
& \times \int_{t^n}^{t^{n+1}} \int_{y_{j-\frac{1}{2}}}^{y_{j+\frac{1}{2}}} \int_{z_{k-\frac{1}{2}}}^{z_{k+\frac{1}{2}}} \mathcal{D}_x \left(\mathbf{v}_h^-(x_{i+\frac{1}{2}}, y, z, t), \right. \\
& \left. \mathbf{v}_h^+(x_{i+\frac{1}{2}}, y, z, t) \right) dz \, dy \, dt,
\end{aligned}
\tag{16}
$$

$$D^y_{i,j+\frac{1}{2},k} = \frac{1}{\Delta t}\frac{1}{\Delta x_i}\frac{1}{\Delta z_k}$$

$$\times \int_{t^n}^{t^{n+1}} \int_{x_{i-\frac{1}{2}}}^{x_{i+\frac{1}{2}}} \int_{z_{k-\frac{1}{2}}}^{z_{k+\frac{1}{2}}} \mathcal{D}_y\big(\mathbf{v}_h^-(x,y_{j+\frac{1}{2}},z,t),$$

$$\mathbf{v}_h^+(x,y_{j+\frac{1}{2}},z,t)\big)\,dz\,dx\,dt, \tag{17}$$

$$D^z_{i,j,k+\frac{1}{2}} = \frac{1}{\Delta t}\frac{1}{\Delta x_i}\frac{1}{\Delta y_j}$$

$$\times \int_{t^n}^{t^{n+1}} \int_{x_{i-\frac{1}{2}}}^{x_{i+\frac{1}{2}}} \int_{y_{j-\frac{1}{2}}}^{y_{j+\frac{1}{2}}} \mathcal{D}_z\big(\mathbf{v}_h^-(x,y,z_{k+\frac{1}{2}},t),$$

$$\mathbf{v}_h^+(x,y,z_{k+\frac{1}{2}},t)\big)\,dy\,dx\,dt. \tag{18}$$

According to this approach, the following path integrals must be prescribed

$$\mathcal{D}_i\big(\mathbf{v}_h^-,\mathbf{v}_h^+\big) = \int_0^1 \mathbf{B}_i\big(\Psi\big(\mathbf{v}_h^-,\mathbf{v}_h^+,s\big)\big)\mathbf{M}\big(\Psi\big(\mathbf{v}_h^-,\mathbf{v}_h^+,s\big)\big)\frac{\partial\Psi}{\partial s}\,ds,$$

$$i \in \{x,y,z\}, \tag{19}$$

where $\Psi(s)$ is a path joining the left and right boundary extrapolated states $\mathbf{v}_h^-$ and $\mathbf{v}_h^+$ in state space of the primitive variables. The simplest option is to use a straight-line segment path

$$\Psi = \Psi\big(\mathbf{v}_h^-,\mathbf{v}_h^+,s\big) = \mathbf{v}_h^- + s\big(\mathbf{v}_h^+ - \mathbf{v}_h^-\big), \quad 0 \le s \le 1. \tag{20}$$

Pragmatic as it is,[b] the choice of the path (20) allows to evaluate the terms $\mathcal{D}_i$ in (19) as

$$\mathcal{D}_i\big(\mathbf{v}_h^-,\mathbf{v}_h^+\big) = \left(\int_0^1 \mathbf{B}_i\big(\Psi\big(\mathbf{v}_h^-,\mathbf{v}_h^+,s\big)\big)\mathbf{M}\big(\Psi\big(\mathbf{v}_h^-,\mathbf{v}_h^+,s\big)\big)\,ds\right)$$

$$\times \big(\mathbf{v}_h^+ - \mathbf{v}_h^-\big), \tag{21}$$

that we compute through a three-point Gauss-Legendre formula (Dumbser et al. 2010; Dumbser and Toro 2011a; Dumbser and Toro 2011b). The computation of the numerical fluxes $\tilde{\mathbf{f}}^i$ in Eq. (11) requires the use of an approximate Riemann solver, see Toro (1999). In this work we have limited our attention to a local Lax-Friedrichs flux (Rusanov flux) and to the Osher-type flux proposed in Dumbser and Toro (2011b), Dumbser and Toro (2011a), Castro et al. (2015). Both of them can be written formally as

$$\tilde{\mathbf{f}}^i = \frac{1}{2}\big(\mathbf{f}^i\big(\mathbf{v}_h^-\big) + \mathbf{f}^i\big(\mathbf{v}_h^+\big)\big) - \frac{1}{2}\mathbf{D}_i\tilde{\mathbf{M}}\big(\mathbf{v}_h^+ - \mathbf{v}_h^-\big),$$

$$i \in \{x,y,z\}, \tag{22}$$

where $\mathbf{D}_i \ge 0$ is a positive-definite dissipation matrix that depends on the chosen Riemann solver. For the Rusanov flux it simply reads

$$\mathbf{D}_i^{\text{Rusanov}} = |s_{\max}|\mathbf{I}, \tag{23}$$

where $|s_{\max}|$ is the maximum absolute value of the eigenvalues admitted by the PDE and $\mathbf{I}$ is the identity matrix. The matrix $\tilde{\mathbf{M}}$ is a *Roe matrix* that allows to write the jumps in the conserved variables in terms of the jump in the primitive variables, i.e.

$$\mathbf{q}_h^+ - \mathbf{q}_h^- = \mathbf{Q}\big(\mathbf{v}_h^+\big) - \mathbf{Q}\big(\mathbf{v}_h^-\big) = \tilde{\mathbf{M}}\big(\mathbf{v}_h^+ - \mathbf{v}_h^-\big). \tag{24}$$

Since $\mathbf{M} = \partial\mathbf{Q}/\partial\mathbf{V}$, the Roe matrix $\tilde{\mathbf{M}}$ can be easily defined by a path integral as

$$\mathbf{Q}\big(\mathbf{v}_h^+\big) - \mathbf{Q}\big(\mathbf{v}_h^-\big) = \int_0^1 \mathbf{M}\big(\Psi\big(\mathbf{v}_h^-,\mathbf{v}_h^+,s\big)\big)\frac{\partial\Psi}{\partial s}\,ds$$

$$= \tilde{\mathbf{M}}\big(\mathbf{v}_h^+ - \mathbf{v}_h^-\big), \tag{25}$$

which in the case of the simple straight-line segment path (20) leads to the expression

$$\tilde{\mathbf{M}} = \int_0^1 \mathbf{M}\big(\Psi\big(\mathbf{v}_h^-,\mathbf{v}_h^+,s\big)\big)\,ds. \tag{26}$$

In the case of the Osher-type flux, on the other hand, the dissipation matrix reads

$$\mathbf{D}_i^{\text{Osher}} = \int_0^1 \big|\mathbf{A}_i\big(\Psi\big(\mathbf{v}_h^-,\mathbf{v}_h^+,s\big)\big)\big|\,ds, \tag{27}$$

with the usual definition of the matrix absolute value operator

$$|\mathbf{A}| = \mathbf{R}|\mathbf{\Lambda}|\mathbf{R}^{-1}, \quad |\mathbf{\Lambda}| = \text{diag}\big(|\lambda_1|,|\lambda_2|,\ldots,|\lambda_v|\big). \tag{28}$$

The path Ψ in Eqs. (27) and (26) is the same segment path adopted in (20) for the computation of the jumps $\mathcal{D}_i$.

2.3 A novel WENO reconstruction in primitive variables

Since we want to compute the time averaged fluxes [cf. Eqs. (11)-(13)] and the space-time averaged sources [cf. Eq. (14)] directly from the primitive variables $\mathbf{V}$, it is necessary to reconstruct a WENO polynomial in primitive variables. However, the underlying finite volume scheme (9) will still advance in time the cell averages of the conserved variables $\bar{\mathbf{Q}}_{ijk}^n$, which are the only known input quantities at the reference time level t^n. Hence, the whole procedure is performed through the following three simple steps:

1. We perform a *first* standard spatial WENO reconstruction of the conserved variables starting from the cell averages $\bar{\mathbf{Q}}_{ijk}^n$. This allows to obtain a reconstructed polynomial $\mathbf{w}_h(x,y,z,t^n)$ in conserved variables valid within each cell.
2. Since $\mathbf{w}_h(x,y,z,t^n)$ is defined at any point inside the cell, we simply *evaluate* it at the cell center in order to obtain the *point value* $\mathbf{Q}_{ijk}^n = \mathbf{w}_h(x_i,y_j,z_k,t^n)$. This conversion from cell averages $\bar{\mathbf{Q}}_{ijk}^n$ to point values

$\mathbf{Q}_{ijk}^n$ is the **main key idea** of our new method, since the simple identity $\mathbf{Q}_{ijk}^n = \bar{\mathbf{Q}}_{ijk}^n$ is valid only up to second order of accuracy! After that, we perform a conversion from the point-values of the conserved variables to the point-values in primitive variables, i.e. we apply Eq. (5), thus obtaining the corresponding primitive variables $\mathbf{V}_{ijk}^n = \mathbf{V}(\mathbf{Q}_{ijk}^n)$ at each cell center. This is the only step in the entire algorithm that needs a conversion from the conservative to the primitive variables.

3. Finally, from the point-values of the primitive variables at the cell centers, we perform a *second* WENO reconstruction to obtain a reconstruction polynomial in *primitive variables*, denoted as $\mathbf{p}_h(x,y,z,t^n)$. This polynomial is then used as the initial condition for the new local space-time DG predictor in primitive variables described in Section 2.4.

As for the choice of the spatial WENO reconstruction, we have adopted a dimension-by-dimension reconstruction strategy, discussed in full details in our previous works (see Dumbser et al. 2013; Dumbser et al. 2014; Zanotti and Dumbser 2015). Briefly, we first introduce space-time reference coordinates $\xi,\eta,\zeta,\tau \in [0,1]$, defined by

$$x = x_{i-\frac{1}{2}} + \xi\,\Delta x_i, \qquad y = y_{j-\frac{1}{2}} + \eta\,\Delta y_j,$$
$$z = z_{k-\frac{1}{2}} + \zeta\,\Delta z_k, \qquad t = t^n + \tau\,\Delta t, \tag{29}$$

and, along each spatial direction, we define a basis of polynomials $\{\psi_l(\lambda)\}_{l=1}^{M+1}$, each of degree M, formed by the $M+1$ Lagrange interpolating polynomials, that pass through the $M+1$ Gauss-Legendre quadrature nodes $\{\mu_k\}_{k=1}^{M+1}$. According to the WENO philosophy, a number of stencils is introduced such that the final polynomial is a data-dependent nonlinear combination of the polynomials computed from each stencil. Here, we use a fixed number N_s of one-dimensional stencils, namely $N_s = 3$ for odd order schemes (even polynomials of degree M), and $N_s = 4$ for even order schemes (odd polynomials of degree M). For example, focusing on the x direction for convenience, every stencil along x is formed by the union of $M+1$ adjacent cells, i.e.

$$S_{ijk}^{s,x} = \bigcup_{e=i-L}^{i+R} I_{ejk}, \tag{30}$$

where $L = L(M,s)$ and $R = R(M,s)$ are the spatial extension of the stencil to the left and to the right.[c]

Now, an important difference emerges depending on whether we are reconstructing the conserved or the primitive variables. In the former case, corresponding to the computation of $\mathbf{w}_h(x,y,z,t^n)$ at step 1 above, we require that the reconstructed polynomial must preserve the *cell-averages* of the *conserved variables* over each element I_{ijk}.

Since the polynomials reconstructed along the x direction can be written as

$$\mathbf{w}_h^{s,x}(x,t^n) = \sum_{r=0}^{M} \psi_r(\xi)\hat{\mathbf{w}}_{ijk,r}^{n,s} := \psi_r(\xi)\hat{\mathbf{w}}_{ijk,r}^{n,s}, \tag{31}$$

the reconstruction equations read

$$\frac{1}{\Delta x_e} \int_{x_{e-\frac{1}{2}}}^{x_{e+\frac{1}{2}}} \mathbf{w}_h^x(x,t^n)\,dx$$
$$= \frac{1}{\Delta x_e} \int_{x_{e-\frac{1}{2}}}^{x_{e+\frac{1}{2}}} \psi_r(\xi(x))\hat{\mathbf{w}}_{ijk,r}^{n,s}\,dx$$
$$= \bar{\mathbf{Q}}_{ejk}^n, \qquad \forall I_{ejk} \in S_{ijk}^{s,x}. \tag{32}$$

Equations (32) provide a system of $M+1$ linear equations for the unknown coefficients $\hat{\mathbf{w}}_{ijk,r}^{n,s}$, which is conveniently solved through linear algebra packages. Once this operation has been performed for each stencil, we construct a data-dependent nonlinear combination of the resulting polynomials, i.e.

$$\mathbf{w}_h^x(x,t^n) = \psi_r(\xi)\hat{\mathbf{w}}_{ijk,r}^n, \quad \text{with } \hat{\mathbf{w}}_{ijk,r}^n = \sum_{s=1}^{N_s} \omega_s \hat{\mathbf{w}}_{ijk,r}^{n,s}. \tag{33}$$

The nonlinear weights ω_s are computed according to the WENO approach (Jiang and Shu 1996) and their explicit expression can be found in Dumbser et al. (2013), Dumbser et al. (2014), Zanotti and Dumbser (2015). The whole procedure must be repeated along the two directions y and z. Hence, although each direction is treated separately, the net effect provides a genuine multidimensional reconstruction. We now proceed with the **key step** of the new algorithm presented in this paper and compute the *point values* of the conserved quantities at the cell centers, simply by *evaluating* the reconstruction polynomials in the barycenter of each control volume:

$$\mathbf{Q}_{ijk}^n = \mathbf{w}_h(x_i, y_j, z_k, t^n). \tag{34}$$

These point values of the conserved quantities $\mathbf{Q}_{ijk}^n$ are now converted into point values of the primitive variables $\mathbf{V}_{ijk}^n$, which requires only a single *cons-to-prim conversion* per cell. In RHD and RMHD, this is one of the most expensive and most delicate parts of the entire algorithm:

$$\mathbf{V}_{ijk}^n = \mathbf{V}(\mathbf{Q}_{ijk}^n). \tag{35}$$

The reconstruction polynomials in primitive variables are spanned by the same basis functions $\psi_r(\xi)$ used for $\mathbf{w}_h$, hence

$$\mathbf{p}_h^{s,x}(x,t^n) = \sum_{r=0}^{M} \psi_r(\xi)\hat{\mathbf{p}}_{ijk,r}^{n,s} := \psi_r(\xi)\hat{\mathbf{p}}_{ijk,r}^{n,s}. \tag{36}$$

According to step 3 listed above, we now require that the reconstructed polynomial must interpolate the *point-values* of the *primitive variables* at the centers of the cells forming each stencil, i.e.

$$\mathbf{p}_h^x(x_e, t^n) = \psi_r(\xi(x_e))\hat{\mathbf{p}}_{ijk,r}^{n,s} = \mathbf{V}_{ejk}^n, \quad \forall I_{ejk} \in \mathcal{S}_{ijk}^{s,x}. \quad (37)$$

The reconstruction equations (37) will also generate a system of $M + 1$ linear equations for the unknown coefficients $\hat{\mathbf{p}}_{ijk,r}^{n,s}$. The rest of the WENO logic applies in the same way, leading to

$$\mathbf{p}_h^x(x, t^n) = \psi_r(\xi)\hat{\mathbf{p}}_{ijk,r}^n, \quad \text{with } \hat{\mathbf{p}}_{ijk,r}^n = \sum_{s=1}^{N_s} \omega_s \hat{\mathbf{p}}_{ijk,r}^{n,s}. \quad (38)$$

We emphasize that thanks to our polynomial WENO reconstruction (instead of the original point-wise WENO reconstruction of Jiang and Shu 1996), the point-value of $\mathbf{w}_h(x, y, z, t^n)$ at each cell center, which is required at step 2 above, is promptly available after evaluating the basis functions at the cell center. In other words, there is no need to perform any special transformation from cell averages to point-values via Taylor series expansions, like in Buchmüller and Helzel (2014), Buchmüller et al. (2015). On the other hand, since the WENO reconstruction is performed twice, once for the conserved variables and once for the primitive variables, we expect that our new approach will become convenient in terms of computational efficiency only for those systems of equations characterized by relations $\mathbf{V}(\mathbf{Q})$ that cannot be written in closed form. In such circumstances, in fact, reducing the number of *cons-to-prim conversions* from $M(M + 1)^{d+1} + d(M + 1)^d$ in d space dimensions (due to the space-time predictor and the numerical flux computation in the finite volume scheme) to just *one single conversion* per cell will compensate for the double WENO reconstruction in space that we must perform. On the contrary, for systems of equations, such as the compressible Euler, for which the *cons-to-prim conversion* is analytic, no benefit will be reported in terms of computational efficiency, but still a significant benefit will be reported in terms of numerical accuracy. All these comments will be made quantitative in Section 3.

2.4 A local space-time DG predictor in primitive variables
2.4.1 Description of the predictor
As already remarked, the computation of the fluxes through the integrals (11)-(13) is more conveniently performed if the primitive variables are available at each space-time quadrature point. In such a case, in fact, no conversion from the conserved to the primitive variables is required. According to the discussion of the previous Section, it is possible to obtain a polynomial $\mathbf{p}_h(x, y, z, t^n)$ in primitive variables at the reference time t^n. This is however

not enough for a high accurate computation of the numerical fluxes, and $\mathbf{p}_h(x, y, z, t^n)$ must be evolved in time, locally for each cell, in order to obtain a polynomial $\mathbf{v}_h(x, y, z, t)$ approximating the solution at any time in the range $[t^n; t^{n+1}]$.

To this extent, we need an operation, to be performed locally for each cell, which uses as input the high order polynomial $\mathbf{v}_h$ obtained from the WENO reconstruction, and gives as output its evolution in time, namely

$$\mathbf{p}_h(x, y, z, t^n) \xrightarrow{LSDG} \mathbf{v}_h(x, y, z, t), \quad t \in [t^n; t^{n+1}]. \quad (39)$$

This can be obtained through an element-local space-time discontinuous Galerkin predictor that is based on the *weak* integral form of Eq. (3). From a mathematical point of view, Eq. (3) is a hyperbolic system in non-conservative form. Therefore, the implementation of the space-time discontinuous Galerkin predictor follows strictly the strategy already outlined in Dumbser et al. (2014) for non-conservative systems. Here we recall briefly the main ideas, focusing on the novel aspects implied by the formulation of Eq. (3). The sought polynomial $\mathbf{v}_h(x, y, z, t)$ is supposed to be expanded in space and time as

$$\mathbf{v}_h = \mathbf{v}_h(\boldsymbol{\xi}, \tau) = \theta_l(\boldsymbol{\xi}, \tau)\hat{\mathbf{v}}_l^n, \quad (40)$$

where the degrees of freedom $\hat{\mathbf{v}}_l^n$ are the unknowns. The space-time basis functions θ_l are given by a dyadic product of the Lagrange interpolation polynomials that pass through the Gauss-Legendre quadrature points, i.e. the tensor-product quadrature points on the hypercube $[0, 1]^{d+1}$, see Stroud (1971). The system (3) is first rephrased in terms of the reference coordinates τ and $\boldsymbol{\xi} = (\xi, \eta, \zeta)$, yielding

$$\frac{\partial \mathbf{V}}{\partial \tau} + \mathbf{C}_1^* \frac{\partial \mathbf{V}}{\partial \xi} + \mathbf{C}_2^* \frac{\partial \mathbf{V}}{\partial \eta} + \mathbf{C}_3^* \frac{\partial \mathbf{V}}{\partial \zeta} = \mathbf{S}^*, \quad (41)$$

with

$$\mathbf{C}_1^* = \frac{\Delta t}{\Delta x_i}\mathbf{C}_1, \qquad \mathbf{C}_2^* = \frac{\Delta t}{\Delta y_j}\mathbf{C}_2,$$
$$\mathbf{C}_3^* = \frac{\Delta t}{\Delta z_k}\mathbf{C}_3, \qquad \mathbf{S}^* = \Delta t\mathbf{M}^{-1}\mathbf{S}. \quad (42)$$

Expression (41) is then multiplied by the piecewise space-time polynomials $\theta_k(\xi, \eta, \zeta, \tau)$ and integrated over the space-time reference control volume, thus providing

$$\int_0^1 \int_0^1 \int_0^1 \int_0^1 \theta_k \frac{\partial \mathbf{v}_h}{\partial \tau} d\boldsymbol{\xi} d\tau$$
$$= \int_0^1 \int_0^1 \int_0^1 \int_0^1 \theta_k \left(\mathbf{S}^* - \mathbf{C}_1^* \frac{\partial \mathbf{v}_h}{\partial \xi}\right.$$
$$\left. - \mathbf{C}_2^* \frac{\partial \mathbf{v}_h}{\partial \eta} - \mathbf{C}_3^* \frac{\partial \mathbf{v}_h}{\partial \zeta}\right) d\boldsymbol{\xi} d\tau, \quad (43)$$

where we have replaced $\mathbf{V}$ with its discrete representation $\mathbf{v}_h$. Integrating the first term by parts in time yields

$$
\int_0^1 \int_0^1 \int_0^1 \theta_k(\boldsymbol{\xi},1)\mathbf{v}_h(\boldsymbol{\xi},1)\,d\boldsymbol{\xi}
$$
$$
-\int_0^1 \int_0^1 \int_0^1 \int_0^1 \left(\frac{\partial}{\partial\tau}\theta_k\right)\mathbf{v}_h(\boldsymbol{\xi},\tau)\,d\boldsymbol{\xi}\,d\tau
$$
$$
= \int_0^1 \int_0^1 \int_0^1 \theta_k(\boldsymbol{\xi},0)\mathbf{p}_h\big(\boldsymbol{\xi},t^n\big)\,d\boldsymbol{\xi}
$$
$$
+\int_0^1 \int_0^1 \int_0^1 \int_0^1 \theta_k\left(\mathbf{S}^* - \mathbf{C}_1^*\frac{\partial\mathbf{v}_h}{\partial\xi}\right.
$$
$$
\left. - \mathbf{C}_2^*\frac{\partial\mathbf{v}_h}{\partial\eta} - \mathbf{C}_3^*\frac{\partial\mathbf{v}_h}{\partial\zeta}\right)d\boldsymbol{\xi}\,d\tau. \tag{44}
$$

Equation (44) is an element-local nonlinear algebraic equation that must be solved locally for each grid-cell in the unknowns $\hat{\mathbf{v}}_l^n$. In practice, we solve the system of equations (44) through a discrete Picard iteration, see Dumbser and Zanotti (2009), Hidalgo and Dumbser (2011), where additional comments about its solution can be found.

2.4.2 An efficient initial guess for the predictor

A proper choice of the initial guess for each of the space-time degrees of freedom $\hat{\mathbf{v}}_l$ can improve the convergence of the Picard process. The easiest strategy is to set $\mathbf{v}_h(\mathbf{x},t) = \mathbf{p}_h(\mathbf{x},t^n)$ i.e. the reconstruction polynomial is simply extended as a constant in time. This is, however, not the best approach. A better strategy for obtaining a good initial guess for the LSDG predictor was presented in Hidalgo and Dumbser (2011), and it is based on the implementation of a MUSCL scheme for the explicit terms, plus a second-order Crank-Nicholson scheme in case stiff source terms are present. In the following, we refer to this version of the initial guess for the LSDG predictor as the MUSCL-CN initial guess. If the source terms are not stiff, however, an even more efficient approach is possible which is based on a space-time extension of multi-level Adams-Bashforth-type ODE integrators. For that purpose, the space-time polynomial denoted by $\mathbf{v}_h^{n-1}(\mathbf{x},t)$ obtained during the previous time step $[t^{n-1},t^n]$ is simply *extrapolated in time* to the new time step $[t^n,t^{n+1}]$ by simple L2 projection:

$$
\int_{I_{ijk}} \int_{t^n}^{t^{n+1}} \theta_k(\mathbf{x},t)\mathbf{v}_h^n(\mathbf{x},t)\,dt\,d\mathbf{x}
$$
$$
= \int_{I_{ijk}} \int_{t^n}^{t^{n+1}} \theta_k(\mathbf{x},t)\mathbf{v}_h^{n-1}(\mathbf{x},t)\,dt\,d\mathbf{x}. \tag{45}
$$

Table 1 CPU time comparison among different versions of the initial guesses for the LSDG predictor

	MUSCL-CN	Adams-Bashforth
$\mathbb{P}_0\mathbb{P}_2$	1.0	0.64
$\mathbb{P}_0\mathbb{P}_3$	1.0	0.75
$\mathbb{P}_0\mathbb{P}_4$	1.0	0.72

The comparison has been performed for the isentropic vortex solution and the numbers have been normalized to the value obtained with the traditional MUSCL-CN initial guess (see Section 2.4.2 for more details).

In terms of the degrees of freedom $\hat{\mathbf{v}}_l^n$ and $\hat{\mathbf{v}}_l^{n-1}$ this relation becomes

$$
\int_0^1 \int_0^1 \int_0^1 \int_0^1 \theta_k(\boldsymbol{\xi},\tau)\theta_l(\boldsymbol{\xi},\tau)\hat{v}_l^n\,dt\,d\boldsymbol{\xi}
$$
$$
= \int_0^1 \int_0^1 \int_0^1 \int_0^1 \theta_k(\boldsymbol{\xi},\tau)\theta_l\big(\boldsymbol{\xi},\tau'\big)\hat{v}_l^{n-1}\,dt\,d\boldsymbol{\xi}, \tag{46}
$$

with $\tau' = 1 + \tau\frac{\Delta t^n}{\Delta t^{n-1}}$ and $\Delta t^{n-1} = t^n - t^{n-1}$.

In the following, we refer to this second version of the initial guess for the LSDG predictor as the Adams-Bashforth (AB) initial guess. In Table 1 we show a comparison among the performances of the LSDG predictor with these two different implementations of the initial guess.

3 Numerical tests with the new ADER-WENO finite volume scheme in primitive variables

In the following we explore the properties of the new ADER-WENO finite volume scheme by solving a wide set of test problems belonging to four different systems of equations: the classical Euler equations, the relativistic hydrodynamics (RHD) and magnetohydrodynamics (RMHD) equations and the Baer-Nunziato equations for compressible two-phase flows. For the sake of clarity, we introduce the notation 'ADER-Prim' to refer to the novel approach of this work for which both the spatial WENO reconstruction and the subsequent LSDG predictor are performed on the primitive variables. On the contrary, we denote the traditional ADER implementation, for which both the spatial WENO reconstruction and the LSDG predictor are performed on the conserved variables, as 'ADER-Cons'. In a few circumstances, we have also compared with the 'ADER-Char' scheme, namely a traditional ADER scheme in which, however, the spatial reconstruction is performed on the characteristic variables. In this Section we focus our attention on finite volume schemes, which, according to the notation introduced in Dumbser et al. (2008a), are denoted as $\mathbb{P}_0\mathbb{P}_M$ methods, where M is the degree of the approximating polynomial. In Section 4 a brief account is given to discontinuous Galerkin methods, referred to as $\mathbb{P}_N\mathbb{P}_N$ methods, for which an ADER-Prim version is also possible.

3.1 Euler equations

First of all we consider the solution of the classical Euler equations of compressible gas dynamics, for which the vectors of the conserved variables $\mathbf{Q}$ and of the fluxes $\mathbf{f}^x$, $\mathbf{f}^y$ and $\mathbf{f}^z$ are given respectively by

$$\mathbf{Q} = \begin{pmatrix} \rho \\ \rho v_x \\ \rho v_y \\ \rho v_z \\ E \end{pmatrix}, \qquad \mathbf{f}^x = \begin{pmatrix} \rho v_x \\ \rho v_x^2 + p \\ \rho v_x v_y \\ \rho v_x v_z \\ v_x(E + p) \end{pmatrix},$$

$$\mathbf{f}^y = \begin{pmatrix} \rho v_y \\ \rho v_x v_y \\ \rho v_y^2 + p \\ \rho v_y v_z \\ v_y(E + p) \end{pmatrix}, \qquad \mathbf{f}^z = \begin{pmatrix} \rho v_z \\ \rho v_x v_z \\ \rho v_y v_z \\ \rho v_z^2 + p \\ v_z(E + p) \end{pmatrix}. \tag{47}$$

Here v_x, v_y and v_z are the velocity components, p is the pressure, ρ is the mass density, $E = p/(\gamma - 1) + \rho(v_x^2 + v_y^2 + v_z^2)/2$ is the total energy density, while γ is the adiabatic index of the supposed ideal gas equation of state, which is of the kind $p = \rho\epsilon(\gamma - 1)$, ϵ being the specific internal energy.

3.1.1 2D isentropic vortex

It is important to assess the convergence properties of the new scheme, in particular comparing with the traditional ADER scheme in conserved and in characteristic variables. To this extent, we have studied the two-dimensional isentropic vortex, see e.g. Hu and Shu (1999). The initial conditions are given by a uniform mean flow, to which a per-turbation is added, such that

$$(\rho, v_x, v_y, v_z, p) = (1 + \delta\rho, 1 + \delta v_x, 1 + \delta v_y, 0, 1 + \delta p), \tag{48}$$

with

$$\begin{pmatrix} \delta\rho \\ \delta v_x \\ \delta v_y \\ \delta p \end{pmatrix} = \begin{pmatrix} (1 + \delta T)^{1/(\gamma-1)} - 1 \\ -(y - 5)\epsilon/2\pi \exp[0.5(1 - r^2)] \\ (x - 5)\epsilon/2\pi \exp[0.5(1 - r^2)] \\ (1 + \delta T)^{\gamma/(\gamma-1)} - 1 \end{pmatrix}. \tag{49}$$

Whatever the perturbation δT in the temperature is, it is easy to verify that there is not any variation in the specific entropy $s = p/\rho^\gamma$, and the flow is advected smoothly and isentropically with velocity $v = (1, 1, 0)$. We have solved this test over the computational domain $\Omega = [0; 10] \times [0; 10]$, assuming

$$\delta T = -\frac{\epsilon^2(\gamma - 1)}{8\gamma\pi^2} \exp\left(1 - r^2\right), \tag{50}$$

with $r^2 = (x - 5)^2 + (y - 5)^2$, vortex strength $\epsilon = 5$ and adiabatic index $\gamma = 1.4$. Table 2 contains the results of our calculation, in which we have compared the convergence properties of three different finite volume ADER schemes: ADER-Prim, ADER-Cons and ADER-Char, obtained with the Osher-type Riemann solver, see Dumbser and Toro (2011b). While all the schemes converge to the nominal order, it is interesting to note that the smallest L_2 error is obtained for the *new* ADER finite volume scheme in *primitive variables*, and that the difference with respect to the

Table 2 L_2 errors of the mass density and corresponding convergence rates for the 2D isentropic vortex problem

2D isentropic vortex problem		ADER-Prim		ADER-Cons		ADER-Char		Theor.
	N_x	L_2 error	L_2 order	L_2 error	L_2 order	L_2 error	L_2 order	
$\mathbb{P}_0\mathbb{P}_2$	100	4.060E-03	-	5.028E-03	-	5.010E-03	-	3
	120	2.359E-03	2.98	2.974E-03	2.88	2.968E-03	2.87	
	140	1.489E-03	2.98	1.897E-03	2.92	1.893E-03	2.92	
	160	9.985E-04	2.99	1.281E-03	2.94	1.279E-03	2.94	
	200	5.118E-04	2.99	6.612E-04	2.96	6.607E-04	2.96	
$\mathbb{P}_0\mathbb{P}_3$	50	2.173E-03	-	4.427E-03	-	5.217E-03	-	4
	60	8.831E-04	4.93	1.721E-03	5.18	2.232E-03	4.65	
	70	4.177E-04	4.85	8.138E-04	4.85	1.082E-03	4.69	
	80	2.194E-04	4.82	4.418E-04	4.57	5.746E-04	4.74	
	100	7.537E-05	4.79	1.605E-04	4.53	1.938E-04	4.87	
$\mathbb{P}_0\mathbb{P}_4$	50	2.165E-03	-	3.438E-03	-	3.416E-03	-	5
	60	6.944E-04	6.23	1.507E-03	4.52	1.559E-03	4.30	
	70	3.292E-04	4.84	7.615E-04	4.43	7.615E-04	4.65	
	80	1.724E-04	4.84	4.149E-04	4.55	4.148E-04	4.55	
	100	5.884E-05	4.82	1.449E-04	4.71	1.448E-04	4.72	

A comparison is shown among the reconstruction in primitive variables (ADER-Prim), in conserved variables (ADER-Cons) and in characteristic variables (ADER-Char). The Osher-type numerical flux has been used.

other two reconstructions increases with the order of the method.

In addition to the convergence properties, we have compared the performances of the Adams-Bashforth version of the initial guess for the LSDG predictor with the traditional version based on the MUSCL-CN algorithm. The comparison has been performed over a 100 × 100 uniform grid. The results are shown in Table 1, from which we conclude that the Adams-Bashforth initial guess is indeed computationally more efficient in terms of CPU time. However, we have also experienced that it is typically less robust, and in some of the most challenging numerical tests discussed in the rest of the paper we had to use the more traditional MUSCL-CN initial guess.

3.1.2 Sod's Riemann problem

We have then solved the classical Riemann problem named after Sod (Sod 1978), assuming an adiabatic index $\gamma = 1.4$, and evolved until $t_{\text{final}} = 0.2$. In spite of the fact that this is a one-dimensional test, we have evolved this problem in two spatial dimensions over the domain $[0,1] \times [-0.2, 0.2]$, using periodic boundary conditions along the passive y direction. In Figure 1 we show the comparison among the solutions obtained with ADER-Prim, ADER-Cons and ADER-Char, together with the exact solution provided in Toro (1999). We have adopted the finite volume scheme at the fourth order of accuracy, namely the $\mathbb{P}_0\mathbb{P}_3$ scheme, in combination with the Rusanov numerical flux and using 400 cells along the x-direction. Although

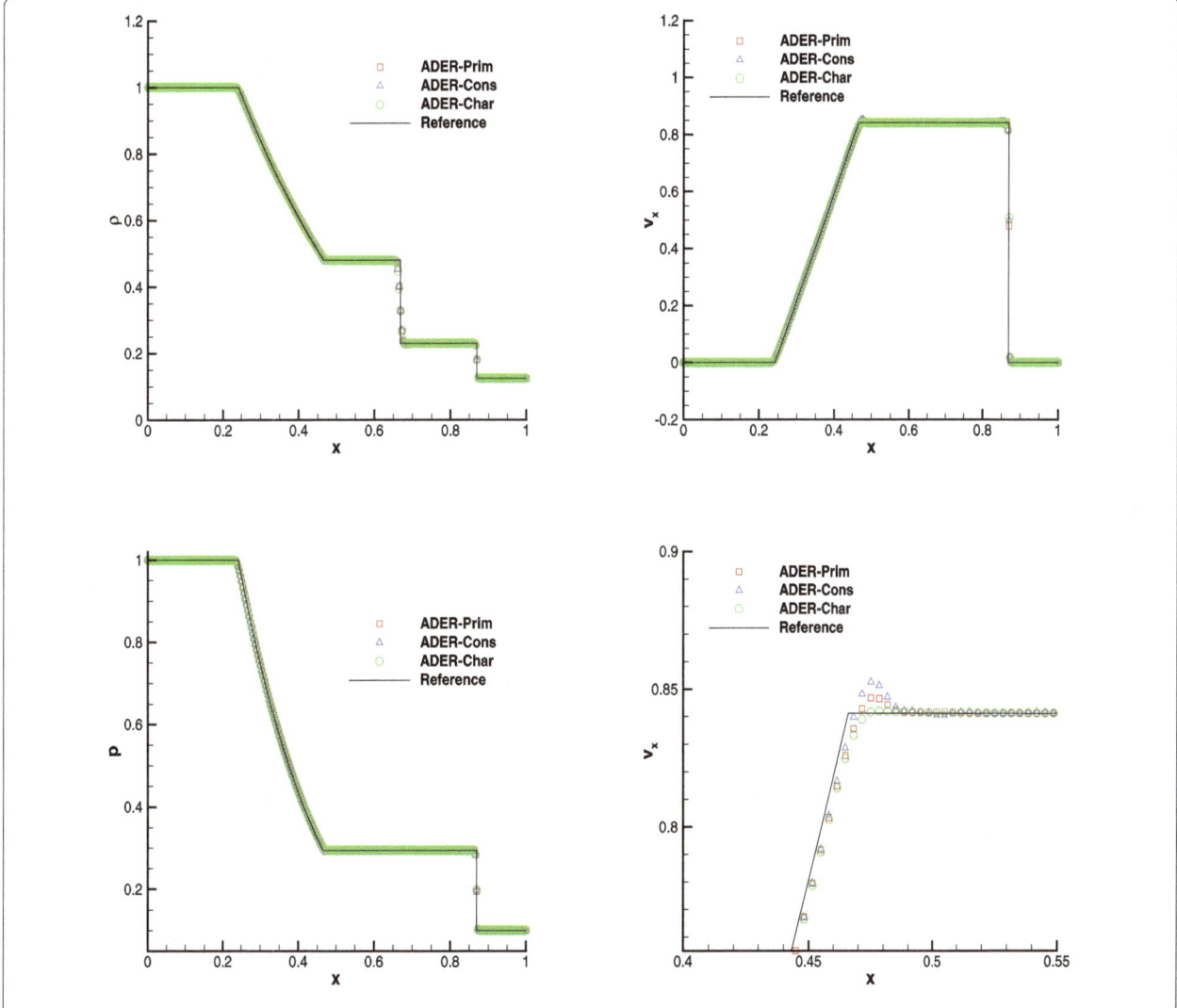

Figure 1 Solution of Sod's Riemann problem with the fourth order ADER-WENO scheme at time $t = 0.2$. The bottom right panel shows a magnification of the velocity at the tail of the rarefaction.

Table 3 CPU time comparison among different ADER implementations for the Sod Riemann problem

	ADER-Prim	ADER-Cons	ADER-Char
$\mathbb{P}_0\mathbb{P}_2$	1.0	0.74	0.81
$\mathbb{P}_0\mathbb{P}_3$	1.0	0.74	0.80
$\mathbb{P}_0\mathbb{P}_4$	1.0	0.77	0.81

The numbers have been normalized to the value obtained with ADER-Prim.

all of the ADER implementations show a very good agreement with the exact solution, a closer look at the tail of the rarefaction, highlighted in the bottom right panel, reveals that the ADER-Cons scheme is actually the worst one, while the solution obtained with ADER-Prim is more similar to the reconstruction in characteristic variables. On the contrary, in terms of CPU-time, ADER-Prim is not convenient for this system of equations because the price paid for performing the double WENO reconstruction in space is not significantly compensated by the reduced number of conversions that are needed from the conserved to the primitive variables. Table 3 reports the CPU times, normalized with respect to the ADER-Prim implementation, for different orders of accuracy, showing that the ADER-Prim scheme is $\sim$25 % slower than the traditional ADER-Cons scheme. As we will see in Table 5 of Section 3.2, the comparison will change in favor of ADER-Prim schemes, when the relativistic equations are solved instead.

3.1.3 Interacting blast waves

The interaction between two blast waves was first proposed by Woodward and Colella (1984) and it is now a standard test for computational fluid dynamics. The initial

conditions are given by

$$(\rho, v_x, p) = \begin{cases} (1.0, 0.0, 10^3) & \text{if } -0.5 < x < -0.4, \\ (1.0, 0.0, 10^{-2}) & \text{if } -0.4 < x < 0.4, \\ (1.0, 0.0, 10^2) & \text{if } 0.4 < x < 0.5, \end{cases} \quad (51)$$

where the adiabatic index is $\gamma = 1.4$. We have evolved this problem in two spatial dimensions over the domain $[-0.6, 0.6] \times [-0.5, 0.5]$, using reflecting boundary conditions in x direction and periodic boundary conditions along the y direction. The results of our calculations, obtained with the $\mathbb{P}_0\mathbb{P}_3$ scheme, are reported in Figure 2, where only the one-dimensional cuts are shown. The number of cells chosen along the x-direction, namely $N_x = 500$, is not particularly large, at least for this kind of challenging problem. This has been intentionally done to better highlight potential differences among the two alternative ADER-Prim and ADER-Cons schemes. As it turns out from the figure, the two methods are very similar in terms of accuracy: the sharp peak in the density at time $t = 0.028$ (left panel) is somewhat better resolved through the ADER-Prim, while the opposite is true for the highest peak at time $t = 0.038$ (right panel). On the overall, however, the two schemes perform equally well for this test.

3.1.4 Double Mach reflection problem

As a representative test for the Euler equations in two space dimensions, we have considered the *double Mach reflection problem*, which implies the interaction of several waves. The dynamics of this problem is triggered by a shock wave propagating towards the right with a Mach number $M = 10$, and intersecting the x-axis at $x = 1/6$ with

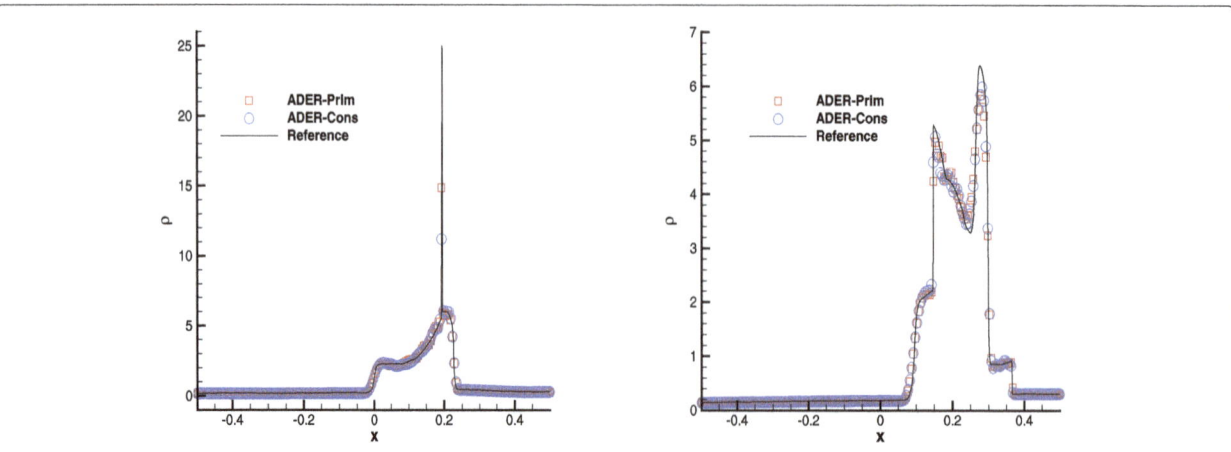

Figure 2 Solution of the interacting Blast-Wave problem at time $t = 0.028$ (left panel) and at time $t = 0.038$ (right panel) obtained with the fourth order ADER-WENO scheme. The computation has been performed over a uniform grid of 500 cells.

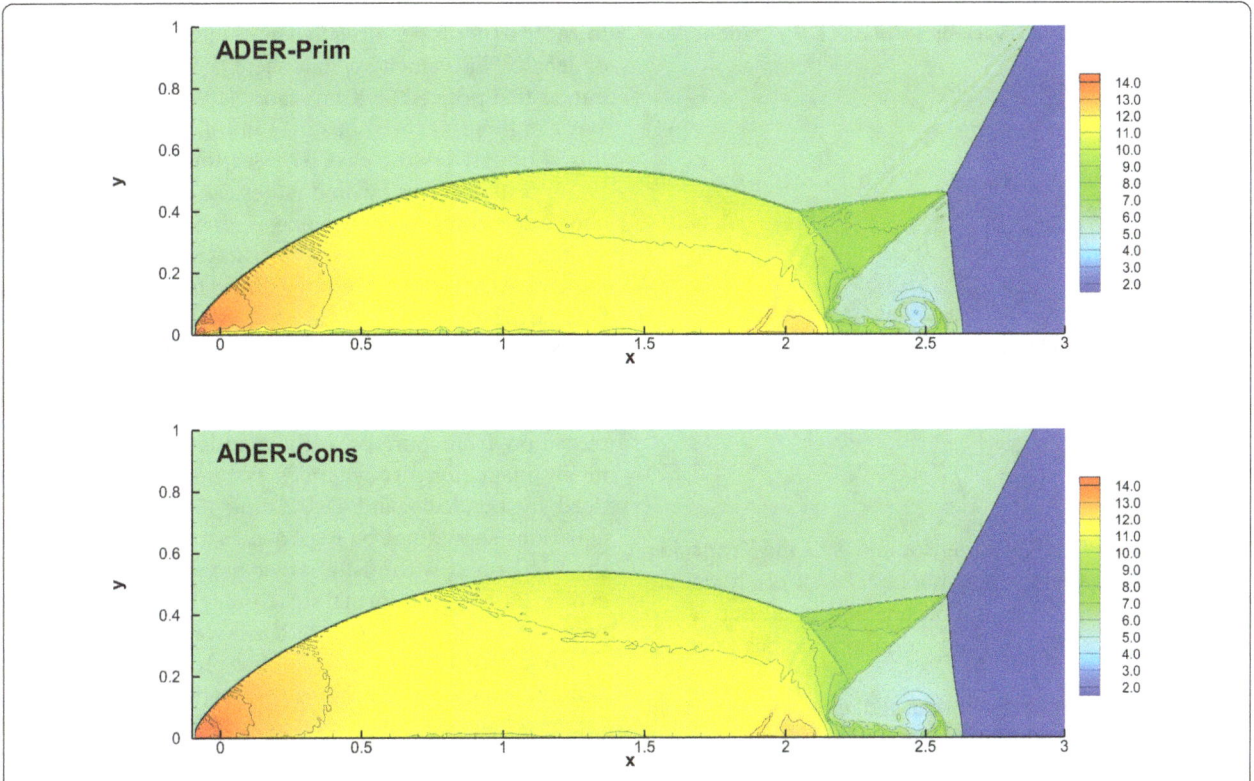

Figure 3 **Double Mach reflection problem at time $t = 0.2$ obtained with the fourth order ADER-WENO scheme and the Rusanov Riemann solver. The computation has been performed over a uniform grid of 1,200 × 300 cells.** Top panel: mass density distribution obtained with ADER-Prim. Bottom panel: mass density distribution obtained with ADER-Cons.

an inclination angle of $\alpha = 60°$. The initial states ahead and behind the shock are fixed after solving the Rankine-Hugoniot conditions, obtaining

$$(\rho, u, v, p)(\mathbf{x}, t = 0)$$

$$= \begin{cases} \frac{1}{\gamma}(8.0, 8.25, 0.0, 116.5), & \text{if } x' < 0.1, \\ (1.0, 0.0, 0.0, \frac{1}{\gamma}), & \text{if } x' \geq 0.1, \end{cases} \quad (52)$$

where $x' = (x - 1/6)\cos\alpha - y\sin\alpha$. The adiabatic index is $\gamma = 1.4$. We fix inflow and outflow boundary conditions on the left side and on the right of the numerical domain, respectively, while on the bottom we have used reflecting boundary conditions. At the top we must impose the exact solution of an isolated moving oblique shock wave with the same shock Mach number $M_s = 10$. We have solved the test over the rectangle $\Omega = [0; 3.0] \times [0; 1]$, covered by a uniform grid composed of 1,200 × 300 cells, using the Rusanov Riemann solver and a fourth order finite volume scheme. The two panels of Figure 3 show the comparison of the solution at time $t = 0.2$ obtained with the ADER-Prim (top panel) and with the ADER-Cons (bottom panel) scheme. The results are very similar in the two cases.

As a tentative conclusion about the performances of ADER-Prim for the Euler equations, we may say that, although it is the most accurate on smooth solutions (see Table 2), and comparable to a traditional ADER with reconstruction in characteristic variables, it is computationally more expensive than ADER-Cons and ADER-Char. Hence, ADER-Prim will rarely become the preferred choice in standard applications for the Euler equations.

3.2 Relativistic hydrodynamics and magnetohydrodynamics

From a formal point of view, the equations of special relativistic hydrodynamics and magnetohydrodynamics can be written in conservative form like the classical Euler equations (see, however, the comments below), namely as in Eq. (1), with the vectors of the conserved variables and of the corresponding fluxes given by

$$\mathbf{Q} = \begin{bmatrix} D \\ S_j \\ U \\ B^j \end{bmatrix}, \quad \mathbf{f}^i = \begin{bmatrix} v^i D \\ W_j^i \\ S^i \\ \epsilon^{jik} E^k \end{bmatrix}, \quad i = x, y, z, \quad (53)$$

where the conserved variables (D, S_j, U, B_j) can be expressed as[d]

$$D = \rho W, \tag{54}$$

$$S_i = \rho h W^2 v_i + \epsilon_{ijk} E_j B_k, \tag{55}$$

$$U = \rho h W^2 - p + \frac{1}{2}\left(E^2 + B^2\right), \tag{56}$$

while the spatial projection of the energy-momentum tensor of the fluid is (Del Zanna et al. 2007)

$$W_{ij} \equiv \rho h W^2 v_i v_j - E_i E_j - B_i B_j$$
$$+ \left[p + \frac{1}{2}\left(E^2 + B^2\right)\right]\delta_{ij}. \tag{57}$$

Here ϵ_{ijk} is the Levi-Civita tensor and δ_{ij} is the Kronecker symbol. We have used the symbol $h = 1 + \epsilon + p/\rho$ to denote the specific enthalpy of the plasma and in all our calculations the usual ideal gas equation of state has been assumed.

The components of the electric and of the magnetic field in the laboratory frame are denoted by E_i and B_i, while the Lorentz factor of the fluid with respect to this reference frame is $W = (1 - v^2)^{-1/2}$. We emphasize that the electric field does not need to be evolved in time under the assumption of infinite electrical conductivity, since it can always be computed in terms of the velocity and of the magnetic field as $\vec{E} = -\vec{v} \times \vec{B}$.

Although formally very similar to the classical gas dynamics equations, their relativistic counterpart present two fundamental differences. The first one is that, while the physical fluxes $\mathbf{f}^i$ of the classical gas dynamics equations can be written analytically in terms of the conserved variables, i.e. $\mathbf{f}^i = \mathbf{f}^i(\mathbf{Q})$, those of the relativistic hydrodynamics (or magnetohydrodynamics) equations need the knowledge of the primitive variables, i.e. $\mathbf{f}^i = \mathbf{f}^i(\mathbf{V})$ for RMHD. The second difference is that, in the relativistic case, the conversion from the conserved to the primitive variables, i.e. the operation $(D, S_j, U, B_j) \rightarrow (\rho, v_i, p, B_i)$, is not analytic, and it must be performed numerically through some appropriate iterative procedure. Since in an ADER scheme such a conversion must be performed in each space-time degree of freedom of the space-time DG predictor and at each Gaussian quadrature point for the computation of the fluxes in the finite volume scheme, we may expect a significant computational advantage by performing the WENO reconstruction and the LSDG predictor directly on the primitive variables. In this way, in fact, the conversion $(D, S_j, U, B_j) \rightarrow (\rho, v_i, p, B_i)$ is required only once at the cell center (see Section 2.3), and not in each space-time degree of freedom of the predictor and at each Gaussian point for the quadrature of the numerical fluxes. We emphasize that the choice of the variables

to reconstruct for the relativistic velocity is still a matter of debate. The velocity v_i may seem the most natural one, but, as first noticed by Komissarov (1999), reconstructing $W v_i$ can increase the robustness of the scheme. However, this is not always the case (see Section 3.2.5 below) and in our tests we have favored either the first or the second choice according to convenience. Concerning the specific strategy adopted to recover the primitive variables, in our numerical code we have used the third method reported in Section 3.2 of Del Zanna et al. (2007). Alternative methods can be found in Noble et al. (2006), Rezzolla and Zanotti (2013).

Finally, there is an important formal change in the transition from purely hydrodynamics systems to genuinely magnetohydrodynamics systems. As already noticed by Londrillo and Del Zanna (2000), the RMHD equations should not be regarded as a mere extension of the RHD ones, with just a larger number of variables to evolve. Rather, their formal structure is better described in terms of a coupled system of conservation laws (the five equations for the dynamics of the plasma) and a set of Hamilton-Jacobi equations, those for the evolution of the vector potential of the magnetic field (Jin and Xin 1998). The different mathematical structure of the RMHD equations reflects the existence of the divergence-free property of the magnetic field, which must be ensured at all times during the evolution. Numerically, we have adopted a simplified and well known approach, which consists of augmenting the system (1) with an additional equation for a scalar field Φ, aimed at propagating away the deviations from $\vec{\nabla} \cdot \vec{B} = 0$. We therefore need to solve

$$\partial_t \Phi + \partial_i B^i = -\kappa \Phi, \tag{58}$$

while the fluxes for the evolution of the magnetic field are also changed, namely $\mathbf{f}^i(B^j) \rightarrow \epsilon^{jik} E^k + \Phi \delta^{ij}$, where $\kappa \in [1; 10]$ in most of our calculations. Originally introduced by Dedner et al. (2002) for the classical MHD equations, this approach has been extended to the relativistic regime by Palenzuela et al. (2009). More information about the mathematical structure of the RMHD equations can be found in Anile (1990), Balsara (2001), Komissarov (1999), Del Zanna et al. (2007), Antón et al. (2010).

In the following, we first limit our attention to a few physical systems for which $B_i = E_i = 0$, hence to relativistic hydrodynamics, and then we consider truly magnetohydrodynamics tests with $B_i \neq 0$.

3.2.1 RHD Riemann problems

Table 4 reports the initial conditions of the two one-dimensional Riemann problems that we have considered, and whose wave-patterns at the final time $t_f = 0.4$ are shown in Figure 4 and Figure 5, respectively. In order to appreciate the differences among the available ADER implementations, we have again solved each problem with

Table 4 Left and right states of the one-dimensional RHD Riemann problems

Problem		γ	ρ	v_x	p	t_f
RHD-RP1	$x > 0$	5/3	1	−0.6	10	0.4
	$x \le 0$		10	0.5	20	
RHD-RP2	$x > 0$	5/3	10^{-3}	0.0	1	0.4
	$x \le 0$		10^{-3}	0.0	10^{-5}	

the three alternative schemes: ADER-Prim, ADER-Cons and ADER-Char. The reference solution, computed as in Rezzolla and Zanotti (2001), is shown too.

In the first Riemann problem, which was also analyzed by Mignone and Bodo (2005), two rarefaction waves are produced, separated by a contact discontinuity. It has been solved through a fourth order $\mathbb{P}_0\mathbb{P}_3$ scheme, using the Ru-

sanov Riemann solver over a uniform grid with 300 cells. As it is clear from Figure 4, the ADER-Prim scheme performs significantly better than the ADER-Cons. In particular, the overshoot and undershoot at the tail of the right rarefaction is absent. In general, the results obtained with ADER-Prim are essentially equivalent to those of ADER-Char, namely when the reconstruction in characteristic variables is adopted. This is manifest after looking at the bottom right panel of Figure 4, where a magnification of the rest mass density at the contact discontinuity is shown. Additional interesting comparisons can be made about the second Riemann problem, which can be found in Radice and Rezzolla (2012), and which is displayed in Figure 5. In this case a third order $\mathbb{P}_0\mathbb{P}_2$ scheme has been used, again with the Rusanov Riemann solver over a uniform grid with 500 cells. The right propagating shock has a strong jump

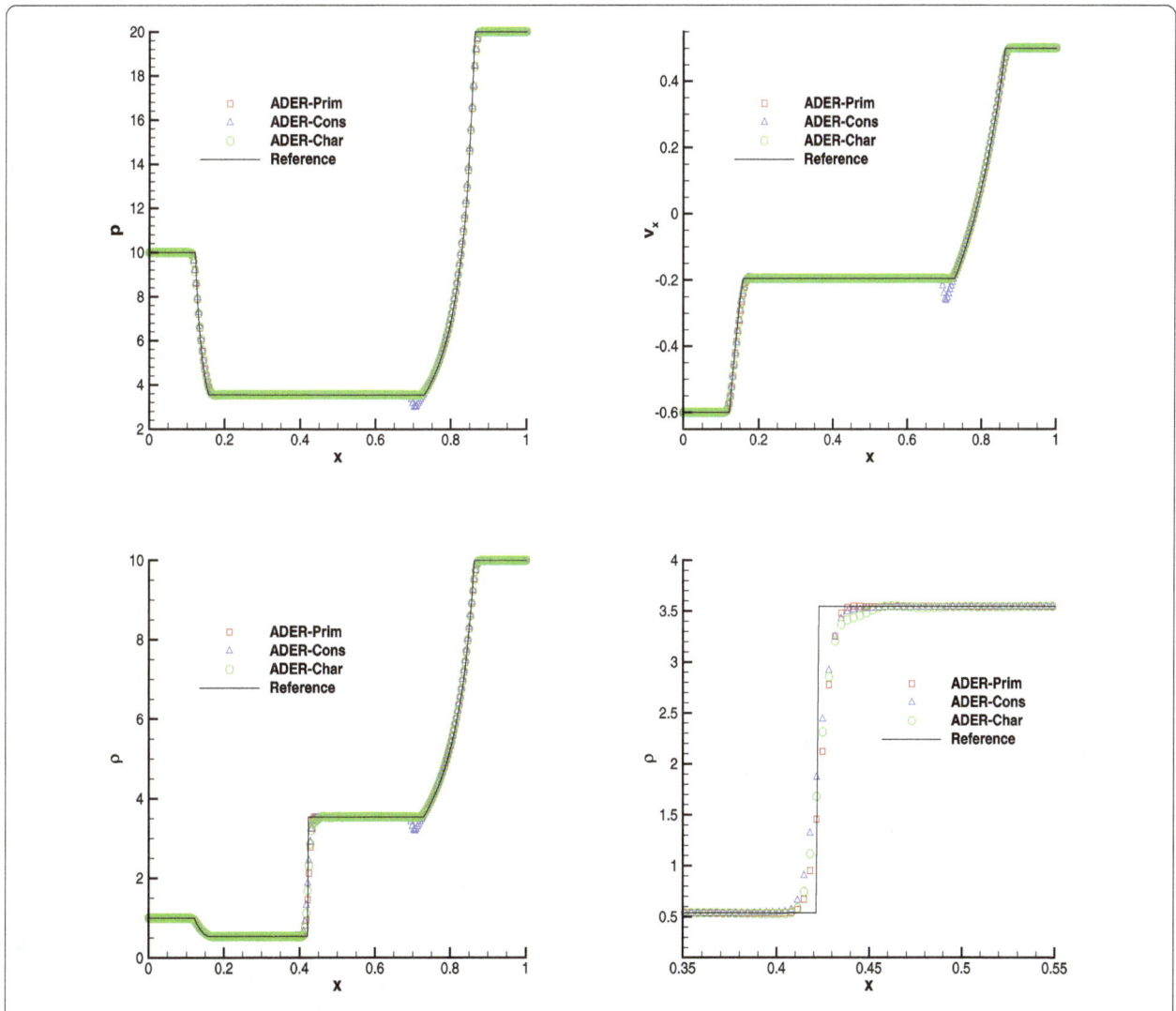

Figure 4 Solution of RHD-RP1 (see Table 4) with the fourth order ADER-WENO scheme at time $t = 0.4$. The bottom right panel shows a magnification around the contact discontinuity.

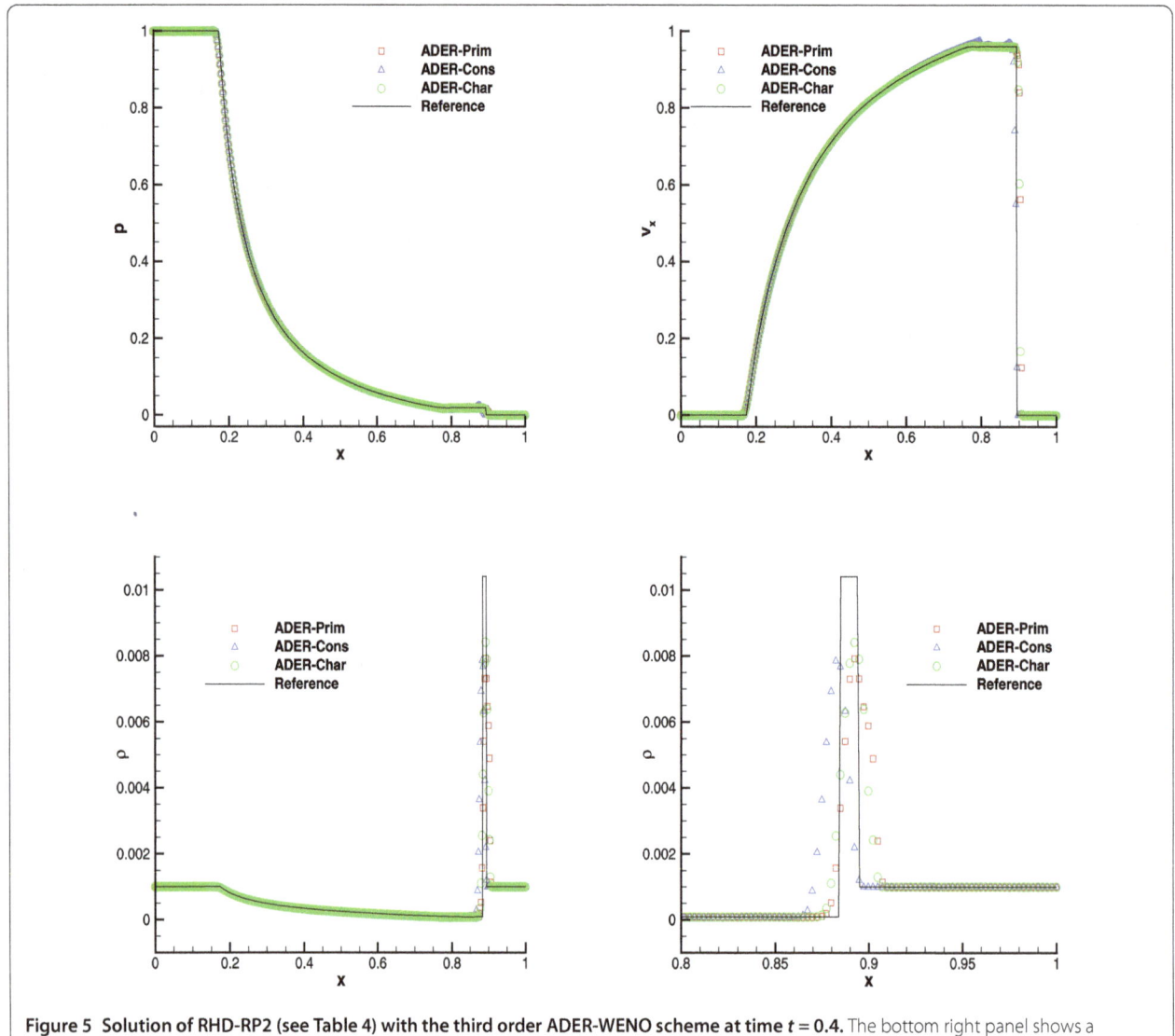

Figure 5 Solution of RHD-RP2 (see Table 4) with the third order ADER-WENO scheme at time *t* = 0.4. The bottom right panel shows a magnification around the right propagating shock.

in the rest mass density, as it is visible from the bottom right panel of the figure, and the position of the shock front is better captured by the two schemes ADER-Prim and ADER-Char.

It is particularly interesting to address the issue of CPU time comparison among different implementations of

Table 5 CPU time comparison among different ADER implementations for the RHD-RP1 problem

	ADER-Prim	ADER-Cons	ADER-Char
$\mathbb{P}_0\mathbb{P}_2$	1.0	1.26	1.40
$\mathbb{P}_0\mathbb{P}_3$	1.0	1.13	1.24
$\mathbb{P}_0\mathbb{P}_4$	1.0	1.04	1.06

The numbers have been normalized to the value obtained with ADER-Prim.

ADER, as already done for the Euler equations. The result of such a comparison, performed for the RHD-RP1 problem, are reported in Table 5, which should be read in synopsis with Table 3. Clearly, ADER-Prim is not only more accurate than ADER-Cons, but it is also more efficient. As anticipated, this is in agreement with our expectations, since in the ADER-Prim implementation a single *cons-to-prim* operation is needed within the cell, rather than at each Gaussian quadrature point and at each space-time degree of freedom. For other tests, see for instance Section 3.2.2, the CPU time reduction implied by ADER-Prim is even more evident, but the numbers shown in Table 5 describe with good fidelity the relative performances of the different ADER in a large number of relativistic tests.

3.2.2 RHD Kelvin-Helmholtz instability

In the relativistic regime, the Kelvin-Helmholtz (KH) instability is likely to be responsible for a variety of physical effects, which are encountered in the dynamics of extragalactic relativistic jets (Bodo et al. 2004; Perucho et al. 2006; Perucho et al. 2007). As an academic test, we simulate the linear growth phase of the KH instability in two spatial dimensions, taking the initial conditions from Mignone et al. (2009) (see also Beckwith and Stone 2011 and Radice and Rezzolla 2012). In particular, the rest-mass density is chosen as

$$\rho = \begin{cases} \rho_0 + \rho_1 \tanh\left[(y - 0.5)/a\right], & y > 0, \\ \rho_0 - \rho_1 \tanh\left[(y + 0.5)/a\right], & y \le 0, \end{cases} \quad (59)$$

with $\rho_0 = 0.505$ and $\rho_1 = 0.495$. Assuming that the shear layer has a velocity $v_s = 0.5$ and a characteristic size $a = 0.01$, the velocity along the x-direction is modulated as

$$v_x = \begin{cases} v_s \tanh\left[(y - 0.5)/a\right], & y > 0, \\ -v_s \tanh\left[(y + 0.5)/a\right], & y \le 0. \end{cases} \quad (60)$$

It is convenient to add a perturbation in the transverse velocity, i.e.

$$v_y = \begin{cases} \eta_0 v_s \sin(2\pi x)\exp[-(y - 0.5)^2/\sigma], & y > 0, \\ -\eta_0 v_s \sin(2\pi x)\exp[-(y + 0.5)^2/\sigma], & y \le 0, \end{cases} \quad (61)$$

where $\eta_0 = 0.1$ is the amplitude of the perturbation, while $\sigma = 0.1$ is its length scale. The adiabatic index is $\gamma = 4/3$ and the pressure is uniform, $p = 1$. The problem has been solved over the computational domain $[-0.5, 0.5] \times [-1, 1]$, covered by a uniform mesh with 200×400 cells, using the $\mathbb{P}_0\mathbb{P}_3$ scheme and the Osher-type numerical flux. Periodic boundary conditions are fixed both in x and in y directions. Figure 6 shows the results of the calculations: in the left, in the central and in the right panels we have reported the solution obtained with the ADER-Prim, with the ADER-Cons and with the ADER-Char scheme, respectively, while the top and the bottom panels correspond to two different times during the evolution, namely $t = 2.0$ and $t = 2.5$. Interestingly, two secondary vortices are visible when the reconstruction is performed in primitive and characteristic variables (see left the right panels), but only one is present in the simulation using the reconstruction in conserved variables. In Zanotti and Dumbser (2015) we have already commented about the elusive character of these details in the solution, which depend both on the resolution and on the Riemann solver adopted. Based on our results, we infer that the ADER-Cons scheme is the most diffusive, while ADER-Prim and ADER-Char seem to produce the same level of accuracy in the solution. However, if we look at the CPU times in the two cases, we find that ADER-Prim is

a factor 2.5 faster than ADER-Cons and a factor 3 faster than ADER-Char, and therefore should be preferred in all relevant applications of RHD.

3.2.3 RMHD Alfvén wave

In Table 2 of Section 3.1.1 we have reported the comparison of the convergence rates among three different implementations of ADER for the Euler equations. We believe it is important to verify the convergence of the new ADER-Prim scheme also for the RMHD equations, which indeed admits an exact, smooth unsteady solution, namely the propagation of a circularly polarized Alfvén wave (see Komissarov 1997; Del Zanna et al. 2007 for a full account). The wave is assumed to propagate along the x direction in a constant density and constant pressure background, say $\rho = p = 1$. The magnetic field, on the other hand, is given by

$$B_x = B_0, \quad (62)$$

$$B_y = \eta B_0 \cos\left[k(x - v_A t)\right], \quad (63)$$

$$B_z = \eta B_0 \sin\left[k(x - v_A t)\right], \quad (64)$$

where $\eta = 1$ is the amplitude of the wave, $B_0 = 1$ is the uniform magnetic field, k is the wave number, while v_A is speed of propagation of the wave. We have solved this problem over the computational domain $\Omega = [0; 2\pi] \times [0; 2\pi]$, using periodic boundary conditions, the Rusanov Riemann solver and the Adams-Bashforth version for the initial guess of the LSDG predictor. We have compared the numerical solution with the analytic one after one period $T = L/v_A = 2\pi/v_A$. Table 6 contains the results of our analysis, showing the L_1 and the L_2 norms of the error of B^y. As apparent from the table, the nominal order of convergence of the new ADER-Prim scheme is recovered with very good accuracy.

3.2.4 RMHD Riemann problems

Riemann problems are very relevant also in RMHD, admitting a larger number of waves than in hydrodynamics. The exact solution was provided by Giacomazzo and Rezzolla (2006) already ten years ago, making them very popular as a precise tool to validate numerical codes. We have selected Test 1 and Test 5 in Table 1 of Balsara (2001), with initial left and right states that are reported in Table 7. Both the tests have been solved using a fourth order ADER-WENO scheme, the Rusanov Riemann solver and over a uniform grid composed of 400 cells. The damping factor for the divergence-cleaning procedure is set to $\kappa = 10$. Figure 7 and Figure 8 allow to compare the exact solution with the results obtained through the ADER-Prim and the ADER-Cons schemes. Especially for RMHD-RP1, the solution obtained with the traditional ADER-Cons scheme is significantly more oscillatory than that produced by ADER-Prim. This is particularly evident in the rest-mass

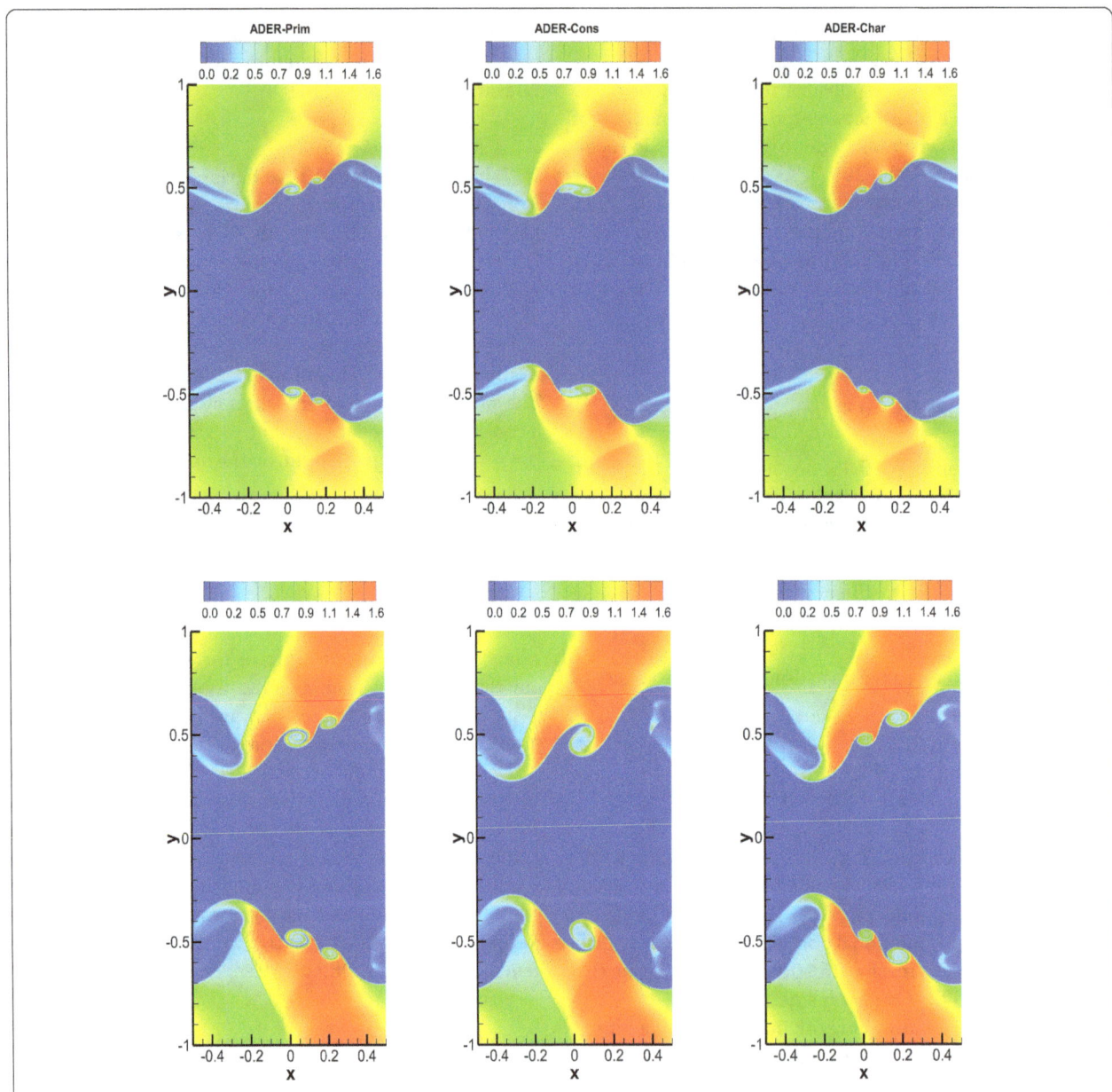

Figure 6 Two-dimensional Kelvin-Helmholtz instability obtained with the $\mathbb{P}_0\mathbb{P}_3$ scheme and with the Osher flux. Left panels: solution with ADER-Prim. Central panels: solution with ADER-Cons. Right panels: solution with ADER-Char. Top panels: solution at $t = 2.0$. Bottom panels: solution at $t = 2.5$.

density and in the velocity v_x. We have here a good indication that the ADER-Prim scheme behaves better than the ADER-Cons scheme when applied to the equations of special relativistic magnetohydrodynamics.

3.2.5 RMHD rotor problem
The relativistic version of the magnetic rotor problem, originally proposed by Balsara and Spicer (1999), has by now become a standard numerical test in RMHD. It describes the evolution of a high density plasma which, at

time $t = 0$, rotates rapidly with angular velocity ω and is surrounded by a low density plasma at rest:

$$\rho = \begin{cases} 10 & \text{for } 0 \leq r \leq 0.1; \\ 1 & \text{otherwise}; \end{cases}$$

$$\omega = \begin{cases} 9.3 & \text{for } 0 \leq r \leq 0.1; \\ 0 & \text{otherwise}; \end{cases} \qquad \mathbf{B} = \begin{pmatrix} 1.0 \\ 0 \\ 0 \end{pmatrix}, \qquad (65)$$

$$p = 1, \qquad \gamma = 4/3.$$

Table 6 L_1 and L_2 errors analysis for the 2D Alfvén wave problem

	N_x	L_1 error	L_1 order	L_2 error	L_2 order	Theor.
2D circularly polarized Alfvén wave						
$\mathbb{P}_0\mathbb{P}_2$	50	5.387E-02	-	9.527E-03	-	3
	60	3.123E-02	2.99	5.523E-03	2.99	
	70	1.969E-02	2.99	3.481E-03	2.99	
	80	1.320E-02	2.99	2.334E-03	2.99	
	100	6.764E-03	3.00	1.196E-03	3.00	
$\mathbb{P}_0\mathbb{P}_3$	50	2.734E-04	-	4.888E-05	-	4
	60	1.153E-04	4.73	2.061E-05	4.74	
	70	5.622E-05	4.66	1.004E-05	4.66	
	80	3.043E-05	4.60	5.422E-06	4.61	
	100	1.108E-05	4.53	1.968E-06	4.54	
$\mathbb{P}_0\mathbb{P}_4$	30	2.043E-03	-	3.611E-04	-	5
	40	4.873E-04	4.98	8.615E-05	4.98	
	50	1.603E-04	4.98	2.846E-05	4.96	
	60	6.491E-05	4.96	1.168E-05	4.88	
	70	3.173E-05	4.64	6.147E-06	4.16	

The errors have been computed with respect to the magnetic field B^y.

Due to rotation, a sequence of torsional Alfvén waves are launched outside the cylinder, with the net effect of reducing the angular velocity of the rotor. We have solved this problem over a computational domain $\Omega = [-0.6, 0.6] \times [-0.6, 0.6]$, discretized by 300×300 numerical cells and using a fourth order finite volume scheme with the Rusanov Riemann solver. No taper has been applied to the initial conditions, thus producing true discontinuities right at the beginning. Figure 9 shows the rest-mass density, the thermal pressure, the relativistic Mach number and the magnetic pressure at time $t = 0.4$. We obtain results which are in good qualitative agreement with those available in the literature (see, for instance, Del Zanna et al. 2003; Dumbser and Zanotti 2009; Loubère et al. 2014 and Kim and Balsara 2014). We emphasize that for this test the reconstruction of the primitive variables v^i turns out to be more robust than that achieved through the reconstruction of the products Wv^i.

3.3 The Baer-Nunziato equations

As a genuinely non-conservative system of hyperbolic equations we consider the Baer-Nunziato model for compressible two-phase flow (see also Baer and Nunziato 1986; Saurel and Abgrall 1999; Andrianov and Warnecke 2004;

Schwendeman et al. 2006; Deledicque and Papalexandris 2007; Murrone and Guillard 2005). In the rest of the paper we define the first phase as the solid phase and the second phase as the gas phase. As a result, we will use the subscripts 1 and s as well as 2 and g as synonyms. Sticking to Baer and Nunziato (1986), we prescribe the interface velocity $\mathbf{v}_I$ and the pressure p_I as $\mathbf{v}_I = \mathbf{v}_1$ and $p_I = p_2$, respectively, although other choices are also possible (Saurel and Abgrall 1999). With these definitions, the system of Baer-Nunziato equations can be cast in the form prescribed by (1) after defining the state vector $\mathbf{Q}$ as

$$\mathbf{Q} = (\phi_1\rho_1, \phi_1\rho_1 v_1^i, \phi_1\rho_1 E_1,$$
$$\phi_2\rho_2, \phi_2\rho_2 v_2^i, \phi_2\rho_2 E_2, \phi_1), \tag{66}$$

where ϕ_k is the volume fraction of phase k, with the condition that $\phi_1 + \phi_2 = 1$. On the other hand, the fluxes $\mathbf{f}^i$, the sources $\mathbf{S}$ and the non-conservative matrices $\mathbf{B}_i$ are expressed by

$$\mathbf{f}^i = \begin{bmatrix} \phi_1\rho_1 v_1^i \\ \phi_1(\rho_1 v_1^i v_1^j + p_1\delta^{ij}) \\ \phi_1 v_1^i(\rho_1 E_1 + p_1) \\ \phi_2\rho_2 v_2^i \\ \phi_2(\rho_2 v_2^i v_2^j + p_2\delta^{ij}) \\ \phi_2 v_2^i(\rho_2 E_2 + p_2) \\ 0 \end{bmatrix},$$
$$\tag{67}$$

$$\mathbf{S} = \begin{bmatrix} 0 \\ -\nu(v_1^i - v_2^i) \\ -\nu\mathbf{v}_1 \cdot (\mathbf{v}_1 - \mathbf{v}_2) \\ 0 \\ -\nu(v_2^i - v_1^i) \\ -\nu\mathbf{v}_1 \cdot (\mathbf{v}_2 - \mathbf{v}_1) \\ \mu(p_1 - p_2) \end{bmatrix},$$

$$\mathbf{B}_i = \begin{pmatrix} 0 & 0 & 0 & 0 & 0 & 0 & 0 & 0 & 0 & 0 & 0 \\ 0 & 0 & 0 & 0 & 0 & 0 & 0 & 0 & 0 & 0 & -p_I\mathbf{e}_i \\ 0 & 0 & 0 & 0 & 0 & 0 & 0 & 0 & 0 & 0 & -p_I v_I^i \\ 0 & 0 & 0 & 0 & 0 & 0 & 0 & 0 & 0 & 0 & 0 \\ 0 & 0 & 0 & 0 & 0 & 0 & 0 & 0 & 0 & 0 & p_I\mathbf{e}_i \\ 0 & 0 & 0 & 0 & 0 & 0 & 0 & 0 & 0 & 0 & p_I v_I^i \\ 0 & 0 & 0 & 0 & 0 & 0 & 0 & 0 & 0 & 0 & v_I^i \end{pmatrix},$$
$$\tag{68}$$

Table 7 Left and right states of the one-dimensional RMHD Riemann problems

Problem		γ	ρ	$(v_x$	v_y	$v_z)$	p	$(B_x$	B_y	$B_z)$	t_f
RMHD-RP1	$x > 0$	2.0	0.125	0.0	0.0	0.0	0.1	0.5	-1.0	0.0	0.4
	$x \le 0$		1.0	0.0	0.0	0.0	1.0	0.5	1.0	0.0	
RMHD-RP2	$x > 0$	5/3	1.0	-0.45	-0.2	0.2	1.0	2.0	-0.7	0.5	0.55
	$x \le 0$		1.08	0.4	0.3	0.2	0.95	2.0	0.3	0.3	

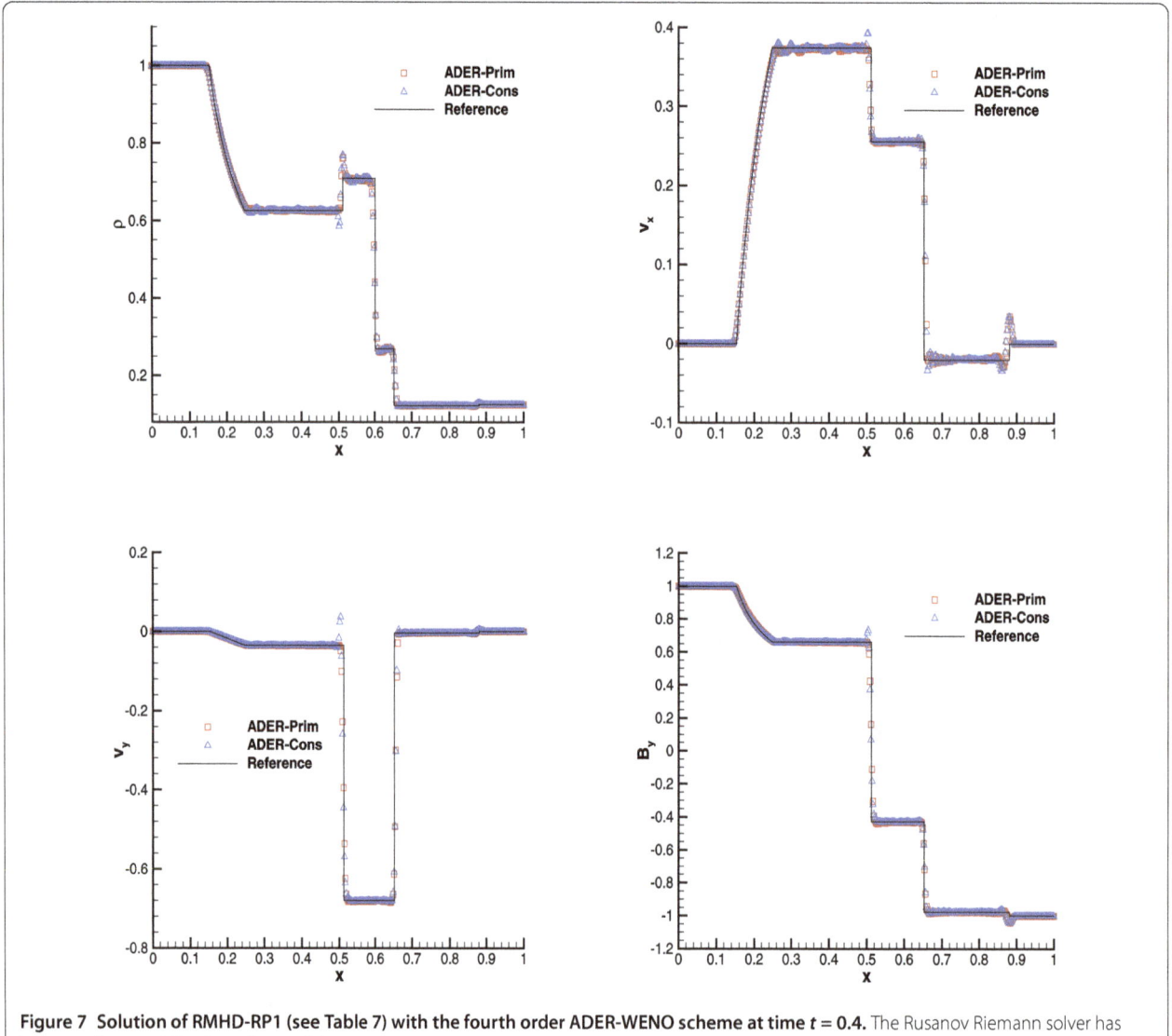

Figure 7 Solution of RMHD-RP1 (see Table 7) with the fourth order ADER-WENO scheme at time *t* = 0.4. The Rusanov Riemann solver has been used over a 400 cells uniform grid.

where $\mathbf{e}_i$ is the unit vector pointing in direction i ($i \in \{x, y, z\}$) and ν and μ are two parameters related to the friction between the phases and to the pressure relaxation.[e]

The equation of state is the so-called stiffened gas equation of state,

$$\epsilon_k = \frac{p_k + \gamma_k \pi_k}{\rho_k (\gamma_k - 1)}, \tag{69}$$

which is a simple modification of the ideal gas EOS and where π_k expresses a reference pressure. For brevity, we have solved this system of equations only for a set of one-dimensional Riemann problems, with initial conditions reported in Table 8. The name of the models, BNRP1, BNRP2, etc., respects the numeration adopted in Dumbser et al. (2010). A reference solution is available for these tests,

and it can be found in Andrianov and Warnecke (2004), Schwendeman et al. (2006), Deledicque and Papalexandris (2007). Each Riemann problem has been solved using a fourth order WENO scheme with 300 cells uniformly distributed over the range $[-0.5; 0.5]$. In Figures 10–14 we have reported the comparison among the solutions obtained with the ADER-Prim, with the ADER-Cons and with the exact solver. In all the tests, with the exception of BNRP2, the ADER-Prim scheme behaves significantly better than the ADER-Cons scheme. On several occasions, such as for v_s and v_g in BNRP1, or for most of the quantities in BNRP5, the solution provided through ADER-Cons manifest evident oscillations, which are instead strongly reduced, or even absent, when the ADER-Prim scheme is used. The CPU time overhead implied by ADER-Prim is comparatively limited, and never larger than $\sim$20 %.

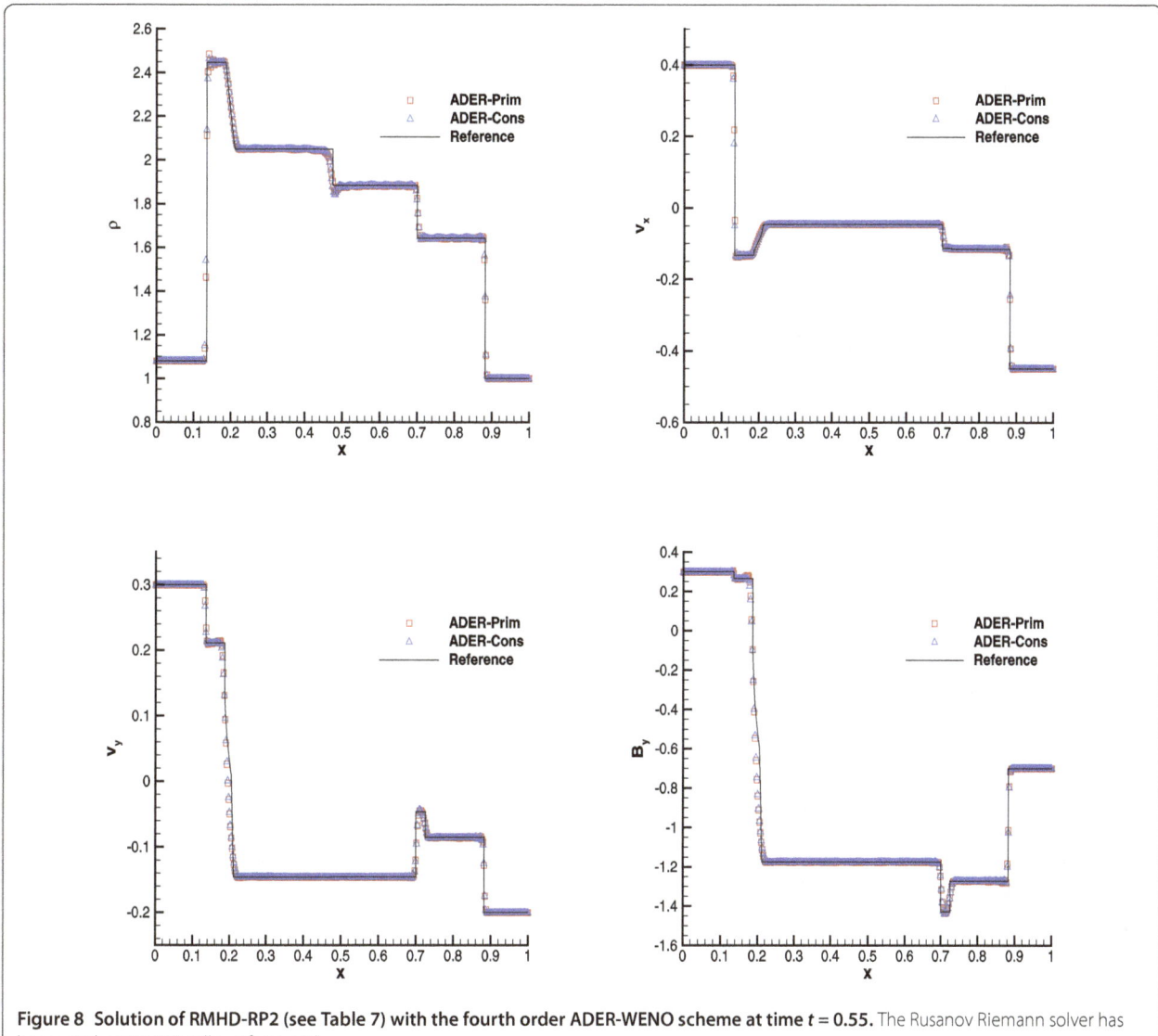

Figure 8 Solution of RMHD-RP2 (see Table 7) with the fourth order ADER-WENO scheme at time *t* = 0.55. The Rusanov Riemann solver has been used over a 400 cells uniform grid.

4 Extension to discontinuous Galerkin and adaptive mesh refinement

Although we have so far concentrated on the implementation of the new ADER-Prim scheme in the context of finite volume methods, the same idea can be extended to discontinuous Galerkin (DG) schemes as well. Incidentally, we note that the interest of computational astrophysics towards DG methods is increasing (Radice and Rezzolla 2011; Teukolsky 2015), and, especially in the relativistic context, they are expected to play a crucial role in the years to come. In a sequence of papers, we have recently developed a class of robust DG schemes which are able to cope even with discontinuous solutions, by incorporating an a posteriori subcell limiter (Dumbser et al. 2014; Zanotti et al. 2015b; Zanotti et al. 2015a). The whole logic

can be briefly summarized as follows. First we assume a *discrete representation* of the solution, in conserved variables, at any given time t^n as

$$\mathbf{u}_h\left(\mathbf{x}, t^n\right) = \sum_{l=0}^{N} \Phi_l(\boldsymbol{\xi})\hat{\mathbf{u}}_l^n = \Phi_l(\boldsymbol{\xi})\hat{\mathbf{u}}_l^n, \quad \mathbf{x} \in T_i, \qquad (70)$$

in which the polynomials

$$\Phi_l(\boldsymbol{\xi}) = \psi_p(\xi)\psi_q(\eta)\psi_r(\zeta) \qquad (71)$$

are built using the spatial Lagrange interpolation polynomials already adopted for the WENO reconstruction. The time evolution of the *degrees of freedom* $\hat{\mathbf{u}}_l^n$ is then obtained after considering the weak form of the governing

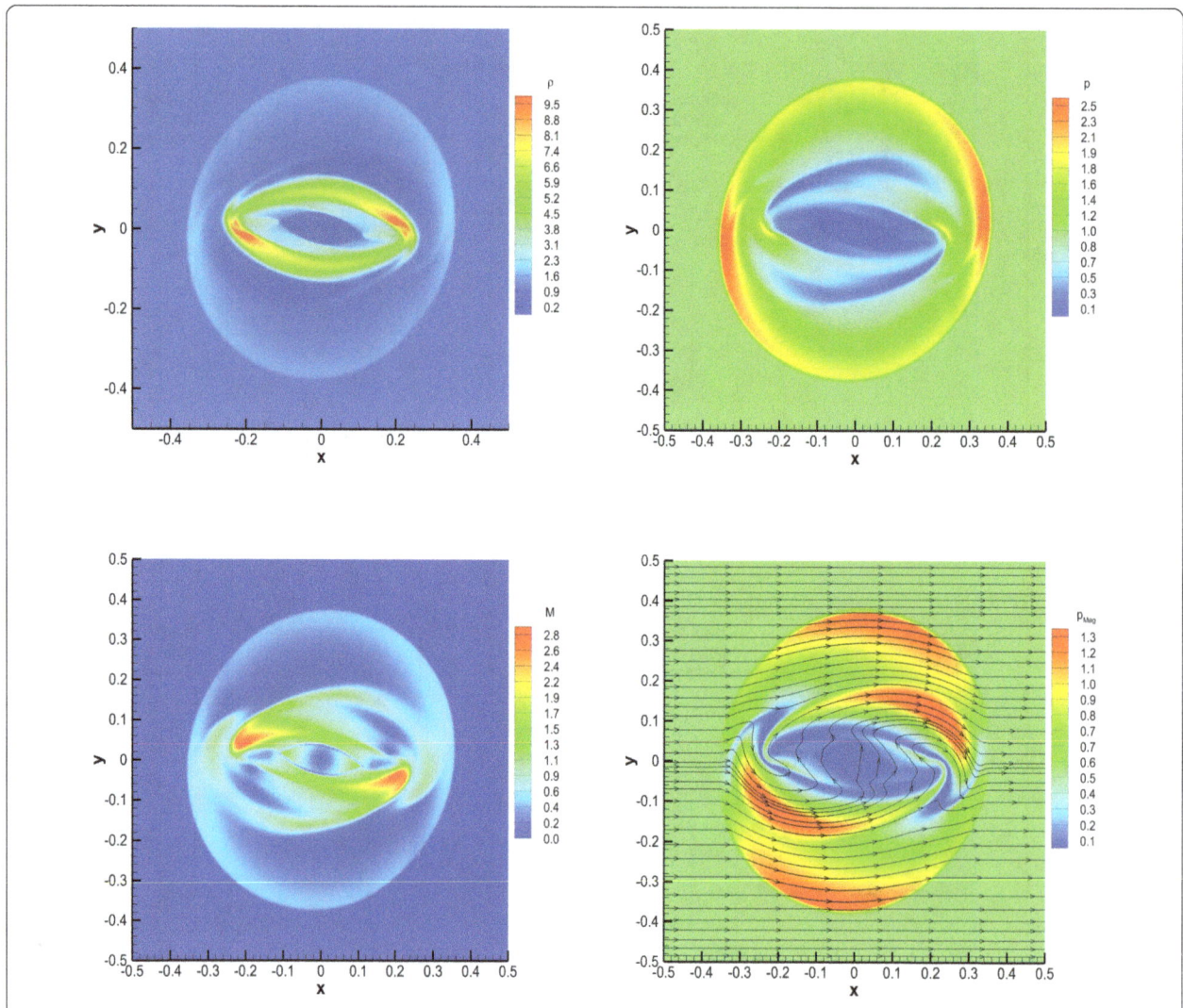

Figure 9 Solution of the RMHD rotor problem at time $t = 0.4$, obtained with the $\mathbb{P}_0\mathbb{P}_3$ scheme on a uniform grid with 300 × 300 cells. Top panels: rest-mass density (left) and thermal pressure (right). Bottom panels: Mach number (left) and magnetic pressure (right).

PDE, which leads to

$$\left(\int_{T_i} \Phi_k \Phi_l \, d\mathbf{x}\right)\left(\hat{\mathbf{u}}_l^{n+1} - \hat{\mathbf{u}}_l^n\right)$$

$$+ \int_{t^n}^{t^{n+1}} \int_{\partial T_i} \Phi_k \left(\tilde{\mathbf{f}}(\mathbf{v}_h^-, \mathbf{v}_h^+) + \frac{1}{2}\mathcal{D}(\mathbf{v}_h^-, \mathbf{v}_h^+)\right) \cdot \mathbf{n} \, dS \, dt$$

$$- \int_{t^n}^{t^{n+1}} \int_{T_i} \nabla \Phi_k \cdot \mathbf{F}(\mathbf{v}_h) \, d\mathbf{x} \, dt$$

$$+ \int_{t^n}^{t^{n+1}} \int_{T_i} \Phi_k \mathbf{B}(\mathbf{v}_h) \cdot \mathbf{M} \nabla \mathbf{v}_h \, d\mathbf{x} \, dt$$

$$= \int_{t^n}^{t^{n+1}} \int_{T_i} \Phi_k \mathbf{S}(\mathbf{v}_h) \, d\mathbf{x} \, dt, \qquad (72)$$

where, just like in Eq. (22), $\tilde{\mathbf{f}}$ denotes a numerical flux function and $\mathcal{D}(\mathbf{v}_h^-, \mathbf{v}_h^+)$ a path-conservative jump term. Obviously, no spatial WENO reconstruction is needed within the DG framework, and the local spacetime DG predictor $\mathbf{v}_h(\mathbf{x}, t)$ entering Eq. (72) will be computed according to the same strategy outlined in Section 2.4.1. T although acting directly over the degrees of freedom $\hat{\mathbf{p}}_l^n$ in primitive variables, which are computed from the degrees of freedom $\hat{\mathbf{u}}_l^n$ in conserved variables simply by

$$\hat{\mathbf{p}}_l^n = \mathbf{V}(\hat{\mathbf{Q}}_l^n), \quad \forall l. \qquad (73)$$

The conversion can be done in such a simple way because we use a *nodal* basis $\Phi_l(\mathbf{x})$. In other words, the degrees of freedom $\hat{\mathbf{u}}_l^n$ in conserved variables are first converted into degrees of freedom $\hat{\mathbf{p}}_l^n$ in primitive variables, which

Table 8 Initial states left (L) and right (R) for the Riemann problems for the Baer-Nunziato equations

	ρ_s	u_s	p_s	ρ_g	u_g	p_g	ϕ_s	t_e
BNRP1 (Deledicque and Papalexandris 2007): $\gamma_s = 1.4$, $\pi_s = 0$, $\gamma_g = 1.4$, $\pi_g = 0$								
L	1.0	0.0	1.0	0.5	0.0	1.0	0.4	0.10
R	2.0	0.0	2.0	1.5	0.0	2.0	0.8	
BNRP2 (Deledicque and Papalexandris 2007): $\gamma_s = 3.0$, $\pi_s = 100$, $\gamma_g = 1.4$, $\pi_g = 0$								
L	800.0	0.0	500.0	1.5	0.0	2.0	0.4	0.10
R	1,000.0	0.0	600.0	1.0	0.0	1.0	0.3	
BNRP3 (Deledicque and Papalexandris 2007): $\gamma_s = 1.4$, $\pi_s = 0$, $\gamma_g = 1.4$, $\pi_g = 0$								
L	1.0	0.9	2.5	1.0	0.0	1.0	0.9	0.10
R	1.0	0.0	1.0	1.2	1.0	2.0	0.2	
BNRP5 (Schwendeman et al. 2006): $\gamma_s = 1.4$, $\pi_s = 0$, $\gamma_g = 1.4$, $\pi_g = 0$								
L	1.0	0.0	1.0	0.2	0.0	0.3	0.8	0.20
R	1.0	0.0	1.0	1.0	0.0	1.0	0.3	
BNRP6 (Andrianov and Warnecke 2004): $\gamma_s = 1.4$, $\pi_s = 0$, $\gamma_g = 1.4$, $\pi_g = 0$								
L	0.2068	1.4166	0.0416	0.5806	1.5833	1.375	0.1	0.10
R	2.2263	0.9366	6.0	0.4890	−0.70138	0.986	0.2	

Values for γ_i, π_i and the final time t_e are also reported.

are then used as initial conditions for the LSDG predictor, i.e.

$$\mathbf{u}_h(\mathbf{x}, t^n) \xrightarrow{Cons2Prim} \mathbf{p}_h(\mathbf{x}, t^n) \xrightarrow{LSDG} \mathbf{v}_h(\mathbf{x}, t),$$
$$t \in [t^n; t^{n+1}]. \tag{74}$$

In those cells in which the main scheme of Eq. (72) fails, either because unphysical values of any quantity are encountered, or because strong oscillations appear in the solution which violate the discrete maximum principle, the computation within the troubled cell goes back to the time level t^n and it proceeds to a complete re-calculation. In practice, a suitable subgrid is generated just within the troubled cell, and a traditional finite volume scheme is used on the subgrid using an alternative data representation in terms of cell averages defined for each cell of the subgrid. This approach and the underlying *a posteriori* MOOD framework have been presented in full details in Clain et al. (2011), Diot et al. (2012), Dumbser et al. (2014), to which we address the interested reader for a deeper understanding.

The resulting ADER-DG scheme in primitive variables can be combined with spacetime adaptive mesh refinement (AMR), in such a way to resolve the smallest details of the solution in highly complex flows. We refer to Zanotti et al. (2015b), Zanotti et al. (2015a) for a full account of our AMR solver in the context of ADER-DG schemes. Here we want to show three representative test cases of the ability of the new ADER-Prim-DG scheme with adaptive mesh refinement, by considering the cylindrical expansion of a blast wave in a plasma with an initially uniform magnetic field (see also Komissarov 1999; Leismann et al. 2005; Del Zanna et al. 2007; Dumbser and Zanotti 2009), as well

as the shock problems of Leblanc, Sedov (1959) and Noh (1987).

4.1 RMHD blast wave problem

At time $t = 0$, the rest-mass density and the pressure are $\rho = 0.01$ and $p = 1$, respectively, within a cylinder of radius $R = 1.0$, while outside the cylinder $\rho = 10^{-4}$ and $p = 5 \times 10^{-4}$. Moreover, there is a constant magnetic field B_0 along the x-direction and the plasma is at rest, while a smooth ramp function between $r = 0.8$ and $r = 1$ modulates the initial jump between inner and outer values, similarly to Komissarov (1999) and Del Zanna et al. (2007).

The computational domain is $\Omega = [-6, 6] \times [-6, 6]$, and the problem has been solved over an initial coarse mesh with 40×40 elements. During the evolution the mesh is adaptively refined using a refinement factor along each direction $\mathfrak{r} = 3$ and two levels of refinement. A simple Rusanov Riemann solver has been adopted, in combination with the $\mathbb{P}_3\mathbb{P}_3$ version of the ADER-DG scheme. On the subgrid we are free to choose any finite volume scheme that we wish, and for this specific test we have found convenient to adopt a second-order TVD scheme. The results for $B_0 = 0.5$ are shown in Figure 15, which reports the rest-mass density, the thermal pressure, the Lorentz factor and the magnetic pressure at time $t = 4.0$. At this time, the solution is composed by an external circular fast shock wave, which is hardly visible in the rest mass density, and a reverse shock wave, which is compressed along the y-direction. The magnetic field is mostly confined between these two waves, as it can be appreciated from the contour plot of the magnetic pressure. The two bottom panels of the figure show the AMR grid (bottom left) and the map of the limiter (bottom right). In the latter we have used the red color to highlight those cells which required the activa-

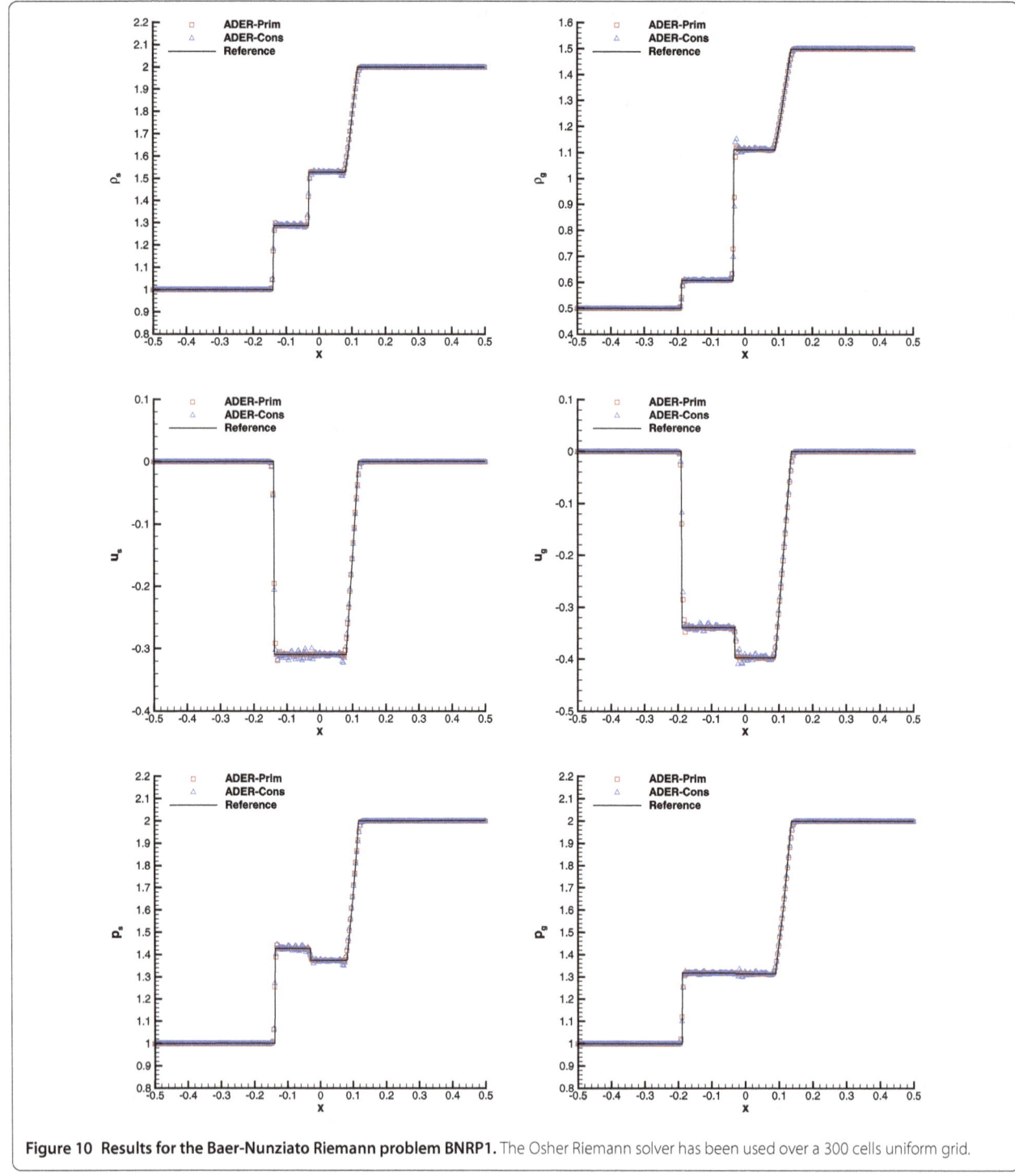

Figure 10 Results for the Baer-Nunziato Riemann problem BNRP1. The Osher Riemann solver has been used over a 300 cells uniform grid.

tion of the limiter over the subgrid, while the blue color is for the regular cells. In practice, the limiter is only needed at the inner shock front, while the external shock front is so weak that the limiter is only occasionally activated. These results confirm the ability of the new ADER-Prim scheme to work also in combination with discontinuous

Galerkin methods, and with complex systems of equations like RMHD.

4.2 Leblanc, Sedov and Noh problem
Here we solve again the classical Euler equations of compressible gas dynamics on a rectangular domain for the Leblanc problem and on a circular domain in the case of

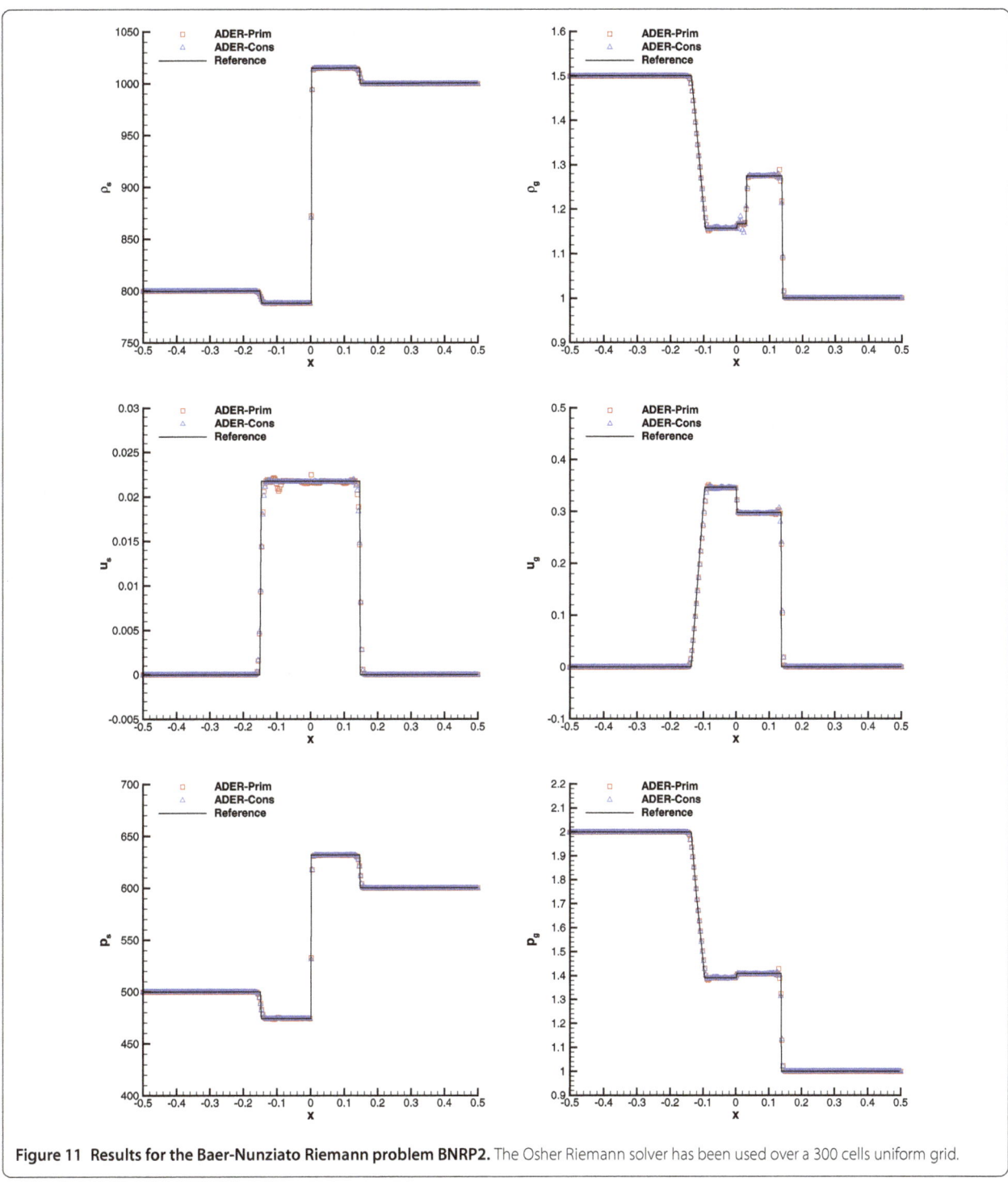

Figure 11 Results for the Baer-Nunziato Riemann problem BNRP2. The Osher Riemann solver has been used over a 300 cells uniform grid.

the shock problems of Sedov and Noh. The initial conditions are detailed in Dumbser et al. (2013), Boscheri et al. (2014b), Boscheri and Dumbser (2014). For the low pressure region that is present in the above test problems, we use $p = 10^{-14}$ for the Leblanc and the Noh problem. The computational results obtained with very high order

ADER-DG $\mathbb{P}_9\mathbb{P}_9$ schemes are depicted in Figures 16, 17 and 18, showing an excellent agreement with the exact solution in all cases, apart from the overshoot in the case of the Leblanc shock tube. We stress that all test problems are extremely severe and therefore clearly demonstrate the robustness of the new approach.

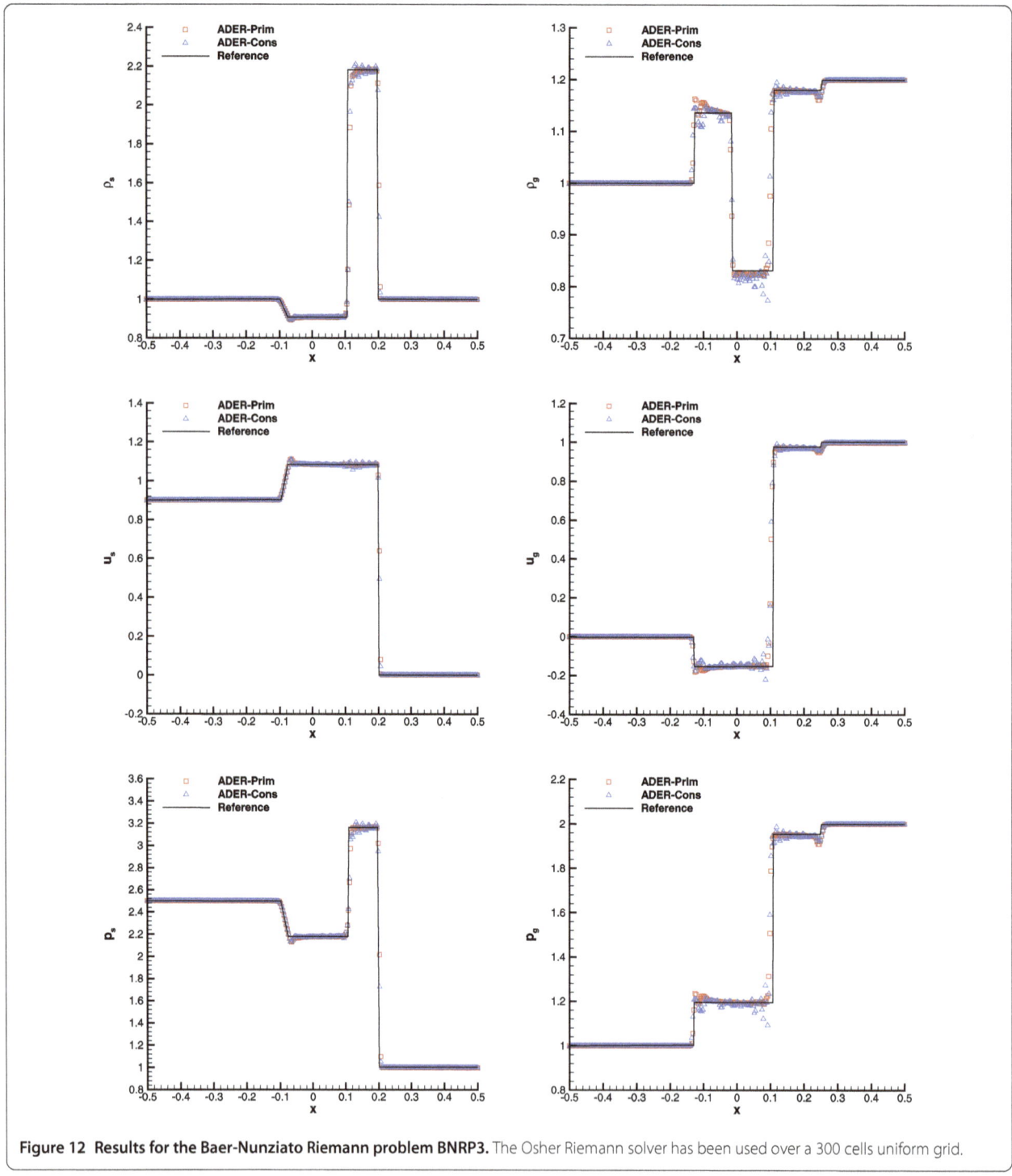

Figure 12 Results for the Baer-Nunziato Riemann problem BNRP3. The Osher Riemann solver has been used over a 300 cells uniform grid.

5 Conclusions

The new version of ADER schemes introduced in Dumbser et al. (2008b) relies on a local space-time discontinuous Galerkin predictor, which is then used for the computation of high order accurate fluxes and sources. This approach has the advantage over classical Cauchy-Kovalewski based

ADER schemes (Toro et al. 2001; Titarev and Toro 2002; Toro and Titarev 2002; Titarev and Toro 2005; Toro and Titarev 2006; Dumbser and Munz 2006; Taube et al. 2007) that it is in principle applicable to general nonlinear systems of conservation laws. However, for hyperbolic systems in which the conversion from conservative to primi-

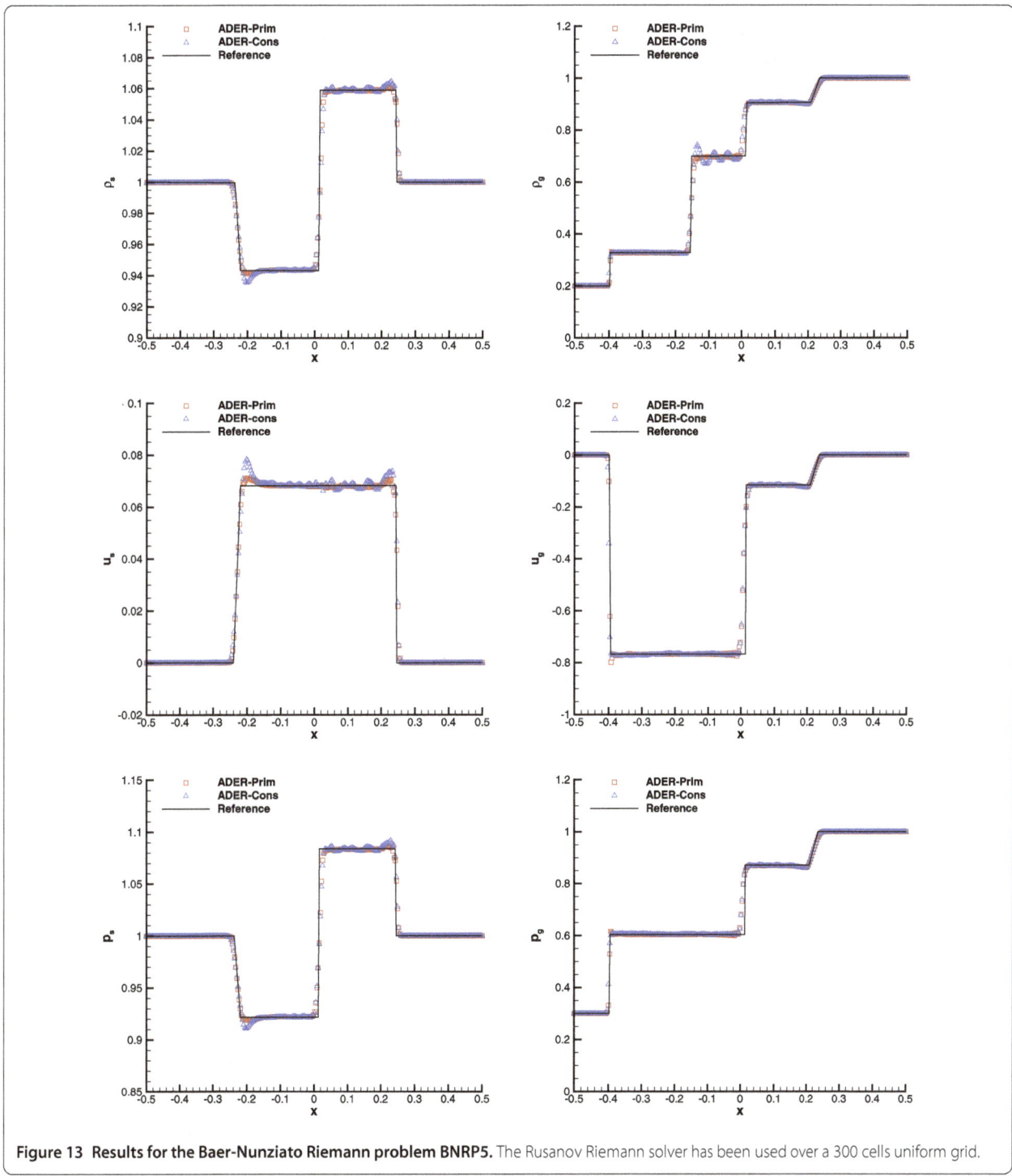

Figure 13 Results for the Baer-Nunziato Riemann problem BNRP5. The Rusanov Riemann solver has been used over a 300 cells uniform grid.

tive variables is not analytic but only available numerically, a large number of such expensive conversions must be performed, namely one for each space-time quadrature point for the integration of the numerical fluxes over the element interfaces and one for each space-time degree of freedom in the local space-time DG predictor.

Motivated by this limitation, we have designed a new version of ADER schemes, valid primarily for finite volume schemes but extendible also to the discontinuous Galerkin finite element framework, in which both the spatial WENO reconstruction and the subsequent local space-time DG predictor act on the primitive variables.

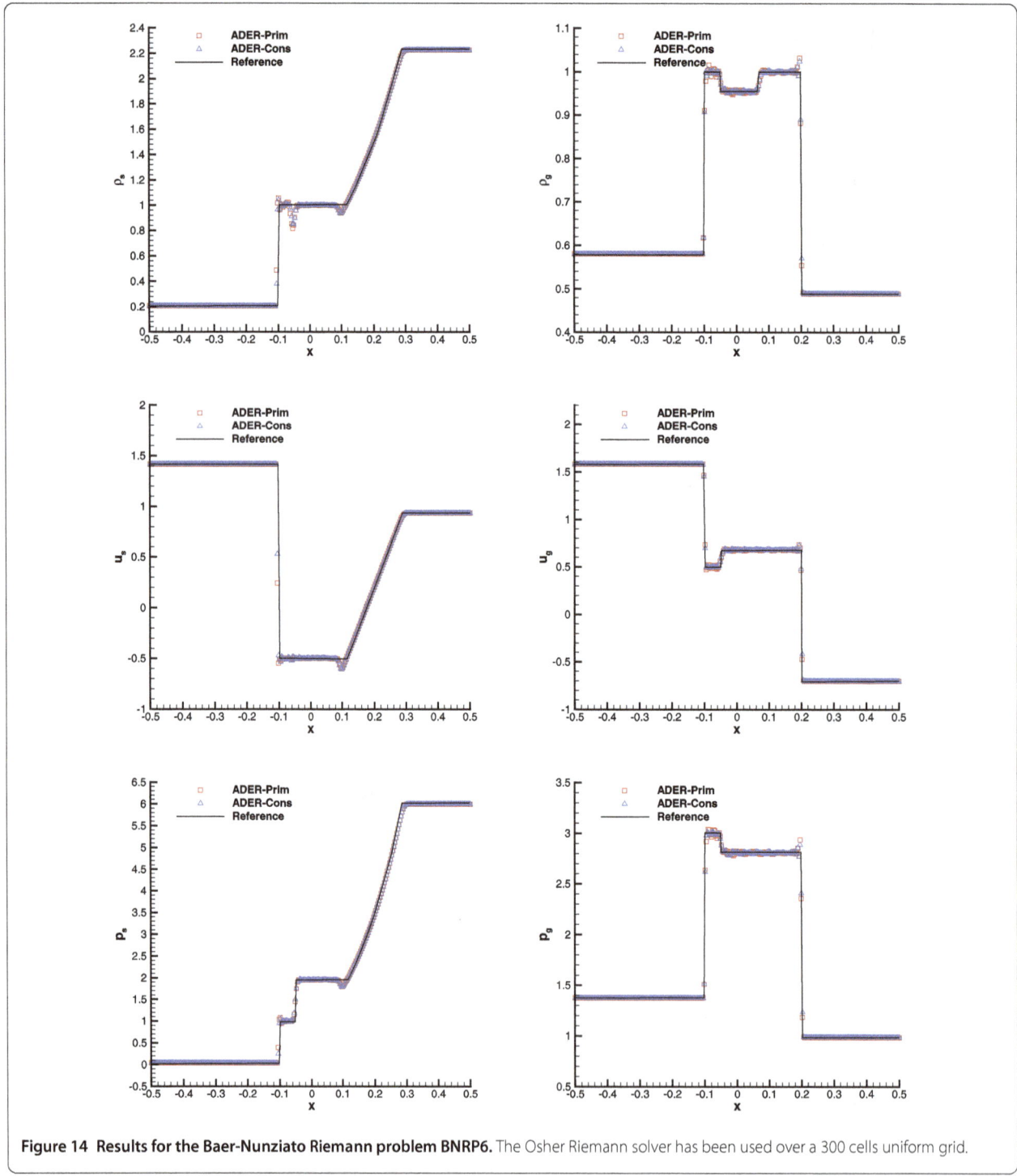

Figure 14 Results for the Baer-Nunziato Riemann problem BNRP6. The Osher Riemann solver has been used over a 300 cells uniform grid.

In the finite volume context this can be done by performing a double WENO reconstruction for each cell. In the first WENO step, piece-wise polynomials of the conserved variables are computed from the cell averages in the usual way. Then, these reconstruction polynomials are simply *evaluated* in the cell centers, in order to obtain *point val-* *ues* of the conserved variables. After that, a single conversion from the conserved to the primitive variable is needed in each cell. Finally, a second WENO reconstruction acts on these point values and provides piece-wise polynomials of the primitive variables. The local space-time discontinuous Galerkin predictor must then be reformulated in a

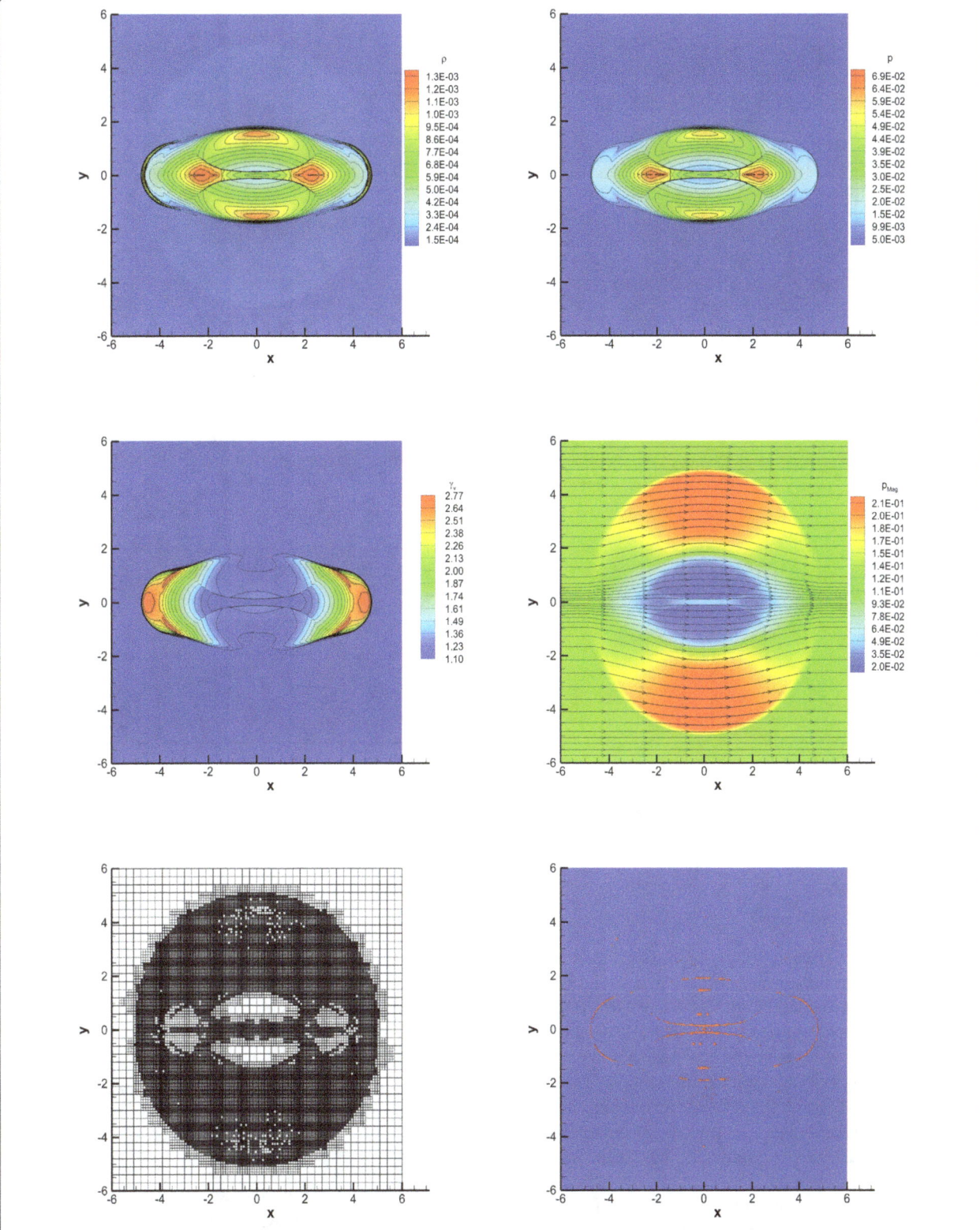

Figure 15 Solution of the RMHD blast wave at time *t* = 4.0, obtained with the ADER-DG $\mathbb{P}_3\mathbb{P}_3$ scheme supplemented with the *a posteriori* second order TVD subcell finite volume limiter. Top panels: rest-mass density (left) and thermal pressure (right). Central panels: Lorentz factor (left) and magnetic pressure (right), with magnetic field lines reported. Bottom panels: AMR grid (left) and limiter map (right) with troubled cells marked in red and regular unlimited cells marked in blue.

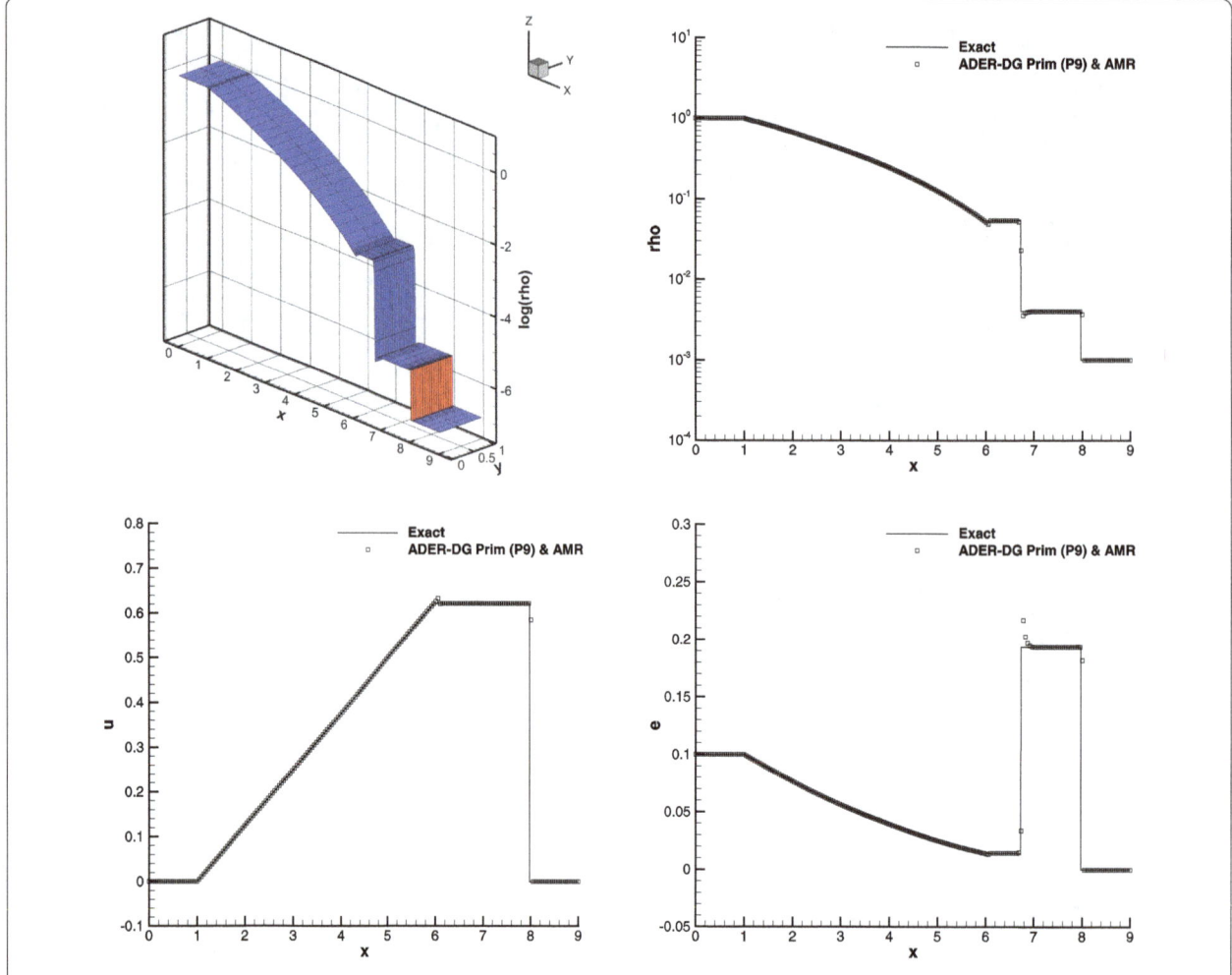

Figure 16 Solution of the Leblanc shock tube problem at time $t = 6.0$, obtained with the ADER-DG $\mathbb{P}_9\mathbb{P}_9$ scheme supplemented with the *a posteriori* second order TVD subcell finite volume limiter. Top left: Troubled cells highlighted in red and unlimited cells in blue. Top right to bottom right: Comparison with the exact solution using a 1D cut through the 2D solution on 200 equidistant sample points for density, velocity and internal energy.

non-conservative fashion, supplying the time evolution of the reconstructed polynomials for the primitive variables.

For all systems of equations that we have explored, classical Euler, relativistic hydrodynamics (RHD) and magnetohydrodynamics (RMHD) and the Baer-Nunziato equations, we have noticed a significant reduction of spurious oscillations provided by the new reconstruction in primitive variables with respect to traditional reconstruction in conserved variables. This effect is particularly evident for the Baer-Nunziato equations. In the relativistic regime, there is also an improvement in the ability of capturing the position of shock waves (see Figure 5). To a large extent, the new primitive formulation provides results that are comparable to reconstruction in characteristic variables.

Moreover, for systems of equations in which the conversion from the conserved to the primitive variables can-

not be obtained in closed form, such as for the RHD and RMHD equations, there is an advantage in terms of computational efficiency, with reductions of the CPU time around $\sim$20 %, or more. We have also introduced an additional improvement, namely the implementation of a new initial guess for the LSDG predictor, which is based on an extrapolation in time, similar to Adams-Bashforth-type ODE integrators. This new initial guess is typically faster than those traditionally available, but it is also less robust in the presence of strong shocks.

We predict that the new version of ADER based on primitive variables will become the standard ADER scheme in the relativistic framework. This may become particularly advantageous for high energy astrophysics, in which both high accuracy and high computational efficiency are required.

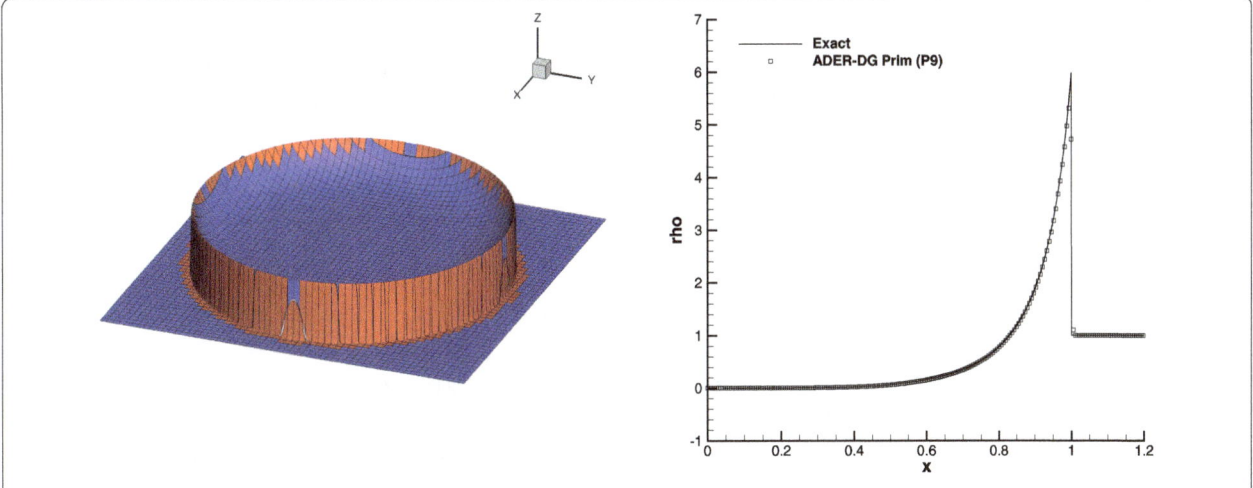

Figure 17 Solution of the Sedov problem at time $t = 1.0$**, obtained with the ADER-DG** $\mathbb{P}_9\mathbb{P}_9$ **scheme supplemented with the** *a posteriori* **second order TVD subcell finite volume limiter.** Left: Troubled cells highlighted in red and unlimited cells in blue. Right: Comparison with the exact solution along the *x*-axis.

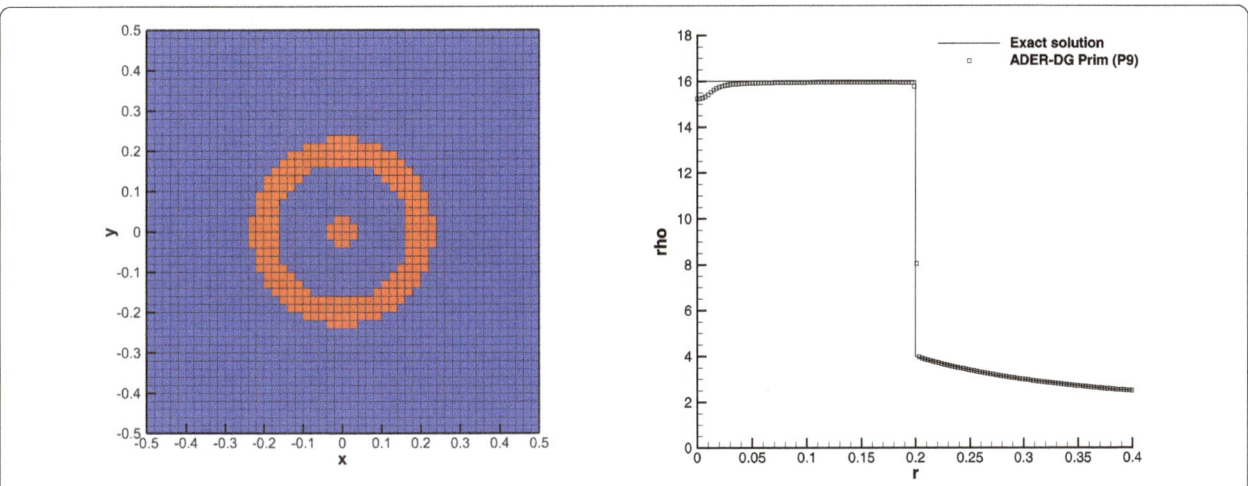

Figure 18 Solution of the Noh problem at time $t = 0.6$**, obtained with the ADER-DG** $\mathbb{P}_9\mathbb{P}_9$ **scheme supplemented with the** *a posteriori* **second order TVD subcell finite volume limiter.** Left: Troubled cells highlighted in red and unlimited cells in blue. Right: Comparison with the exact solution along the *x*-axis.

Competing interests
The authors declare that they have no competing interests.

Authors' contributions
All authors contributed equally to the writing of this paper. All authors read and approved the final manuscript.

Acknowledgements
The research presented in this paper was financed by (i) the European Research Council (ERC) under the European Union's Seventh Framework Programme (FP7/2007-2013) with the research project *STiMulUs*, ERC Grant agreement no. 278267 and (ii) it has received funding from the European Union's Horizon 2020 Research and Innovation Programme under grant agreement no. 671698 (call FETHPC-1-2014, project *ExaHyPE*). We are grateful to Bruno Giacomazzo and Luciano Rezzolla for providing the numerical code for the exact solution of the Riemann problem in RMHD. We would also like to acknowledge PRACE for awarding access to the SuperMUC supercomputer based in Munich (Germany) at the Leibniz Rechenzentrum (LRZ), and ISCRA, for awarding access to the FERMI supercomputer based in Casalecchio (Italy).

Endnotes
a   Since we adopt Cartesian coordinates, $\mathbf{f}^x(\mathbf{Q})$, $\mathbf{f}^y(\mathbf{Q})$, $\mathbf{f}^z(\mathbf{Q})$ express the fluxes along the *x*, *y* and *z* directions, respectively.
b   See Müller et al. (2013) for more sophisticated paths.
c   See Appendix A of Zanotti and Dumbser (2015) for a graphical representation.
d   We note that, since the spacetime is flat and we are using Cartesian coordinates, the covariant and the contravariant components of spatial vectors can be used interchangeably, namely $A_i = A^i$, for the generic vector $\vec{A}$.
e   In the tests below ν and μ are both set to zero.

References

Andrianov, N, Warnecke, G: The Riemann problem for the Baer-Nunziato two-phase flow model. J. Comput. Phys. **212**, 434-464 (2004)

Anile, AM: Relativistic Fluids and Magneto-fluids. Cambridge University Press, Cambridge (1990)

Antón, L, Miralles, JA, Martí, JM, Ibáñez, JM, Aloy, MA, Mimica, P: Relativistic magnetohydrodynamics: renormalized eigenvectors and full wave decomposition Riemann solver. Astron. Astrophys. Suppl. Ser. **188**, 1-31 (2010)

Baer, MR, Nunziato, JW: A two-phase mixture theory for the deflagration-to-detonation transition (DDT) in reactive granular materials. Int. J. Multiph. Flow **12**, 861-889 (1986)

Balsara, D: Total variation diminishing scheme for relativistic magnetohydrodynamics. Astrophys. J. Suppl. Ser. **132**, 83-101 (2001)

Balsara, D, Spicer, D: A staggered mesh algorithm using high order Godunov fluxes to ensure solenoidal magnetic fields in magnetohydrodynamic simulations. J. Comput. Phys. **149**, 270-292 (1999)

Balsara, DS, Meyer, C, Dumbser, M, Du, H, Xu, Z: Efficient implementation of ADER schemes for Euler and magnetohydrodynamical flows on structured meshes - speed comparisons with Runge-Kutta methods. J. Comput. Phys. **235**, 934-969 (2013)

Beckwith, K, Stone, JM: A second-order Godunov method for multi-dimensional relativistic magnetohydrodynamics. Astrophys. J. Suppl. Ser. **193**(1), 6 (2011)

Bodo, G, Mignone, A, Rosner, R: Kelvin-Helmholtz instability for relativistic fluids. Phys. Rev. E **70**(3), 036304 (2004)

Boscheri, W, Dumbser, M: Arbitrary-Lagrangian-Eulerian one-step WENO finite volume schemes on unstructured triangular meshes. Commun. Comput. Phys. **14**, 1174-1206 (2013)

Boscheri, W, Dumbser, M: A direct arbitrary-Lagrangian-Eulerian ADER-WENO finite volume scheme on unstructured tetrahedral meshes for conservative and nonconservative hyperbolic systems in 3D. J. Comput. Phys. **275**, 484-523 (2014)

Boscheri, W, Balsara, DS, Dumbser, M: Lagrangian ADER-WENO finite volume schemes on unstructured triangular meshes based on genuinely multidimensional HLL Riemann solvers. J. Comput. Phys. **267**, 112-138 (2014a)

Boscheri, W, Dumbser, M, Balsara, DS: High order Lagrangian ADER-WENO schemes on unstructured meshes - application of several node solvers to hydrodynamics and magnetohydrodynamics. Int. J. Numer. Methods Fluids **76**, 737-778 (2014b)

Bucciantini, N, Del Zanna, L: General relativistic magnetohydrodynamics in axisymmetric dynamical spacetimes: the X-ECHO code. Astron. Astrophys. **528**, A101 (2011)

Buchmüller, P, Helzel, C: Improved accuracy of high-order WENO finite volume methods on Cartesian grids. J. Sci. Comput. **61**(2), 343-368 (2014)

Buchmüller, P, Dreher, J, Helzel, C: Finite volume WENO methods for hyperbolic conservation laws on Cartesian grids with adaptive mesh refinement. Appl. Math. Comput. (2015). doi:10.1016/j.amc.2015.03.078

Castro, MJ, Gallardo, JM, Parés, C: High-order finite volume schemes based on reconstruction of states for solving hyperbolic systems with nonconservative products. Applications to shallow-water systems. Math. Comput. **75**, 1103-1134 (2006)

Castro, MJ, Gallardo, JM, López, JA, Parés, C: Well-balanced high order extensions of Godunov's method for semilinear balance laws. SIAM J. Numer. Anal. **46**, 1012-1039 (2008a)

Castro, MJ, LeFloch, PG, Muñoz-Ruiz, ML, Parés, C: Why many theories of shock waves are necessary: convergence error in formally path-consistent schemes. J. Comput. Phys. **227**(17), 8107-8129 (2008b)

Castro, MJ, Gallardo, JM, Marquina, A: Approximate Osher-Solomon schemes for hyperbolic systems. Appl. Math. Comput. (2015). doi:10.1016/j.amc.2015.06.104

Clain, S, Diot, S, Loubère, R: A high-order finite volume method for systems of conservation laws - multi-dimensional optimal order detection (MOOD). J. Comput. Phys. **230**(10), 4028-4050 (2011)

Dal Maso, G, LeFloch, PG, Murat, F: Definition and weak stability of nonconservative products. J. Math. Pures Appl. **74**, 483-548 (1995)

Dedner, A, Kemm, F, Kröner, D, Munz, CD, Schnitzer, T, Wesenberg, M: Hyperbolic divergence cleaning for the MHD equations. J. Comput. Phys. **175**, 645-673 (2002)

Del Zanna, L, Bucciantini, N, Londrillo, P: An efficient shock-capturing central-type scheme for multidimensional relativistic flows. II. Magnetohydrodynamics. Astron. Astrophys. **400**, 397-413 (2003)

Del Zanna, L, Zanotti, O, Bucciantini, N, Londrillo, P: ECHO: a Eulerian conservative high-order scheme for general relativistic magnetohydrodynamics and magnetodynamics. Astron. Astrophys. **473**, 11-30 (2007)

Deledicque, V, Papalexandris, MV: An exact Riemann solver for compressible two-phase flow models containing non-conservative products. J. Comput. Phys. **222**, 217-245 (2007)

Diot, S, Clain, S, Loubère, R: Improved detection criteria for the multi-dimensional optimal order detection (MOOD) on unstructured meshes with very high-order polynomials. Comput. Fluids **64**, 43-63 (2012)

Dumbser, M: Arbitrary high order PNPM schemes on unstructured meshes for the compressible Navier-Stokes equations. Comput. Fluids **39**, 60-76 (2010)

Dumbser, M: A diffuse interface method for complex three-dimensional free surface flows. Comput. Methods Appl. Mech. Eng. **257**, 47-64 (2013)

Dumbser, M, Boscheri, W: High-order unstructured Lagrangian one-step WENO finite volume schemes for non-conservative hyperbolic systems: applications to compressible multi-phase flows. Comput. Fluids **86**, 405-432 (2013)

Dumbser, M, Munz, C: Building blocks for arbitrary high order discontinuous Galerkin schemes. J. Sci. Comput. **27**, 215-230 (2006)

Dumbser, M, Toro, EF: A simple extension of the Osher Riemann solver to non-conservative hyperbolic systems. J. Sci. Comput. **48**, 70-88 (2011a)

Dumbser, M, Toro, EF: On universal Osher-type schemes for general nonlinear hyperbolic conservation laws. Commun. Comput. Phys. **10**, 635-671 (2011b)

Dumbser, M, Zanotti, O: Very high order PNPM schemes on unstructured meshes for the resistive relativistic MHD equations. J. Comput. Phys. **228**, 6991-7006 (2009)

Dumbser, M, Balsara, DS, Toro, EF, Munz, C-D: A unified framework for the construction of one-step finite volume and discontinuous Galerkin schemes on unstructured meshes. J. Comput. Phys. **227**, 8209-8253 (2008a)

Dumbser, M, Enaux, C, Toro, EF: Finite volume schemes of very high order of accuracy for stiff hyperbolic balance laws. J. Comput. Phys. **227**, 3971-4001 (2008b)

Dumbser, M, Castro, M, Parés, C, Toro, EF: ADER schemes on unstructured meshes for non-conservative hyperbolic systems: applications to geophysical flows. Comput. Fluids **38**, 1731-1748 (2009)

Dumbser, M, Hidalgo, A, Castro, M, Parés, C, Toro, EF: FORCE schemes on unstructured meshes II: non-conservative hyperbolic systems. Comput. Methods Appl. Mech. Eng. **199**, 625-647 (2010)

Dumbser, M, Uuriintsetseg, A, Zanotti, O: On arbitrary-Lagrangian-Eulerian one-step WENO schemes for stiff hyperbolic balance laws. Commun. Comput. Phys. **14**, 301-327 (2013)

Dumbser, M, Zanotti, O, Hidalgo, A, Balsara, DS: ADER-WENO finite volume schemes with space-time adaptive mesh refinement. J. Comput. Phys. **248**, 257-286 (2013)

Dumbser, M, Hidalgo, A, Zanotti, O: High order space-time adaptive ADER-WENO finite volume schemes for non-conservative hyperbolic systems. Comput. Methods Appl. Mech. Eng. **268**, 359-387 (2014)

Dumbser, M, Zanotti, O, Loubère, R, Diot, S: A posteriori subcell limiting of the discontinuous Galerkin finite element method for hyperbolic conservation laws. J. Comput. Phys. **278**, 47-75 (2014)

Giacomazzo, B, Rezzolla, L: The exact solution of the Riemann problem in relativistic MHD. J. Fluid Mech. **562**, 223-259 (2006)

Hidalgo, A, Dumbser, M: ADER schemes for nonlinear systems of stiff advection-diffusion-reaction equations. J. Sci. Comput. **48**, 173-189 (2011)

Hu, C, Shu, C: Weighted essentially non-oscillatory schemes on triangular meshes. J. Comput. Phys. **150**, 97-127 (1999)

Jiang, G, Shu, C: Efficient implementation of weighted ENO schemes. J. Comput. Phys. **126**, 202-228 (1996)

Jin, S, Xin, Z: Numerical passage from systems of conservation laws to Hamilton-Jacobi equations, and relaxation schemes. SIAM J. Numer. Anal. **35**(6), 2385-2404 (1998)

Kim, J, Balsara, DS: A stable HLLC Riemann solver for relativistic magnetohydrodynamics. J. Comput. Phys. **270**, 634-639 (2014)

Komissarov, SS: On the properties of Alfvén waves in relativistic magnetohydrodynamics. Phys. Lett. A **232**, 435-442 (1997)

Komissarov, SS: A Godunov-type scheme for relativistic magnetohydrodynamics. Mon. Not. R. Astron. Soc. **303**, 343-366 (1999)

Leismann, T, Antón, L, Aloy, MA, Müller, E, Martí, JM, Miralles, JA, Ibáñez, JM: Relativistic MHD simulations of extragalactic jets. Astron. Astrophys. **436**, 503-526 (2005)

Londrillo, P, Del Zanna, L: High-order upwind schemes for multidimensional magnetohydrodynamics. Astrophys. J. **530**, 508-524 (2000)

Loubère, R, Dumbser, M, Diot, S: A new family of high order unstructured MOOD and ADER finite volume schemes for multidimensional systems of hyperbolic conservation laws. Commun. Comput. Phys. **16**, 718-763 (2014)

Mignone, A, Bodo, G: An HLLC Riemann solver for relativistic flows - I. Hydrodynamics. Mon. Not. R. Astron. Soc. **364**, 126-136 (2005)

Mignone, A, Ugliano, M, Bodo, G: A five-wave Harten-Lax-van Leer Riemann solver for relativistic magnetohydrodynamics. Mon. Not. R. Astron. Soc. **393**(4), 1141-1156 (2009)

Müller, LO, Parés, C, Toro, EF: Well-balanced high-order numerical schemes for one-dimensional blood flow in vessels with varying mechanical properties. J. Comput. Phys. **242**, 53-85 (2013)

Muñoz, ML, Parés, C: Godunov method for nonconservative hyperbolic systems. Math. Model. Numer. Anal. **41**, 169-185 (2007)

Munz, CD: On the construction and comparison of two-step schemes for the Euler equations. Notes Numer. Fluid Mech. **14**, 195-217 (1986)

Murrone, A, Guillard, H: A five equation reduced model for compressible two phase flow problems. J. Comput. Phys. **202**, 664-698 (2005)

Noble, SC, Gammie, CF, McKinney, JC, Del Zanna, L: Primitive variable solvers for conservative general relativistic magnetohydrodynamics. Astrophys. J. **641**, 626-637 (2006)

Noh, W: Errors for calculations of strong shocks using an artificial viscosity and an artificial heat flux. J. Comput. Phys. **72**(1), 78-120 (1987)

Palenzuela, C, Lehner, L, Reula, O, Rezzolla, L: Beyond ideal MHD: towards a more realistic modelling of relativistic astrophysical plasmas. Mon. Not. R. Astron. Soc. **394**, 1727-1740 (2009)

Pares, C: Numerical methods for nonconservative hyperbolic systems: a theoretical framework. SIAM J. Numer. Anal. **44**(1), 300-321 (2006)

Parés, C, Castro, MJ: On the well-balance property of Roe's method for nonconservative hyperbolic systems. Applications to shallow-water systems. Math. Model. Numer. Anal. **38**, 821-852 (2004)

Pareschi, L, Puppo, G, Russo, G: Central Runge-Kutta schemes for conservation laws. SIAM J. Sci. Comput. **26**(3), 979-999 (2005)

Perucho, M, Lobanov, AP, Martí, J-M, Hardee, PE: The role of Kelvin-Helmholtz instability in the internal structure of relativistic outflows. The case of the jet in 3C 273. Astron. Astrophys. **456**, 493-504 (2006)

Perucho, M, Hanasz, M, Martí, J-M, Miralles, J-A: Resonant Kelvin-Helmholtz modes in sheared relativistic flows. Phys. Rev. E **75**(5), 056312 (2007)

Pidatella, R, Puppo, G, Russo, G, Santagati, P: Semi-conservative schemes for conservation laws. J. Sci. Comput. (2016, to appear)

Radice, D, Rezzolla, L: Discontinuous Galerkin methods for general-relativistic hydrodynamics: formulation and application to spherically symmetric spacetimes. Phys. Rev. D **84**(2), 024010 (2011)

Radice, D, Rezzolla, L: THC: a new high-order finite-difference high-resolution shock-capturing code for special-relativistic hydrodynamics. Astron. Astrophys. **547**, A26 (2012)

Rezzolla, L, Zanotti, O: An improved exact Riemann solver for relativistic hydrodynamics. J. Fluid Mech. **449**, 395-411 (2001)

Rezzolla, L, Zanotti, O: Relativistic Hydrodynamics. Oxford University Press, Oxford (2013)

Saurel, R, Abgrall, R: A multiphase Godunov method for compressible multifluid and multiphase flows. J. Comput. Phys. **150**, 425-467 (1999)

Schwendeman, DW, Wahle, CW, Kapila, AK: The Riemann problem and a high-resolution Godunov method for a model of compressible two-phase flow. J. Comput. Phys. **212**, 490-526 (2006)

Sedov, LI: Similarity and Dimensional Methods in Mechanics (1959)

Sod, GA: A survey of several finite difference methods for systems of nonlinear hyperbolic conservation laws. J. Comput. Phys. **27**, 1-31 (1978)

Stroud, A: Approximate Calculation of Multiple Integrals. Prentice-Hall, Englewood Cliffs (1971)

Taube, A, Dumbser, M, Balsara, D, Munz, C: Arbitrary high order discontinuous Galerkin schemes for the magnetohydrodynamic equations. J. Sci. Comput. **30**, 441-464 (2007)

Teukolsky, SA: Formulation of discontinuous Galerkin methods for relativistic astrophysics. arXiv:1510.01190

Titarev, V, Toro, E: ADER: arbitrary high order Godunov approach. J. Sci. Comput. **17**(1-4), 609-618 (2002)

Titarev, VA, Toro, EF: ADER schemes for three-dimensional nonlinear hyperbolic systems. J. Comput. Phys. **204**, 715-736 (2005)

Toro, E, Titarev, V: Solution of the generalized Riemann problem for advection-reaction equations. Proc. R. Soc. Lond. **458**, 271-281 (2002)

Toro, E, Millington, R, Nejad, L: Towards very high order Godunov schemes. In: Toro, E (ed.) Godunov Methods. Theory and Applications, pp. 905-938. Kluwer/Plenum Academic Publishers, New York (2001)

Toro, EF: Riemann Solvers and Numerical Methods for Fluid Dynamics, 2nd edn. Springer, Berlin (1999)

Toro, EF, Hidalgo, A: ADER finite volume schemes for nonlinear reaction-diffusion equations. Appl. Numer. Math. **59**, 73-100 (2009)

Toro, EF, Titarev, VA: Derivative Riemann solvers for systems of conservation laws and ADER methods. J. Comput. Phys. **212**(1), 150-165 (2006)

Woodward, P, Colella, P: The numerical simulation of two-dimensional fluid flow with strong shocks. J. Comput. Phys. **54**, 115-173 (1984)

Zanotti, O, Dumbser, M: Numerical simulations of high Lundquist number relativistic magnetic reconnection. Mon. Not. R. Astron. Soc. **418**, 1004-1011 (2011)

Zanotti, O, Dumbser, M: High order numerical simulations of the Richtmyer-Meshkov instability in a relativistic fluid. Phys. Fluids **27**(7), 074105 (2015)

Zanotti, O, Dumbser, M: A high order special relativistic hydrodynamic and magnetohydrodynamic code with space-time adaptive mesh refinement. Comput. Phys. Commun. **188**, 110-127 (2015)

Zanotti, O, Roedig, C, Rezzolla, L, Del Zanna, L: General relativistic radiation hydrodynamics of accretion flows - I. Bondi-Hoyle accretion. Mon. Not. R. Astron. Soc. **417**, 2899-2915 (2011)

Zanotti, O, Fambri, F, Dumbser, M: Solving the relativistic magnetohydrodynamics equations with ADER discontinuous Galerkin methods, a posteriori subcell limiting and adaptive mesh refinement. Mon. Not. R. Astron. Soc. **452**, 3010-3029 (2015a)

Zanotti, O, Fambri, F, Dumbser, M, Hidalgo, A: Space-time adaptive ADER discontinuous Galerkin finite element schemes with a posteriori sub-cell finite volume limiting. Comput. Fluids **118**, 204-224 (2015b)

9

Entropy-limited hydrodynamics: a novel approach to relativistic hydrodynamics

Federico Guercilena[1*], David Radice[2,3] and Luciano Rezzolla[1,4]

Abstract

We present entropy-limited hydrodynamics (ELH): a new approach for the computation of numerical fluxes arising in the discretization of hyperbolic equations in conservation form. ELH is based on the hybridisation of an unfiltered high-order scheme with the first-order Lax-Friedrichs method. The activation of the low-order part of the scheme is driven by a measure of the locally generated entropy inspired by the artificial-viscosity method proposed by Guermond et al. (J. Comput. Phys. 230(11):4248-4267, 2011, doi:10.1016/j.jcp.2010.11.043). Here, we present ELH in the context of high-order finite-differencing methods and of the equations of general-relativistic hydrodynamics. We study the performance of ELH in a series of classical astrophysical tests in general relativity involving isolated, rotating and nonrotating neutron stars, and including a case of gravitational collapse to black hole. We present a detailed comparison of ELH with the fifth-order monotonicity preserving method MP5 (Suresh and Huynh in J. Comput. Phys. 136(1):83-99, 1997, doi:10.1006/jcph.1997.5745), one of the most common high-order schemes currently employed in numerical-relativity simulations. We find that ELH achieves comparable and, in many of the cases studied here, better accuracy than more traditional methods at a fraction of the computational cost (up to ~50% speedup). Given its accuracy and its simplicity of implementation, ELH is a promising framework for the development of new special- and general-relativistic hydrodynamics codes well adapted for massively parallel supercomputers.

Keywords: flux limiters; entropy limited hydrodynamics; numerical methods; central schemes

1 Introduction

Large-scale general-relativistic hydrodynamical numerical simulations have been shown to be a very powerful tool for the study of astrophysical systems involving compact objects such as black holes and neutron stars (Font 2008; Shibata and Taniguchi 2011; Rezzolla and Zanotti 2013; Martí and Müller 2015; Shibata 2016; Baiotti and Rezzolla 2016; Paschalidis 2016). The realisation of such simulations requires dealing with very different physical, mathematical and computational issues. One of the most challenging of such issues, and one that could lead to significant differences on the outcome of resolution-limited simulations, is the choice of the numerical method for the solution of the hydrodynamics equations (one such difference would be e.g., the dephasing of the gravitational waveforms in

binary neutron star merger simulations, where different numerical schemes can lead to different dynamics of the matter bulk, see e.g., Baiotti et al. 2011; Read et al. 2013; Radice et al. 2014a; Thierfelder et al. 2011; Bernuzzi et al. 2012).

The most commonly used methods in this context are collectively known as high-resolution shock-capturing (HRSC) techniques (see Rezzolla and Zanotti 2013 for an introduction). Belonging to this class, which contains both finite-differences and finite-volumes schemes, are e.g., slope limiting methods (e.g., Roe 1985), the piece-wise parabolic method (PPM) (Colella and Woodward 1984; Martí and Müller 1996), the fifth-order monotonicity preserving (MP5) method (Suresh and Huynh 1997), essentially/weighted non-oscillatory (ENO/WENO) methods (Harten et al. 1987; Liu et al. 1994; Jiang and Shu 1996; De Pietri et al. 2016; Bernuzzi and Dietrich 2016) and many others.

HRSC methods are very effective in dealing with shocks and suppressing spurious oscillations, and have been em-

[*]Correspondence: guercilena@th.physik.uni-frankfurt.de
[1]Institut für Theoretische Physik, Goethe Universität, Max-von-Laue-Str. 1, Frankfurt am Main, 60438, Germany
Full list of author information is available at the end of the article

ployed with varying degree of success in astrophysical simulations. In recent times much work has gone into improving these schemes (e.g., by innovative mesh refinement techniques such as in DeBuhr et al. 2015) or moving beyond them; one promising alternative paradigm is that of discontinuous Galerkin methods (see, e.g., Radice and Rezzolla 2011; Bugner et al. 2016; Zanotti et al. 2015; Kidder et al. 2016; Miller and Schnetter 2016). Many of such schemes, however, potentially suffer from a few shortcomings: (i) they are complex to derive and implement, or to extend and modify (e.g., to increase the formal order of accuracy); (ii) they often depend on many coefficients that require some degree of optimisation (e.g., the WENO methods); (iii) they can lead to load imbalance in parallel implementations as a result of their complexity.

In this work we propose a different approach that is able to address some of these shortcomings (especially the points (i) and (iii) previously mentioned), which we refer to as 'entropy-limited hydrodynamics' (ELH) and formulate it in a finite-differences framework. The underlying concept is relatively straightforward: the hydrodynamical fluxes are computed using an unfiltered, high-order stencil, to which a contribution from a low-order, stable numerical flux (the Lax-Friedrichs flux) is added in order to ensure stability. To determine which gridpoints are in need of the low-order contribution, we employ a 'shock detector', which not only marks region of the computational domain requiring the limiter, but also determines the relative weights of the high and low-order fluxes.

The use of a hybrid numerical flux to achieve both accuracy and stability places our method in the class of flux-limiting schemes (see, e.g., the classification in Leveque 1992), which have long been a feature in the panorama of numerical schemes for hydrodynamics. In this context the Lax-Friedrichs flux is a common choice for low-order methods, being monotone and dissipative (a different example of combining high- and low-order methods, in the context of the reconstruction method, would be e.g., Tchekhovskoy et al. 2007).

To drive the activation of the Lax-Friedrichs method, a criterion to flag generically problematic points of the computational domain is needed. Such a criterion is offered by the entropy viscosity function described by Guermond et al. (we refer primarily to Guermond et al. 2011, but see also Guermond and Pasquetti 2008; Zingan et al. 2013), in which the local production of entropy is used to identify shocks. Since entropy is produced only in the presence of shocks, this results in a stable method able to recover high-order in regions of smooth flow. We extend the definition of the entropy viscosity from the classical to the relativistic case, and employ it to drive the flux limiting scheme rather than as a weight to additional viscous terms in the hydrodynamical equations. As a result, and in contrast to the approach by Guermond et al. (2011), we do not modify

the underlying equations of relativistic hydrodynamics by introducing additional entropy-related terms.

In the following we describe the method and the details of our implementation, then report the results of tests we conducted in order to gauge its behaviour against a standard HRSC method, namely, MP5 (Suresh and Huynh 1997). The paper is structured as follows: in Section 2 we briefly summarise the equations of relativistic hydrodynamics and the finite-differences framework we employ. In Section 3 the ELH method and its implementation are described, while the results of the numerical tests are collected in Section 4. We present our conclusions and outlook in Section 5.

In the following we use the spacetime signature $(-, +, +, +)$, with Greek indices running from 0 to 3 and Latin indices from 1 to 3. We also employ the Einstein convention for the summation over repeated indices. Unless otherwise stated, all quantities are expressed in a geometrized system of units in which $c = G = 1$.

2 Relativistic hydrodynamics: theory and numerics overview

2.1 Relativistic hydrodynamics

We summarize here the mathematical framework of relativistic hydrodynamics. In the interest of simplicity, the discussion is limited to special relativity, while the general-relativistic case, which is relevant for the neutron-star tests of Section 4.2, can be found in Appendix 1.

Since most of our interest is in modelling neutron-star matter, we assume it to be described by a perfect fluid, therefore with a corresponding energy-momentum tensor given by

$$T_{\mu\nu} = \rho h u_\mu u_\nu + p \eta_{\mu\nu}, \tag{1}$$

where $\eta_{\mu\nu}$ is the Minkowski metric, ρ is the rest-mass density, u^μ is the fluid four-velocity, p is the pressure, $h = 1 + \epsilon + p/\rho$ is the specific enthalpy and ϵ the specific internal energy (Rezzolla and Zanotti 2013). The equations of motion for the fluid are the conservation of the stress-energy tensor

$$\partial_\mu T^{\mu\nu} = 0, \tag{2}$$

and conservation of rest mass

$$\partial_\mu \left(\rho u^\mu \right) = 0. \tag{3}$$

These two sets of equations are closed by an equation of state (EOS) $p = p(\rho, \epsilon)$, and we here assume a simple ideal-fluid (or Gamma-law) EOS:

$$p = \rho \epsilon (\Gamma - 1), \tag{4}$$

with Γ the adiabatic index.

Equations (2) and (3) can be cast in conservation form and therefore written symbolically as (Martí and Müller 2015)

$$\partial_t \boldsymbol{U} + \partial_i \boldsymbol{F}^i = \boldsymbol{S}, \tag{5}$$

by defining the conserved variables $\boldsymbol{U}$ as

$$\boldsymbol{U} := \begin{pmatrix} D \\ S_j \\ \tau \end{pmatrix}$$

$$:= \begin{pmatrix} \rho W \\ \rho h W^2 v_j \\ \rho h W^2 - p - \rho W \end{pmatrix}. \tag{6}$$

These are functions of the 'primitive' variables (ρ, v_i, p, ϵ). In these expressions and in the following the fluid three-velocity measured by the normal (or Eulerian) observers is defined as $v^i := u^i/W$ and the Lorentz factor is $W := (1 - v^i v_i)^{-\frac{1}{2}} = u^t$. The fluxes $\boldsymbol{F}^i$ are

$$\boldsymbol{F}^i = \begin{pmatrix} v^i D \\ S_j v^i + p \delta_j^i \\ S^i - D v^i \end{pmatrix}, \tag{7}$$

where δ_j^i is the Kronecker delta.

Note that the source functions $\boldsymbol{S}$ are identically zero in special relativity, but this is no longer the case in a generic spacetime, where metric-dependent terms appear both in the fluxes and sources (see Appendix 1).

2.2 Numerical methods

The ELH method proposed here has been implemented in the code WhiskyTHCEL as a variant of the WhiskyTHC code (Radice and Rezzolla 2012; Radice et al. 2014a, 2014b) based on the Einstein Toolkit (Löffler et al. 2012; Zilhão and Löffler 2013; Einstein Toolkit 2010). WhiskyTHC implements both finite-difference and finite-volume methods applied to a characteristic-variables decomposition with a Lax-Friedrichs flux-splitting for upwinding. It also crucially provides a positivity preserving limiter to cope with large rest-mass density jumps, e.g., as those appearing across the surface of compact stars. In the following we summarise the main components of the underlying algorithm and refer the interested reader to Radice and Rezzolla (2012) and Radice et al. (2014a) for a more detailed description.

Given a discrete mesh with Δ being the spatial grid spacing, the finite-difference algorithm we employ provides an estimate for the right-hand-side of an evolution equation in flux-conservative form as

$$\partial_t U|_i = -\frac{f_{i+1/2} - f_{i-1/2}}{\Delta} + S_i, \tag{8}$$

i.e., as a difference between numerical fluxes at the cell interfaces, plus the sources contribution (to simplify the notation, here U is any one of the components of (6)).

The numerical fluxes $f_{i\pm1/2}$ are obtained via a reconstruction operator, i.e., an operator yielding a high-order approximation of a generic function at a given point using its volume averages (see, e.g., Leveque 1992; Rezzolla and Zanotti 2013; Martí and Müller 2015). Out of the variety of reconstruction operators available in the literature and implemented in WhiskyTHC, we focus here on the fifth-order and seventh-order unfiltered stencils

$$^5\mathcal{S}^- := \frac{1}{60}(2f_{i-2} - 13f_{i-1} + 47f_i$$
$$+ 27f_{i+1} - 3f_{i+2}) \tag{9}$$

and

$$^7\mathcal{S}^- := -\frac{1}{140}f_{i-3} + \frac{5}{84}f_{i-2} - \frac{101}{420}f_{i-1} + \frac{319}{420}f_i$$
$$+ \frac{107}{210}f_{i+1} - \frac{19}{210}f_{i+2} + \frac{1}{105}f_{i+3}, \tag{10}$$

returning the value of the flux at $x_{i+1/2}$, and which we refer to as U5 and U7, respectively (we have only written here the left-biased operators $\mathcal{S}^-$, since the right-biased ones $\mathcal{S}^+$ are symmetric).[a]

Furthermore, we select the MP5 scheme as our benchmark against which to test the properties of the EL method. MP5 is built on top of the U5 stencil, but the resulting fluxes are limited so as to preserve monotonicity near discontinuities (Suresh and Huynh 1997; Mignone et al. 2010; Radice and Rezzolla 2012). MP5 offers a good compromise between robustness and accuracy and it has been successfully employed in several realistic scenarios in which it also achieved high convergence-order (Radice et al. 2014a, 2014b; Radice et al. 2015). It has therefore become a *de facto* standard in our work, hence we use it as a reference.

To ensure the stability of the scheme, the reconstruction must be appropriately upwinded, i.e., a right- (left-) biased operator has to be applied to the left- (right-) going part of the flux. We therefore split the flux f in a right-going flux f^+ and a left-going one f^-, so that $f = f^+ + f^-$. The splitting is performed using the Lax-Friedrichs or Rusanov flux splitter (Shu 1997), i.e.,

$$f^{\pm} := f(U) \pm \kappa U, \quad \kappa := \max|f'(U)|, \tag{11}$$

where the maximum is taken over the stencil of the reconstruction operator.

The reconstruction procedure outlined above can be applied on each equation in the system (5) (this is also called a components split) or to its local characteristic variables (in which case it is referred to as a characteristics split).

A further ingredient in our algorithm is a so-called 'positivity-preserving' limiter (Hu et al. 2013; Radice et al. 2014a). The basic idea is to split the numerical flux in Eq. (8) in two contributions, as:

$$f_{i+1/2} = \theta f_{i+1/2}^{HO} + (1 - \theta) f_{i+1/2}^{LF}, \tag{12}$$

where f^{HO} is the original high-order flux, while the Lax-Friedrichs flux f^{LF} has the standard form

$$f_{i+1/2}^{LF} := \frac{1}{2}(f_i + f_{i+1}) - \frac{\kappa}{2}(U_i - U_{i+1}) \tag{13}$$

and κ is defined as in Eq. (11). For a single Euler step, the result of the evolution of U can be explicitly written as

$$U_i^{n+1} = \frac{1}{2}\left(U_i^n + 2\lambda f_{i-1/2}\right) + \frac{1}{2}\left(U_i^n - 2\lambda f_{i+1/2}\right), \tag{14}$$

where the fluxes are defined as in Eq. (12) and λ depends on the maximum propagation speed of the system as well as the CFL factor. The value of θ is defined as the one that makes both terms of Eq. (14) positive. Applied to the continuity equation this guarantees that the density never becomes negative (see Wu 2017 for a way to generally ensure the physicality of the fluid state-vector in a generic spacetime).[b]

We note that this algorithm does not free us from having to employ an artificial floor (or atmosphere) to treat (ideally) vacuum regions: these are filled with a uniform and tenuous fluid with rest-mass density ρ_{atmo}. Whenever in the subsequent evolution the rest-mass density of a grid-point falls below the floor value ρ_{atmo}, it is reset to the floor value, its three-velocity is set to zero and the specific internal energy is set to the corresponding value coming from the EOS. In neutron star simulations we fix the floor at $\rho_{atmo} = 10^{-16}\, M_\odot^{-2}$, i.e. the typical value of ρ_{atmo}/ρ_{max} is $\sim 10^{-13}$.

Details of the algorithms we employ to evolve the spacetime and couple it to the fluid evolution are given in Appendix 2.

The last step in the algorithm is the actual time evolution. Since after the spatial discretization, the original PDEs to be solved are in the form of a coupled systems of ODEs, this is taken care of in a method-of-lines (MOL) fashion by means of a fixed step Runge-Kutta time integrator. We employ either the standard fourth-order Runge-Kutta method (RK4) or a third-order (RK3, see Gottlieb et al. 2009) method with strong stability preserving (SSP) properties.

3 Entropy-limited hydrodynamics

3.1 Description of the scheme

The scheme we propose consists of two building blocks: (1) a function detecting shocks; (2) a limiter scheme of the high-order fluxes. We will start discussing the latter.

As customary in flux-limiting schemes, we modify the high-order approximation of the flux by combining (or 'hybridising') it with a (local) Lax-Friedrichs flux contribution as in (12).

Of course, the hybridisation of the high-order flux with the Lax-Friedrichs should be activated only in regions of the flow that are problematic. In order to flag such regions we introduce a regularisation function that we refer to as the 'viscosity' ν. Hence, we *redefine* the parameter $\theta \in [0, 1]$ in Eq. (12) in terms of the quantity

$$\theta := \min\left[\tilde{\theta}, 1 - \frac{1}{2}(\nu_i + \nu_{i+1})\right], \tag{15}$$

so that the contribution of the Lax-Friedrichs flux grows linearly with the viscosity. The value of the coefficient $\tilde{\theta}$ is the one mentioned in the previous section to guarantee the positivity of the rest-mass density. With the choice (15) for the limiting coefficient θ, additional dissipation is inserted when ν attains large values as well as in near-vacuum regions. On the other hand, in regions where the flow is smooth and away from near vacuum, θ is close to unity, ensuring the use of the high-order flux and preserving the high accuracy of the method.

We still need to associate the viscosity ν to some property of the flow. To this scope we take inspiration from the work of Guermond and collaborators (Guermond et al. 2011) and associate ν to the specific entropy s. In general, the precise functional form of s will depend on the EOS, but for the simpler case of a perfect fluid with an ideal-fluid EOS (cf., Eq. (4)), the specific entropy can be shown to be equal to (apart from constant multiplicative factors) to (Rezzolla and Zanotti 2013)

$$s = \log\left(\frac{\epsilon}{\rho^{\Gamma-1}}\right). \tag{16}$$

Of course, the specific entropy must satisfy the second law of thermodynamics, so that we can introduce the *entropy residual*, or entropy-production rate, $\mathcal{R}$ as

$$\mathcal{R} := \partial_\mu\left(\rho s u^\mu\right) \geq 0. \tag{17}$$

As a result, the computation of the entropy residual $\mathcal{R}$, effectively represents the first step in defining the viscosity and hence the root to limiter parameter θ.

The expected behaviour of the entropy residual is that it cannot decrease in time and that is spatially restricted to very small regions in the neighbourhood of shocks, ideally expressed a delta function peaked at the location $\boldsymbol{x}_s$ of shocks, i.e., $\mathcal{R} \propto \delta(\boldsymbol{x} - \boldsymbol{x}_s)$. A physical justification for this latter expectation is rather simple to motivate. Euler equations generally apply to perfect fluids, and while they can capture non-ideal features (i.e., shocks), the description of the latter is only approximate. As long as the flow is smooth and the perfect-fluid approximation holds, all phenomena are reversible and there can be no production of entropy. However, in those regimes where the perfect-fluid approximation breaks down and non-ideal effects appear, namely, at the location of shocks, the entropy production is nonzero and the entropy jumps locally to a higher value. Since shocks are regions of dimension $N-1$ in spatial manifolds with N spatial dimensions, the entropy residual $\mathcal{R}$ must be a Dirac delta peaked at shock locations for it to provide a finite contribution.

To seal the strict connection between the entropy viscosity ν_e and the entropy residual and to embody the property that this quantity should be a function of the spatial discretization, we define it as

$$\nu_e := c_e \Delta |\mathcal{R}|. \tag{18}$$

The absolute value is used for the entropy residual since the inequality (18) is not guaranteed to be satisfied during the numerical integration. In fact, $\mathcal{R}$, having to approximate a delta function, is expected to show an oscillatory behaviour with potential negative values in practical numerical applications. In expression (18), Δ is the spacing of the mesh, c_e is a positive tunable constant with dimensions of $[\text{time}]^3 \times [\text{temperature}] \times [\text{mass}]^{-1}$, so that ν_e is effectively dimensionless. We note that despite ν_e not having the dimensions of a viscosity, we still refer to as the 'entropy viscosity', mostly for convenience and in analogy with the very similar quantity defined in Guermond et al. (2011).

An additional benefit of this definition is the ability of the resulting scheme in differentiating automatically between shocks and contact discontinuities. This follows from the fact that at contact discontinuities there is no entropy production and therefore the viscosity there would be zero as well (Rezzolla and Zanotti 2013).

A potential problem of the definition (18) is that it can lead to an unbounded value since the entropy residual $\mathcal{R}$ is not physically upper limited. In our case the value of θ cannot however exceed unity, and so the viscosity must not exceed this value as well. To enforce this requirement and cut-off excessively large values of the entropy viscosity we set the entropy viscosity to be used in the limiter (15) as

$$\nu := \min[\nu_e, c_{\max}], \tag{19}$$

where ν_e is given by Eq. (18). Here, $c_{\max}$ is a tunable dimensionless coefficient, which, together with the other tunable coefficient c_e, we have assumed to be equal to one in all of the tests presented here. As we will comment in Section 4, the results are not very sensitive to the choices made for these coefficients.

3.2 Numerical implementation

In our numerical implementation we compute the entropy residual (17) by first rewriting its definition in a way that involves only derivatives of the specific entropy s

$$\mathcal{R} = \partial_\mu (s \rho u^\mu) = s \partial_\mu (\rho u^\mu) + \rho u^\mu \partial_\mu s = \rho u^\mu \partial_\mu s, \tag{20}$$

where the continuity equation (3) was used to obtain the final expression in (20). By expressing the 4-velocity u^μ in terms of the fluid three-velocity v^i, we finally write the residual as

$$\mathcal{R} = \rho W (\partial_t s + v^i \partial_i s). \tag{21}$$

The spatial derivatives in Eq. (21) are approximated with a standard centered finite-difference stencil of order $p+1$, where p is the order of the stencil used to approximate the physical fluxes. This restriction arises from the need to ensure that the viscosity converges to zero fast enough not to spoil the overall convergence of the scheme at the nominal order. The time derivative in (21) is also approximated by finite differencing. In particular, at every iteration we use the current value of the specific entropy and the values at the two previous timesteps to compute a second-order approximation of $\partial_t s$ via a one-sided stencil, i.e., as

$$(\partial_t s)^n = \frac{1}{2\Delta t} (3s^n - 4s^{n-1} + s^{n-2}) + \mathcal{O}((\Delta t)^2). \tag{22}$$

A few remarks are useful at this point. First, the time derivative of the specific entropy in Eq. (21) is computed with a low-order method and this could in principle be a limiting factor for the convergence properties of the overall scheme. In practice, however, we find that the space discretization error dominates over the error on the time derivative in the tests we have performed, so that the scheme achieves high-order convergence as expected despite the use of a low-order approximation for $\partial_t s$. Second, the high-order flux f^{HO} is computed component by component. In fact since the reconstruction operators U5 and U7 ((9) and (10)) are linear, they commute with the matrices used to perform the characteristic decomposition, and there is therefore no difference in this case between component-by-component and characteristic decomposition. This contributes (along with other intrinsic differences in the formulation of the schemes) to a significant speed-up of the code (up to ~50%, depending on the setup of the grid on the computing nodes) with respect to MP5, since there is no need to compute the system eigenvectors and apply the resulting matrix. By contrast, the MP5 reconstruction is nonlinear and does not commute with the characteristic decomposition. As a result, when using MP5 we always switch to characteristic variables, since this is known to reduce spurious numerical oscillations in high-order methods (Suresh and Huynh 1997).

Two further operations are applied on the viscosity before it is used in Eq. (15). Firstly, since the viscosity is found to be very close to zero in regions of very low rest-mass density, we improve the behaviour close to atmosphere values by simply setting the viscosity to some small and constant value ν_v at a given point x_{ijk} whenever the rest-mass density at the given point, and at all nearest neighbours, is below a certain threshold ρ_v, i.e., if

$$\rho_{i+l,j+m,k+n} < \rho_v \quad \forall l, m, n = -1, 0, 1$$

then

$$\nu_{ijk} = \nu_v.$$

In practice, therefore, (19) needs to be slightly revised and the expression for the entropy viscosity we actually implement in the numerical code is

$$\nu := \begin{cases} \nu_v & \text{if } \rho < \rho_v, \\ \min[\nu_e, \nu_{max}] & \text{elsewhere.} \end{cases} \quad (23)$$

In all of the numerical tests presented we have used $\nu_v = 10^{-12}$ and $\rho_v = 10^{-11} \, M_\odot^{-2}$ (i.e., the threshold is 5 orders of magnitude larger then the atmosphere floor, $\rho_{atmo} = 10^{-16} \, M_\odot^{-2}$). Secondly, and following the original implementation in Guermond et al. 2011, we introduce a smoothing step which removes unwanted oscillations in the viscosity. This is accomplished by applying a five-point stencil of the form

$$\bar{\nu}_{ijk} := \sum_{l=-2}^{2} \sum_{m=-2}^{2} \sum_{n=-2}^{2} a_l a_m a_n \nu_{i+l,j+m,k+n}, \quad (24)$$

where the coefficients a_l have values $a_0 = 0.58$, $a_{\pm 1} = 0.06$ and $a_{\pm 2} = 0.15$. This stencil in Eq. (24) is constructed so to approximate the convolution of the viscosity with a Gaussian kernel of characteristic cutoff length scale 4 times the grid spacing, in such a way that the residual of the transfer function of the target filter and of its approximation is minimised over a broad range of wavelengths (see Sagaut and Grohens 1999 for details).

In addition to dampening oscillations in the viscosity mentioned in the previous section, the necessity of the smoothing step stems from the fact that the viscosity is computed once at the beginning of every new timestep before its value is used in (15). Since the viscosity is kept constant during the series of Runge-Kutta internal steps, it 'lags behind' in time with respect to the solution. This issue is addressed by the smoothing procedure, but in practice we have found that this does not represent a problem in our tests. The smoothing (24) also prevents the viscosity to plunge to very small values where it should instead be

non negligible. This can happen, e.g., close to stellar surfaces as a result of oscillations in the solution. The application of the smoothing operator removes this problem by joining seamlessly the values of the viscosity in the neighbouring points.

4 Numerical tests

We report in this section the results of some of the tests obtained with the ELH method described in the previous sections. In all tests we compare the ELH results with those obtained using the monotonicity preserving, fifth-order scheme (MP5). In particular, unless otherwise stated, we couple the ELH method to the fifth-order U5 stencil (9), to make a fair and sensible comparison between methods of the same order. In few cases, however, we will also report results obtained with the seventh-order stencil U7 (10). We will refer to the corresponding schemes as to EL5 and EL7, respectively. Finally, it is useful to remark that in all of the following tests no attempt was made to tune the coefficients c_e and c_{max} introduced in Section 3.1, and that have been set to unity here. Despite this very simple choice, the ELH method is stable and accurate in all cases considered, as the following sections make clear. At the same time, we consider it possible (if not likely) that the results could be further improved after a careful exploration of the changes in the solution upon a change of c_e and c_{max}; we will leave this exploration to a future work.

4.1 Special-relativistic tests

We begin with a series of mostly one-dimensional tests, performed in special-relativistic hydrodynamics, so that the metric $g_{\mu\nu}$ is fixed to the flat Minkowski metric $\eta_{\mu\nu}$ and no spacetime evolution is performed. Also, since we are mostly interested in the behaviour of the scheme in realistic astrophysical applications, we focus on just a handful of one-dimensional test cases: a smooth nonlinear wave and three shock-tube tests.

4.1.1 Smooth nonlinear wave

We first show the accuracy of the scheme in the case of a smooth solution and measure rigorously its convergence order so as to show that the entropy-driven limiter does not spoil the convergence properties of the high-order method upon which it is built. This test has been discussed in Radice and Rezzolla (2012) (which adapted it from Zhang and MacFadyen 2006). In short, we consider a one-dimensional, large-amplitude, smooth, nonlinear wave with initial rest-mass density profile given by

$$\rho_0(x) = \begin{cases} 1 + \exp[-1/(1 - x^2/L^2)] & \text{if } |x| < 1, \\ 1 & \text{elsewhere,} \end{cases} \quad (25)$$

where $L = 0.3$. The initial data employs a polytropic EOS, $p = K\rho^{\tilde{\gamma}}$, with $K = 100$ and $\tilde{\gamma} = 5/3$, and we then evolve

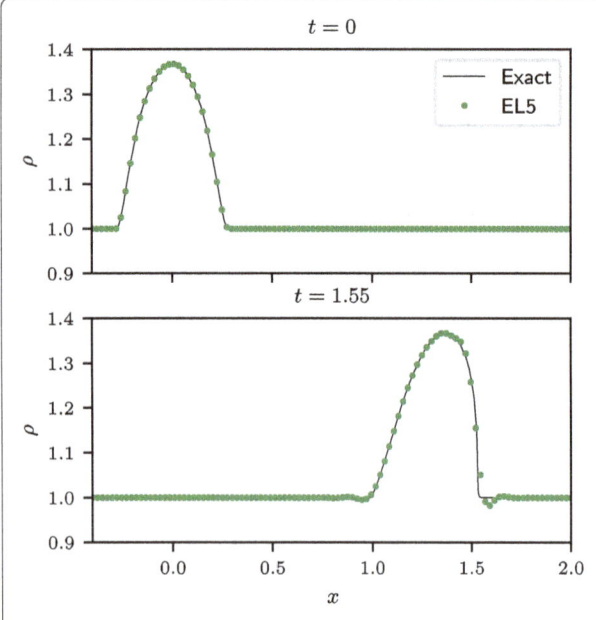

Figure 1 Smooth wave test: rest-mass density profiles. Rest-mass density profiles for the smooth nonlinear wave test. The EL5 data shown corresponds to the coarsest resolution of 100 gridpoints over the domain.

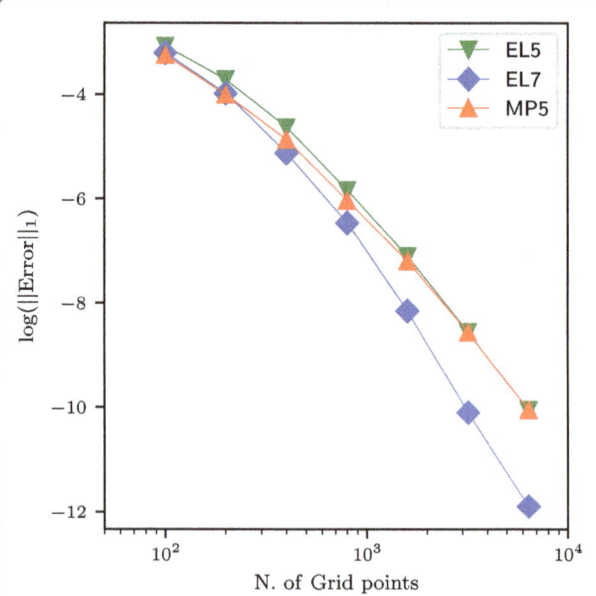

Figure 2 Smooth wave test: L_1-norm of error. L_1-norm of the error on the rest-mass density for the smooth nonlinear-wave test at time $t = 0.8$.

it with the ideal-fluid EOS (4) with $\Gamma = 5/3$. Since in this test discontinuities are absent (so that $\tilde{\gamma} = \Gamma$) and there are no stability issues, we use as time integrator the standard fourth-order RK4 method with a timestep of ~ 0.13 times the grid spacing.

The analytic solution of the test (shown in Figure 1 with a black solid line) is represented by a wave profile propagating towards the right and 'steepening' in the direction of its motion. At time $t_c \simeq 1.6$, the wave develops a shock, leading to a sharp discontinuity. Up to the formation of the caustic, the analytic solution can be computed using the method of characteristics (Anile 1990) on a Lagrangian grid. We obtain an accurate enough approximation by computing it on a very fine grid of 10^5 gridpoints and interpolating the solution using cubic splines on the Eulerian grid. This solution, which we refer to as the 'exact' solution, is then used as the reference against which the numerical solutions are compared.

We perform this test with both the EL5 and EL7 schemes of the ELH method to validate that high-order schemes can be employed with great ease in our approach by simply swapping a lower-order stencil for a higher-order one; this operation is far more demanding in standard finite-volume HRSC schemes.

Figure 2 shows the L_1-norm of the error with respect to the analytic solution at time $t = 0.8$ for the various schemes and at various resolutions. The latter are parametrized by the number of gridpoints used on the x axis, and we

have considered seven different resolutions, each twice as fine as the preceding one, going from 100 gridpoints up to 6,400. The different lines in Figure 2 show that at the lowest resolutions all schemes show very similar errors, MP5 being the most accurate by a small margin. As the resolution is increased, however, the gap in accuracy between EL5 and MP5 decreases and disappears at very high resolutions. The error curve of EL7, being a higher-order scheme, decreases much more rapidly with resolution, so that its error at the highest resolution of 6,400 gridpoints is two orders of magnitude lower than for the fifth-order schemes.

We also compute the convergence order of the various schemes using the data at resolutions of 1,600 and 3,200 gridpoints and after comparing it with the 'exact' solution. The result is shown in Figure 3 as a function of time to the development of the shock. The computed order should be equal to the nominal order of each scheme as long as the solution is smooth, gradually degrading to first order as the caustic is approached. Every scheme matches this description, in particular EL5, whose convergence order is almost exactly five. EL7 similarly appears to saturate just below its nominal convergence order of seven. Deviations from the nominal convergence order of each scheme are due to contaminations from other error sources, which become increasingly significant at high resolution; these are: the truncation error due to the time-integrator, the accuracy of the inversion from conservative to primitive variables,

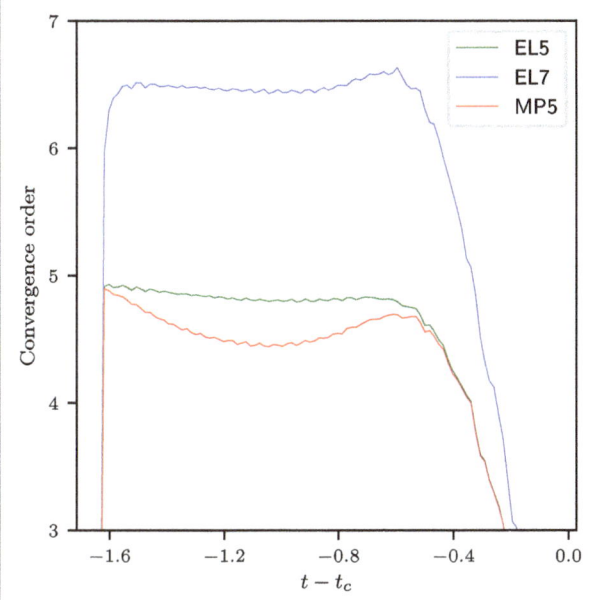

Figure 3 Smooth wave test: convergence order. Convergence order computed on the smooth nonlinear-wave test as a function of time to caustic formation.

or the low-order approximation for the evolution of the entropy (cf., Eq. (22)). What is relevant here is that the EL method does not interfere with the convergence properties of the underlying stencil and can exploit their accuracy.

4.1.2 Shock-tube tests

We choose as a first shock-tube test the special-relativistic version of the classical Sod test (Sod 1978). In this case, the adiabatic index for both the polytropic initial data EOS and the ideal-fluid evolution EOS is $\Gamma = 1.4$ and the right (R) and left (L) initial states are

$$(\rho_R, v_R, p_R) = (0.125, 0, 0.1),$$
$$(\rho_L, v_L, p_L) = (1, 0, 1). \tag{26}$$

The analytic solution consists in a left-going rarefaction wave and a right-going shock wave separated by a right-going contact discontinuity. We perform the test with a variety of spatial resolutions ranging from $\Delta x = 0.01$ to $\Delta x = 3.125 \times 10^{-4}$, and a timestep $\Delta t = 0.1\,\Delta x$.

Figure 4 shows the test results at time $t = 0.6$ for both the EL5 and MP5 schemes at resolution $\Delta x = 1.25 \times 10^{-3}$. Both schemes capture the main features of the solution, with the shocks being captured within $\sim$3 gridpoints, as are the constant states in the pressure and velocity. However, the EL5 scheme displays some oscillations downstream of the shock as well as over- and undershoots around the location of the discontinuities and in the transition between the rarefaction wave and the surrounding flat regions, while the MP5 scheme is able to resolve the solution avoiding such artefacts. This is not surprising since MP5 is a monotonicity preserving scheme (the number of local maxima and minima cannot increase by effect of this method, therefore over- and undershoots cannot occur by construction)

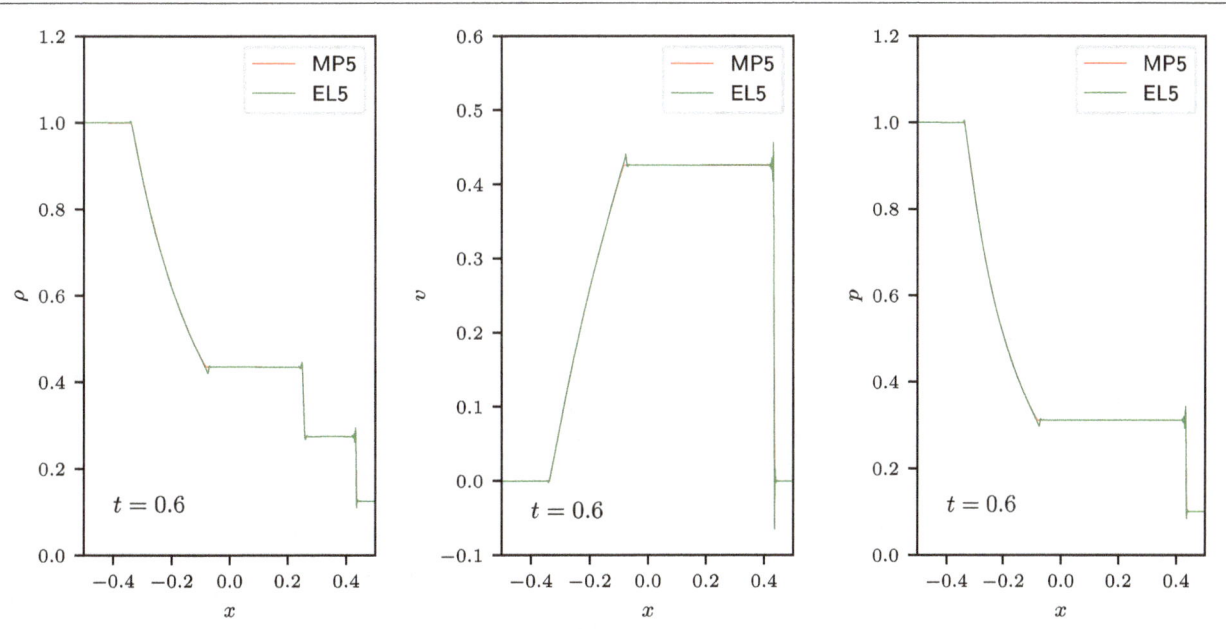

Figure 4 Sod test: solution profiles. Profiles of the rest-mass density (left), velocity (center) and pressure (right) for the special-relativistic Sod test at $t = 0.6$. The solution is computed on a grid of 800 points. The EL5 scheme correctly captures the features of the solution despite oscillations at the discontinuities.

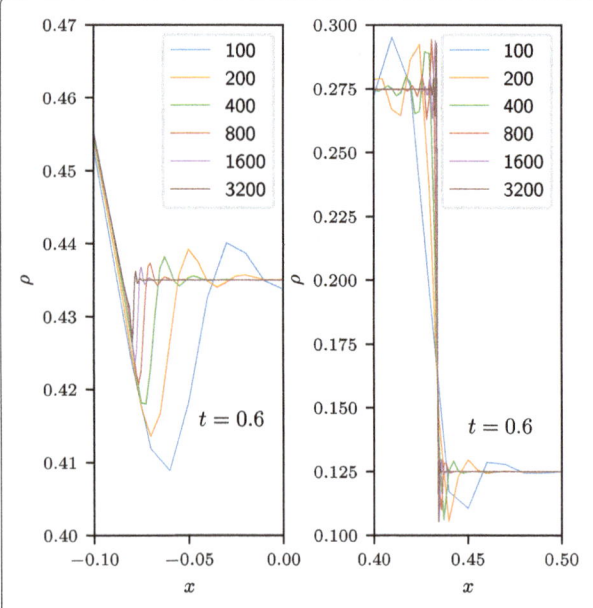

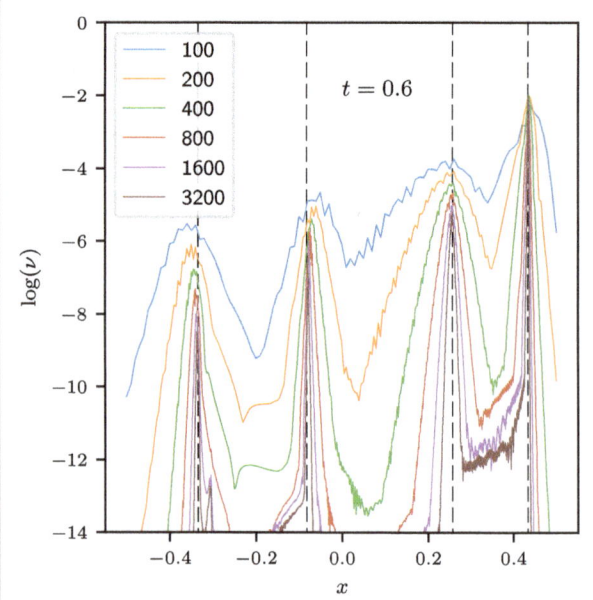

Figure 5 **Sod test: solution at different resolutions.** Rest-mass density profiles, zoomed on the right edge of the rarefaction wave (left) and on the shock (right) for the special-relativistic Sod test at $t = 0.6$, computed with EL5 at different resolutions (parametrized by the number of points on the x axis). The oscillations in the solution can be seen converging away with resolution.

Figure 6 **Sod test: viscosity at different resolutions.** Profiles of the viscosity in logarithmic scale for the special-relativistic Sod test at $t = 0.6$, computed with EL5 at different resolutions (parametrized by the number of points on the x axis). The four peaks correspond to the four different features of the solution, i.e., from left to right, the edges of the rarefaction wave, the contact discontinuity and the shock (vertical dashed lines highlight their location). As the resolution increases, they tend to delta functions.

while EL5 is not. We should remark, however, that this property of MP5 is valid only for scalar equations in one spatial dimension, and it basically accounts for the differences in the behaviour of the two schemes. It should also be stressed that the EL5 scheme is indeed stable and that the oscillations that are present in the solution converge away with resolution (see Figure 5).

The behaviour of the viscosity is displayed in Figure 6. It presents four well distinct peaks, each corresponding to the four nonlinear waves generated by the Riemann problem and corresponding to the edges of the rarefaction wave, where the solution is continuous but non-smooth, of the contact discontinuity and of the shock. The viscosity is higher in correspondence with the latter, and decreases by several orders of magnitude for the other three. It can also be seen clearly how the peaks in the viscosity sharpen as the resolution is increased, mirroring the decreasing size of the aforementioned features (see Figure 5), and seemingly tending towards a delta function at infinite resolution, as expected.

The second shock-tube test we select is a more extreme 'blast-wave' test (Martí and Müller 2003). In this case, the adiabatic index used is $\Gamma = 5/3$ and the right and left initial states are

$$(\rho_R, v_R, p_R) = \left(10^{-3}, 0, 1\right),$$
$$(\rho_L, v_L, p_L) = \left(10^{-3}, 0, 10^{-5}\right).$$

(27)

The exact solution consists in a right-going shock wave, followed by a contact discontinuity and a left-going rarefaction wave. We employ the same resolutions and time-step choices as for the Sod test.

Figure 7 is similar to Figure 4, but relative to the blast-wave test at time $t = 0.4$. We should remark that this is a very extreme test (the pressure has an initial jump of five orders of magnitude) in which the contact discontinuity and shock wave move at essentially the same speed, yielding a very narrow constant rest-mass density state between the two. The oscillations in the EL5 scheme data are in this case more severe than in the Sod shock-tube test, especially around the shock location. As a result, the solution with the EL5 scheme tends to a general decrease of the pressure between the rarefaction wave and the shock wave, whose relative value is however of $\lesssim 7\%$ at most; the MP5 scheme performs better and has a relative error in pressure that is $\sim 1\%$. In both cases, the agreement with the exact solution improves with resolution.

Finally, we perform a three-dimensional shock-tube problem, involving non-grid-aligned shocks i.e., the relativistic-explosion test. The initial data in this case is given by

$$\begin{cases} (\rho, v_i, p) = (1, 0, 1) & \text{if } r \leq 0.4, \\ (\rho, v_i, p) = (0.125, 0, 0.1) & \text{otherwise}, \end{cases}$$

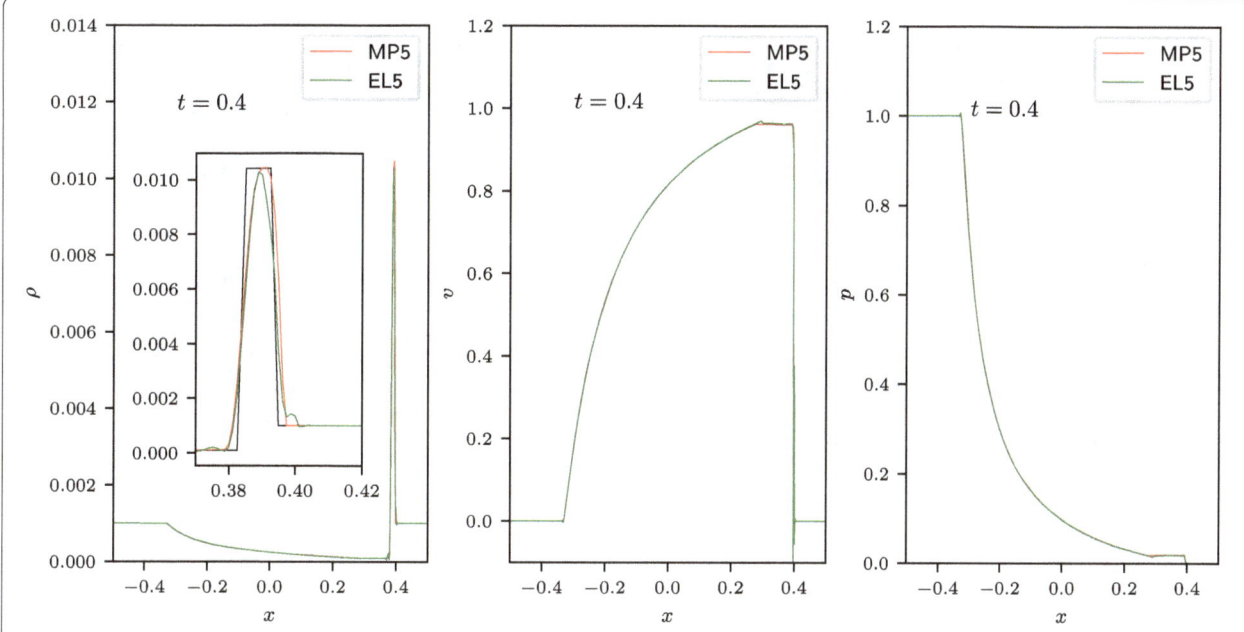

Figure 7 Blast-wave test: solution profiles. Profiles of the rest-mass density (left), velocity (center) and pressure (right) for the special-relativistic blast-wave test at $t = 0.4$. The solution is computed on a grid of 800 points. The inset in the density panel magnifies the blast wave, showing also the exact solution in black. Inversion failures due to oscillations when using the EL5 scheme spoil the quality of the solution.

where r is the distance from the origin. The computational domain is a cube of side 1 centered on the origin, and we use a grid spacing of $\Delta x = 0.01$ and a timestep $\Delta t = 0.1 \, \Delta x$. The adiabatic index for this test is again $\Gamma = 1.4$. The feature of the solution are similar to those of the Sod test i.e., an ingoing rarefaction wave and an outgoing shock, separated by an outgoing contact discontinuity. Note however that because of the spherical symmetry of the test (compared to the planar symmetry in the Sod case), the regions at the two sides of the contact discontinuity are no longer constant states in rest-mass density, velocity and pressure, but display a smooth radial dependence.

Figure 8 shows the rest-mass density for this test at time $t = 0.25$, on the (x, y) plane as well as on the x axis. Both EL5 and MP5 perform very similarly, with differences being barely noticeable in the two-dimensional plot. The curves on the x axis reveal that while both schemes capture the features of the solution, as in the Sod test, the EL5 scheme is slightly more oscillatory.

Overall, these shock-tube tests demonstrate how the entropy-driven hybridisation of the high-order stencil is sufficient to stabilise the scheme even for discontinuous initial data and it is remarkable that such a simple scheme can achieve good accuracy.

4.2 Three-dimensional general-relativistic tests: neutron stars

We next test the EL5 scheme against a series of three-dimensional tests mostly based on the evolution of single,

isolated neutron stars in general relativity (with the exception of grazing-collision test of Section 4.2.6). In each test we employ for the evolution the ideal-fluid EOS (4) with $\Gamma = 2$. The neutron star initial data is constructed using a polytropic EOS $p = K\rho^{\tilde{\gamma}}$ also with $\tilde{\gamma} = 2$ and $K = 100 \, M_\odot^{-2}$.

4.2.1 Isolated star in the Cowling approximation

The first test we perform is the evolution of a stable non-rotating (or TOV, from Tolmann-Oppenheimer-Volkoff) neutron star in a fixed spacetime (i.e., adopting the Cowling approximation) with the goal of assessing the properties of the EL5 scheme over long timescales. Despite its conceptual simplicity (a TOV is just a static solution of the Einstein-Euler equations) the test can be rather challenging. This is because in this test the location of the stellar surface, which is the hardest feature to simulate due to the steep gradient in the hydrodynamics variables, is essentially stationary; as a result, errors can accumulate and grow, affecting the accuracy of the simulation. This behaviour is to be contrasted with the typical situation encountered when evolving inspiralling binary neutron stars, where the stellar surfaces move very supersonically with respect to the floor and most of the errors at the surface are absorbed into the shocks.

For this test we build and evolve a TOV model with central rest-mass density $1.28 \times 10^{-3} \, M_\odot^{-2}$, yielding a (baryon) rest mass of $1.5 \, M_\odot$ and a radius of $\sim 10 \, M_\odot$. We perform the test on a single refinement level with outer boundaries

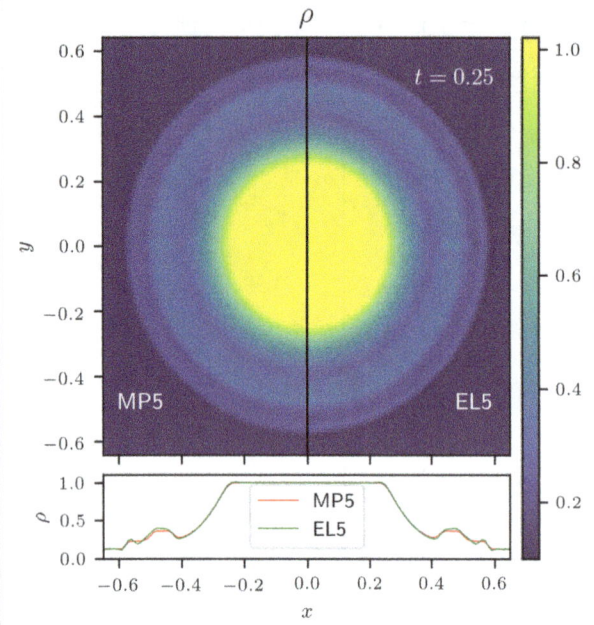

Figure 8 Relativistic-explosion test: density on (x, y) plane.
Rest-mass density for the relativistic-explosion test at time $t = 0.25$. In the top panel the distribution on the (x, y) plane is plotted, MP5 on the left side and EL5 on the right. In the bottom panel, the rest mass density is plotted on the x axis. Both schemes capture very well the solution.

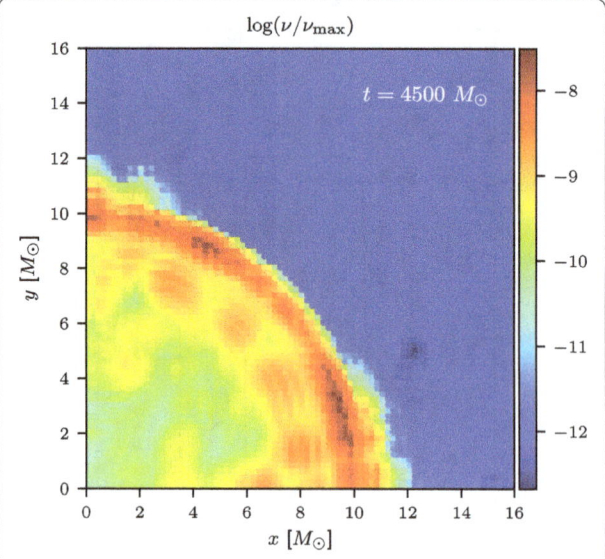

Figure 9 Cowling TOV test: viscosity on (x, y) plane.
Two-dimensional viscosity distribution relative to its upper limit on the equatorial (x, y) plane at time $t = 4,500\ M_\odot$ for the Cowling TOV test. The viscosity peaks at the stellar surface, identified as a shock by the scheme, and drops in the interior.

placed at $16\ M_\odot$ and a resolution of $\Delta^i = 0.2\ M_\odot \simeq 0.3$ km. The timestep is set to 0.15 times the grid spacing, and the time integrator is RK3.

Figure 9 reports the distribution of the viscosity on the equatorial plane, which clearly shows a local annular peak around the location of the stellar surface, where the hydrodynamical variables experience the most violent variations, leading to large values of the viscosity. In the external low-density fluid, the viscosity is set to a small constant value almost everywhere as detailed in Section 3.2. The inner part of the neutron star is expected to be isentropic, as it consists of a shock-free perfect fluid. Indeed, in the stellar interior the viscosity is nonzero but also 10^2 to 10^3 times smaller than at the surface and does not significantly affect the evolution. Quite generally, these features of the viscosity profile are typical in all the tests we considered, whenever a sharp matter/vacuum interface is present.

The general behaviour of the EL5 scheme, when compared to the MP5 scheme, is well illustrated by Figure 10, showing the rest-mass density distribution on the equatorial plane for the two schemes (the left part of the panel, i.e., for $x < 0$, refers to the MP5 scheme, while the right part, i.e., for $x > 0$, to the EL5 scheme[c]). As it can be seen from the figure, both schemes accurately capture the solution in the stellar interior, but significant differences arise at the surface and in the exterior. The MP5 scheme shows a rather diffusive behaviour, with a smooth transition to

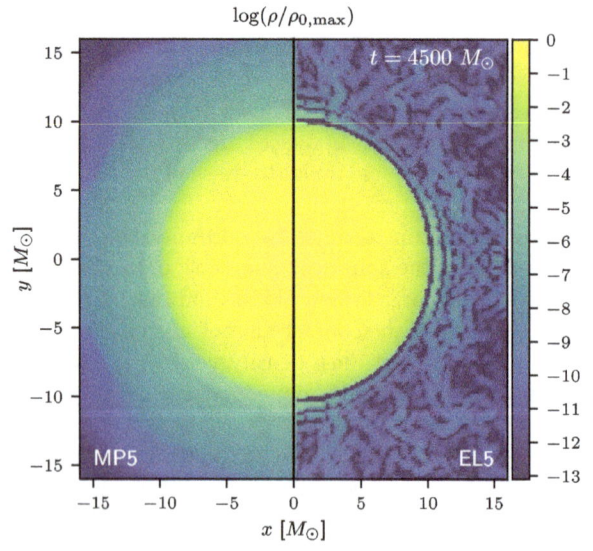

Figure 10 Cowling TOV test: density on (x, y) plane.
Two-dimensional rest-mass density distribution relative to the initial data maximum value on the equatorial (x, y) plane at time $t = 4,500\ M_\odot$ for the Cowling TOV test; the left part of the panel (i.e., $x < 0$) refers to the MP5 scheme, while the right (i.e., $x > 0$) part to the EL5 scheme. Oscillations are visible with the EL5 scheme at the stellar surface, but the exterior fluid is visibly less dense than in the MP5 case.

the external 'vacuum' (i.e., to a region close to the rest-mass density floor) and extended low-density tails. The EL5 scheme, on the other hand, produces a sharper edge.

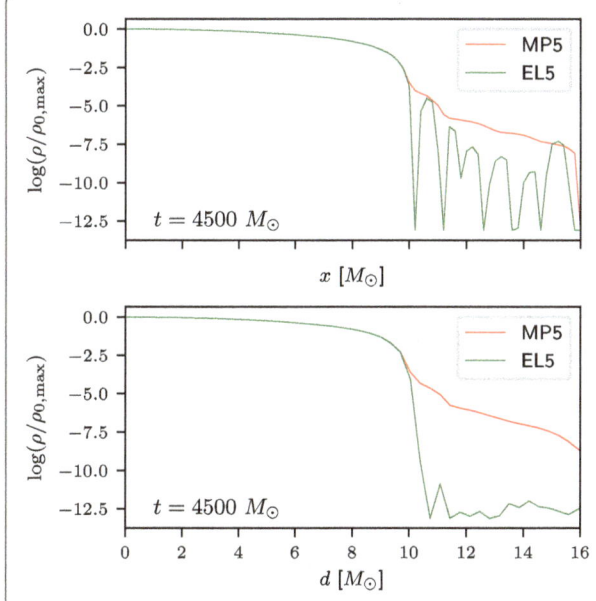

Figure 11 Cowling TOV test: 1D density profiles. One-dimensional rest-mass density profiles in the x (top) and d direction (bottom) at time $t = 4{,}500\ M_\odot$ relative to the initial data maximum value for the Cowling TOV test. The oscillations seen in the EL5 data are a direction dependent artefact, absent in the diagonal direction.

Oscillations in the solution can be seen just outside of the star, resulting in shell-like structures around the surface, which are particularly noticeable in the coordinate axes directions. The stellar exterior is much closer to the vacuum with the EL5 scheme and, in contrast to MP5, it also displays small-scale dynamics at very low densities.

The properties of the oscillations present in the solution computed with the EL5 scheme are made clearer in Figure 11, which shows the rest-mass density profiles along different radial cuts. Along the x direction, the oscillations in the EL5 data have large amplitude and a similar behaviour is observed along the y and z axes. On the other hand, on the three-dimensional diagonal (i.e., along the $x = y = z$ line), the EL5 scheme manages to capture the sharp transition between the stellar interior and the outside vacuum almost perfectly, without significant oscillations or other artefacts. By contrast, the use of the MP5 scheme leads to smooth, rest-density profiles that are only slowly decaying in all directions.[d]

The direction-dependent behaviour shown in Figure 11 for the EL5 scheme is due to the well-known anisotropy of the phase error common to finite-differencing schemes (Vichnevetsky and Bowles 1982; Lele 1992). The MP5 scheme is able to mask this behaviour, but at the price of sacrificing the ability to sharply define stellar surface. We expect that the performance of the EL5 scheme could be improved through the use of multidimensional stencils (i.e., employing a multidimensional interpolation in the re-

construction step), as opposed to the current approach in which the stencil is simply oriented in the direction of the flux to be reconstructed.

The quantitative differences between the two schemes are better captured in Figure 12, where the evolution of the total rest mass and of the central rest-mass density are shown. We recall that the total rest mass (or baryon mass), is defined as

$$M := \int \rho W \sqrt{\gamma}\, d^3 x, \tag{28}$$

where the integral is performed over the whole computational domain. From the continuity equation (35) follows that it should be conserved in absence of a net total flow of matter in or out of the domain. The numerical schemes we employ are conservative (see e.g., Leveque 1992), and therefore preserve the value of the rest mass to the one determined by the initial data. Nonetheless, violations of this conservation can take place in at least three different ways. First, winds originating at the stellar surface (physically, as e.g., in binary neutron star merger, or spuriously as in a stationary case such as the present one) can yield a net loss of mass when they reach the outer boundary and leave the computational domain. Second, matter can be spuriously created or destroyed, in a way that is hard to control, because of floating-point or interpolation errors at the boundaries of refinement levels (this is not the case for this particular test clearly, since we employ a single grid, but it is relevant for the following ones). Finally, when a value of the density is floored mass is spuriously created or destroyed. It is therefore important to characterize the interplay between the numerical scheme and these grid related effects.

The left panel of Figure 12 shows deviations, in absolute value, of the rest mass from the initial value for the two schemes. The EL5 scheme is evidently much better at conserving mass in this test than MP5, leading to a cumulative deviation of $\sim 10^{-7}\ M_\odot$ which is almost three orders of magnitude smaller than the MP5 value.

The central rest-mass density also undergoes an evolution (right panel of Figure 12), with oscillations triggered by the treatment of the stellar surface. Both schemes perform at a similar level of accuracy, with relative variations from the initial value no greater than about 0.3% (even though spurious peaks are present in both data series at various times). The short term behaviour of the two schemes is noticeably different, and the frequency content in the two data series appears different, with the MP5 scheme seeming to show more pronounced high-frequency modes. However, at later times both schemes appear to relax and oscillations decrease significantly in amplitude.

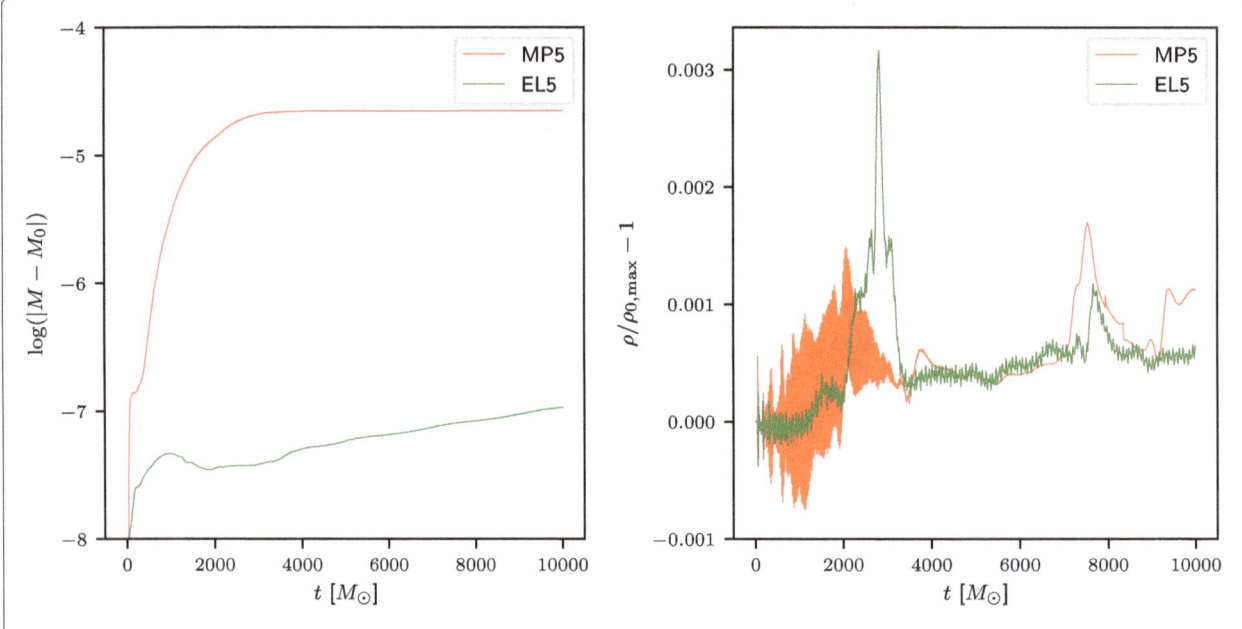

Figure 12 Cowling TOV test: total mass and central density evolution. Deviation of the total rest mass (left panel) and central rest-mass density (right panel) from the initial values for the Cowling TOV test.

4.2.2 Isolated star in a dynamical spacetime

We then proceed to relax the Cowling approximation and test the entropy-limited method coupled with a dynamically evolved spacetime. As first step, we evolve the same initial data for a isolated stable star as in the previous section (i.e., with central density $1.28 \times 10^{-3} M_\odot^{-2}$, baryon mass of $1.5 M_\odot$ and radius $\sim 10 M_\odot$). We perform the test on a grid consisting of three refinement levels centered on the star with sides lengths 16, 32 and 60 $M_\odot$ from finest to coarsest, and with a constant refinement factor of 2. The spatial resolution of the innermost and finest level is set to $\Delta^i = 0.2 M_\odot \simeq 0.3$ km, and the timestep to 0.15 times the grid spacing. This factor is largest possible to guarantee the positivity of the rest-mass density (see discussion in Section 3.1 and Radice et al. 2014a for details). The atmosphere value of the density is set to $10^{-16} M_\odot^{-2}$, that is, almost 13 orders of magnitude smaller than the maximum value. As a time integrator we select the third-order SSP Runge-Kutta RK3. Unless stated differently, we employ the same grid setup for each one of the following single star tests.

Figure 13 shows the distribution of rest-mass density on the equatorial plane for this test, again with the MP5 and EL5 schemes shown on the left and right parts of the panel, respectively. It can be appreciated how the MP5 scheme produces rest-mass tails which are even more dense and extended than in the Cowling case, making the near vacuum solution obtained by the EL5 scheme all the more striking.

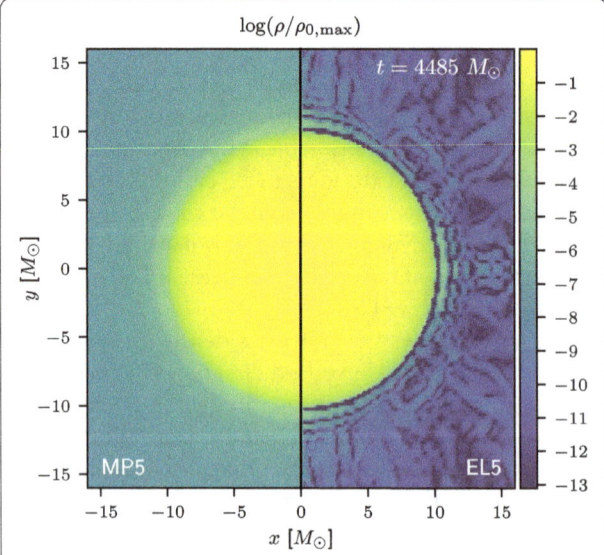

Figure 13 Dynamical TOV test: density on (x, y) plane. Two-dimensional rest-mass density distribution relative to the initial data maximum value on the equatorial (x, y) plane at time $t = 4{,}485 M_\odot$ for the dynamical TOV test. The matter tails are even more extended in MP5 case compared with the Cowling test, EL5 instead preserves its behaviour at the stellar surface and exterior.

Another difference from the Cowling test can be seen in the conservation of the rest mass (left panel of Figure 14). In this case too, the EL5 scheme is able to conserve the initial value to an accuracy roughly two orders of magnitude

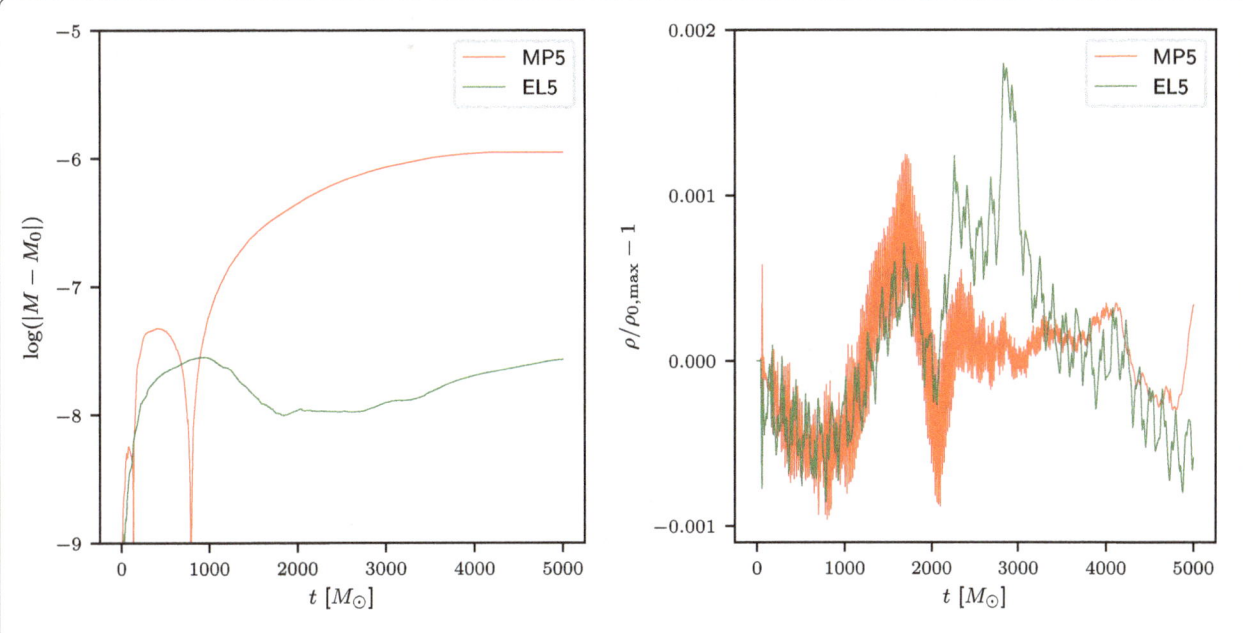

Figure 14 Dynamical TOV test: total mass and central density evolution. Deviation of the total rest mass (left panel) and central rest-mass density (right panel) from the initial values for the dynamical TOV test.

better than the MP5 scheme. Furthermore, it is interesting to notice how the behaviour of the EL5 scheme is much more smooth and predictable; MP5 by contrast displays both spurious losses and gains of mass, which lead to the zero crossings clearly visible in the figure. This is due to interpolation errors arising during the restriction and prolongation operations between different refinement levels. These errors are more severe with MP5 due to the presence of long tails of low density matter in the stellar exterior, as we checked by varying the extent of the refinement levels. In contrast, the EL5 scheme is less affected since the exterior of the star (especially away from the coordinate axes) is nearly vacuum.

The evolution of the central rest-mass density, as shown in the right panel of Figure 14, is similar to the one shown in the previous section for the Cowling approximation, with both schemes varying no more than 0.2% from the initial value, but with MP5 displaying oscillations at much higher frequency.

To further investigate this point, we compute the power spectral density (PSD) of the density evolution, in order to quantitatively gauge the differences between the two schemes. The PSD is computed over the first 5,000 $M_\odot$ of data and with the use of a Hann window function. Before computing the PSD, any linearly growing component of the signal is removed via a least-squares fit.

The PSDs for both schemes are shown in Figure 15 along with the oscillation frequencies of this stellar model as computed perturbatively following the methods discussed in Yoshida and Eriguchi (2001), Takami et al. (2011), and

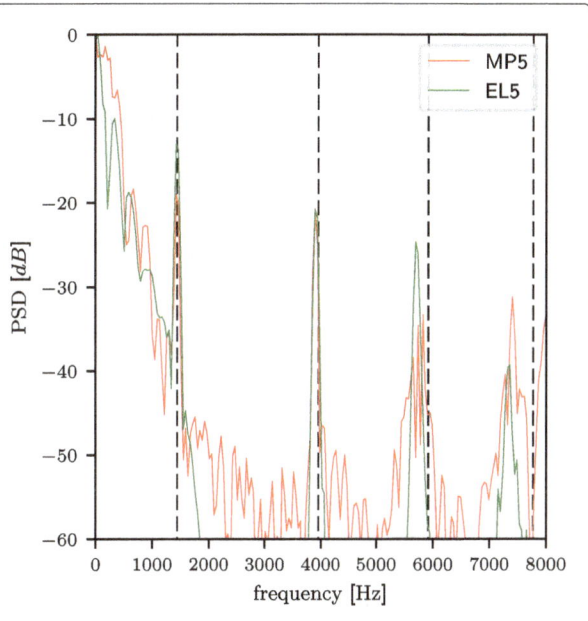

Figure 15 Dynamical TOV test: PSD of central density. PSD of the central rest-mass density evolution and physical eigenfrequencies of the stellar model for the dynamical TOV test. Note the good agreement between the first eigenfrequencies and the peaks in the data.

shown as vertical dashed lines. For both schemes, the PSD is dominated by a low-frequency component due the well-known secular changes in the central rest-mass density (Font et al. 2002) and that disappear with resolution.

However, peaks are clearly visible above the noise. The lowest-frequency peaks correspond to the fundamental oscillation mode of the star and its first overtone, while the following ones are higher overtones and are progressively more offset from the corresponding perturbative eigenfrequencies. The peaks in the EL5 data appear to be more clearly identifiable and less broad than in the MP5 case. Above ~8,000 Hz, and not shown in Figure 15, the MP5 scheme shows significant high-frequency components, clearly visible in the first part of the corresponding curve in Figure 14, but with a rather disordered spectrum. These same frequencies are instead greatly suppressed in the EL5 scheme. Overall, also this test highlights that the EL5 scheme captures quite well the physical behaviour of the system as expected from perturbative methods and is free from some of the artefacts which appear instead in the evolution with the MP5 scheme.

4.2.3 Perturbed isolated star

The next test we perform is a slight modification of previous setup, i.e., we evolve the same isolated neutron star model, but applying a small velocity perturbation to the initial solution. The perturbation consists of a radially outgoing velocity growing linearly in radius to a maximum value of 0.005.

We employ this scenario, more realistic than the simple smooth-wave test of Section 4.1.1, to measure the convergence order of the EL5 and MP5 methods. We performed three sets of simulations at resolutions 0.24, 0.12 and 0.06 $M_\odot$ on the finest level, extracting the evolution of the rest-mass density over time from each one. The initial perturbation is added so that the density evolution is not dominated by the truncation error, but displays a cleaner behaviour. Otherwise, as the resolution is increased, the density evolution would show additional high-frequency modes, which would make the dependence on resolution discontinuous, making it difficult to compute the instantaneous convergence order.

Using the values of the L_1-norm of the rest-mass density over the domain at the three resolutions we also computed the instantaneous convergence order $M_\odot$ as shown in Figure 16. Because this is the instantaneous convergence order and because the underlying system is oscillating, the curves are somewhat noisy (especially for MP5); however, when taking the running average, both schemes generally show a convergence order just below three, consistent with the results in Radice et al. (2014a, 2014b). It is also however apparent how EL5 maintains a fairly constant order of convergence through time, while the behaviour of MP5 is more irregular, especially at later times.

While the hydrodynamics schemes are both formally fifth-order accurate, other parts of the algorithm operate at different degrees of accuracy. In particular, the time integrator is third-order accurate, which most likely accounts

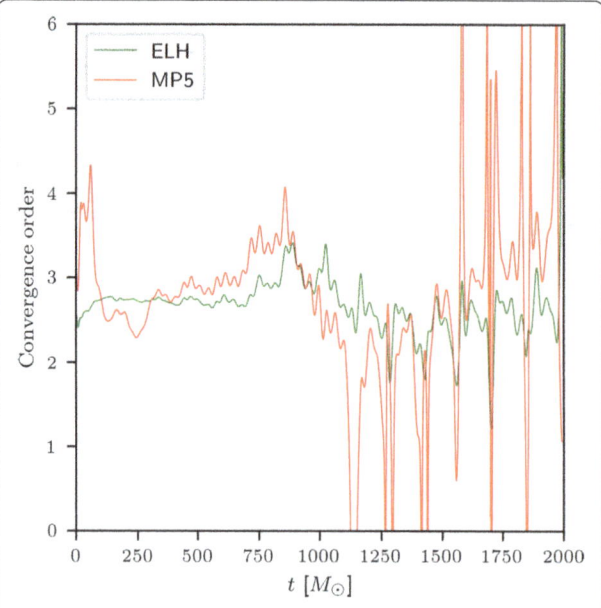

Figure 16 Perturbed TOV test: convergence order. Instantaneous convergence order measured in a perturbed TOV simulation, computed from the L_1-norm of the rest-mass density.

for the convergence order being closer to three than to five. The result is also consistent with the ones found for the MP5 scheme in Radice et al. (2014a, 2014b), Radice et al. (2015). Overall, this test highlights how both the ELH and MP5 schemes perform fairly consistently over time, with no major loss of accuracy.

4.2.4 Migration test

Another important test in our series is the migration of a TOV star moving from a solution on the unstable branch of equilibrium solutions to a stable one. We recall that for any given EOS, increasingly massive but stable TOV models can be constructed by considering increasingly large values of the central rest-mass density. This can continue until a maximum mass is reached, at which point, an increase of the central rest-mass density corresponds to a decrease of the mass of the star. Models on this second branch of the mass/central-rest-mass density curve are unstable, and if a perturbation is present will evolve to either a stable configuration or collapse to a black hole. This is precisely the physical scenario that the migration test simulates: we construct a model on the unstable branch of the mass/density curve and force its 'migration' to a stable configuration by applying a suitable velocity perturbation.

This is a common test for numerical relativity codes (see, e.g., Font et al. 2002; Baiotti et al. 2003, 2005; Cordero-Carrión et al. 2009; Thierfelder et al. 2011), and has been studied in detail in Radice et al. (2010). In particular, we build a nonrotating stellar model on the unstable branch of the equilibrium solutions and with central rest-mass density of $7 \times 10^{-3}\ M_\odot^{-2}$ (yielding a total rest mass of 1.6 $M_\odot$

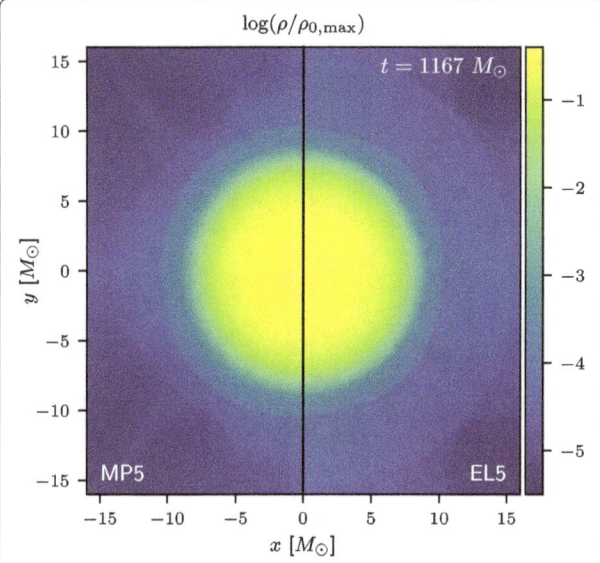

Figure 17 Migration test: density on (x, y) plane. Two-dimensional rest-mass density distribution relative to the initial data maximum value on the equatorial (x, y) plane at time $t = 1,167\,M_\odot$ for the migration test. Virtually no difference can be detected in the two schemes behaviour.

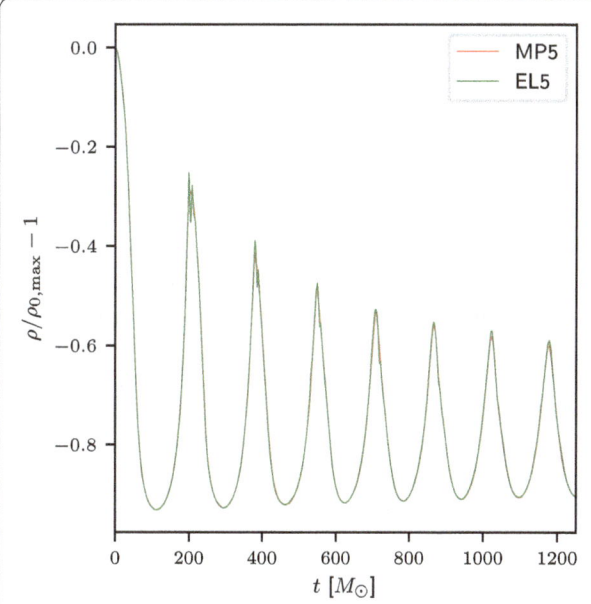

Figure 18 Migration test: central density evolution. Central rest-mass density in the migration test. The agreement between the two schemes is apparent over the whole evolution, apart from high-frequency modes at the maxima in EL5 data.

and a radius of $6\,M_\odot$). The migration is then triggered by injecting a radially outgoing velocity perturbation where the velocity grows linearly in radius, reaching a maximum value of 0.01. The star then undergoes a series of violent expansions and contractions as it migrates to the stable branch and then settles on the new equilibrium. During each contraction and expansion strong shocks are formed, and the shocked matter is ejected at large velocities.

In Figure 17 we show the rest-mass density distribution on the equatorial plane for both schemes and during one of the contractions of the star, just before the central rest-mass density reaches a maximum (cf., Figure 18). The snapshot clearly shows that both the EL5 and MP5 schemes produce almost identical results for this test. This is not surprising and mainly due to the matter outflow driven by the stellar oscillations, which rapidly fills the domain and removes the sharp feature of the stellar surface, which is the most problematic structure to resolve and the main difference in the two schemes.

The evolutions of the maximum rest-mass density are shown in Figure 18, where they are reported as normalized to the initial value. The agreement between the two schemes is extremely good during the entire evolution and the main difference between the two solutions is the presence of some high-frequency modes near the maxima of the density in the EL5 data. Such oscillations are the result of inward-propagating shock waves generated in the outer layers of the star during the contraction phase. Figure 19 shows a magnification of the behaviour of the maximum

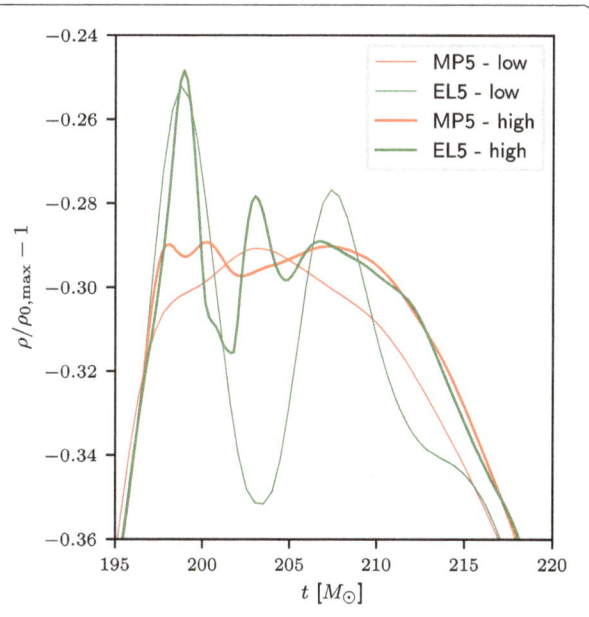

Figure 19 Migration test: central density evolution at different resolutions. Magnification of the central rest-mass density evolution around the first contraction of the migrating star. The low-resolution data (thin curves) corresponds to Figure 18 and to a grid spacing of $0.2\,M_\odot$. The high-resolution data (thick curves) corresponds to a grid spacing of $0.086\,M_\odot$. The high-frequency modes present in the EL5 data at low resolution persist at high resolution in both schemes.

rest-mass density at the peak of the first contraction, comparing not only the two schemes but also the evolutions with two different resolutions. At high resolution, both the MP5 and the EL5 scheme show small-scale and high-frequency oscillations that are less pronounced in the low-resolution data. Interestingly, these oscillations are essentially smoothed out in the low-resolution run of the MP5 scheme, while they are very visible in the low-resolution EL5 run. This seems to indicate that the two schemes tend, with increasing resolution, towards a solution where the small-scale oscillations are present and therefore physically correct and not a numerical artefact. Finally, as the evolution progresses, the contraction/expansion phases become less and less violent as part of the kinetic energy is converted into internal energy, thereby leading to milder and milder shocks, and the high-frequency oscillations in the central rest-mass density all but disappear.

4.2.5 Isolated rotating neutron star

As the last test case for a stable (or metastable) isolated relativistic stars we consider the evolution of a rapidly and uniformly rotating star. More precisely, we set up axisymmetric initial data relative to a uniformly rotating neutron star governed by a polytropic EOS with $K = 100\ M_\odot^{-2}$ and $\Gamma = 2$, having a central rest-mass density of $1.28 \times 10^{-3}\ M_\odot^{-2}$ and a polar to equatorial axis ratio of 0.8 using the RNS code (Stergioulas and Friedman 1995). This results in a star with total rest mass 1.6 $M_\odot$, radius 10 $M_\odot$, rotation frequency $f = 673.2$ Hz (about 60% of the mass shedding frequency) and dimensionless angular momentum $J/M^2 = 0.46$. Also in this case the spacetime is evolved in time despite the solution being stationary.

In Figure 20 we report again the rest-mass density distribution on the equatorial plane for both schemes and at time $t = 4,300\ M_\odot$, that is, after about 14 rotation periods. The figure clearly illustrates that that both schemes evolve the rotating star with no noticeable problems and that, as already seen in the case of nonrotating stars, the part of the domain exterior to the stellar surface rapidly fills with matter. Also in this case, the behaviour of the two methods in the low-density regions is rather different, with the MP5 scheme yielding to a volume which is filled by uniform but comparatively higher-density material, while the EL5 scheme produces a stellar exterior which has lower-density matter but with small-scale condensations (cf., Figures 10 and 13 for the equivalent behaviour in the absence of rotation).

However, in contrast with the behaviour seen in the case of nonrotating stars, the dynamics of the low-density material in the stellar exterior results in a degradation of the conservation of mass for the EL5 scheme, as shown in the left panel of Figure 21 (cf., the left panels of Figures 12 and 14 for the equivalent behaviour in the absence of rotation). The deviation of the total rest-mass density from its initial

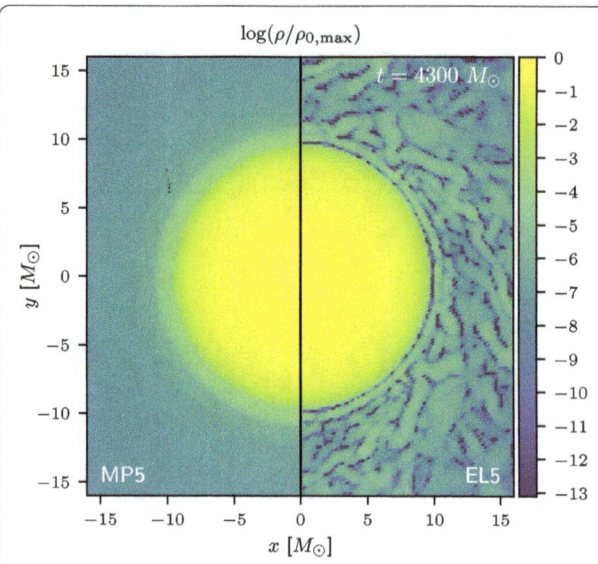

Figure 20 Rotating star test: density on (x, y) plane.
Two-dimensional rest-mass density distribution relative to the initial data maximum value on the equatorial (x, y) plane at time $t = 4,300\ M_\odot$ for the rotating star test. The behaviour of the star exterior is dynamic and chaotic with EL5 as compared with MP5.

value is more than one order of magnitude larger for the EL5 scheme than for MP5 and reaches values of $\sim 10^{-5}\ M_\odot$. This is the result of failures in the conversion from the conserved variables to the primitive ones, triggered by oscillations in the solution: the solution can be evolved to an unphysical state, and in this case the rest-mass density could reach values below the atmosphere floor value; if so, the conversion routine resets the affected cells to the atmosphere value, thereby spuriously creating mass. We speculate that most of these failures result from the large tangential velocity that is acquired by the shell-like distribution of matter that builds up in the case of the EL5 scheme and that is present already in the nonrotating case. While rather innocuous in the absence of rotation, this shell of matter can fling material to large distances (but within the computational domain) and lead to a much more chaotic dynamics of the fluid in the low-density regions (see the discussion in Section 3.2.3 of Radice et al. 2014a).

To assess the impact of the fluid dynamics in the stellar exterior we report the evolution of the central rest-mass density for the two schemes in the right panel of Figure 21. We find that the low-density fluctuations appearing in the stellar exterior with the EL5 scheme do not impact the solution in the stellar interior, with the low-frequency central density oscillations essentially being in phase for the two schemes. Also quite apparent is that the EL5 scheme yields rather constant-amplitude oscillations and this should contrasted with the MP5 scheme, where the oscillations are comparatively larger in the first

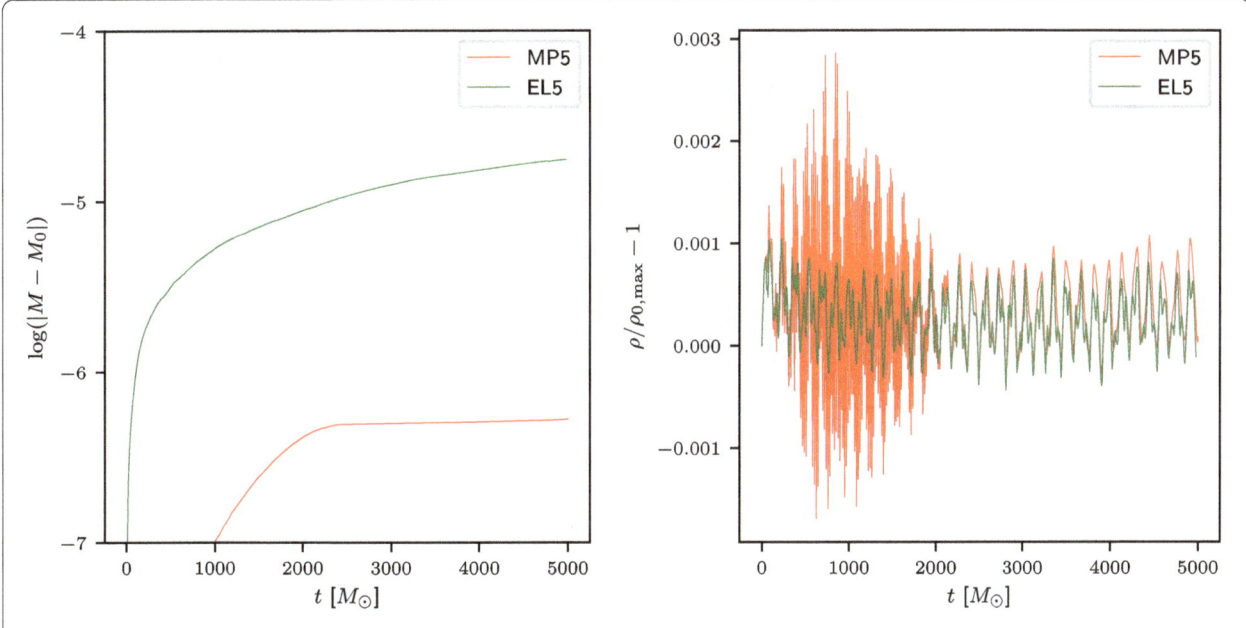

Figure 21   Rotating star test: total mass and central density evolution. Deviation of the total rest mass (left panel) and central rest-mass density (right panel) from the initial values for the rotating-star test.

~2,000$M_\odot$ of the evolution. In both cases, however, the oscillations are extremely small and below 0.1%.

4.2.6   Grazing collision of neutron stars

We further test the ELH scheme in another truly dynamical test: the motion across the numerical grid of two neutron stars in a grazing collision. This is a setup that is very similar to that of a binary-neutron star system in quasi-circular orbit, the most obvious difference being the initial momenta of the two stars do not result in a quasi-circular orbit and that the initial fluid velocity can be taken to be arbitrary. In practice, the initial data is set up by generating two identical TOV models (the same as considered in Sections 4.2.1 and 4.2.2), superimposing the two data sets on the computational grid and imparting suitable initial momenta resulting in a small, but nonzero, impact parameter. Clearly, such initial data is valid only as a first approximation since the stars are not in the hydrostatic equilibrium corresponding to the binary system and the intial metric and extrinsic curvature do not reflect a solution of the Einstein constraints equations.

These violations lead to initial oscillations in the evolution (see Kastaun et al. 2013; Tsatsin and Marronetti 2013 for a more detailed discussion of a more sophisticated setup in which the stars are also subject to a spin up) which can however be reduced significantly by setting the initial distance of the two stars to a rather large value. More importantly, however, these oscillations do not interfere with the main goal of this test, namely, that of validating the ability of the ELH scheme to preserve sharply the features

of the stellar surface also when the star moves across the numerical grid.

More in detail, the star centers are set at positions $(x_1, y_1, z_1) = (50, -50, 0)$ and $(x_2, y_2, z_2) = (-50, 50, 0)$ in units of $M_\odot$, i.e., symmetric with respect to the grid center on the (x, y) plane and at a distance of ~141 $M_\odot$. The initial 3-velocities are $(v_1^x, v_1^y, v_1^z) = (0, -0.1, 0)$ and $(v_2^x, v_2^y, v_2^z) = (0, 0.1, 0)$ respectively. We evolve the system on a cubic grid of radius 512 $M_\odot$, but employ reflection symmetry boundary conditions across the (x, y) plane and 180 degrees rotation symmetry boundary conditions across the (y, z) plane to reduce the computational cost. The grid structure consists of two identical box-in-box refinement levels hierarchies with refinement factor 2, each centered on a star and consisting of 5 cubic levels with radii $12, 25, 50, 100, 200$ $M_\odot$, plus the coarse base level with radius 512 $M_\odot$, so that the grid spacing in the innermost refinement level is $\Delta^i = 0.2$ $M_\odot \simeq 0.3$ km. The refinement levels moved to track the positions of the stars during the evolution (see also Radice et al. 2016 for further details on the initial data and grid structure). We set again $\Delta t = 0.15\,\Delta x$.

The evolution consists in the two stars initially traversing the grid in the x direction and approaching each other, then bending their trajectories as in a gravitational scattering process; we do not follow the dynamics of the process after the first fly-by. Figure 22 shows the rest-mass density distribution on the (x, y) plane for this test. The snapshots of one of the two stars are taken at time $t = 768$ $M_\odot$, when the two stars are past the point of closest approach and are

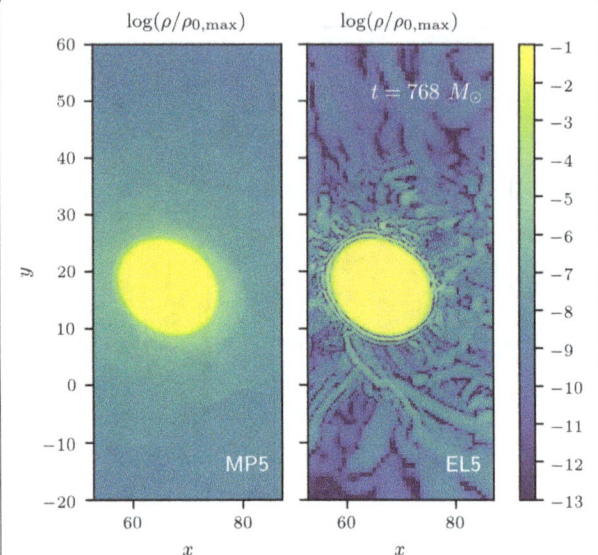

Figure 22 Grazing collision test: density on (x, y) plane.
Two-dimensional rest-mass density distribution relative to the initial data maximum value on the equatorial (x, y) plane for the grazing-collision tests at time t = 768 $M_\odot$, i.e., after the point of closest approach.

a dynamical spacetime: EL5 shows a sharper star surface with respect to MP5, as well as an 'emptier' surrounding region, while the bulk of the star itself is very well resolved by both schemes.

The conservation of the total rest mass (left panel of Figure 23) is instead similar to the rotating-star case, thus showing a better performance by the MP5 scheme. Note however that the differences between the schemes are far smaller in this case, less than one order of magnitude. Furthermore, in contrast with the preceding tests, the grid structure in this case is much more complicated as well as dynamically updated to track the stars. Interpolation errors at the refinement level boundaries play therefore a greater role in the conservation of rest mass.

The evolution of the rest-mass density at the star centers, as shown in the right panel of Figure 23, is very similar for both schemes. There is an initial sudden increase in the density of about 4% with respect to the initial value, due to the evolution scheme bringing the star in hydrostatic equilibrium from the initial state. The density then oscillates around this new value, due to perturbations that in this case are not only induced by the violation of the constraint equations but also by the gravitational interaction. Overall, both schemes reproduce well all of these effects and show a very good agreement.

4.2.7 Gravitational collapse to a black hole

As a final test we evolve the violently dynamical collapse of a star to a black hole. This is also a common numerical-relativity benchmark (see, e.g., Font et al. 2002; Baiotti

escaping. One can appreciate the deformation due to the boost, the acquired spin angular momentum and the tidal gravitational interaction. From the hydrodynamics point of view the behaviour of the MP5 and EL5 schemes is consistent with the previous tests, in particular the TOV in

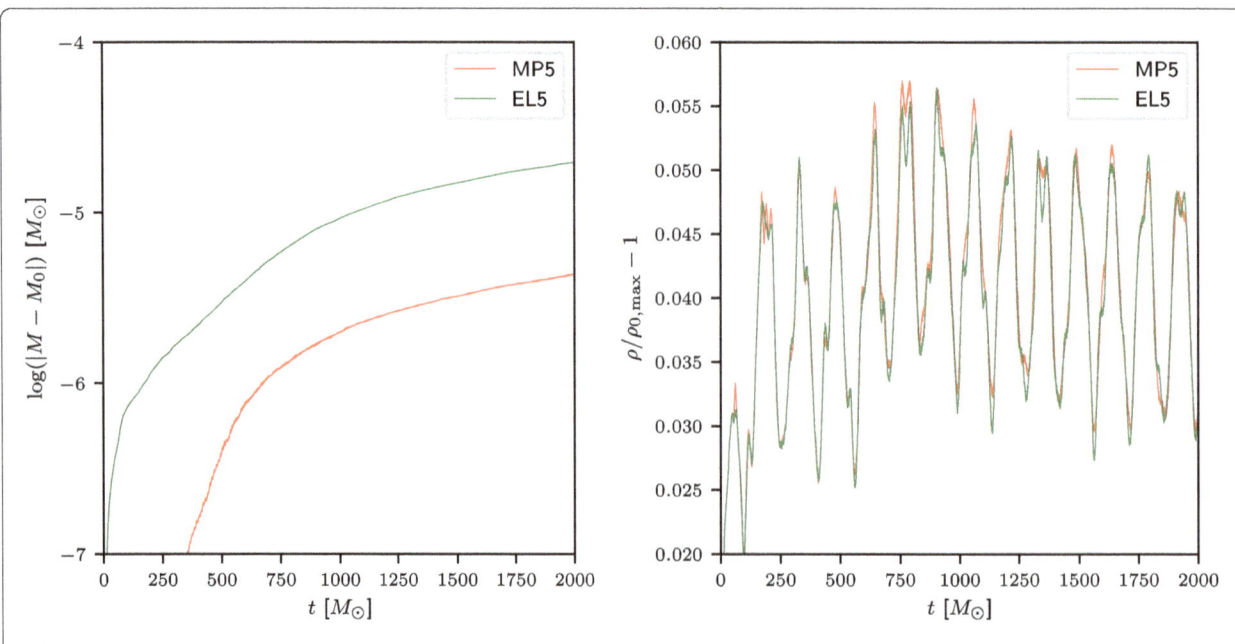

Figure 23 Grazing collision test: total mass and central density evolution. Deviation of the total rest mass (left panel) and central rest-mass density (right panel) from the initial values for the grazing-collision test.

et al. 2005; Baiotti and Rezzolla 2006; Thierfelder et al. 2010), which allows us to validate ELH in the presence of a physical singularity and of an apparent horizon. More specifically, we consider a nonrotating star with central rest-mass density $8 \times 10^{-3} \, M_{\odot}^{-2}$, corresponding to a baryon mass of $1.5 \, M_{\odot}$ and radius $6 \, M_{\odot}$, and initiate the collapse with a velocity perturbation analogous to the one used in the migration test, but with the opposite sign, i.e., radially ingoing.

We define the time of black-hole formation as the instant at which an apparent horizon is first detected on the numerical domain, which, given the chosen setup, happens at $t \simeq 48 \, M_{\odot}$. Since we use singularity avoiding slicing conditions, we do not need to excise the interior spacetime of the black hole (Baiotti and Rezzolla 2006; Baiotti et al. 2007; Thierfelder et al. 2010). At the same time, we set the hydrodynamical variables to their atmosphere values inside a surface with the same shape as the apparent horizon, but radius $r = 0.9 \, r_{\mathrm{AH}}$ in every angular direction, r_{AH} being the radius of the apparent horizon. This 'hydrodynamic excision' is not strictly necessary as our code can handle the collapse without it, regardless of the scheme we employ. However, we have observed that its use improves the accuracy of the subsequent evolution, most notably, it improves the behaviour of the rest-mass density and we therefore choose to employ it.

Figure 24 shows a snapshot of the rest-mass density on the equatorial plane just after the collapse. The central area of uniform low density is where the hydrodynamical variables have been set to atmosphere inside the horizon, and in this plot appears of identical size and shape for the two codes. The areas outside the horizon are instead filled with matter which has been spuriously ejected from the outer layers of the star star during the collapse. The rest-mass density is evidently higher in the case of the EL5 scheme, due to a slightly higher proportion of matter ejected during the collapse, which is in turn triggered by larger oscillations around the stellar surface for the EL5 scheme. However, in both cases the total rest mass outside the horizon is tiny, $\sim 10^{-6} \, M_{\odot}$ for the EL5 scheme and $\sim 10^{-9} \, M_{\odot}$ for MP5, and thus dynamically essentially irrelevant on the properties of the solution. Furthermore, most of this matter is gravitationally bound and hence accretes back onto the newly formed black hole, forming streams of infalling matter. This is particularly evident along the coordinate directions, where the numerical viscosity of the high-order finite-differences stencil is smaller, independently of the scheme employed.

The very close agreement between the two solutions is summarised in Figure 25, where the evolution of the central rest-mass density, minimum lapse and L_2-norm of the Hamiltonian constraint is plotted. The discontinuities in the curves at about $50 \, M_{\odot}$ correspond to the time of collapse and arise because we exclude points inside the horizon in the calculation of extrema and norms. In each panel before the formation of the apparent horizon, the curves

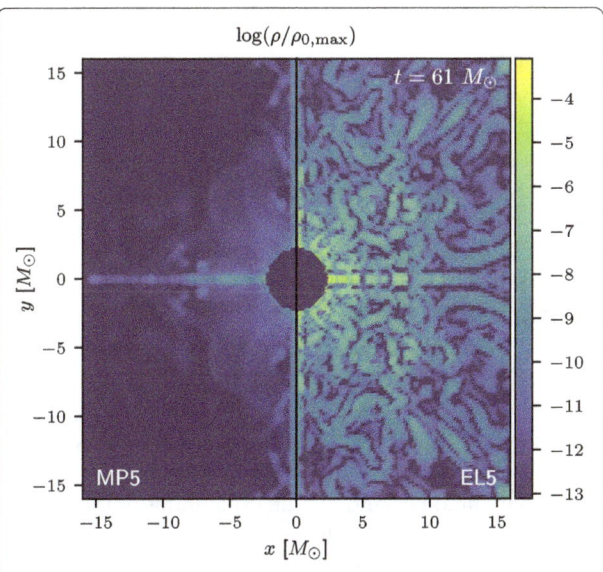

Figure 24 BH collapse test: density on (x, y) plane.
Two-dimensional rest-mass density distribution relative to the initial data maximum value on the equatorial (x, y) plane at time $t = 61 M_{\odot}$ for the collapse test. Streams of matter ejected during collapse and accreting back onto the black hole are clearly visible and more prominent in EL5 data.

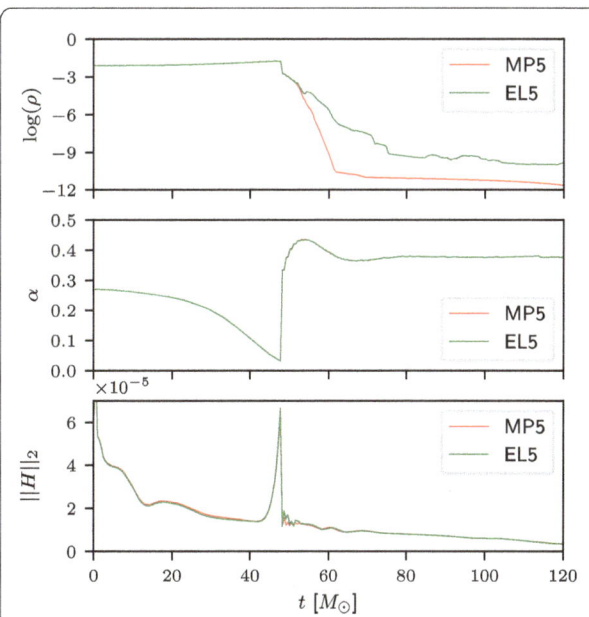

Figure 25 BH collapse test: density, lapse and Hamiltonian constraint evolution. Maximum rest-mass density (top), minimum lapse (middle) and L_2-norm of the Hamiltonian constraint (top) for the stellar-collapse test. Note that the violation of the Hamiltonian constrain will grow on longer timescales as it is typical of BSSNOK evolutions (Alic et al. 2013).

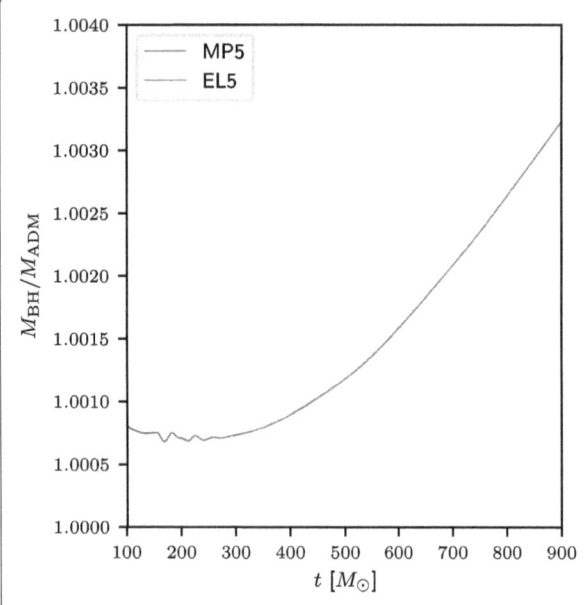

Figure 26 BH collapse test: BH mass. Ratio of the apparent-horizon mass to the ADM mass in the stellar-collapse test. Note that the growth is actually very small and is amplified here to show the difference between the two schemes.

for the solution of the classical equations of hydrodynamics (Guermond et al. 2011), but it is also importantly different in that it does not require any change in the equations of relativistic hydrodynamics.

To assess the robustness and accuracy of our new method, we have discussed its implementation in the WhiskyTHCEL code, which exploits the finite-difference capabilities of the WhiskyTHC code, and tested its validity with an extensive series of tests, comparing the results of ELH with those obtained with another well-tested and high-order HRSC scheme: the fifth-order monotonicity-preserving MP5 method (Suresh and Huynh 1997). Overall, we have found that the scheme is stable and able to cope with shocks and discontinuities, both in classical test such as shock-tube tests, as well as in realistic astrophysical simulations.

Under all of these conditions the scheme has been found to be stable and to yield accuracy that is comparable, if not better, of that of the MP5 method. In some tests involving stars that are nonrotating or not moving across the grid, it also offers definite advantages, such as a sharper resolution of the surface/vacuum interface. At the same time, however, it also shows a less good conservation of the rest mass for stars that are rotating or moving across the computational domain (the opposite is true for stationary nonrotating stars, where the new method conserves rest mass more accurately). Quite surprisingly, all of the results presented here were obtained without any fine tuning of the two arbitrary coefficients that enter the definition of the scheme. Finally, thanks to its linearity and simplicity, the ELH method can also offer advantages in terms of performance. In our tests, we have found EL5 to be ∼50% faster than MP5, even though our implementation was not particularly optimized. For instance, a definite advantage of ELH, which we did not exploit, is that it can be easily vectorised. At the same time, we remark that the exact speed-up that can be achieved with ELH depends also on external factors, such as the grid setup and number of ghost zones, which may vary for different applications. An interesting development in this sense would be the use of this scheme in a discontinuous Galerkin framework, whose superior scalability properties should decouple the performance of the ELH method from the grid setup.

The work presented here could be improved in at least two ways. Firstly, the already good capturing properties of steep gradients as those given by the stellar surface, could be further enhanced and the full capabilities of the scheme further exploited by coupling it to truly multidimensional stencils. Second, the two free coefficients that appear in the method, and that we have here set to unity for simplicity, could potentially be tuned to optimise some of the features of the solution. Both of these aspects will be explored in future work.

In conclusion, we have shown that entropy-limited hydrodynamics is a robust, stable, and accurate alternative

corresponding to the EL5 and MP5 schemes are essentially on top of each other (the largest differences being of the order of 0.3%, 0.6%, and 4.7% for each plot, respectively), showing the very good agreement in the evolution between the two schemes. Note also that after the apparent horizon formation and due to our approach of 'hydrodynamic excision', the upper panel of Figure 25 shows the maximum of the density in the exterior of the horizon rather than the central density. The disagreement in the EL5 and MP5 curves relates therefore to the tiny amount of residual matter outside of the black hole, and as such it has no dynamical impact on the black-hole solution.

A final confirmation of the equivalence between the two numerical solutions comes from the comparison of the two black-hole masses computed using the dynamical horizon formalism (Ashtekar and Krishnan 2003) and shown in Figure 26. As can be seen from the figure, we find a very close agreement between the two schemes.

5 Conclusions

We have presented a new high-order numerical method for the solution of the Euler equations of general-relativistic hydrodynamics that we name 'entropy-limited hydrodynamics' (ELH). The scheme is of the flux-limiting type, where a high-order numerical flux is combined with a stable low-order method, namely the Lax-Friedrichs flux. The flux-limiting is activated and driven by a shock indicator based on a measure of the entropy generated by the solution. Such a special and general-relativistic method is inspired by the entropy-viscosity method proposed recently to commonly employed HRSC schemes. Its performance reaches the level of accuracy and stability necessary to apply it to realistic astrophysical simulations. Given these encouraging prospects, work is already in progress to apply this method to realistic simulations of binaries involving neutron stars and black holes.

Appendix 1: General relativistic hydrodynamics formulation

In this section we summarize the general relativistic formulation of hydrodynamics, which forms the theoretical background for the neutron star tests presented above; as well as providing details on the ELH scheme in the general relativistic case, when they differ from the discussion of Section 3.

A.1 3 + 1 decomposition

We adopt the usual 3 + 1 decomposition of spacetime (see Alcubierre 2008; Bona et al. 2009; Baumgarte and Shapiro 2010; Gourgoulhon 2012; Rezzolla and Zanotti 2013; Shibata 2016 for a detailed discussion), in which the spacetime is decomposed into spacelike hypersurfaces with normal $n^\mu = \alpha^{-1}(1, -\beta^i)$, where α is the lapse function and β^i are the components of the shift vector. Within this formalism, the spacetime metric $g_{\mu\nu}$ is written as

$$
\begin{aligned}
ds^2 &= g_{\mu\nu}\, dx^\mu\, dx^\nu \\
&= -\alpha^2\, dt^2 + \gamma_{ij}\bigl(dx^i + \beta^i\, dt\bigr)\bigl(dx^j + \beta^j\, dt\bigr),
\end{aligned} \tag{29}
$$

where γ_{ij} is the spatial three-metric, which together with the extrinsic curvature $K_{ij} = -\frac{1}{2}\mathcal{L}_n\gamma_{ij}$, $\mathcal{L}_n$ being the Lie derivative along n^μ, fully determines the geometry of each leaf of the foliation.

The matter content of the spacetime is described through its energy-momentum tensor $T_{\mu\nu}$, which within the 3 + 1 split can be decomposed in its timelike, spacelike and mixed parts as (see e.g., Rezzolla and Zanotti 2013)

$$
E := n^\mu n^\nu T_{\mu\nu}, \tag{30}
$$

$$
S_i := -\gamma_i^{\,\mu} n^\nu T_{\mu\nu}, \tag{31}
$$

$$
S_{ij} := \gamma_i^{\,\mu} \gamma_j^{\,\nu} T_{\mu\nu}, \tag{32}
$$

where $\gamma_\nu^{\,\mu} := \delta_\nu^\mu + n_\nu n^\mu$ is the projection operator orthogonal to n_μ, hence yielding purely spatial tensors.

A.2 Relativistic hydrodynamics

The general relativistic generalization of the energy-momentum tensor (1) consists in replacing the Minkowski metric $\eta_{\mu\nu}$ with the generic metric $g_{\mu\nu}$, i.e.

$$
T_{\mu\nu} = \rho h u_\mu u_\nu + p g_{\mu\nu}. \tag{33}
$$

The corresponding equations of motion for the fluid are the 'conservation' of the stress-energy tensor

$$
\nabla_\mu T^{\mu\nu} = 0, \tag{34}
$$

and conservation of rest mass

$$
\nabla_\mu\bigl(\rho u^\mu\bigr) = 0, \tag{35}
$$

which differ from their special relativistic counterparts (2) and (3) by the use of the ∇ operator, i.e. the covariant derivative constructed from the metric (29).

Equations (34) and (35) are solved in conservation form (Eq. (5)), where the 'conserved' variables $\boldsymbol{U}$ are defined as

$$
\begin{aligned}
\boldsymbol{U} &:= \sqrt{\gamma}\begin{pmatrix} D \\ S_j \\ \tau \end{pmatrix} \\
&:= \sqrt{\gamma}\begin{pmatrix} \rho W \\ \rho h W^2 v_j \\ \rho h W^2 - p - \rho W \end{pmatrix}.
\end{aligned} \tag{36}
$$

This definitions differ from (6) by the multiplicative factor γ, i.e. the determinant of the 3-metric γ_{ij}. The fluxes and sources, containing metric-dependent terms, are given by

$$
\boldsymbol{F}^i = \sqrt{\gamma}\begin{pmatrix} (\alpha v^i - \beta^i)D \\ \alpha S_j^i - \beta^i S_j \\ \alpha(S^i - D v^i) - \beta^i \tau \end{pmatrix}, \tag{37}
$$

and

$$
\boldsymbol{S} = \sqrt{\gamma}\begin{pmatrix} 0 \\ \frac{1}{2}\alpha S^{lm}\partial_j\gamma_{lm} + S_k\partial_j\beta^k - E\partial_j\alpha \\ \alpha S^{ij}K_{ij} - S^k\partial_k\alpha \end{pmatrix}. \tag{38}
$$

This formulation is known as the 'Valencia formulation' and was first proposed by (Banyuls et al. 1997). Note that the sources terms for the momentum and energy equations are non-vanishing, which corresponds to the fact that the momentum and energy of the fluid are not independently conserved, but the coupling of the fluid to the spacetime and vice versa has to be taken into account (see e.g., Rezzolla and Zanotti 2013; Shibata 2016; Baumgarte and Shapiro 2010 for details). The fluid three-velocity measured by the normal observers is defined as

$$
v^i := \frac{1}{\alpha}\left(\frac{u^i}{u^t} + \beta^i\right) \tag{39}
$$

which also contains metric-dependent terms, and the Lorentz factor is $W := (1 - v^i v_i)^{-\frac{1}{2}} = \alpha u^t$. We also have used the fact that $\sqrt{-g} = \alpha\sqrt{\gamma}$.

A.3 ELH scheme extension to GR

The general relativistic formulation of the second law of thermodynamics, and therefore of the entropy residual $\mathcal{R}$, is

$$\mathcal{R} := \nabla_\mu \left(\rho s u^\mu \right) \geq 0, \tag{40}$$

once again swapping the partial derivative ∂ for the covariant one ∇. This gets rewritten as

$$\begin{aligned}
\mathcal{R} &= \nabla_\mu \left(s\rho u^\mu \right) \\
&= s\nabla_\mu \left(\rho u^\mu \right) + \rho u^\mu \nabla_\mu s \\
&= \rho u^\mu \nabla_\mu s = \rho u^\mu \partial_\mu s,
\end{aligned} \tag{41}$$

which again involves only derivatives of the specific entropy s, and where the continuity equation (35) was used to remove the first term on the second line. The 4-velocity u^μ can again be written in terms of the fluid three-velocity v^i, but it will also contain the lapse and shift (see Eq. (39)), so that the final form of the residual in the 3 + 1 split is

$$\mathcal{R} = \frac{\rho W}{\alpha} \left[\partial_t s + \left(\alpha v^i - \beta^i \right) \partial_i s \right]. \tag{42}$$

Appendix 2: Spacetime evolution

The evolution of the spacetime is of course determined by Einstein's equations

$$R_{\mu\nu} - \frac{1}{2} R g_{\mu\nu} = 8\pi T_{\mu\nu}, \tag{43}$$

where $R_{\mu\nu}$ is the Ricci tensor and $R := R_\mu^\mu$ the Ricci scalar. The energy-momentum tensor $T_{\mu\nu}$ given by (33) appears in the right-hand-side of the equations, coupling the spacetime to the and fluid and vice versa.

We employ the common BSSNOK formulation (Shibata and Nakamura 1995; Baumgarte and Shapiro 1999; Brown 2009) of the Einstein equations (extensions to the CCZ4 (Alic et al. 2012) system are also in progress), along with the standard 1+log slicing condition and Gamma driver to specify the evolution of the lapse and shift, respectively (see Baiotti and Rezzolla 2016 for a review of these gauges). Given the following definitions for the evolved fields

$$\phi = \frac{1}{12} \ln(\gamma), \tag{44a}$$

$$K = \gamma^{ij} K_{ij}, \tag{44b}$$

$$\tilde{\gamma}_{ij} = e^{-4\phi} \gamma_{ij}, \tag{44c}$$

$$\tilde{A}_{ij} = e^{-4\phi} \left(K_{ij} - \frac{1}{3} \gamma_{ij} K \right), \tag{44d}$$

$$\tilde{\Gamma}^i = \tilde{\gamma}^{jk} \tilde{\Gamma}^i_{jk}, \tag{44e}$$

where $\tilde{\Gamma}^i_{jk}$ are the Christoffel symbols computed from the conformal metric $\tilde{\gamma}_{ij}$, the BSSNOK and gauge equations take the form:

$$\partial_\perp \phi = \frac{1}{6} \partial_k \beta^k - \frac{1}{6} \alpha K, \tag{45a}$$

$$\partial_\perp \tilde{\gamma}_{ij} = -2\alpha \tilde{A}_{ij} - \frac{2}{3} \tilde{\gamma}_{ij} \partial_k \beta^k, \tag{45b}$$

$$\begin{aligned}
\partial_\perp K = {}& \alpha \left(\tilde{A}_{ij} \tilde{A}^{ij} + \frac{1}{3} K^2 \right) - \gamma^{ij} \nabla_i \nabla_j \alpha \\
&+ 4\pi \left(S^k_k + E \right),
\end{aligned} \tag{45c}$$

$$\begin{aligned}
\partial_\perp \tilde{A}_{ij} = {}& e^{-4\phi} \left[\alpha (R_{ij} - 8\pi S_{ij}) - \nabla_i \nabla_j \alpha \right]^{\mathrm{TF}} \\
&- \frac{2}{3} \tilde{A}_{ij} \partial_k \beta^k + \alpha \left(K \tilde{A}_{ij} - 2\tilde{A}_{ik} \tilde{A}^k_j \right),
\end{aligned} \tag{45d}$$

$$\begin{aligned}
\partial_\perp \tilde{\Gamma}^i = {}& \tilde{\gamma}^{kl} \partial_k \partial_l \beta^i + \frac{2}{3} \tilde{\gamma}^{jk} \tilde{\Gamma}^i_{jk} \partial_l \beta^l \\
&+ \frac{1}{3} \tilde{\nabla}^i \left(\partial_k \beta^k \right) - 2\tilde{A}^{ik} \partial_k \alpha + 2\alpha \tilde{A}^{kl} \tilde{\Gamma}^i_{kl} \\
&+ 12\alpha \tilde{A}^{ik} \partial_k \phi - \frac{4}{3} \alpha \tilde{\nabla}^i K - 16\pi \alpha \tilde{\gamma}^{ij} S_j,
\end{aligned} \tag{45e}$$

$$\partial_t \alpha = -2\alpha K + \beta^k \partial_k \alpha, \tag{45f}$$

$$\partial_t \beta^i = \frac{3}{4} B^i + \beta^k \partial_k \beta^i, \tag{45g}$$

$$\partial_t B^i = \partial_t \tilde{\Gamma}^i - \eta B^i + \beta^k \partial_k B^i, \tag{45h}$$

where the operator $\partial_\perp$ stands for $\partial_t - \mathcal{L}_{\boldsymbol{\beta}}$, i.e., the derivative with respect to coordinate time minus the Lie derivative along the shift, and the notation $[\cdots]^{\mathrm{TF}}$ indicates terms that are made trace free with respect to the conformal metric. The covariant derivatives ∇ and $\tilde{\nabla}$ are constructed from the physical and covariant three metric respectively, and $\eta \sim 1/M$ (M being the mass of the system).

The three dimensional Ricci tensor R_{ij} is split in two parts, $R_{ij} = \tilde{R}^\phi_{ij} + \tilde{R}_{ij}$, the first involving the conformal factor ϕ and the second the derivatives of the conformal metric $\tilde{\gamma}_{ij}$:

$$\begin{aligned}
\tilde{R}^\phi_{ij} = {}& \phi^{-2} \left[\phi \left(\tilde{\nabla}_i \tilde{\nabla}_j \phi + \tilde{\gamma}_{ij} \tilde{\nabla}^k \tilde{\nabla}_k \phi \right) \right. \\
&\left. - 2\tilde{\gamma}_{ij} \tilde{\nabla}^k \phi \tilde{\nabla}_k \phi \right],
\end{aligned} \tag{46a}$$

$$\begin{aligned}
\tilde{R}_{ij} = {}& -\frac{1}{2} \tilde{\gamma}^{lm} \partial_l \partial_m \tilde{\gamma}_{ij} + \tilde{\gamma}_{k(i} \partial_{j)} \tilde{\Gamma}^k \\
&+ \tilde{\Gamma}^k \tilde{\Gamma}_{(ij)k} + \tilde{\gamma}^{lm} \left[2\tilde{\Gamma}^k_{l(i} \tilde{\Gamma}_{j)km} + \tilde{\Gamma}^k_{im} \tilde{\Gamma}_{kjl} \right].
\end{aligned} \tag{46b}$$

In integrating these equations a constrained approach is used, i.e., we enforce the constraints $\det \tilde{\gamma}_{ij} = 1$ and $\mathrm{tr} \tilde{A}_{ij} = 0$ at every step.

The spacetime evolution is taken care of by the McLachlan code (Brown et al. 2009) to evolve the spacetime variables in the BSSNOK formulation. McLachlan approximates the equations using standard central finite-difference operators with upwinding of the shift advection terms and Kreiss-Oliger dissipation (Kreiss and Oliger 1973) to ensure stability. It supports up to eighth-order operators, however since the major source of errors in our simulations is the hydrodynamical part, we restrict ourselves here to fourth-order accuracy for the spacetime.

Acknowledgements

We thank Kentaro Takami for providing the stellar oscillation eigenfrequencies, while Erik Schnetter, Ian Hinder, and Massimiliano Leoni for useful discussions. The simulations were performed on the SuperMUC cluster at the LRZ in Garching, on the LOEWE cluster in CSC in Frankfurt, on the HazelHen cluster at the HLRS in Stuttgart and on the Caltech compute cluster Zwicky.

Funding

This research is supported in part by the ERC synergy grant 'BlackHoleCam: Imaging the Event Horizon of Black Holes' (Grant No. 610058), by 'NewCompStar', COST Action MP1304, by the LOEWE-Program in the Helmholtz International Center (HIC) for FAIR, and by the European Union's Horizon 2020 Research and Innovation Programme (Grant 671698) (call FETHPC-1-2014, project ExaHyPE). FG is supported by HIC for FAIR and the graduate school HGS-HIRe. DR gratefully acknowledges support from the Schmidt Fellowship and the Max-Planck/Princeton Center (MPPC) for Plasma Physics (NSF PHY-1523261).

Competing interests

The authors declare that they have no competing interests.

Authors' contributions

The implementation of the EL method in `WhiskyTHC` and all of the tests were performed by FG. The project was initiated by DR and LR, and was closely supervised by both.

Author details

[1]Institut für Theoretische Physik, Goethe Universität, Max-von-Laue-Str. 1, Frankfurt am Main, 60438, Germany. [2]Institute for Advanced Study, 1 Einstein Dr., Princeton, NJ 08540, USA. [3]Department of Astrophysical Sciences, Princeton University, 4 Ivy Lane, Princeton, NJ 08544, USA. [4]Frankfurt Institute for Advanced Studies, Ruth-Moufang-Str. 1, Frankfurt am Main, 60438, Germany.

Endnotes

[a] Note that the operators (9) and (10) return approximations of the function h defined by $f_i =: \int_{x_{i-1/2}}^{x_{i+1/2}} h(x') \, dx'$ at $x_{i+1/2}$. In this sense, they act on volume averages, the point-wise flux being the volume average of h. The values $h_{i\pm1/2}$ should appear in (8) instead of $f_{i\pm1/2}$. We have here simplified the notation, but a full discussion can be found in Radice and Rezzolla (2012).

[b] Further details can be found in Radice et al. (2014a). Here, θ is computed following the algorithm in Section 2.2 of Hu et al. (2013), but replacing the rest-mass density ρ with its conserved relativistic counterpart D.

[c] The use of a higher-order stencil in the EL approach, e.g., EL7, does not yield to improvements in the solution; the treatment of the low-density regions is far more delicate and the mass conservation is degraded.

[d] Of course, for both schemes the amount of rest-mass outside the star is minute, being only 10^{-7} of the initial rest-mass for the EL5 scheme and $\sim 10^{-5}$ for the MP5 scheme.

References

Alcubierre, M: Introduction to 3 + 1 Numerical Relativity. Oxford University Press, Oxford (2008). doi:10.1093/acprof:oso/9780199205677.001.0001

Alic, D, Bona-Casas, C, Bona, C, Rezzolla, L, Palenzuela, C: Conformal and covariant formulation of the Z4 system with constraint-violation damping. Phys. Rev. D **85**(6), 064040 (2012). doi:10.1103/PhysRevD.85.064040, arXiv:1106.2254

Alic, D, Kastaun, W, Rezzolla, L: Constraint damping of the conformal and covariant formulation of the Z4 system in simulations of binary neutron stars. Phys. Rev. D **88**(6), 064049 (2013). doi:10.1103/PhysRevD.88.064049, arXiv:1307.7391

Anile, AM: Relativistic Fluids and Magneto-Fluids. Cambridge University Press, Cambridge (1990)

Ashtekar, A, Krishnan, B: Dynamical horizons and their properties. Phys. Rev. D **68**, 104030 (2003). arXiv:gr-qc/0308033

Baiotti, L, Damour, T, Giacomazzo, B, Nagar, A, Rezzolla, L: Accurate numerical simulations of inspiralling binary neutron stars and their comparison with effective-one-body analytical models. Phys. Rev. D **84**(2), 024017 (2011). doi:10.1103/PhysRevD.84.024017, arXiv:1103.3874

Baiotti, L, Hawke, I, Montero, P, Rezzolla, L: A new three-dimensional general-relativistic hydrodynamics code. In: Capuzzo-Dolcetta, R (ed.) Computational Astrophysics in Italy: Methods and Tools, vol. 1, p. 210. MSAlt, Trieste (2003)

Baiotti, L, Hawke, I, Montero, PJ, Löffler, F, Rezzolla, L, Stergioulas, N, Font, JA, Seidel, E: Three-dimensional relativistic simulations of rotating neutron-star collapse to a Kerr black hole. Phys. Rev. D **71**(2), 024035 (2005). doi:10.1103/PhysRevD.71.024035, arXiv:gr-qc/0403029

Baiotti, L, Hawke, I, Rezzolla, L: On the gravitational radiation from the collapse of neutron stars to rotating black holes. Class. Quantum Gravity **24**, 187-206 (2007). arXiv:gr-qc/0701043

Baiotti, L, Hawke, I, Rezzolla, L, Schnetter, E: Gravitational-wave emission from rotating gravitational collapse in three dimensions. Phys. Rev. Lett. **94**, 131101 (2005). arXiv:gr-qc/0503016

Baiotti, L, Rezzolla, L: Challenging the paradigm of singularity excision in gravitational collapse. Phys. Rev. Lett. **97**, 141101 (2006). arXiv:gr-qc/0608113

Baiotti, L, Rezzolla, L: Binary neutron-star mergers: a review of Einstein's richest laboratory. arXiv:1607.03540 (2016)

Banyuls, F, Font, JA, Ibáñez, JM, Martí, JM, Miralles, JA: Numerical 3 + 1 general-relativistic hydrodynamics: a local characteristic approach. Astrophys. J. **476**, 221-231 (1997). doi:10.1086/303604

Baumgarte, TW, Shapiro, SL: Numerical integration of Einstein's field equations. Phys. Rev. D **59**(2), 024007 (1999). doi:10.1103/PhysRevD.59.024007, arXiv:gr-qc/9810065

Baumgarte, TW, Shapiro, SL: Numerical Relativity: Solving Einstein's Equations on the Computer. Cambridge University Press, Cambridge (2010). doi:10.1017/cbo9781139193344

Bernuzzi, S, Dietrich, T: Gravitational waveforms from binary neutron star mergers with high-order WENO schemes in numerical relativity. arXiv:1604.07999 (2016)

Bernuzzi, S, Nagar, A, Thierfelder, M, Brügmann, B: Tidal effects in binary neutron star coalescence. Phys. Rev. D **86**(4), 044030 (2012). doi:10.1103/PhysRevD.86.044030, arXiv:1205.3403

Bona, C, Palenzuela-Luque, C, Bona-Casas, C: Elements of Numerical Relativity and Relativistic Hydrodynamics: From Einstein's Equations to Astrophysical Simulations. Lecture Notes in Physics. Springer, Berlin (2009). http://books.google.co.uk/books?id=KgPGHaCUaAYC

Brown, D, Diener, P, Sarbach, O, Schnetter, E, Tiglio, M: Turduckening black holes: an analytical and computational study. Phys. Rev. D **79**, 044023 (2009). doi:10.1103/PhysRevD.79.044023, arXiv:0809.3533 [gr-qc]

Brown, DJ: Covariant formulations of Baumgarte, Shapiro, Shibata, and Nakamura and the standard gauge. Phys. Rev. D **79**(10), 104029 (2009). doi:10.1103/PhysRevD.79.104029

Bugner, M, Dietrich, T, Bernuzzi, S, Weyhausen, A, Brügmann, B: Solving 3D relativistic hydrodynamical problems with weighted essentially nonoscillatory discontinuous Galerkin methods. Phys. Rev. D **94**(8), 084004 (2016). doi:10.1103/PhysRevD.94.084004, arXiv:1508.07147

Colella, P, Woodward, PR: The piecewise parabolic method (PPM) for gas-dynamical simulations. J. Comput. Phys. **54**(1), 174-201 (1984). doi:10.1016/0021-9991(84)90143-8

Cordero-Carrión, I, Cerdá-Durán, P, Dimmelmeier, H, Jaramillo, JL, Novak, J, Gourgoulhon, E: Improved constrained scheme for the Einstein equations:

an approach to the uniqueness issue. Phys. Rev. D **79**(2), 024017 (2009). doi:10.1103/PhysRevD.79.024017, arXiv:0809.2325

De Pietri, R, Feo, A, Maione, F, Löffler, F: Modeling equal and unequal mass binary neutron star mergers using public codes. Phys. Rev. D **93**(6), 064047 (2016). doi:10.1103/PhysRevD.93.064047, arXiv:1509.08804

DeBuhr, J, Zhang, B, Anderson, M, Neilsen, D, Hirschmann, EW: Relativistic hydrodynamics with wavelets. arXiv:1512.00386 (2015)

Einstein Toolkit: open software for relativistic astrophysics (2010). http://einsteintoolkit.org

Font, JA: Numerical hydrodynamics and magnetohydrodynamics in general relativity. Living Rev. Relativ. **11**, 7 (2008). doi:10.12942/lrr-2008-7

Font, JA, Goodale, T, Iyer, S, Miller, M, Rezzolla, L, Seidel, E, Stergioulas, N, Suen, W-M, Tobias, M: Three-dimensional numerical general relativistic hydrodynamics. II. Long-term dynamics of single relativistic stars. Phys. Rev. D **65**(8), 084024 (2002). doi:10.1103/PhysRevD.65.084024, arXiv:gr-qc/0110047

Gottlieb, S, Ketcheson, D, Shu, C-W: High order strong stability preserving time discretizations. J. Sci. Comput. **38**, 251-289 (2009). doi:10.1007/s10915-008-9239-z

Gourgoulhon, E: 3 + 1 Formalism in General Relativity. Lecture Notes in Physics, vol. 846. Springer, Berlin (2012). doi:10.1007/978-3-642-24525-1

Guermond, JL, Pasquetti, R: Entropy-based nonlinear viscosity for Fourier approximations of conservation laws. C. R. Math. Acad. Sci. **346**, 801-806 (2008). doi:10.1016/j.crma.2008.05.013

Guermond, JL, Pasquetti, R, Popov, B: Entropy viscosity method for nonlinear conservation laws. J. Comput. Phys. **230**(11), 4248-4267 (2011). doi:10.1016/j.jcp.2010.11.043

Harten, A, Engquist, B, Osher, S, Chakravarthy, SR: Uniformly high order accurate essentially non-oscillatory schemes III. J. Comput. Phys. **71**, 231-303 (1987). doi:10.1016/0021-9991(87)90031-3

Hu, XY, Adams, NA, Shu, C-W: Positivity-preserving method for high-order conservative schemes solving compressible Euler equations. J. Comput. Phys. **242**, 169-180 (2013). doi:10.1016/j.jcp.2013.01.024, arXiv:1203.1540

Jiang, G-S, Shu, C-W: Efficient implementation of weighted eno schemes. J. Comput. Phys. **126**, 202-228 (1996)

Kastaun, W, Galeazzi, F, Alic, D, Rezzolla, L, Font, JA: Black hole from merging binary neutron stars: how fast can it spin? Phys. Rev. D **88**(2), 021501 (2013). doi:10.1103/PhysRevD.88.021501, arXiv:1301.7348

Kidder, LE, Field, SE, Foucart, F, Schnetter, E, Teukolsky, SA, Bohn, A, Deppe, N, Diener, P, Hébert, F, Lippuner, J, Miller, J, Ott, CD, Scheel, MA, Vincent, T: SpECTRE: A task-based discontinuous Galerkin code for relativistic astrophysics. arXiv:1609.00098 (2016)

Kreiss, HO, Oliger, J: Methods for the Approximate Solution of Time Dependent Problems. GARP Publication Series, vol. 10, Geneva (1973)

Lele, SK: Compact finite difference schemes with spectral-like resolution. J. Comput. Phys. **103**, 16-42 (1992)

Leveque, RJ: Numerical Methods for Conservation Laws. Birkhäuser, Basel (1992)

Liu, X-D, Osher, S, Chan, T: Weighted essentially non-oscillatory schemes. J. Comput. Phys. **115**, 200-212 (1994). doi:10.1006/jcph.1994.1187

Löffler, F, Faber, J, Bentivegna, E, Bode, T, Diener, P, Haas, R, Hinder, I, Mundim, BC, Ott, CD, Schnetter, E, Allen, G, Campanelli, M, Laguna, P: The Einstein Toolkit: a community computational infrastructure for relativistic astrophysics. Class. Quantum Gravity **29**(11), 115001 (2012). doi:10.1088/0264-9381/29/11/115001, arXiv:1111.3344

Martí, JM, Müller, E: Extension of the piecewise parabolic method to one-dimensional relativistic hydrodynamics. J. Comput. Phys. **123**, 1-14 (1996). doi:10.1006/jcph.1996.0001

Martí, JM, Müller, E: Numerical hydrodynamics in special relativity. Living Rev. Relativ. **6**, 7 (2003). doi:10.12942/lrr-2003-7

Martí, JM, Müller, E: Grid-based methods in relativistic hydrodynamics and magnetohydrodynamics. Living Rev. Comput. Astrophys. **1**, 3 (2015). doi:10.1007/lrca-2015-3

Mignone, A, Tzeferacos, P, Bodo, G: High-order conservative finite difference GLM-MHD schemes for cell-centered MHD. J. Comput. Phys. **229**, 5896-5920 (2010). doi:10.1016/j.jcp.2010.04.013, arXiv:1001.2832

Miller, JM, Schnetter, E: An operator-based local discontinuous Galerkin method compatible with the BSSN formulation of the Einstein equations. arXiv:1604.00075 (2016)

Paschalidis, V: General relativistic simulations of compact binary mergers as engines of short gamma-ray bursts. arXiv:1611.01519 (2016)

Radice, D, Galeazzi, F, Lippuner, J, Roberts, LF, Ott, CD, Rezzolla, L: Dynamical mass ejection from binary neutron star mergers. Mon. Not. R. Astron. Soc. **460**, 3255-3271 (2016). doi:10.1093/mnras/stw1227, arXiv:1601.02426

Radice, D, Rezzolla, L: Discontinuous Galerkin methods for general-relativistic hydrodynamics: formulation and application to spherically symmetric spacetimes. Phys. Rev. D **84**(2), 024010 (2011). doi:10.1103/PhysRevD.84.024010, arXiv:1103.2426

Radice, D, Rezzolla, L: THC: a new high-order finite-difference high-resolution shock-capturing code for special-relativistic hydrodynamics. Astron. Astrophys. **547**, 26 (2012). doi:10.1051/0004-6361/201219735, arXiv:1206.6502

Radice, D, Rezzolla, L, Galeazzi, F: High-order fully general-relativistic hydrodynamics: new approaches and tests. Class. Quantum Gravity **31**(7), 075012 (2014a). doi:10.1088/0264-9381/31/7/075012, arXiv:1312.5004

Radice, D, Rezzolla, L, Galeazzi, F: Beyond second-order convergence in simulations of binary neutron stars in full general-relativity. Mon. Not. R. Astron. Soc. Lett. **437**, 46-50 (2014b). doi:10.1093/mnrasl/slt137, arXiv:1306.6052

Radice, D, Rezzolla, L, Galeazzi, F: High-order numerical-relativity simulations of binary neutron stars. In: Numerical Modeling of Space Plasma Flows: ASTRONUM-2014. ASP Conference Series, vol. 498, pp. 121-126 (2015). arXiv:1502.00551

Radice, D, Rezzolla, L, Kellerman, T: Critical phenomena in neutron stars: I. Linearly unstable nonrotating models. Class. Quantum Gravity **27**(23), 235015 (2010). doi:10.1088/0264-9381/27/23/235015, arXiv:1007.2809.

Read, JS, Baiotti, L, Creighton, JDE, Friedman, JL, Giacomazzo, B, Kyutoku, K, Markakis, C, Rezzolla, L, Shibata, M, Taniguchi, K: Matter effects on binary neutron star waveforms. Phys. Rev. D **88**(4), 044042 (2013). doi:10.1103/PhysRevD.88.044042, arXiv:1306.4065

Rezzolla, L, Zanotti, O: Relativistic Hydrodynamics. Oxford University Press, Oxford (2013). doi:10.1093/acprof:oso/9780198528906.001.0001

Roe, PL: Some contributions to the modelling of discontinuous flows. In: Lee, RL, Sani, RL, Shih, TM, Gresho, PM (eds.) Large-Scale Computations in Fluid Mechanics, pp. 163-193 (1985)

Sagaut, P, Grohens, R: Discrete filters for large eddy simulation. Int. J. Numer. Methods Fluids **31**, 1195-1220 (1999). doi:10.1103/PhysRevD.79.104029

Shibata, M: Numerical Relativity. World Scientific, Singapore (2016). doi:10.1142/9692

Shibata, M, Nakamura, T: Evolution of three-dimensional gravitational waves: harmonic slicing case. Phys. Rev. D **52**, 5428-5444 (1995). doi:10.1103/PhysRevD.52.5428

Shibata, M, Taniguchi, K: Coalescence of black hole-neutron star binaries. Living Rev. Relativ. **14**, 6 (2011). doi:10.12942/lrr-2011-6

Shu, CW: Essentially non-oscillatory and weighted essentially non-oscillatory schemes for hyperbolic conservation laws. Lecture notes ICASE report 97-65; NASA CR-97-206253, NASA Langley Research Center. http://ntrs.nasa.gov/archive/nasa/casi.ntrs.nasa.gov/19980007543_1998045663.pdf (1997)

Sod, GA: A survey of several finite difference methods for systems of nonlinear hyperbolic conservation laws. J. Comput. Phys. **27**, 1-31 (1978). doi:10.1016/0021-9991(78)90023-2

Stergioulas, N, Friedman, JL: Comparing models of rapidly rotating relativistic stars constructed by two numerical methods. Astrophys. J. **444**, 306-311 (1995). doi:10.1086/175605, arXiv:astro-ph/9411032

Suresh, A, Huynh, HT: Accurate monotonicity-preserving schemes with Runge-Kutta time stepping. J. Comput. Phys. **136**(1), 83-99 (1997). doi:10.1006/jcph.1997.5745

Takami, K, Rezzolla, L, Yoshida, S: A quasi-radial stability criterion for rotating relativistic stars. Mon. Not. R. Astron. Soc. **416**, 1-5 (2011). doi:10.1111/j.1745-3933.2011.01085.x, arXiv:1105.3069

Tchekhovskoy, A, McKinney, JC, Narayan, R: WHAM: a WENO-based general relativistic numerical scheme? I. Hydrodynamics. Mon. Not. R. Astron. Soc. **379**(2), 469-497 (2007). doi:10.1111/j.1365-2966.2007.11876.x

Thierfelder, M, Bernuzzi, S, Brügmann, B: Numerical relativity simulations of binary neutron stars. Phys. Rev. D **84**(4), 044012 (2011). doi:10.1103/PhysRevD.84.044012, arXiv:1104.4751

Thierfelder, M, Bernuzzi, S, Hilditch, D, Bruegmann, B, Rezzolla, L: The trumpet solution from spherical gravitational collapse with puncture gauges. Phys. Rev. D **83**, 064022 (2010). arXiv:1012.3703

Tsatsin, P, Marronetti, P: Initial data for neutron star binaries with arbitrary spins. Phys. Rev. D **88**, 064060 (2013). doi:10.1103/PhysRevD.88.064060

Vichnevetsky, R, Bowles, JB: Fourier Analysis of Numerical Approximations of Hyperbolic Equations. SIAM, Philadelphia (1982)

Wu, K: Design of provably physical-constraint-preserving methods for general relativistic hydrodynamics. Phys. Rev. D **95**, 103001 (2017). arXiv:1610.06274

Yoshida, S, Eriguchi, Y: Quasi-radial modes of rotating stars in general relativity. Mon. Not. R. Astron. Soc. **322**, 389-396 (2001)

Zanotti, O, Fambri, F, Dumbser, M: Solving the relativistic magnetohydrodynamics equations with ADER discontinuous Galerkin methods, a posteriori subcell limiting and adaptive mesh refinement. Mon. Not. R. Astron. Soc. **452**, 3010-3029 (2015). doi:10.1093/mnras/stv1510, arXiv:1504.07458

Zhang, W, MacFadyen, AI: RAM: a relativistic adaptive mesh refinement hydrodynamics code. Astrophys. J. Suppl. Ser. **164**, 255-279 (2006)

Zilhão, M, Löffler, F: An introduction to the Einstein toolkit. Int. J. Mod. Phys. A **28**, 40014 (2013). doi:10.1142/S0217751X13400149, arXiv:1305.5299

Zingan, V, Guermond, JL, Morel, J, Popov, B: Implementation of the entropy viscosity method with the discontinuous Galerkin method. Comput. Methods Appl. Mech. Eng. **253**, 479-490 (2013). doi:10.1016/j.cma.2012.08.018

10

GPU-enabled particle-particle particle-tree scheme for simulating dense stellar cluster system

Masaki Iwasawa[1,2*], Simon Portegies Zwart[2] and Junichiro Makino[1,3]

Abstract

We describe the implementation and performance of the P^3T (Particle-Particle Particle-Tree) scheme for simulating dense stellar systems. In P^3T, the force experienced by a particle is split into short-range and long-range contributions. Short-range forces are evaluated by direct summation and integrated with the fourth order Hermite predictor-corrector method with the block timesteps. For long-range forces, we use a combination of the Barnes-Hut tree code and the leapfrog integrator. The tree part of our simulation environment is accelerated using graphical processing units (GPU), whereas the direct summation is carried out on the host CPU. Our code gives excellent performance and accuracy for star cluster simulations with a large number of particles even when the core size of the star cluster is small.

PACS Codes: 95.10.Ce; 98.10.+z

Keywords: methods: N-body simulations

1 Background

Direct N-body simulation has been the most useful tool for the study of the evolution of collisional stellar systems such as star clusters and the center of the galaxy (Aarseth 1963). The force calculations, of which the cost is $O(N^2)$, are the most compute-intensive part of direct N-body simulations. Barnes and Hut (1986) developed a scheme which reduces the calculation cost to $O(N \log N)$ by constructing the tree structure and evaluating the multipole expansions. Dehnen (2002, 2014) developed a scheme to reduce the calculation cost to $O(N)$ by combining the fast multipole method (Greengard and Rokhlin 1987) and the tree code. Recently, the graphical processing units (GPU), which is a device originally developed for rendering the graphical image, started to be used for scientific simulations. The tree code is also implemented on GPUs and it is much faster than it is on CPUs (Gaburov et al. 2010; Bédorf et al. 2012). Bédorf et al. (2014) parallelized the tree

code on GPUs and showed good scalability up to 18,600 GPUs. They also simulated the Milky Way Galaxy with N of up to 242 billion and reported that the average calculation time per iteration on 18,600 GPUs was 4.8 seconds.

The tree schemes are widely used for collisionless system simulations. However, for collisional system simulations, the use of the tree code has been very limited. One reason might be that a collisional stellar system spans a wide range in timescales. Thus it is essential that each particle has its own integration timestep. This scheme is called the individual timestep or the block timestep (McMillan 1986). However, when we use the tree code and the block timestep together, the tree structure is reconstructed at every block timestep, because the positions of integrated particle are updated. The cost of the usual complete reconstruction of the tree is $O(N \log N)$ and not negligible.

To reduce the cost of the reconstruction of the tree, McMillan and Aarseth (1993) introduced local reconstruction of tree. They demonstrated a good performance, but there seems to be no obvious way to parallelize their scheme.

---

*Correspondence: masaki.iwasawa@riken.jp
[1]RIKEN Advanced Institute for Computational Science,
Minatojima-minamimachi, Chuo-ku, Kobe, Japan
[2]Sterrewacht Leiden, P.O. Box 9513, 2300 RA, The Netherlands

Recently, Oshino et al. (2011) introduced another approach to combine the tree code and the block timesteps which they called the P^{3}T scheme. This scheme is based on the idea of Hamiltonian splitting (Kinoshita et al. 1991; Wisdom and Holman 1991; Duncan et al. 1998; Chambers 1999; Brunini and Viturro 2003; Fujii et al. 2007; Moore and Quillen 2011). In the P^{3}T scheme, the Hamiltonian of the system is split into short-range and long-range parts and they are integrated with different integrators. The long-range part is evaluated with the tree code and is integrated using the leapfrog scheme with a shared timestep. The short range part is evaluated with direct summation and integrated using the fourth-order Hermite scheme (Makino and Aarseth 1992) with the block timesteps. They investigated the accuracy and the performance of the P^{3}T scheme for planetary formation simulations and showed that the P^{3}T scheme achieves high performance.

In this paper, we present the implementation of the P^{3}T scheme on GPUs and report its accuracy and performance for star cluster simulations. We found that the P^{3}T scheme demonstrates a very good performance for star cluster simulations, even when the core of the cluster becomes small.

The structure of this paper is as follows. In Section 2, we briefly describe the P^{3}T scheme. In Section 3, we report the accuracy and performance of the P^{3}T scheme. We summarize these results in Section 4.

2 Methods

2.1 Formulation

In this section, we describe the P^{3}T scheme. The Hamiltonian H of a gravitational N-body system is given by

$$H = \sum_i^N \frac{|\boldsymbol{p}_i|^2}{2m_i} - \sum_i^N \sum_{i<j}^N \frac{Gm_im_j}{s_{ij}}, \tag{1}$$

$$s_{ij} = \sqrt{|\boldsymbol{q}_{ij}|^2 + \epsilon^2}, \tag{2}$$

$$\boldsymbol{q}_{ij} = \boldsymbol{q}_i - \boldsymbol{q}_j, \tag{3}$$

where $\boldsymbol{p}_i$, m_i and $\boldsymbol{q}_i$ are momentum, mass and position of the particle i, respectively. To avoid the singularity of the $1/r$ potential, we use the Plummer softening ϵ (Aarseth 1963). With the P^{3}T scheme, H is split into H_{hard} and H_{soft} as follows (Oshino et al. 2011):

$$H = H_{\text{hard}} + H_{\text{soft}}, \tag{4}$$

$$H_{\text{hard}} = \sum_i^N \frac{|\boldsymbol{p}_i|^2}{2m_i} - \sum_i^N \sum_{i<j}^N \frac{m_im_j}{s_{ij}} \left[1 - W(s_{ij}) \right], \tag{5}$$

$$H_{\text{soft}} = -\sum_i^N \sum_{i<j}^N \frac{m_im_j}{s_{ij}} W(s_{ij}). \tag{6}$$

Here $W(s_{ij})$ is a smooth transition function. A suitable form of $W(s_{ij})$ should be zero when a distance between two particles is smaller than the inner cutoff radius r_{in} and should be unity if the distance is larger than the outer cutoff radius r_{cut}. This splitting is introduced by Chambers (1999) to avoid undesirable energy error from close encounters between particles. Similar splitting has been used with P^{3}M (Particle-Particle Particle-Mesh) scheme, in which the long-range part of the interaction is evaluated by using FFT (Hockney and Eastwood 1981).

Forces derived from H_{hard} and H_{soft} are given by

$$\boldsymbol{F}_{\text{hard},i} = -\frac{\partial H_{\text{hard}}}{\partial \boldsymbol{q}_i} = -\sum_{j\neq i}^N \frac{m_im_j}{s_{ij}^3} \left(1 - K(s_{ij})\right) \boldsymbol{q}_{ij}, \tag{7}$$

$$\boldsymbol{F}_{\text{soft},i} = -\frac{\partial H_{\text{soft}}}{\partial \boldsymbol{q}_i} = -\sum_{j\neq i}^N \frac{m_im_j}{s_{ij}^3} K(s_{ij}) \boldsymbol{q}_{ij}, \tag{8}$$

$$K(s_{ij}) = W(s_{ij}) - s_{ij} \frac{dW(s_{ij})}{ds_{ij}}. \tag{9}$$

We call $K(s_{ij})$ the cutoff function.

The tree algorithm is used for the evaluation of $\boldsymbol{F}_{\text{soft},i}$ to reduce the calculation cost.

The formal solution of the equation of motion for the phase space coordinate $\boldsymbol{w} = (\boldsymbol{q}, \boldsymbol{p})$ at time $t + \delta t$ for the given Hamiltonian H is

$$\boldsymbol{w}(t + \delta t) = e^{\delta t\{,H\}} \boldsymbol{w}(t) = e^{\delta t\{,H_{\text{soft}}+H_{\text{hard}}\}} \boldsymbol{w}(t). \tag{10}$$

Here the braces $\{,\}$ stand for the Poisson bracket. In the P^{3}T scheme, we use the second order approximation;

$$\boldsymbol{w}(t + \delta t) = e^{\delta t/2\{,H_{\text{soft}}\}} e^{\delta t\{,H_{\text{hard}}\}} e^{\delta t/2\{,H_{\text{soft}}\}} \boldsymbol{w}(t) + O(\delta t^3). \tag{11}$$

Here, the formal solution for the H_{soft} term is the simple velocity kick, since H_{soft} contains the potential only. We numerically integrate the H_{hard} term, since it cannot be solved analytically. We use the fourth-order Hermite scheme with the block timestep (Makino and Aarseth 1992). The fourth-order integrator requires $K(s_{ij})$ to be three-times differentiable with respect to position. We use the following formula:

$$K(x) = \begin{cases} 0 & (x < 0), \\ -20x^7 + 70x^6 - 84x^5 + 35x^4 & (0 \leq x < 1), \\ 1 & (1 \leq x), \end{cases} \tag{12}$$

$$x = \frac{y - \gamma}{1 - \gamma}, \tag{13}$$

$$y = \frac{s_{ij}}{r_{\text{cut}}}, \tag{14}$$

$$\gamma = \frac{r_{\text{in}}}{r_{\text{cut}}}. \tag{15}$$

This $K(x)$ is the lowest-order polynomial which satisfies the requirement that derivatives up to the third order is zero for $x = 0$ and 1 (i.e. the highest-order term of the lowest-order polynomial is the seventh, because there are eight boundary conditions at $x = 0$ and $x = 1$).

In Figure 1, we plot $K(y)$ (top panel) and forces (bottom panel) with $\gamma = 0.1$. According to Oshino et al. (2011), Chambers (1999), $K(y)$ with $\gamma = 0.1$, is smooth enough to be integrated. Thus, for all calculations, we use $\gamma = 0.1$. The functional form of $W(y;\gamma)$ is given by

$$W(y;\gamma)$$

$$= \begin{cases} \frac{7(\gamma^6 - 9\gamma^5 + 45\gamma^4 - 60\gamma^3 \log\gamma - 45\gamma^2 + 9\gamma - 1)}{3(\gamma-1)^7}y & (y < \gamma), \\ G(y;\gamma) + (1 - G(1;\gamma))y & (\gamma \le y < 1), \\ 1 & (1 \le y), \end{cases} \tag{16}$$

$$G(y;\gamma) = \left(-10/3y^7 + 14(\gamma + 1)y^6 - 21(\gamma^2 + 3\gamma + 1)y^5 \right.$$

$$+ (35(\gamma^3 + 9\gamma^2 + 9\gamma + 1)/3)y^4$$

$$- 70(\gamma^3 + 3\gamma^2 + \gamma)y^3$$

$$+ 210(\gamma^3 + \gamma^2)y^2 - 140\gamma^3 y \log(y)$$

$$\left. + (\gamma^7 - 7\gamma^6 + 21\gamma^5 - 35\gamma^4)\right)/(1-\gamma)^7. \tag{17}$$

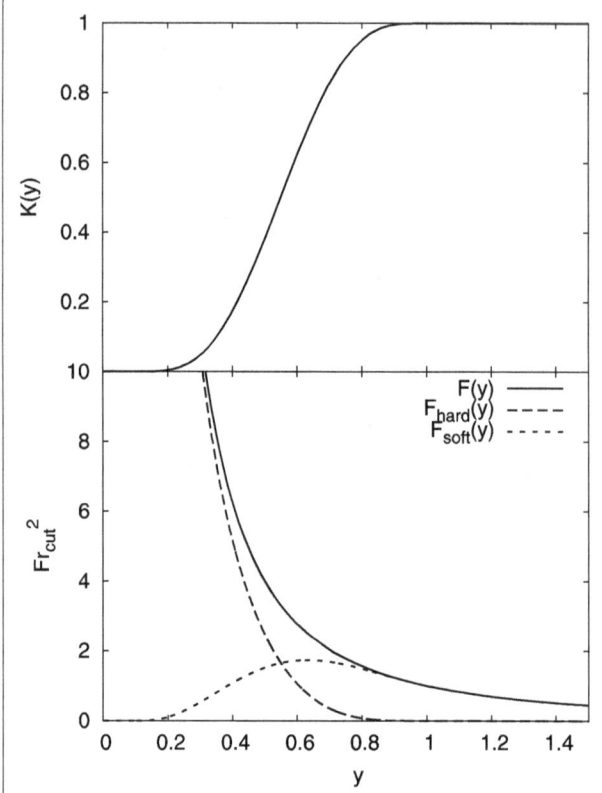

Figure 1 The cutoff function $K(y)$ (top) and the forces (bottom) as functions of $y = s_{ij}/r_{\text{cut}}$.

With the P^{3}T scheme, the time integration proceeds as follows:

(1) At time t, by using the tree code, calculate the acceleration due to H_{soft}, $\boldsymbol{a}_{\text{soft},i}$, and construct a list of all particles which come within r_{cut} from particle i for Δt_{soft}. Here, Δt_{soft} is the timestep for the soft Hamiltonian.

(2) Update the velocities of all particles with $\boldsymbol{v}_{\text{new},i} = \boldsymbol{v}_{\text{old},i} + (1/2)\Delta t_{\text{soft}}\boldsymbol{a}_{\text{soft},i}$.

(3) Integrate all particles to time $t + \Delta t_{\text{soft}}$ under H_{hard}, using the neighbour list and the fourth order Hermite integrator with the block timesteps.

(4) Calculate the acceleration due to H_{soft} at new time $t + \Delta t_{\text{soft}}$ and update the velocity

(5) Go back to step 2.

For the timestep criterion for the block timestep, we use the following form (Oshino et al. 2011).

$$\Delta t_i = \min\left(\eta \sqrt{\frac{\sqrt{|\boldsymbol{a}_i^{(0)}|^2 + a_0^2}|\boldsymbol{a}_i^{(2)}| + |\boldsymbol{a}_i^{(1)}|^2}{|\boldsymbol{a}_i^{(0)}||\boldsymbol{a}_i^{(3)}| + |\boldsymbol{a}_i^{(2)}|^2}}, \Delta t_{\text{max}}\right), \tag{18}$$

$$a_0 = \alpha \frac{m}{r_{\text{cut}}^2}. \tag{19}$$

Here η is the accuracy parameter of the timestep and its typical value is 0.1. Δt_{max} is the maximum timestep which should be smaller than Δt_{soft}, $\boldsymbol{a}_i^{(n)}$ is the nth time derivative of the acceleration of particle i, a_0 is a constant introduced to prevent Δt_i from becoming too small when the distance to the nearest neighbor is close to r_{cut} and α is a parameter to control a_0. In this case, the acceleration from H_{hard} becomes very small and there is no need to use very small Δt_i. According to Oshino et al. (2011), when we choose $\alpha \le 1$, α hardly affects the energy error. Thus we set $\alpha = 0.1$ for all simulations.

In our Hermite implementation, $\boldsymbol{a}_i^{(2)}$ and $\boldsymbol{a}_i^{(3)}$ are derived using interpolation of $\boldsymbol{a}_i^{(0)}$ and $\boldsymbol{a}_i^{(1)}$, and as a consequence we cannot use equation (18) for the first step. We use:

$$\Delta t_i = \min\left(\eta_s \sqrt{\frac{|\boldsymbol{a}_i^{(0)}|^2 + a_0^2}{|\boldsymbol{a}_i^{(1)}|^2}}, \Delta t_{\text{max}}\right). \tag{20}$$

This criterion dose not contain the 2nd and 3rd time derivatives of the acceleration. To prevent the timestep derived by equation (20) from becoming too large, we set η_s to be the one-tenth of η for all simulation in this paper.

We summarize all accuracy parameters in Table 1.

Table 1 Symbols and definitions for the accuracy parameters of the P^3T scheme

α	timestep softening. For all runs $\alpha = 0.1$
γ	ratio of inner and outer cutoff radius (r_{in}/r_{cut}). For all runs $\gamma = 0.1$
Δr_{buff}	width of the buffer shell. $\Delta r_{buff} = 3\sigma \Delta t_{soft}$, as a standard value
Δt_{soft}	timestep of the soft part. $\Delta t_{soft} = (1/256)(N/16K)^{-1/3}$, as a standard value
Δt_{max}	maximum timestep of the hard part. $\Delta t_{max} = \Delta t_{soft}/4$, as a standard value
ϵ	plummer softening length. $\epsilon = (4/N)$, as a standard value
η	accuracy parameter for timestep criterion. $\eta = 0.1$, as a standard value
r_{cut}	outer cutoff radius of smooth transition functions W and K. $r_{cut} = 4\Delta t_{soft}$, as a standard value
r_{in}	inner cutoff radius of smooth transition functions W and K ($r_{in} = \gamma r_{cut}$)
θ	opening criterion for tree. $\theta = 0.4$, as a standard value

2.2 Implementation on GPUs

Even with the Barnes-Hut tree algorithm, obtaining $\boldsymbol{F}_{soft,i}$ is still costly and dominates the total calculation time (Oshino et al. 2011). To accelerate this part, we use GPUs, by modifying the `sequoia` library (Bédorf, Gaburov and Portegies Zwart, submitted to ComAC), on which the high-performance tree code for parallel GPUs `Bonsai` (Bédorf et al. 2012) is based. Our library calculates the long range forces on all particles, $\boldsymbol{F}_{soft,i}$ by the Barnes-Hut tree algorithm (up to the quadrupole moment). On the other hand, we calculate $\boldsymbol{F}_{hard,i}$ on the host computer. The library also returns, for each particle, the list of particles within the distance h from it. We use this list of neighbors to calculate $\boldsymbol{F}_{hard,i}$. The value of h should be sufficiently larger than r_{cut} to guarantee that the particles which are not on the list of the neighbors of particle i do not enter the sphere of the radius r_{cut} around particle i during the time interval Δt_{soft}.

We call the sphere with a radius of r_{cut} the neighbor sphere and the shell between the sphere with a radius of h and the neighbor sphere the buffer shell. The particles of which the nearest neighbor is outside the sphere with radius h are considered isolated and the particles on the list of neighbors are considered neighbor particles. We denote the width of the buffer shell as Δr_{buff} (i.e. $h = r_{cut} + \Delta r_{buff}$).

The compute procedures of our implementation of the P^3T scheme on GPU is as follows:

(1) Evaluate long range forces on all particles $\boldsymbol{F}_{soft,i}$ using GPU.
(2) Particles are divided into two groups; isolated and non-isolated, by using the neighbour list made on GPU.
(3) For non-isolated particles, $\boldsymbol{F}_{hard,i}$ are calculated on the host computer.
(4) All particles receive a velocity kick through $\boldsymbol{F}_{soft,i}$ for $\Delta t_{soft}/2$.
(5) Isolated particles are drifted by $\boldsymbol{r}_i \leftarrow \boldsymbol{r}_i + \Delta t_{soft}\boldsymbol{v}_i$.
(6) Non-isolated particles are integrated with the fourth-order Hermite scheme for Δt_{soft}.
(7) Evaluate $\boldsymbol{F}_{soft,i}$ and make the neighbour list in the same way as in step 1-2.

(8) All particles obtain the velocity kick again for $\Delta t_{soft}/2$.
(9) go back to step 3.

3 Results
3.1 Accuracy and performance

We performed a number of test calculations using the P^3T scheme on GPUs, to study its accuracy and performance. In this section, we describe the result of these tests. For most of them we adopted a Plummer model (Plummer 1911) with 128K (hereafter K = 2^{10}) equal-mass particles as the initial condition. We use the so-called N-body unit or Heggie unit, in which total mass M = 1, the gravitational constant G = 1 and total energy E = $-1/4$ (Heggie and Mathieu 1986). To avoid the singularity of the gravitational potential, we use the Plummer softening and set $\epsilon = 4/N$. Since this value is a typical separation of a hard binary in the N-body unit, we can follow the evolution of the system up to the moment of the core collapse.

Note, in this paper, we use the energy errors as an indicator of the accuracy of the scheme. However, energy conservation dose not guarantee accuracy of simulations (though it is necessary). Thus we will perform realistic simulations in Section 3.2 and check the statistical character of stellar systems by comparing the results with the Hermite scheme, which is widely used in collisional stellar system simulations. As we will see later, for simulations of the core collapse of the star cluster, when the relative energy error is $\lesssim 10^{-3}$ at the moment of the core collapse, the behavior of the core collapse with the P^3T scheme agrees with that with the Hermite scheme very well.

3.1.1 Accuracy

With the P^3T scheme, we have six accuracy parameters. First, we discuss how each parameter controls the accuracy of the P^3T scheme. Finally, we describe the accumulation of the energy error in a long-term integration. To measure energy errors accurately, we calculate potential energies by the direct summation instead of the tree code for all runs in this paper.

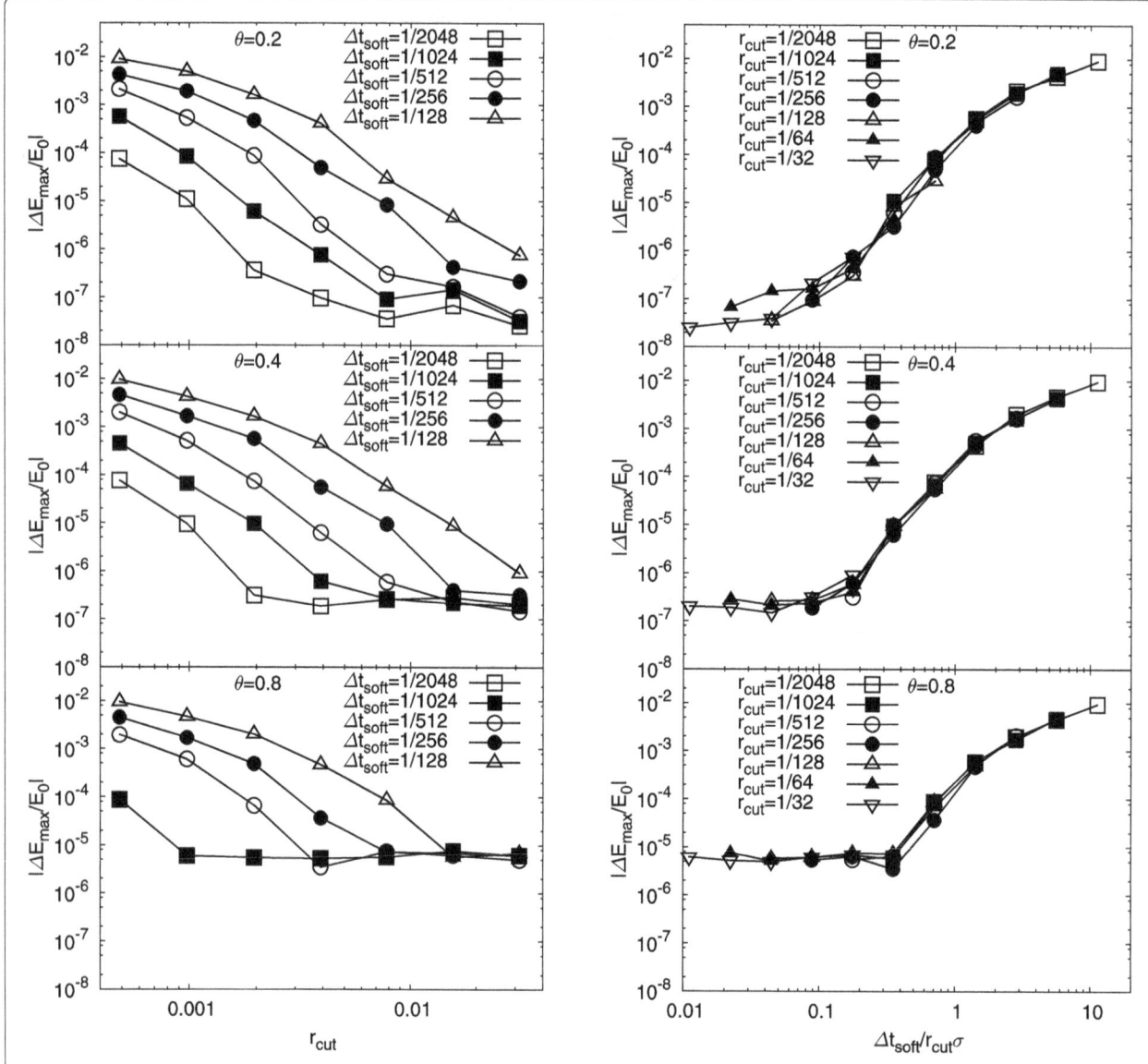

Figure 2 Maximum relative energy errors as functions of r_{cut} (left) and $\Delta t_{soft}/r_{cut}\sigma$ (right). Top, middle and bottom panels show the results for $\theta = 0.2, 0.4$ and 0.8, respectively. For all runs, we use $\eta = 0.1$, $\Delta t_{max} = \Delta t_{soft}/4$ and $\Delta r_{buff} = 3\sigma\Delta t_{soft}$.

Effect of r_{cut}, Δt_{soft} and θ In Figure 2, we present the maximum relative energy error $|\Delta E_{max}/E_0|$ over 10 N-body time units as a function of r_{cut} and Δt_{soft} for several different values of the opening criterion of the tree, θ. Here ΔE_{max} is the maximum energy error and E_0 is the initial energy. We choose $\eta = 0.1$, $\Delta t_{max} = \Delta t_{soft}/4$ and $\Delta r_{buff} = 3\sigma\Delta t_{soft}$, where σ is the global three dimensional velocity dispersion and we adopt $\sigma = 1/\sqrt{2}$.

We can see that the error is smaller for smaller θ, smaller Δt_{soft}, or larger r_{cut}. Roughly speaking, the error depends on two terms, $\Delta t_{soft}/r_{cut}\sigma$ and θ. If $\Delta t_{soft}/r_{cut}\sigma$ is large, it determines the error. In this regime, the error is dominated by the truncation error of the leapfrog integrator. If it is

small enough, θ determines the error, in other words, the tree force error dominates the total error. Even for a very small value of θ like 0.2, the tree force error dominates if $\Delta t_{soft}/r_{cut}\sigma \lesssim 0.05$.

In Figure 3, we plot the maximum energy error as a function of θ. We use the same η, Δt_{max} and Δr_{buff} as in Figure 2. For the runs with $r_{cut} = 1/256$ and $\Delta t_{soft} = 1/512$, the energy error does not drop below 10^{-6} because the error of the leapfrog integrator is larger than the tree force error. In an chaotic system like the model used in our simulations such energy error is sufficient to warrant a scientifically reliable result (Portegies Zwart and Boekholt 2014). On the other hand, for the run with $r_{cut} = 1/128$ and

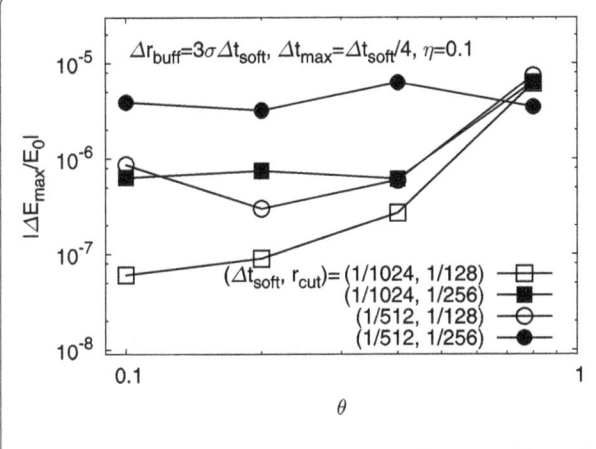

Figure 3   Maximum relative energy error as a function of θ. For all runs, we use $\eta = 0.1$, $\Delta t_{max} = \Delta t_{soft}/4$ and $\Delta r_{buff} = 3\sigma \Delta t_{soft}$.

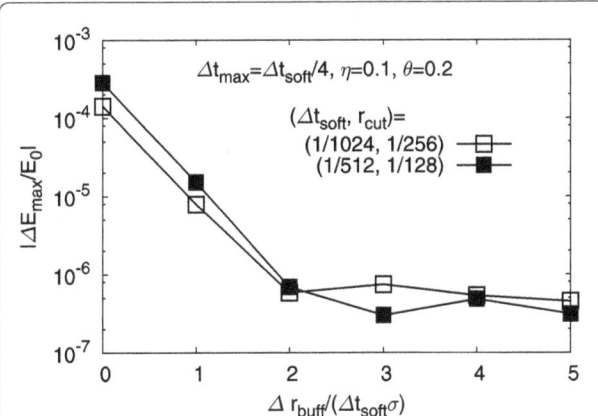

Figure 4   Maximum relative energy error as a function of Δr_{buff} in unit of $\Delta t_{soft}\sigma$. Here σ is the global three dimensional velocity dispersion of the system ($= 1/\sqrt{2}$). For all runs, we use $\eta = 0.1$, $\Delta t_{soft} = 1/512$, $t_{max} = \Delta t_{soft}/4$, $\theta = 0.1$ and $\Delta r_{buff} = 3\sigma \Delta t_{soft}$.

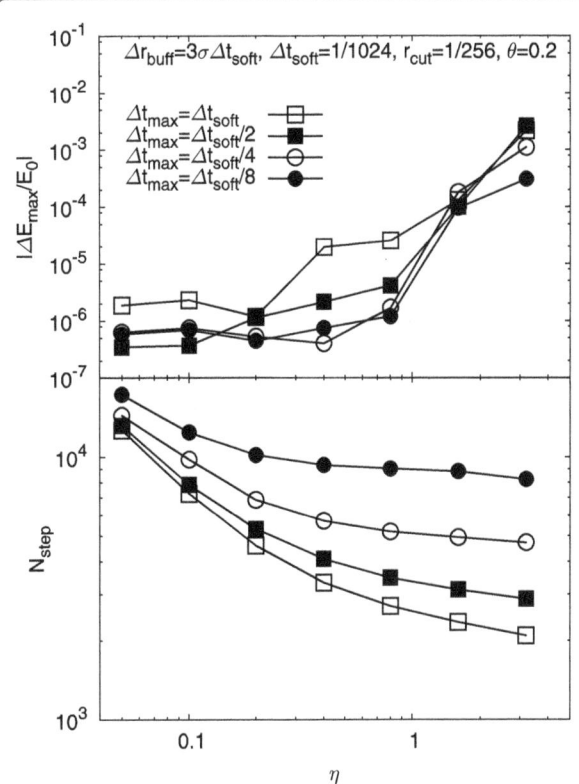

Figure 5   Maximum relative energy error and the steps for the Hermite part against η. Top and bottom panels show the maximum relative energy error and the steps for the Hermite part par particle par unit time against η, respectively.

$\Delta t_{soft} = 1/1,024$, integration error is smaller than the tree force error.

Effect of Δr_{buff}   In Figure 4, we show the maximum relative energy error as a function of Δr_{buff} for the runs with $\Delta t_{max} = \Delta t_{soft}/4$, $\eta = 0.1$, $\theta = 0.2$, for $(\Delta t_{soft}, r_{cut}) = (1/512, 1/128)$ and $(1/1,024, 1/256)$. The energy error is almost constant for $\Delta r_{buff} \gtrsim 2\Delta t_{soft}\sigma$, which indicates that the energy error for $\Delta r_{buff} < 2\Delta t_{soft}\sigma$ is caused by particles that are initially outside the buffer shell (with radius $r_{cut} + \Delta r_{buff}$) and plunge into the neighbour sphere (with radius r_{cut}) during the timestep Δt_{soft}. We can prevent this by adopting $\Delta r_{buff} \gtrsim 2\Delta t_{soft}\sigma$.

Effect of Δt_{max} and η   The maximum relative energy errors over 10 N-body time units are shown in the top panel of Figure 5 as a function of η and the number of steps for

the Hermite part (per particle per unit time, N_{step}) are presented in the bottom panel. The energy errors go down as η decrease until $\eta \sim 0.2$. For $\eta \lesssim 0.2$, the errors hardly depend on Δt_{max}.

Long term integration   In Figure 6, we show the time evolution of the relative energy error until $T = 500$. We compare the accuracy of our P^{3}T scheme with two other schemes, the direct fourth-order Hermite scheme and the leapfrog scheme with the Barnes-Hut tree code. The calculations with the direct Hermite scheme are performed by using the Sapporo library on GPU (Gaburov and Harfst 2009), and the calculations with the leapfrog scheme are performed by using the Bonsai library on GPU (Bédorf et al. 2012). The energy error of the P^{3}T scheme behaves like a random walk whereas that of the leapfrog and the Hermite schemes grow monotonically. In the right-hand panels of Figure 6, we show the same evolution of the error as in the left panels, but time is plotted with a logarithmic scale. This allows us to realize that the error growth of Hermite and tree schemes are linear, whereas the error in the P^{3}T scheme grows as $\propto T^{1/2}$. This latter proportionality is caused by the short-term error of the P^{3}T scheme,

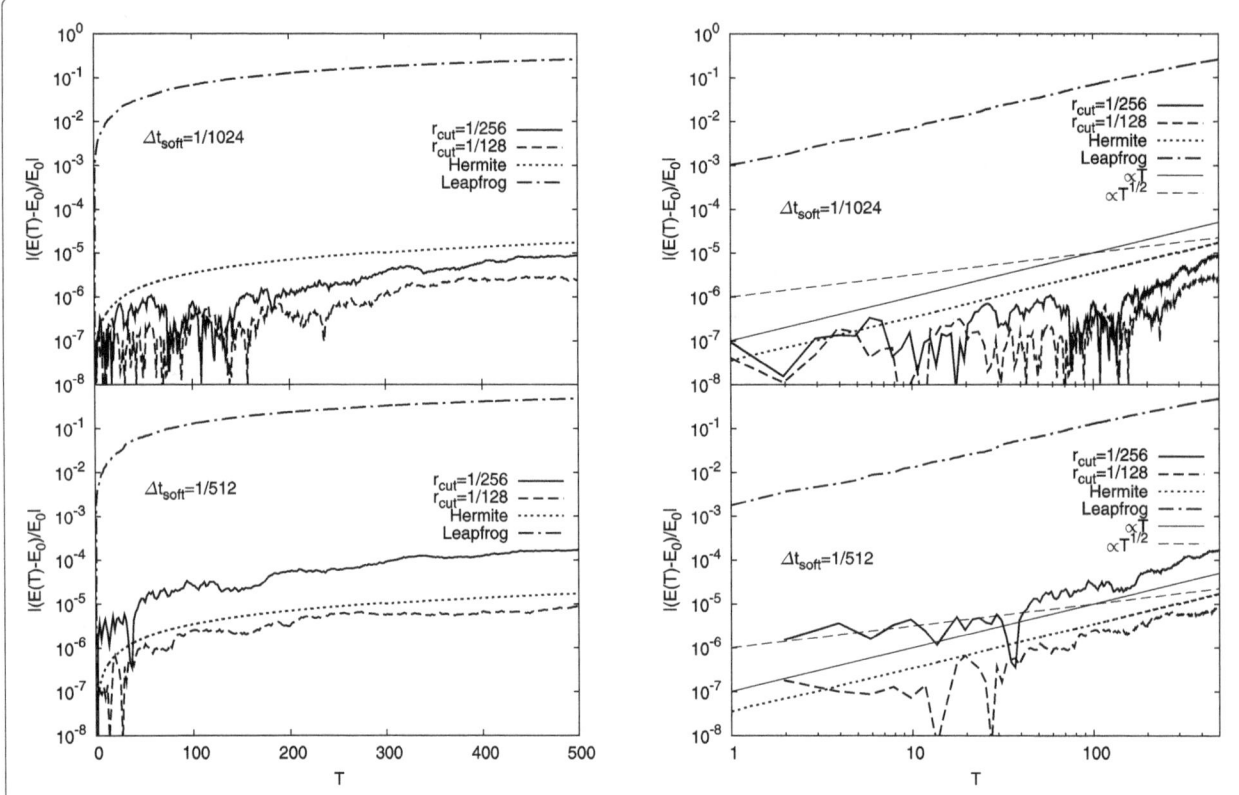

Figure 6 Evolution of relative energy errors with various schemes. We use $\Delta t_{\max} = \Delta t_{\mathrm{soft}}/4$, $\Delta r_{\mathrm{buff}} = 3\sigma\,\Delta t_{\mathrm{soft}}$ for all runs and $\theta = 0.4$ for the tree code and $\eta = 0.1$ for the Hermite scheme. In left and right panels, the x-axes are linear and logarithmic scales, respectively. Thin curves in right panels are proportional to T (solid) and $T^{1/2}$ (dashed).

which is dominated by the randomly changing tree-force error. For long-term integration the P³T scheme conserves energy better than the Hermite or leapfrog schemes.

3.1.2 Calculation cost

In this section, we discuss the calculation cost of the P³T scheme and its dependence on the number of particles N, required accuracy, and other parameters.

First, we construct a simple theoretical model of the dependence of the calculation cost on parameters of the integration scheme such as N, Δt_{soft}, θ and r_{cut}. Finally, we derive the optimal set of parameters from the model and compare this model with the result of the numerical tests. We found that the calculation cost per unit time is proportional to $N^{4/3}$.

Theoretical model The calculation cost for the force evaluations in P³T is split into the tree part and the Hermite part. For the tree part, the calculation cost of evaluating forces for all particles per tree step is proportional to $O(\theta^{-3}N\log N)$. Since we use constant timestep for the tree part, the calculation costs of the integration of particles per unit time for the tree part is proportional to $O(\theta^{-3}N\log N/\Delta t_{\mathrm{soft}})$.

For the Hermite part, since each particle has its own neighbour particles and timesteps, the number of interactions for all particles per unit timestep is given by

$$N_{\mathrm{int,hard}} = \sum_{i}^{N} N_{\mathrm{ngh},i}N_{\mathrm{step},i} \tag{21}$$

$$\sim \sum_{i}^{N} 4\pi/3(r_{\mathrm{cut}} + \Delta r_{\mathrm{buff}})^3 n_i\langle\Delta t_i\rangle^{-1} \tag{22}$$

$$\propto N^2(r_{\mathrm{cut}} + \Delta r_{\mathrm{buff}})^3\langle\langle\Delta t\rangle\rangle^{-1}. \tag{23}$$

Here $N_{\mathrm{ngh},i}$ is the number of the neighbour particles around particle i, $N_{\mathrm{step},i}$ is the number of timesteps required to integrate particle i for one unit time, n_i is the local density around particle i, $\langle\Delta t_i\rangle$ is the average timestep of particle i over one unit time and $\langle\langle\Delta t\rangle\rangle$ is the average of $\langle\Delta t_i\rangle$ over all particles. Here we assume n_i is constant within the radius of $r_{\mathrm{cut}} + \Delta r_{\mathrm{buff}}$ around particle i.

Next we express the $\langle\langle\Delta t\rangle\rangle$ as a function of N and r_{cut}. To simplify the discussion, we define the timestep of the particle through the relative position and velocity from its nearest neighbour particle; $\langle\langle\Delta t\rangle\rangle \propto r_{\mathrm{NN}}/v_{\mathrm{NN}}$, where r_{NN}

and v_{NN} are the relative position and the velocity of the nearest neighbour particle. We can replace v_{NN} to the velocity dispersion σ. Thus average timestep is given by

$$\langle\langle\Delta t\rangle\rangle \propto r_{NN}/v_{NN} \sim r_{NN}/\sigma. \tag{24}$$

To further simplify the derivation we assume that the number density of particles in the system is uniform. If r_{cut} is larger than the mean inter-particle distance $\langle r\rangle$ (i.e. if most particles have neighbour particles), the average timestep is roughly given by

$$\langle\langle\Delta t\rangle\rangle \sim \min\left(\eta\frac{R}{\sigma}N^{-1/3}, \Delta t_{max}\right), \tag{25}$$

where R is the typical size of the system. In this case, the average timestep depend only on N (does not depend on r_{cut}).

If r_{cut} is small compared to $\langle r\rangle$, most particles are isolated and most of the non-isolated particles have only one neighbour particle. In this case, $\langle\langle\Delta t\rangle\rangle$ is given by

$$\langle\langle\Delta t\rangle\rangle \sim \min\left(\eta\frac{r_{cut}}{\sigma}, \Delta t_{max}\right). \tag{26}$$

In Figure 7 we show the number of steps per particle per unit time N_{step} for a plummer sphere as a function of N (top panel) and as a function of r_{cut} (bottom panel). In the top panel, we can see that N_{step} is roughly proportional to $N^{1/3}$ for large N (i.e. $\langle r\rangle$ is small). On the other hand when N is small N_{step} is almost constant because $\langle r\rangle$ is large (see equation (26)).

The bottom panel of Figure 7 shows that all curves eventually approach to constant values for both of large and small r_{cut}. For large r_{cut}, the timesteps of the non-isolated particles are determined by N, not by r_{cut} (see equation (25)), whereas for small values of r_{cut} the non-isolated particles have a timesteps Δt_{max}. This is because most neighbouring particles are in the buffer shell and not in the neighbour sphere. For runs with $\Delta t_{soft} = 1/2,048, 1/1,024$ and $1/512$, we can see bumps of N_{step} at $r_{cut} \sim 1/512$ due to the dependence on r_{cut} shown in equation (26).

Using above discussions, the number of interactions for all particles per unit time of the Hermite part $N_{int,hard}$ and the tree part $N_{int,soft}$ are given by

$$N_{int,hard} \propto \begin{cases} N^{7/3}(r_{cut} + \Delta r_{buff})^3 & \text{(for } r_{cut} \gg \langle r\rangle), \\ N^2(r_{cut} + \Delta r_{buff})^3 & \text{(for } r_{cut} \ll \langle r\rangle), \end{cases} \tag{27}$$

$$N_{int,soft} \propto \theta^{-3}N\log N/\Delta t_{soft}, \tag{28}$$

Optimal set of accuracy parameters   In this section, we derive the optimal values of r_{cut} and Δt_{soft} from the point of view of the balance of the calculation costs between the

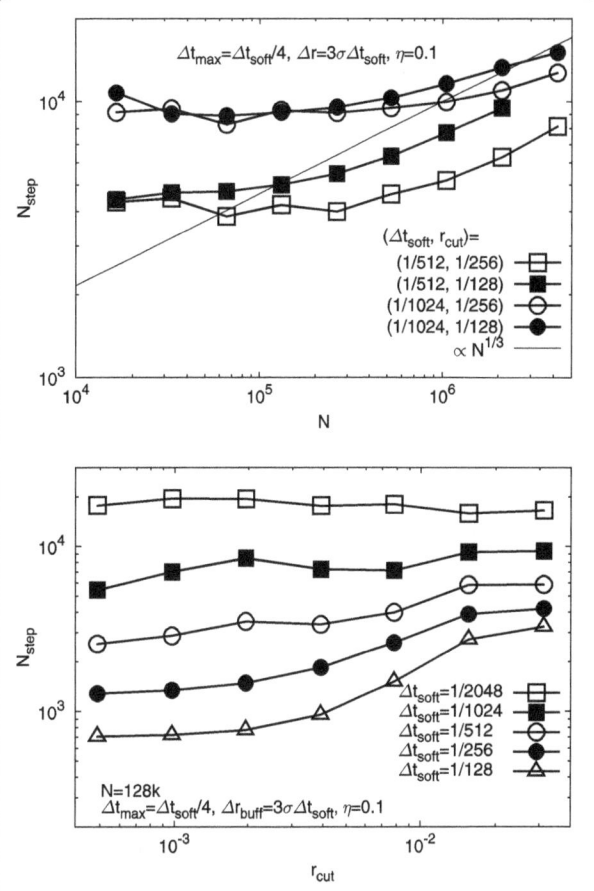

Figure 7 Number of steps of non-isolated particles as functions of N (top) and r_{cut} (bottom). Bottom panel shows the result of the runs with $N = 128$k. For all runs, we chose $\eta = 0.1$, $\Delta r_{buff} = 3\sigma\Delta t_{soft}$ and $\Delta t_{max} = \Delta t_{soft}/4$.

tree and the Hermite parts, in other words we express r_{cut} and Δt_{soft} as functions of N such that $N_{int,hard}/N_{int,soft}$ is independent of N. Following the discussion in Section 3.1.1 and because the energy errors can be controlled through $\Delta t_{soft}/r_{cut}$ and $\Delta t_{soft}/\Delta r_{buff}$, r_{cut} and Δr_{buff} should be proportional to Δt_{soft}.

The requirements are met for $N_{int,hard} \propto N^{7/3}(r_{cut} + \Delta r_{buff})^3$ (or $N^2(r_{cut} + \Delta r_{buff})^3$), $\Delta t_{soft} \propto N^{-1/3}$ and $r_{cut} \propto N^{-1/4}$ and both $N_{int,hard}$ and $N_{int,soft}$ are proportional to $N^{4/3}$ (or $N^{5/4}$). Here we have neglected the $\log N$ dependence in the tree part.

This is illustrated in Figure 8, where we plot $N_{int,hard}$ for a plummer sphere as a function of N. Following above discussions, we use the N-dependent tree timestep: $\Delta t_{soft} = (1/256)(N/16K)^{-1/3}$ and $N_{int,hard}$ as well as $N_{int,soft}$ are proportional to $N^{4/3}$.

In Figures 9 and 10, we plot the wall-clock time of execution T_{cal} and the maximum relative energy errors $|\Delta E_{max}/E_0|$ for the time integration for 10 N-body units against N. Figure 9 shows the results of the runs with

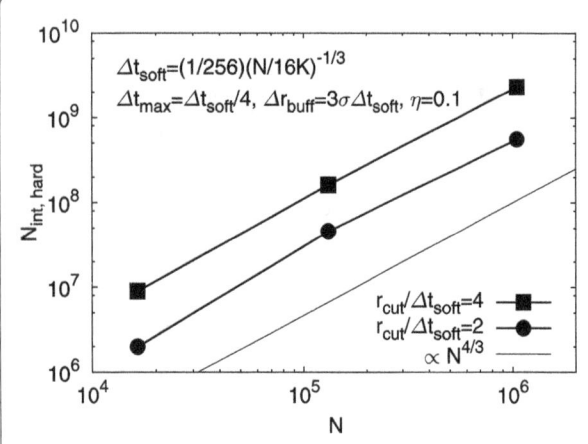

Figure 8 Number of interactions for all particles per unit time as a function of N. For all runs, we use $\Delta r_{\text{buff}} = 3\sigma \Delta t_{\text{soft}}$, $\eta = 0.1$, $\Delta t_{\text{max}} = \Delta t_{\text{soft}}/4$ and $\Delta t_{\text{soft}} = (1/256)(N/16\text{K})^{-1/3}$.

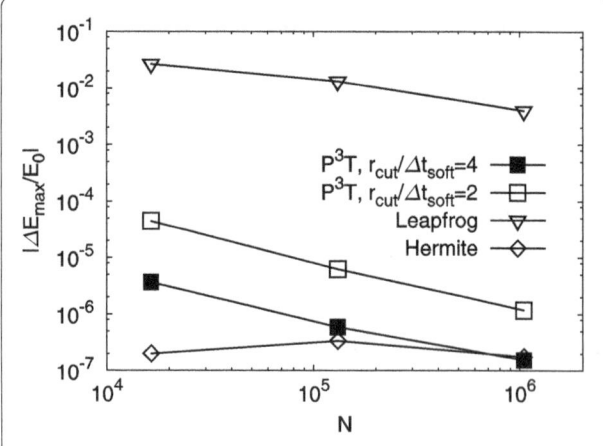

Figure 10 Maximum relative energy errors over 10 N-body time units. All runs are the same as those in Figure 9.

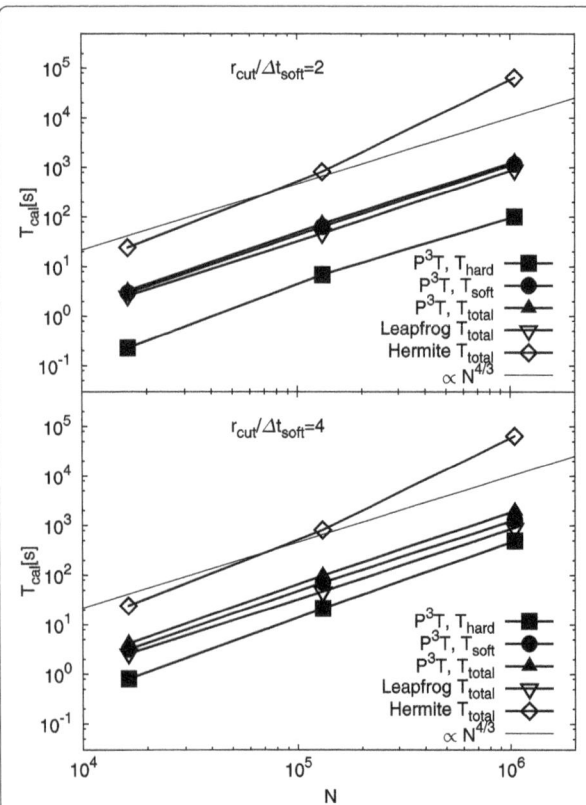

Figure 9 Wall-clock time of execution as a function of N. Top (bottom) panel shows the results of the runs with $r_{\text{cut}}/\Delta t_{\text{soft}} = 2(4)$. We use $\theta = 0.4$, $\eta = 0.1$, $\Delta r_{\text{buff}} = 3\sigma \Delta t_{\text{soft}}$ and $\Delta t_{\text{soft}} = (1/256)(N/16\text{K})^{-1/3}$.

$r_{\text{cut}}/\Delta t_{\text{soft}} = 2$ (top panel) and 4 (bottom panel). All runs in these figures are carried out on NVIDIA GeForce GTX680[a] GPU and Intel Core i7-3770K CPU. For each run, we use one CPU core and one GPU card.

We also perform the simulations using the direct Hermite integrator with the same η and the standard tree code with the same θ and Δt_{soft}. These calculations are performed with the Sapporo GPU library (Gaburov and Harfst 2009) and a standard tree code with the same θ and Δt_{soft} using the Bonsai GPU library (Bédorf et al. 2012). The calculation time for our P^3T implementation is also proportional to $N^{4/3}$, as we presented above, while for the Hermite integrator it is proportional to $N^{7/3}$. The P^3T scheme is faster than the direct Hermite integrator for $N > 16\text{K}$ and when $N = 1\text{M}$ ($M = 2^{20}$), the P^3T scheme is about 50 times faster than the direct Hermite scheme. The pure tree code is slightly faster than the P^3T scheme, but the integration errors are worse by several orders of magnitude (see Figures 6 and 10).

3.2 Examples of practical applications
In Sections 3.1.1 and 3.1.2, we presented a detailed discussion on the accuracy and performance of our P^3T scheme. However, we performed simple simulations, where the stellar systems are in the dynamical equilibrium. In this section, we study the performance of our P^3T scheme when applied to more realistic, or more difficult, simulations by comparing the results of the Hermite scheme. In Section 3.2.1, we discuss the case of the simulation of star clusters up to core collapse. In Section 3.2.2, we discuss the case of a galaxy model with massive central black hole binary.

3.2.1 Star cluster down to core collapse
In this section, we discuss the performance of our P^3T scheme for the simulation of the core collapse of a star cluster. First, we describe the initial condition and parameters of the integration scheme. Next, we compare the calculation results obtained by the P^3T and Hermite schemes, and finally, the calculation speed.

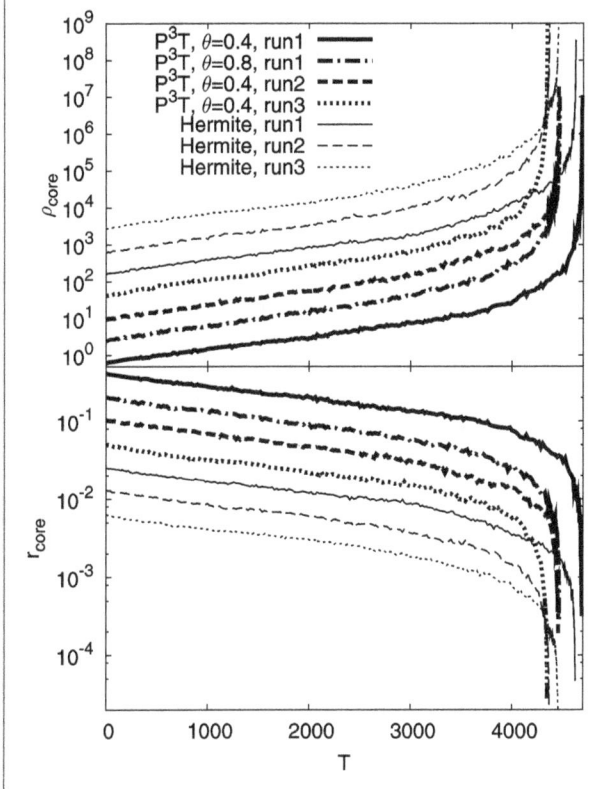

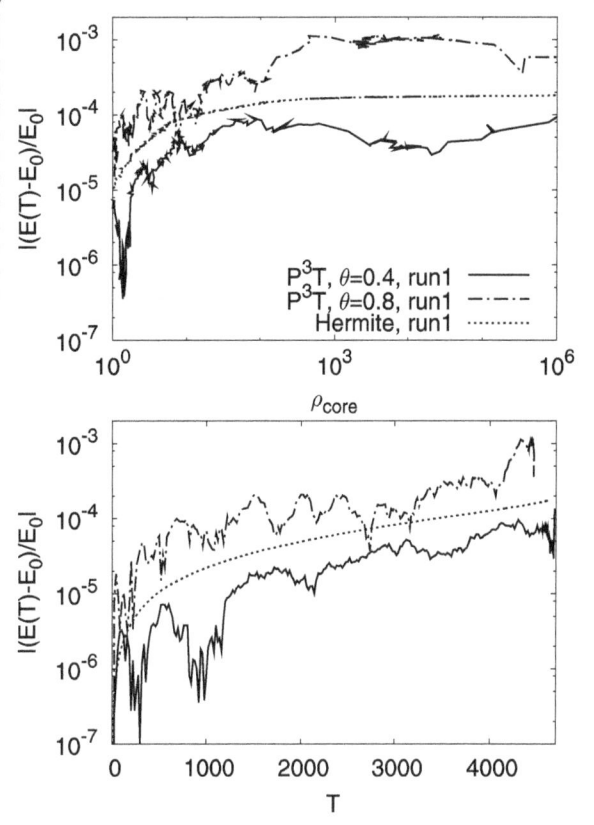

Figure 11 **Time evolution of the core density (top) and the core radius (bottom).** Thick and thin curves show the results of the P^3T and Hermite scheme, respectively. The curves for different runs are vertically shifted by a factor of 8 (top) and 2 (bottom).

Figure 12 **Relative energy error as functions of ρ_{core} (top) and time (bottom).**

Initial conditions We apply the P^3T scheme to the evolution of a star cluster consisting of 16K stars to the moment of the core collapse (Lynden-Bell and Eggleton 1980). We use an equal-mass plummer model as an initial density profile and we adopt $\eta = 0.1$. We apply the Plummer softening $\epsilon = 4/N = 1/4,096$. The simulations are terminated when the core number-density exceeds 10^6, at which point the mean interparticle distance in the core is comparable to ϵ. Next, we set θ. We must choose θ so that the tree force error is smaller than the force due to the two-body relaxation. Hernquist et al. (1993) pointed out that, for $\theta = 0.5$ with monopole and quadrupole, the tree-force error is much smaller than the force due to the two-body relaxation. Thus we choose $\theta = 0.4$ with quadrupole as a standard model. For comparison, we also perform a run with $\theta = 0.8$.

To resolve the motions of the particles in the core, we impose Δt_{soft} to be smaller than $1/128$ of the dynamical time of the core ($\sim \sqrt{3\pi/16\rho_{core}}$, where ρ_{core} is the core density). To reduce the calculation cost for the Hermite part we require $r_{cut} \propto \rho_{core}^{-1/3}$ and set the initial value of $r_{cut} = 1/64$. We also change $\Delta r_{buff} = 3\sigma_{core}\Delta t_{soft}$, where σ_{core} is the velocity dispersion in the core, and $\Delta t_{max} = \Delta t_{soft}/4$, as Δt_{soft} and

σ_{core} are changing. Here, to calculate ρ_{core} and σ_{core}, we use the formula proposed by Casertano and Hut (1985). The same simulation is repeated using the fourth-order Hermite scheme with the block timesteps with the same value of $\eta = 0.1$.

Results In Figure 11 we present the evolution of the core densities ρ_{core} (top panel) and the core radii r_{core} (bottom panel) for P^3T and Hermite schemes. For each scheme, we perform three runs, changing the initial random seed for generating the initial conditions of the Plummer model. The behaviors of the cores for all runs are similar. The differences between two schemes are smaller than run-to-run variations.

Figure 12 shows the relative energy errors of the runs with the same initial seed as functions of the core density (top panel) and the time (bottom panel). The energy errors of the runs with P^3T scheme change randomly, whereas those of the Hermite code grow monotonically. As a result, the P^3T scheme with $\theta = 0.4$ conserves energy better than the Hermite scheme in the long run. The errors for the P^3T scheme with $\theta = 0.8$ is slightly worse than that of the Hermite scheme, but the behavior of the core are sim-

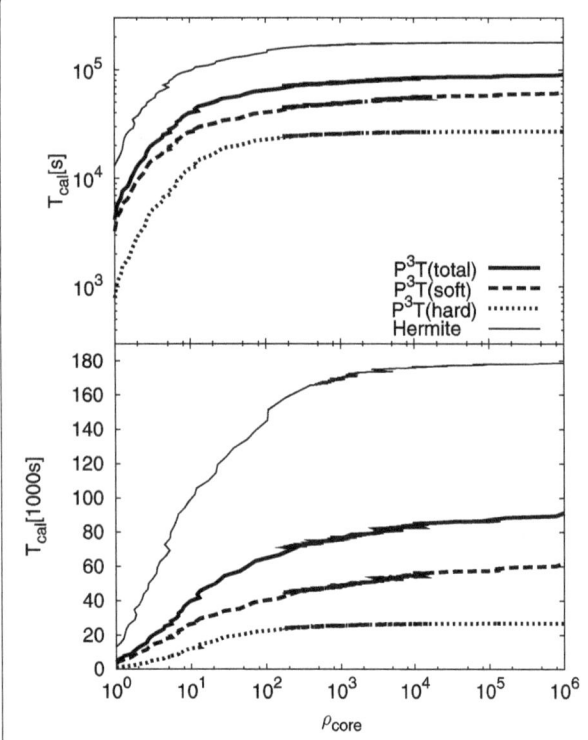

Figure 13 Wall-clock time of execution as functions of ρ_{core}. In top and bottom panels, the y-axes are logarithmic and linear scales, respectively.

ilar with other runs. Thus the choice of $\theta = 0.4$ is enough to follow the core collapse simulations.

Calculation speed Figure 13 shows the calculation time of the P^3T scheme ($\theta = 0.4$) and Hermite scheme on GPU. As shown in this figure, the calculation time of the P^3T scheme is dominated by the tree (soft) part calculation.

Initially the P^3T scheme is much faster than the Hermite scheme, but after the time when $\rho_{\text{core}} \sim 10^4$, the P^3T scheme is slightly slower than the Hermite scheme because in the P^3T scheme, Δt_{soft} is proportional to $\rho_{\text{core}}^{-1/2}$. However, even for the P^3T scheme, the CPU time spent after ρ_{core} reaches 10^4 is small. As a result, the calculation time to the moment of the core collapse of the P^3T scheme is smaller than that of the Hermite scheme by a factor of two.

3.2.2 Orbital evolution of SMBH binary

In this section, we also discuss the performance of the P^3T scheme applied to simulations of a galaxy with a supermassive black hole (SMBH) binary. First, we describe the initial conditions and parameters of the integration scheme. Next, we compare the calculation results obtained by the P^3T and Hermite schemes, and finally, the calculation speed.

Initial conditions and methods We use the Plummer model with $N = 16\text{K}, 128\text{K}$ and 256K as the initial galaxy model. Two SMBH particles with a mass of 1% of that of the galaxy are placed at the positions $(\pm 0.5, 0.0, 0.0)$ with the velocities $(0.0, \pm 0.5, 0.0)$. We use three values for the cut off radius with respect to three different kinds of interactions. For the interaction between field stars (FSs), we set $r_{\text{cut,FS–FS}} = 1/256$. For the interaction between SMBHs, the force is not split and $F_{\text{soft}} = 0$. In other words, the force between SMBHs is integrated with the pure Hermite scheme. We set the cut off radius between SMBH and FS $r_{\text{cut,BH–FS}} = 1/32$ which is large enough that Δt_{soft} is smaller than the Kepler time of a particle in orbit around the SMBH binary at a distance of $r_{\text{cut,BH–FS}}$. We use the Plummer softening $\epsilon = 10^{-4}$ for the interactions between FS-FS and FS-SMBH. For the SMBH-SMBH interaction, we do not use the softening. The accuracy parameter of timestep criterion for FS η_{FS} is 0.1, and for SMBH η_{BH} is 0.03. We adopt $\Delta r_{\text{buff}} = 3\sigma \Delta t_{\text{soft}}$, $\Delta t_{\text{max}} = \Delta t_{\text{soft}}/4$ and $\theta = 0.4$.

We use $\Delta t_{\text{soft}} = 1/1{,}024$ at $T = 0$, and as the binary becomes harder, we decrease Δt_{soft} to suppress the aliasing error of the binary. As a standard model, we set Δt_{soft} to be less than half of the Kepler time of the SMBH binary t_{kep}. Only for $N = 128\text{K}$, we also perform two other runs, where $\Delta t_{\text{soft}} < t_{\text{kep}}/4$ and t_{kep}.

We also perform the same simulations by the Hermite scheme with the same η_{FS} and η_{BH}.

Results Figure 14 shows the evolution of the semi-major axis (top panel) and eccentricity (middle panel) of the SMBH binary and the relative energy error (bottom panel) as functions of time for our standard models ($\Delta t_{\text{soft}} < t_{\text{kep}}/2$). The behaviors of the semi-major axis of the SMBH binary for the runs with the same N agree very well. The hardening rate of the binary depends on N because of the loss-cone refilling through the two-body relaxation (Begelman et al. 1980; Makino and Funato 2004; Berczik et al. 2005). The evolution of the eccentricity has large variation, because this evolution is sensitive to small N fluctuation (Merritt et al. 2007). In the cases of $N = 16\text{K}$ with the Hermite scheme, the relative energy error increases dramatically after $T = 150$ because the binding energy and the eccentricity of the binary are very high.

Figure 15 is the same as Figure 14 but for several different values of Δt_{soft}. Thick solid, dashed and dotted curves indicate the results for $\Delta t_{\text{soft}} < t_{\text{kep}}/4$, $t_{\text{kep}}/2$ and t_{kep}, respectively. The orbital parameters show similar behaviors for all runs. The absolute value of the energy errors of P^3T runs ($\sim 10^{-5}$) are small compared with the binding energy of SMBH binary, which is roughly 0.05.

Calculation speed Figure 16 shows the calculation time for runs for several different values of N with $\Delta t_{\text{soft}} < t_{\text{kep}}/2$.

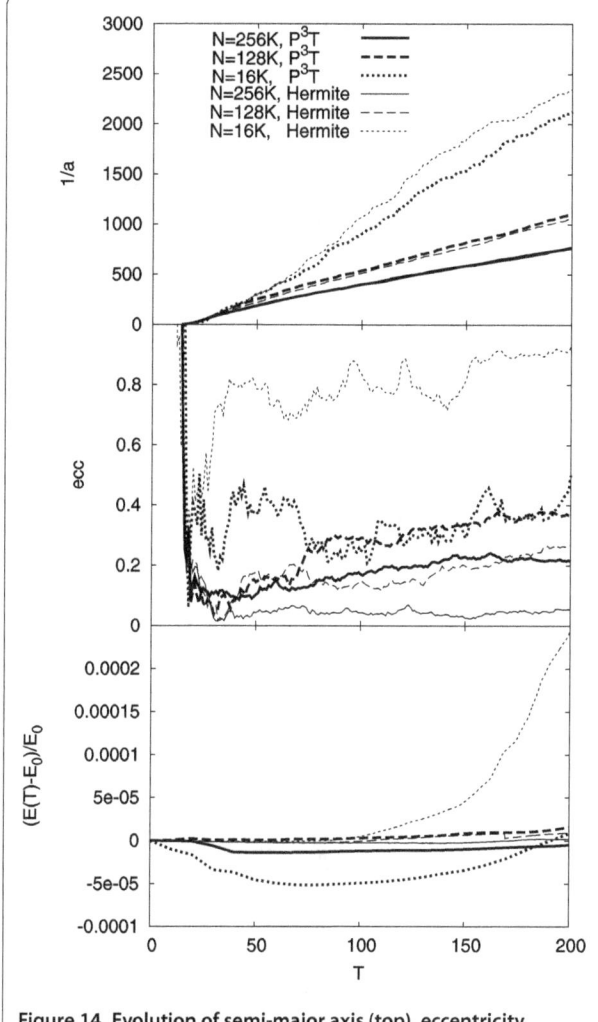

Figure 14 Evolution of semi-major axis (top), eccentricity (middle) of the SMBH binary and energy error (bottom) for several different values of N.

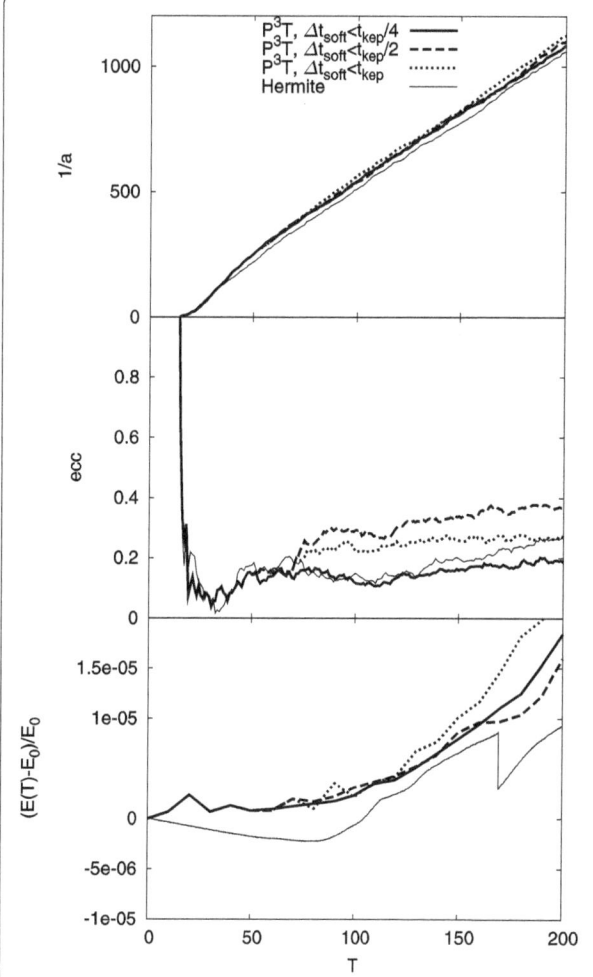

Figure 15 Evolution of semi-major axis (top), eccentricity (middle) of the SMBH binary and energy error (bottom) for the several different value of Δt_{soft}. Thick solid, dashed and dotted curves show the results of P^{3}T scheme with Δt_{soft} is less than $t_{kep}/4$, $t_{kep}/2$ and t_{kep}, respectively.

Initially, the P^{3}T scheme is much faster than the Hermite scheme. As the SMBH binary becomes harder, the P^{3}T scheme slows down more significantly than the direct Hermite scheme does. We can see that T_{cal} of the Hermite scheme is roughly proportional to a^{-1} for $a^{-1} > 300$, whereas that of the P^{3}T scheme is roughly proportional to $a^{-5/2}$, because Δt_{soft} is proportional to the Kepler time of the binary ($\propto a^{3/2}$). However, the calculation time for all runs with the P^{3}T scheme is shorter than that with the Hermite scheme by $a = 1/800$. We can also confirm that as we use more N, the ratio of the calculation time of the P^{3}T scheme to the Hermite scheme become larger. The reason why the P^{3}T scheme becomes slower for large a^{-1} is simply that we force the timestep of all particles to be smaller than the orbital period of the SMBH binary. For the Hermite scheme, we do not put such constraint. Thus, in the Hermite scheme, particles far away from the SMBH have the timestep much larger than the orbital period of the SMBH

binary. This large timestep can cause accuracy problem (Nitadori and Makino 2008). With P^{3}T, it is possible to apply the perturbation approximation to F_{soft} between the SMBH binary and other particles. Such a treatment should improve the accuracy and speed of the P^{3}T scheme when the SMBH binary becomes very hard.

In Figure 17, we plot the calculation time of the hard and soft parts for the standard model with $N = 128k$. We can see that the soft parts dominate the calculation time.

In Figure 18, we compare the calculation time for the runs with various Δt_{soft} ($< t_{kep}, t_{kep}/2, t_{kep}/4$). Since the most of the calculation time is spent after the binary becomes hard, the calculation time strongly depends on the criterion of the Δt_{soft}. From Figure 15, the evolution of the orbital parameters for all runs with the P^{3}T scheme are

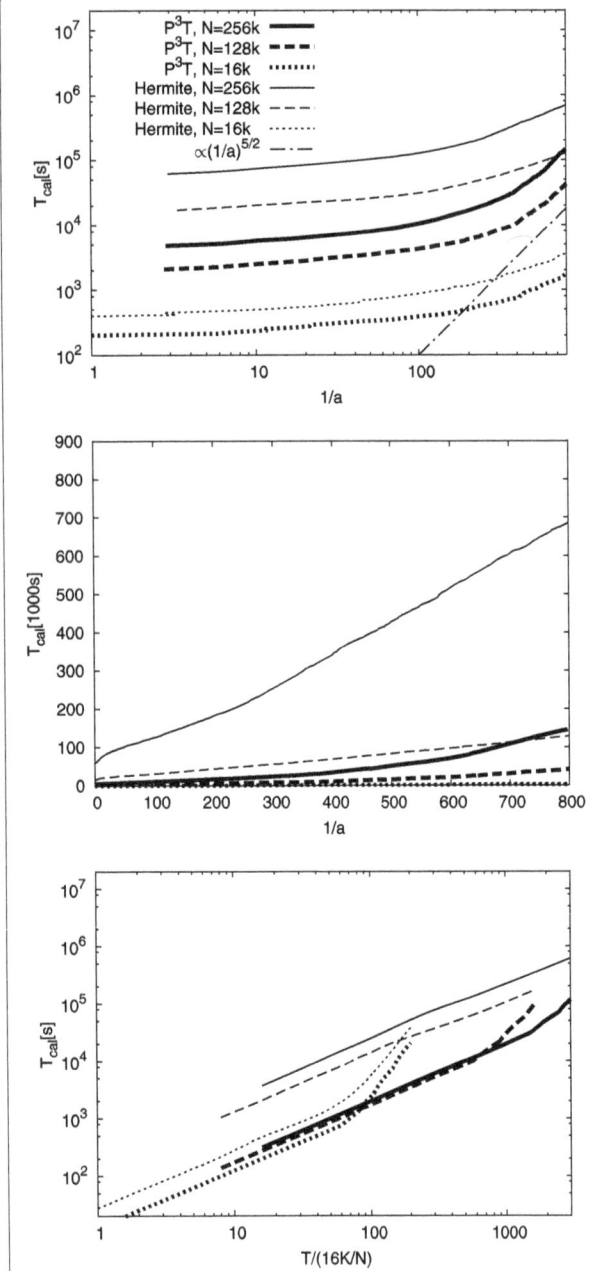

Figure 16 Wall-clock times as a function of 1/a (top and middle) and the system time of the simulations (bottom) for several different values of N. In top and middle panels, the x- and y-axis are logarithmic and linear scales, respectively. In the bottom panel, the x-axis is scaled by N/16K.

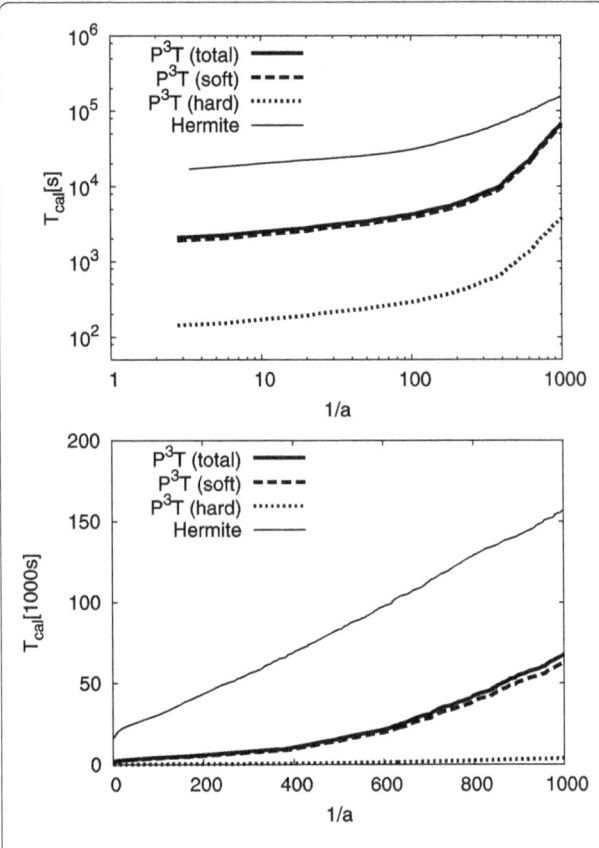

Figure 17 Wall-clock times as a function of 1/a. In top and bottom panels, the x- and y-axis are logarithmic and linear scales, respectively.

similar for various Δt_{soft} criteria. Thus we could choose larger $\Delta t_{soft} \gtrsim t_{kep}$ after the binary formation.

4 Conclusions

We have described the implementation and performance of the P^3T scheme for simulating dense stellar systems. In our implementation, the tree part is accelerated using

GPU. The accuracy and performance of the P^3T scheme can be controlled through six parameters: Δr_{cut}, Δr_{buff}, Δt_{soft}, Δt_{max}, η and θ. We find that $\Delta r_{buff} \gtrsim 2\sigma \Delta t_{soft}$ is a good choice to prevent non-neighbour particles from entering the neighbour sphere. The integration errors can be controlled through $\Delta t_{soft}/\Delta r_{cut}\sigma$. For $\theta = 0.2$, if we set Δt_{soft} to be less than $0.05\Delta r_{cut}/\sigma$, the integration error is smaller than the tree force error. For the Hermite part, if we choose $\eta \lesssim 0.2$, the errors hardly depend on Δt_{max}.

From the point of view of the balance of the calculation costs between the tree and Hermite parts, we derive the optimal set of accuracy parameters, and find that the calculation cost is proportional to $N^{4/3}$.

The P^3T scheme is suitable for simulating large N stellar clusters with a high density contrast, such as star clusters or galactic nuclei. We demonstrate the efficiency of the code and show that it is able to integrate N-body systems to the moment of the core collapse. We also performed the simulations of the galaxy with the SMBH binary and found that the P^3T scheme can be applied to these simulations.

Finally, we discuss the possibilities of implementing two important effects on star cluster evolution in P^3T. The first is the effect of a tidal field which dramatically changes the

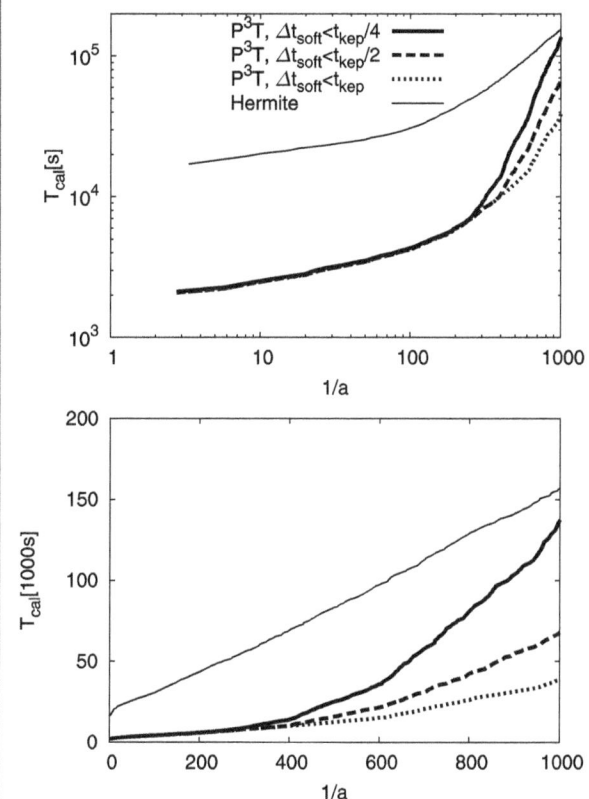

Figure 18 Wall-clock time as a function of $1/a$ for several different values of Δt_{soft}. In top and bottom panels, the x- and y-axis are logarithmic and linear scales, respectively.

collapse time and the evaporation time of a star cluster. The tidal field effect can be included in the soft part.

The other is an effect of stellar-mass binaries. A stellar-mass binary plays an important role in halting the core collapse. In this paper, we introduce the Plummer softening and neglect these binary effect. However, we could treat these effects by integrating stellar-mass binaries in the hard part.

Our P^3T code is incorporated in the AMUSE frameworks and free for use (Portegies Zwart et al. 2013; Pelupessy et al. 2013).

Competing interests
The authors declare that they have no competing interests.

Authors' contributions
All authors, MI, SPZ and JM conceived of the study. MI developed the code, performed all simulations and drafted the manuscript. SPZ and JM helped to draft the manuscript. All authors read and approved the final manuscript.

Author details
[1]RIKEN Advanced Institute for Computational Science, Minatojima-minamimachi, Chuo-ku, Kobe, Japan. [2]Sterrewacht Leiden, P.O. Box 9513, Leiden, 2300 RA, The Netherlands. [3]Earth-Life Science Institute, Tokyo Institute of Technology, Ookayama, Meguro-ku, Tokyo, Japan.

Acknowledgements
We are grateful to Jeroen Bédorf, for preparations of the GPU cluster and GPU library. We also thank to Shoichi Oshino, Daniel Caputo and Keigo Nitadori for stimulating discussion, and to Edwin van der Helm for carefully reading the manuscript. This work was supported by NWO (grants VICI [#639.073.803], AMUSE [#614.061.608] and LGM [# 612.071.503]), NOVA and the LKBF.

Endnote
[a] GTX680 does not have ECC (Error Check and Correct) memories. However, as we will see later, we do not observe any large energy error in any of our runs, which means the hardware error does not affect our result. Betz et al. (2014) performed Molecular Dynamics simulations, in order to investigate the rate of bit-flip error events. They observed a single bit-flip error event in about 4,700 GPU*hours without ECC and conclude that the bit-flip error is exceedingly rare.

References
Aarseth, SJ: Dynamical evolution of clusters of galaxies, I. Mon. Not. R. Astron. Soc. **126**, 223-255 (1963)

Barnes, J, Hut, P: A hierarchical $O(N \log N)$ force-calculation algorithm. Nature **324**, 446-449 (1986)

Dehnen, W: A hierarchical $O(N)$ force calculation algorithm. J. Comput. Phys. **179**, 27-42 (2002)

Dehnen, W: A fast multipole method for stellar dynamics. Comput. Astrophys. Cosmol. **1**, 1 (2014)

Greengard, L, Rokhlin, V: A fast algorithm for particle simulations. J. Comput. Phys. **73**, 325-348 (1987)

Gaburov, E, Bédorf, J, Portegies Zwart, S: Gravitational tree-code on graphics processing units: implementation in CUDA. Proc. Comput. Sci. **1**, 1119-1127 (2010)

Bédorf, J, Gaburov, E, Portegies Zwart, S: A sparse octree gravitational N-body code that runs entirely on the GPU processor. J. Comput. Phys. **231**, 2825-2839 (2012)

Bédorf, J, Gaburov, E, Fujii, MS, Nitadori, K, Ishiyama, T, Portegies Zwart, S: 24.77 Pflops on a gravitational tree-code to simulate the Milky Way Galaxy with 18600 GPUs. In: SC'14 Proceedings of the International Conference for High Performance Computing, Networking, Storage and Analysis, pp. 54-65. IEEE Press, Piscataway (2014). doi:10.1109/SC.2014.10

McMillan, SLW: The vectorization of small-n integrators. In: Hut, P, McMillan, SLW (eds.) The Use of Supercomputers in Stellar Dynamics. Lecture Notes in Physics, vol. 267, pp. 156-161. Springer, Berlin (1986)

McMillan, SLW, Aarseth, SJ: An $O(N \log N)$ integration scheme for collisional stellar systems. Astrophys. J. **414**, 200-212 (1993)

Oshino, S, Funato, Y, Makino, J: Particle-particle particle-tree: a direct-tree hybrid scheme for collisional N-body simulations. Publ. Astron. Soc. Jpn. **63**, 881-892 (2011)

Kinoshita, H, Yoshida, H, Nakai, H: Symplectic integrators and their application to dynamical astronomy. Celest. Mech. Dyn. Astron. **50**, 59-71 (1991)

Wisdom, J, Holman, M: Symplectic maps for the n-body problem. Astron. J. **102**, 1528-1538 (1991)

Duncan, MJ, Levison, HF, Lee, MH: A multiple time step symplectic algorithm for integrating close encounters. Astron. J. **116**, 2067-2077 (1998)

Chambers, JE: A hybrid symplectic integrator that permits close encounters between massive bodies. Mon. Not. R. Astron. Soc. **304**, 793-799 (1999)

Brunini, A, Viturro, HR: A tree code for planetesimal dynamics: comparison with a hybrid direct code. Mon. Not. R. Astron. Soc. **346**, 924-932 (2003)

Fujii, M, Iwasawa, M, Funato, Y, Makino, J: BRIDGE: a direct-tree hybrid N-body algorithm for fully self-consistent simulations of star clusters and their parent galaxies. Publ. Astron. Soc. Jpn. **59**, 1095-1106 (2007)

Moore, A, Quillen, AC: QYMSYM: a GPU-accelerated hybrid symplectic integrator that permits close encounters. New Astron. **16**, 445-455 (2011)

Makino, J, Aarseth, SJ: On a Hermite integrator with Ahmad-Cohen scheme for gravitational many-body problems. Publ. Astron. Soc. Jpn. **44**, 141-151 (1992)

Hockney, RW, Eastwood, JW: Computer Simulation Using Particles (1981)

Plummer, HC: On the problem of distribution in globular star clusters. Mon. Not. R. Astron. Soc. **71**, 460-470 (1911)

Heggie, DC, Mathieu, RD: Standardised units and time scales. In: Hut, P, McMillan, SLW (eds.) The Use of Supercomputers in Stellar Dynamics. Lecture Notes in Physics, vol. 267, pp. 233-235. Springer, Berlin (1986)

Portegies Zwart, S, Boekholt, T: On the minimal accuracy required for simulating self-gravitating systems by means of direct N-body methods. Astrophys. J. Lett. **785**, L3 (2014)

Gaburov, E, Harfst, S, Portegies Zwart, S: SAPPORO: a way to turn your graphics cards into a GRAPE-6. New Astron. **14**, 630-637 (2009)

Betz, RM, DeBardeleben, NA, Walker, RC: An investigation of the effects of hard and soft errors on graphics processing unit-accelerated molecular dynamics simulations. Concurr. Comput., Pract. Exp. **26**, 2134-2140 (2014)

Lynden-Bell, D, Eggleton, PP: On the consequences of the gravothermal catastrophe. Mon. Not. R. Astron. Soc. **191**, 483-498 (1980)

Hernquist, L, Hut, P, Makino, J: Discreteness noise versus force errors in N-body simulations. Astrophys. J. Lett. **402**, L85 (1993)

Casertano, S, Hut, P: Core radius and density measurements in N-body experiments connections with theoretical and observational definitions. Astrophys. J. **298**, 80-94 (1985)

Begelman, MC, Blandford, RD, Rees, MJ: Massive black hole binaries in active galactic nuclei. Nature **287**, 307-309 (1980)

Makino, J, Funato, Y: Evolution of massive black hole binaries. Astrophys. J. **602**, 93-102 (2004)

Berczik, P, Merritt, D, Spurzem, R: Long-term evolution of massive black hole binaries. II. Binary evolution in low-density galaxies. Astrophys. J. **633**, 680-687 (2005)

Merritt, D, Mikkola, S, Szell, A: Long-term evolution of massive black hole binaries. III. Binary evolution in collisional nuclei. Astrophys. J. **671**, 53-72 (2007)

Nitadori, K, Makino, J: Sixth- and eighth-order Hermite integrator for N-body simulations. New Astron. **13**, 498-507 (2008)

Portegies Zwart, S, McMillan, SLW, van Elteren, E, Pelupessy, I, de Vries, N: Multi-physics simulations using a hierarchical interchangeable software interface. Comput. Phys. Commun. **183**, 456-468 (2013)

Pelupessy, FI, van Elteren, A, de Vries, N, McMillan, SLW, Drost, N, Portegies Zwart, SF: The astrophysical multipurpose software environment. Astron. Astrophys. **557**, A84 (2013)

PERMISSIONS

LIST OF CONTRIBUTORS

Tjarda Boekholt and Simon Portegies Zwart
Leiden Observatory, Leiden University, PO Box 9513, Leiden, 2300 RA, The Netherlands

Brian Punsly
1415 Granvia Altamira, Palos Verdes Estates, CA 90274, USA.
ICRANet, Piazza della Repubblica 10, Pescara, 65100, Italy

Dinshaw Balsara, Jinho Kim and Sudip Garain
Physics Department, University of Notre Dame du Lac, 225 Nieuwland Science Hall, Notre Dame, IN 46556, USA

Andreas Bleuler and Romain Teyssier
Institute for Computational Science, University of Zurich, Zurich, CH-8057, Switzerland

Sébastien Carassou
Institute for Computational Science, University of Zurich, Zurich, CH-8057,Switzerland
Institut d'Astrophysique de Paris, 98bis boulevard Arago, Paris,75014, France

Davide Martizzi
Institute for Computational Science, University of Zurich, Zurich, CH-8057, Switzerland
Department of Astronomy, University of California, Berkeley,CA 94720-3411, USA

David Radice, Sean M Couch and Christian D Ott
TAPIR, Walter Burke Institute for Theoretical Physics, California Institute of Technology, E California Blvd, Pasadena, CA 91125, USA

Alex Rimoldi, Simon Portegies Zwart and Elena Maria Rossi
Leiden Observatory, Leiden University, Niels Bohrweg 2, Leiden, 2333 CA, The Netherlands

Harshitha Menon, Lukasz Wesolowski, Gengbin Zheng, Pritish Jetley, Laxmikant Kale
Department of Computer Science, University of Illinois at Urbana-Champaign, Urbana, USA

Thomas Quinn and Fabio Governato
Department of Astronomy, University of Washington, Seattle,USA

Oliver Porth, Hector Olivares, Yosuke Mizuno and Ziri Younsi
Institute for Theoretical Physics, Max-von-Laue-Str. Frankfurt am Main, 60438, Germany

Luciano Rezzolla
Institute for Theoretical Physics, Max-von-Laue-Str. 1, Frankfurt am Main, 60438, Germany
Frankfurt Institute for Advanced Studies, Ruth-Moufang-Straße 1, Frankfurt am Main, D-60438, Germany

Monika Moscibrodzka and Heino Falcke
Department of Astrophysics/IMAPP, Radboud University Nijmegen, P.O. Box 9010, Nijmegen, 65008, The Netherlands

Michael Kramer
Max-Planck-Institut für Radioastronomie, Auf dem Hügel 69, Bonn, D-53121, Germany

Olindo Zanotti and Michael Dumbser
Laboratory of Applied Mathematics, Department of Civil, Environmental and Mechanical Engineering, University of Trento, Via Mesiano 77, Trento, 38123, Italy

Federico Guercilena
Institut für Theoretische Physik, Goethe Universität, Max-von-Laue-Str. 1, Frankfurt am Main, 60438, Germany

David Radice
Institute for Advanced Study, 1 Einstein Dr., Princeton, NJ 08540, USA
Department of Astrophysical Sciences, Princeton University

Luciano Rezzolla
Institut für Theoretische Physik, Goethe Universität, Max-von-Laue-Str. 1, Frankfurt am Main, 60438, Germany
Ivy Lane, Princeton, NJ 08544, USA. 4Frankfurt Institute for Advanced Studies, Ruth-Moufang-Str. 1, Frankfurt am Main, 60438, Germany

Iwasawa
RIKEN Advanced Institute for Computational Science, Minatojima-minamimachi, Chuo-ku, Kobe, Japan
Sterrewacht Leiden, P.O.Box 9513, Leiden, 2300 RA, The Netherlands

Simon Portegies Zwart
Sterrewacht Leiden, P.O.Box 9513, Leiden, 2300 RA, The Netherlands

Junichiro Makino
RIKEN Advanced Institute for Computational Science, Minatojima-minamimachi, Chuo-ku, Kobe, Japan
Earth-Life Science Institute, Tokyo Institute of Technology, Ookayama, Meguro-ku, Tokyo, Japan

Index

www.ingramcontent.com/pod-product-compliance
Lightning Source LLC
Chambersburg PA
CBHW080406190526
45161CB00003B/147